CHEMISTRY Explorer

STUDENT EDITION

USER GUIDE

LOGAL Software, Inc.

PRENTICE HALL, Upper Saddle River, NJ 07458

Printed in the United States of America

10 9 8 7 6 5 4 3 2 1

Prentice-Hall International (UK) Limited, *London*
Prentice-Hall of Australia Pty. Limited, *Sydney*
Prentice-Hall Canada, Inc. *Toronto*
Prentice-Hall Hispanoamericana, S.A., *Mexico*
Prentice-Hall of India Private Limited, *New Delhi*
Prentice-Hall of Japan, Inc. *Tokyo*
Simon & Schuster Asia Pte. Ltd., *Singapore*
Editora Prentice-Hall do Brasil, Ltda., *Rio de Janeiro*

Table of Contents

Chapter 1:
What is *Student Edition Chemistry Explorer*?

Welcome to the *Student Edition Chemistry Explorer*.

In order to make the best use of this product, you probably will find the following information quite useful.

1. Overview of the *Explorer* Series

The purpose of the *Explorer* series is to enhance student discovery and learning through models and simulations which are both educationally stimulating and scientifically accurate. *Explorer* products exist in the fields of chemistry, biology and physics.

■ Models and Simulations

Models are aids to understanding. In science, a model is a representation of specified phenomena. When such a model is stated in mathematical terms, the model can be used to produce simulations, or special instances of a particular model, given certain conditions. The conditions can then be changed to produce a different simulation, and the instances can be compared.

For example, the *Chemical Equilibrium* model can be used to represent a wide variety of phenomena by generating many specific simulations, such as endothermic and exothermic reactions or titration.

■ The *Student Edition Chemistry Explorer*

The *Student Edition Chemistry Explorer* can be thought of as a "simulation processor." Just as a word processor can be used to produce many written documents, each *Explorer* product can be used to generate and save many simulation lab files dealing with the topics covered by its model.

Each model is based on the same *Explorer* simulation software engine. This engine is responsible for doing many things, such as allowing you to manipulate the model with various "tools," run the simulations and present the resulting data.

Because all *Explorer* models are based on this same software engine, each model has the same "feel," i.e., the same human-computer interface. As a result, once you have learned to use one *Explorer* model, it becomes much easier to learn to use other *Explorer* models.

The *Student Edition Chemistry Explorer* is intended to give you a tool that you can use to explore further the phenomena that you are studying in class and to prepare presentations. There are sample labs included, together with accompanying text to explain the steps for creating and using them.

■ A Quick Look At How *Explorer* Works

For every model defined by the *Explorer* system, there is a predefined set of **input variables**. You can set up different simulations by changing the values of these variables. The results of the new values are reflected when you run the simulation.

Every *Explorer* model also has predefined **output variables**. Output data may be simply the value of some or all of these variables, or expressions involving them. *Explorer* provides a variety of methods for recording or displaying output data. For example, output data can be sent to a window for presentation in the form of line graphs, pie charts, or bar charts. You can choose which expressions are graphed, set numerical limits for their values, and set the scale.

For a given simulation, the model takes a set of input values and computes a new set of output values. On the next cycle of the simulation, these outputs become inputs to the model, which uses them to generate a new set of outputs. The process continues until you halt the simulation or until the simulation halts automatically.

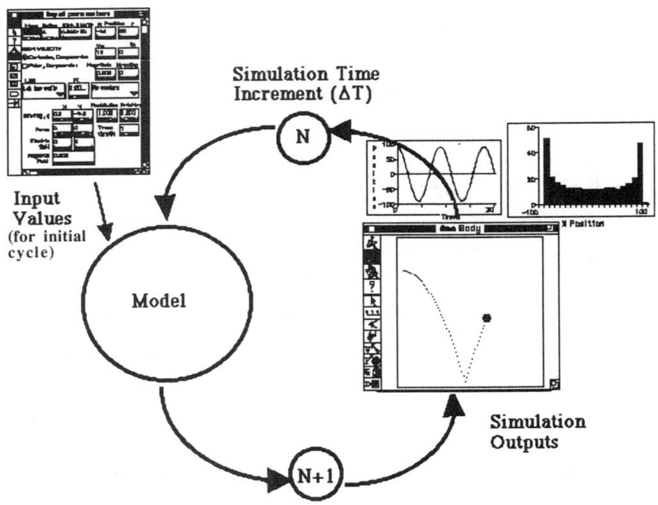

Note: The time between cycles is called the time increment, or Δt. For the *Student Edition Chemistry Explorer* the Δt has been preset to typical values.

2. How to Use The Disks and This Guide

Installing the Software

> **NOTE:** If your computer has a RAM-resident virus protection application running, you must deactivate the application and restart the computer before attempting to install *Explorer*. After you have completed the software installation, you may need to instruct the virus protection program that a new application has been installed.

A. Macintosh

Insert Disk #1 into the disk drive. Double-click the Installer icon. Click the "Continue" button on the opening screen.

To install the entire *Student Edition Chemistry Explorer* package, choose the Easy Install option and click the "Install" button. The software will automatically be installed into a folder called Explorer.

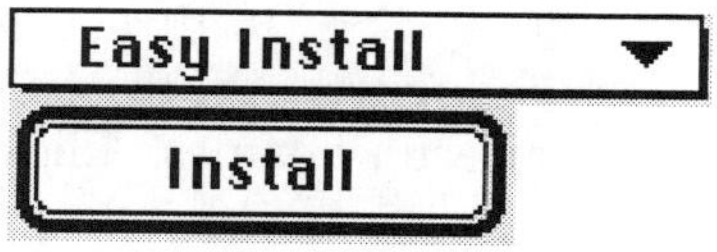

If you want to install it into a different folder, click the "Select Folder..." button and indicate the desired folder.

Note that you need at least 4.9 Mb available to install the entire product on your hard disk. If you have already installed a complete *Run Time Chemistry Explorer* package into the same directory, you need only an additional 2 Mb to install the *Student Edition Chemistry Explorer.*

To install selected elements of the product, choose the Custom Install option.

A menu offering the basic application, the chemistry models, and the *Student Edition* lab files appears. Click the check box of the software that you want to install. To determine the size requirements and the contents of a given element, click its Get Info icon.

To delete one of the elements, choose the Custom Remove option. Then click the check box of the element whose files you want to remove. Note that you cannot delete the models in this way because they may be in use by a *Run Time Explorer* product.

B. Windows

Insert Disk #1 into the disk drive.

From the Windows Program Manager File menu, "Run" option, type

a:\setup (If the disk is in a drive other than a:, type the appropriate drive.)

Follow the instructions given on the monitor.

The default directory is C:\EXPLR_RT. To change the path, enter the desired drive and directory names. Click the Continue button.

A program group called *Student Edition Explorer* will be added to the Program Manager.

Note that you need at least 8.5 Mb available to install the entire product on your hard disk. If you have already installed a complete *Run Time Chemistry Explorer* package into the same directory, you need only an additional 2 Mb to install the *Student Edition Chemistry Explorer.*

The Lab Files

- **Macintosh:** All of the *Student Edition* lab files provided with this package are in a folder called SE Chem Labs, which is automatically installed inside the product folder.

- **Windows:** The *Student Edition* lab files provided with this package are divided into subdirectories by model. The subdirectories are automatically created inside the product directory.

The prepared lab files and templates are locked, so you cannot perform an ordinary Save operation on them. When you are ready to save modifications that you have made to one of these files, use the Save As option, and give the file any name that you want. The original file remains for you to refer to during future work.

The Guide

- ***Chapter 1: What is the Student Edition Chemistry Explorer*** ?

  This chapter gives you a brief introduction to *Explorer*: modeling and simulations, the *Student Edition*, how *Explorer* works, and installation instructions.

- ***Chapter 2: Explorer Tutorial — Heating a Gas***

  The tutorial in Chapter 2 is a detailed series of six lessons that take you from a general template to a finished exploration on heating a gas, based on the *Gas Laws* model. The lessons explain how to plan and use the exploration, making use of all the basic *Explorer* design and analysis tools.

  It is highly recommended that you work through the tutorial before working on the templates or on a design of your own.

■ *Chapter 3: Lab Templates*

Each model is accompanied by an introductory lab file and four template files to help you get started:

— First Look (on disk only), presenting an introductory tour through each model

— Manipulation Template (on disk and in book), focusing on how to use the model window tools to solve problems

— Analysis Template (on disk and in book), explaining the considerations in choosing *Explorer* analysis tools for specific problems

— Design Template (on disk and in book), presenting outlined problems for independent work

— General Template (on disk only), serving as a starting point for your own lab designs

It is recommended that you begin work on each model by first working through the relevant First Look lab file. The templates are independent of one another and you can work through them in any order.

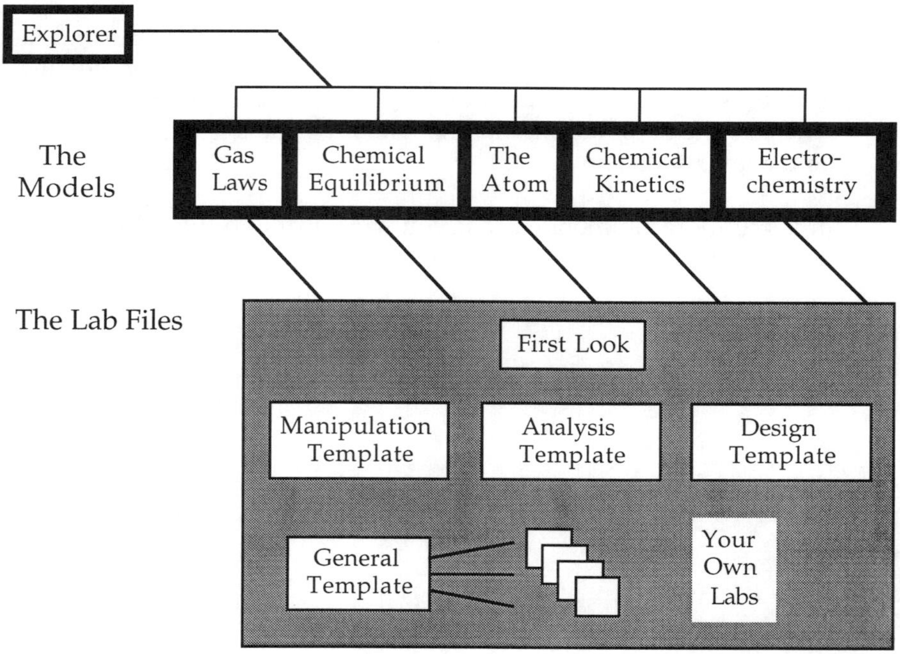

- *Appendices: Reference Guide*

  The reference guide is divided into two appendices:

  — A detailed reference on the *Explorer* tools and spreadsheet

  — A model-by-model description of tools, input and output parameters, and model reactions available in the *Student Edition Chemistry Explorer*.

The skills required for scientific investigation are useful skills that can help you in many areas of life. Learning with the *Explorer* series can be a process of discovery. We hope you find this process both enjoyable and intellectually rewarding!

Chapter 2: Explorer Tutorial — Heating a Gas (*Gas Laws* Model)

This tutorial is designed to familiarize you with the *Explorer* tools. It presents a typical problem you can analyze with *Explorer* and shows you recommended steps for designing your answer. Your final result will be a presentation that you can use to show how you answer the problem. You will then be ready to analyze other problems.

You will be working with the *Gas Laws* model, which is one of five models in the *Chemistry Explorer* series. The *Gas Laws* model simulates the behavior of a gas enclosed in a cylinder with a moveable piston. The gas can be set to be an ideal gas, one of a number of real gases, or a generic real gas. You can determine a set of initial conditions and a process to arrive at a final state. The process may be constant volume, constant pressure, constant temperature, or adiabatic, enabling you to simulate experiments such as those of Boyle, Charles, and Gay-Lussac. You can find and measure deviations from ideal gas behavior for real gases under different conditions of pressure, temperature, or volume.

The model makes measurements of variables such as pressure, temperature, volume, number of moles, work done, and heat out during the simulation.

You can familiarize yourself with this model by working through the *Gas Laws First Look* lab file provided on disk. (This file is called *1stlook4.lab* in Windows.)

This tutorial assumes a basic knowledge of gas laws. While the focus of the tutorial is to teach you to work with *Explorer* and not to teach you chemistry, you may find it useful to review your chemistry textbook on gas laws.

How to Use the Tutorial

The tutorial is built around the following problem:

a. *A gas is confined to a cylinder with a moveable piston under constant atmospheric pressure. What changes will occur if the gas is heated from 298K to 500K?*

b. *Describe the energy changes the gas undergoes while being heated.*

This tutorial is divided into six sections:

Section 1: Addressing the Problem — Using the control and model tools

In this section you will learn to:

- open lab file windows
- access the work window tools and use some of them
- insert input value boxes and format them
- specify different processes and set initial values in the model window
- operate the simulation

Section 2: Displaying Outputs — Using the work window tools

In this section you will learn to:

- display numerical output from the simulation
- label displays
- make a line graph and a bar chart for displaying outputs

Section 3: Creating Buttons

In this section you will learn to:

- create a button
- use the button to see its effect

Section 4: The Spreadsheet — Dynamic sampling, summary sampling, plots

In this section you will learn to:

- generate a summary spreadsheet of data collected from multiple simulations
- generate a dynamic spreadsheet for collecting data during a single simulation
- plot spreadsheet results

Section 5: Finishing Touches — Screen design features

In this section you will learn to:

- name and arrange work windows in the Windows menu
- change text font and style
- color coordinate elements on the screen

Section 6: Extension — Effect of number of moles

In this section you are presented with an additional question that you can explore with your simulation setup.

The tutorial starts from the general template for the *Gas Laws* model. Each section guides you through steps in a particular area of lab construction, building on the previous section. You should save your work after each section, and begin the next section with the same file. When you have finished all six sections of the tutorial, you will have a lab file suitable for a class presentation. You will also understand the basics of the *Explorer* system.

The tutorial uses the following conventions:

- Indicates steps in analyzing the problem

» Indicates general tasks

➊➋➌ etc. Indicate steps you should perform to complete a given task

Section 1: Addressing the Problem

The Problem

a. *A gas is confined to a cylinder with a moveable piston under constant atmospheric pressure. What changes will occur if the gas is heated from 298K to 500K?*

b. *Describe the energy changes the gas undergoes while being heated.*

- There are four variables that define the state, or condition, of a gas:

  T, the temperature

  P, the pressure

  V, the volume

  n, the number of moles (the quantity of matter)

- In addition, depending on the process type, there might be changes in thermodynamic data:

  ΔE, changes in internal energy of the system

  ΔH, changes in enthalpy of the system

  W, the work associated with the expansion/compression of a gas

Keeping all this in mind, you are going to learn how you can explore a heating process similar to the problem above, and the changes in the related parameters using the *Explorer* tool.

Launching the Software

» Open *Student Edition Chemistry Explorer* by double-clicking its icon.

» In the dialog box that opens, choose the lab file *Gas Laws Template* (Mac, SE Chem Labs folder) or *tmplate4.lab* (Windows, in the \gas_laws subdirectory).

Note: If you are unfamiliar with the *Gas Laws* model, it is recommended that you work through the *Gas Laws First Look* file (*gas_laws**1stlook4.lab* in Windows).

Using a Template Window

- Let's introduce the parameters concerned with the process — the pressure, temperature, and the equation involved — into a work window.

» The Gas Laws Template contains a single work window:

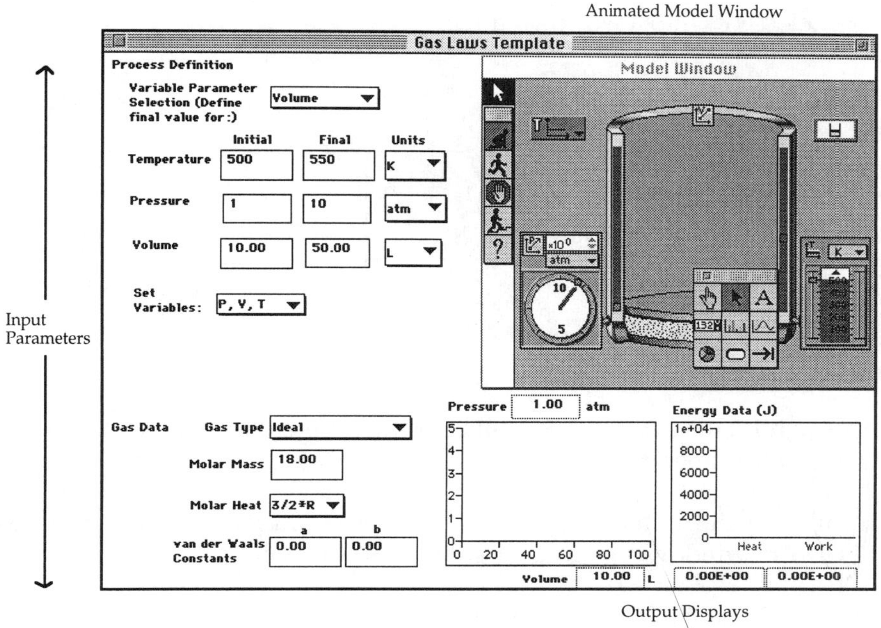

This work window already contains *Gas Laws* model input variables available for your use. (You could work directly in this window to experiment with the initial values in order to set up the simulation.) For most simulations, you will only use some of these variables.

There are specific inputs you will want to access during the simulation. For the given problem, you will want to change the gas conditions, so you will want to create a new work window with the appropriate inputs. Later you will add output displays and buttons to this window as well.

To create your new work window, follow these steps:

» From the Windows menu, choose the New Work Window option. A new, untitled work window is opened for you.

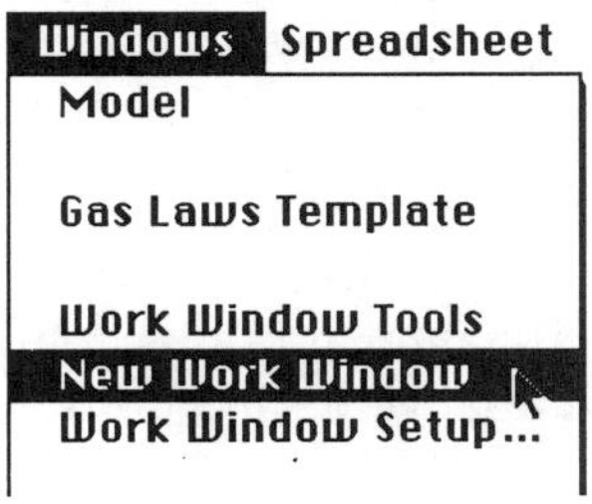

» Also from the Windows menu, choose the Work Window Tools option. This option opens a palette containing tools for introducing elements into a work window:

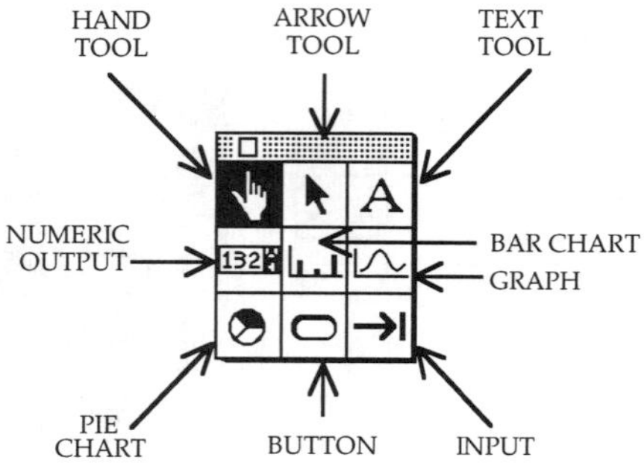

Throughout the tutorial, you will be using different tools from this palette. The detailed operations of the tools are explained in Appendix 1.

In order to copy items from the Gas Laws Template window into your new work window, you must use the ARROW tool.

❶ Click the ARROW tool to select it.

❷ Reopen the *Gas Laws* Template window by choosing it from the Windows menu:

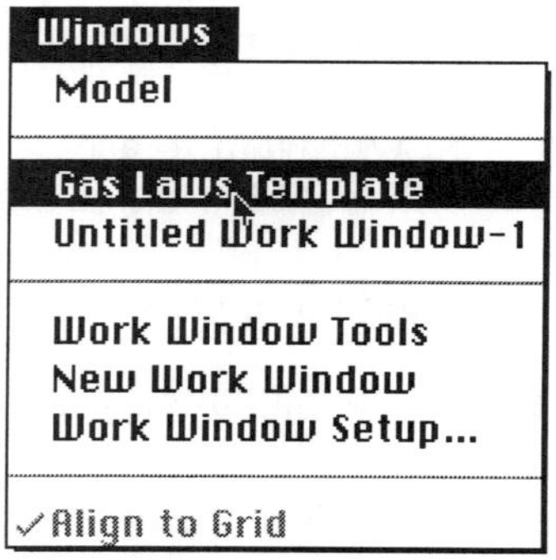

❸ With the ARROW tool, click the pressure initial value input box in order to copy it to your new work window. Small drag handles mark the corners of the item to indicate that it is selected.

You can hold down the Shift key and click additional items to select them as well. You will probably want to copy the following input boxes and their labels: pressure initial value and units; temperature initial and final values and units; volume initial value and units; Set Variables option.

❹ From the Edit menu, choose the Copy option to place these items on the Clipboard.

❺ Open your new work window from the Windows menu. Note that *Explorer* has automatically named it "Untitled Work Window-1" and added it to the Windows menu. From the Edit menu, choose the Paste option to place these items in the window.

You can use the ARROW tool to move or resize the items that you have just pasted.

When you copy the factors, remember to copy the text and units that go with the factors. (You can change the text if you want, or even delete it!)

There are two shortcuts for selecting multiple items: you can select all items in a work window by first activating the ARROW tool and then choosing the Select All option from the Edit menu. You can select specific items by clicking in a blank space on the screen and dragging across the desired items. Any item crossed by the "rectangle" created by the mouse will be selected when you release the mouse button.

Inserting Inputs with the INPUT Tool

The Template Window itself was built using the INPUT tool and the TEXT tool. You should know how to use these tools in case you accidentally delete an item from the Template Window, and then want to reinsert it.

» For example, suppose you have deleted the Initial Temperature input value box from the template, planning to set the value directly in the Model Window. Then you decide that you prefer to set the temperature's initial value in the work window so that you can set a more exact value than possible in the Model Window.

» For the sake of this exercise, use the ARROW tool to select the Initial Temperature input value box. Choose the Cut option from the Edit menu.

» Select the INPUT tool in the Work Window Tools palette. Place the cursor anywhere in your work window. Hold down the mouse button. A pop-up list of available inputs appears. Drag down to **Initial Condition Values**, drag across to **Temperature,** and release the button.

The **Initial Conditions—Temperature** input box is inserted into the window. If the value is too long to fit into the box, it will appear as an asterisk (*). You can drag a corner of the box to increase its size.

You can control the number of significant digits that are displayed in this input value box.

❶ Select the box with the ARROW tool.

❷ Choose the Format option from the Edit menu.

The following dialog box opens:

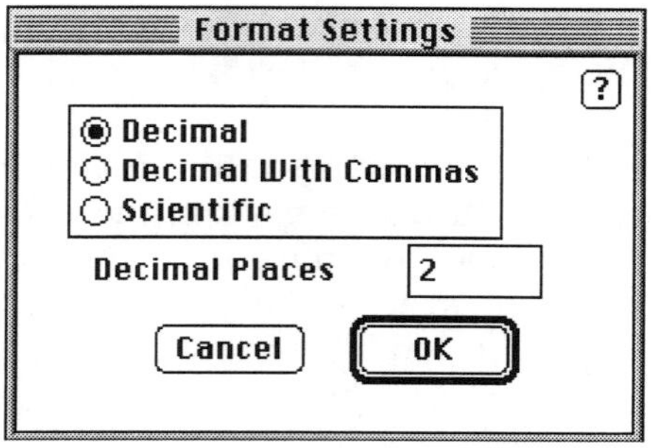

❸ Type 0 into the Decimal Places field.

❹ Click OK to accept the setting.

Next you should label the input box so that you know what variable it controls.

❺ Select the TEXT tool in the Work Window Tools palette. Place the cursor near the new input box. Click to create a text field, and then type a label for the input box that you have just inserted. To "finish" your text entry, click outside the text field or choose the ARROW tool.

You can use the ARROW tool to place or resize the box or the label. You can use the Font, Size and Style options in the Text menu to format it. You can also color it from the list in the Colors menu.

Setting Limits on an Input

- For this particular lab, the pressure for a given run is constant, and throughout the problem we will choose pressures no higher than 3 atm. Therefore, we want to limit the possible values of the initial pressure value box to a range of 0–3 atm.

❶ Select the ARROW tool. First double-click the Initial Pressure input value box. Its definition dialog box opens.

❷ Click in the Min. Lab Limits field and type 0. Click in the Max. Lab Limits field and type 3. Your dialog box should appear as below:

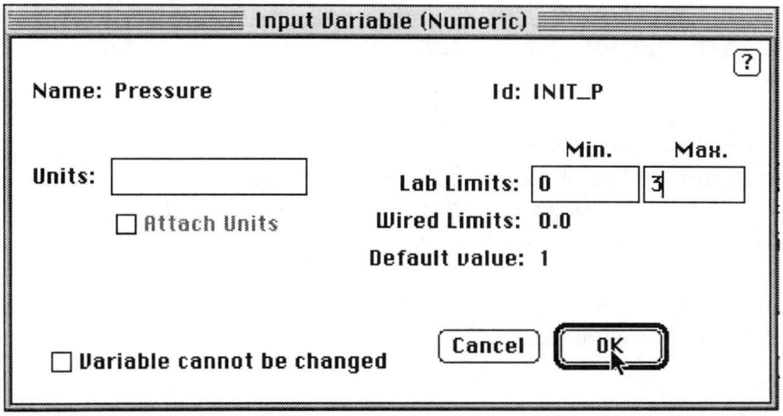

❸ Click OK to accept.

When an input variable has lab limits (defined for the specific file) or wired limits (preprogrammed in the model), its input box automatically appears with a scroll bar that enables you to scroll from the minimum to the maximum value.

Now when you use the scroll bar to set the initial pressure value, the values scroll from 0 to 3. If you type a number larger than 3 into the input value box, an alert informs you that the value is outside of the limits.

Note that once you have changed the parameter's limits, it is limited in the Model Window as well. (You will change values in the Model Window next.) **Although you can drag the indicator on the pressure gauge beyond 3 atm, the model resets itself to this upper limit.**

However you decide to build the window to answer the problem, your final window should look something like the one we show here:

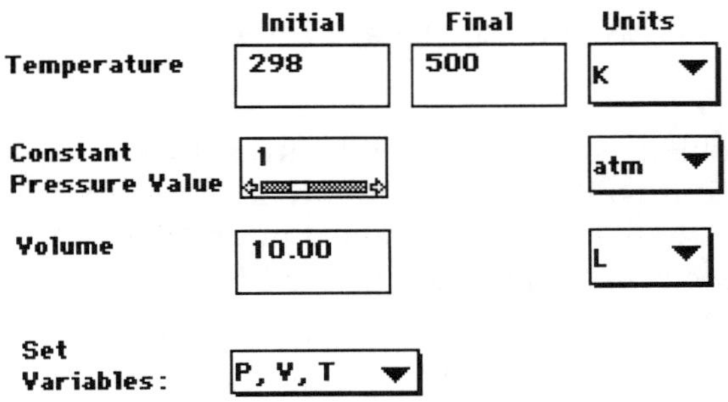

Figure 1

Notice that since for this problem the pressure remains constant for each simulation run, we have retitled Initial Pressure Value to Constant Pressure Value.

Note also that since the problem does not consider the number of moles n of the confined gas, we choose the Set Variable option for P, V, and T for this problem. The *Gas Laws* model will automatically calculate the number of moles to use for the simulation.

At this stage your window may appear somewhat untidy. Use the ARROW tool to arrange the items in the window.

Setting Up the Model Window

Now let's look at the Model Window. If you are unfamiliar with the elements in this window, refer to the *Gas Laws First Look* lab file (*gas_laws**1stlook4.lab* in Windows) or to Appendix 2 of this guide.

You are building a presentation to simulate a constant pressure process.

» Use the PROCESS TYPE menu to select the constant pressure process.

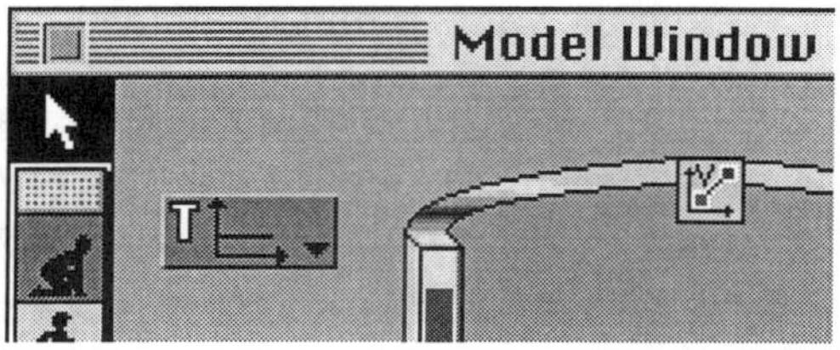

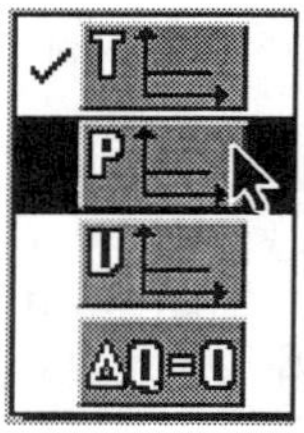

Using Inputs in the Model Window

You have put several input value boxes in the work window. An alternative way of setting certain input values is via controls in the Model Window.

Back to the problem: you were asked to investigate a gas under a constant pressure of 1 atm while the temperature is increased from 298 to 500K. Translating this to the model's inputs, it means that the pressure defines the initial conditions and the temperature is the changing parameter.

» Click the slider on the pressure control to activate it:

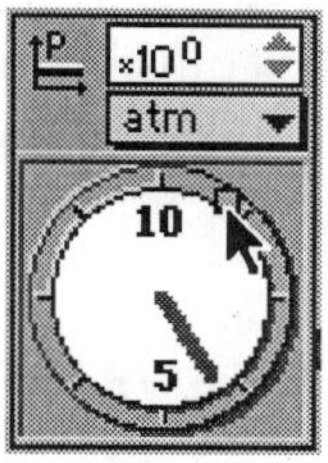

The slider should be green when active. Drag it to a value of 1 atm, if this is not currently the value (displayed at the top of the window). Note that when you click within the pressure control, a readout appears at the top of the Model Window giving the current pressure. You will use this readout later on!

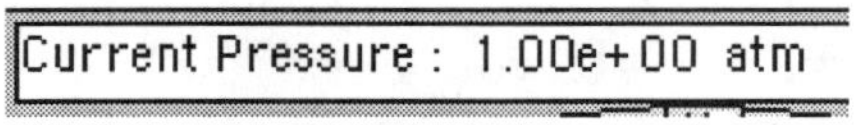

» Next click the small graph in the temperature control:

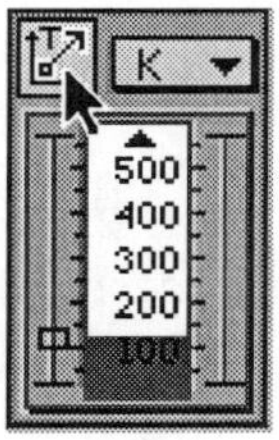

This graph indicates that you will set initial *and* final values for the temperature range, using the green lever for the initial temperature and the red one for the final temperature. In general, to change the temperature range by increments of 500 click one of the arrows in the control.

Then to set a temperature value in increments of 10 degrees, drag the given slider on the side of the temperature control:

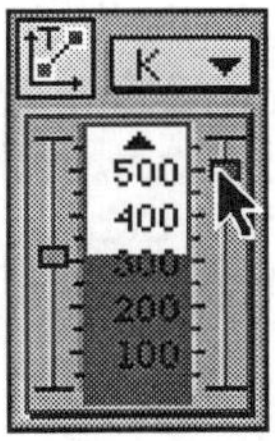

Note that you can change the units of the temperature using the pop-up menu at the top of the control. For the given problem, leave the units as Kelvin. 

Since the initial temperature required by the problem is 298K, use the input value box in the work window to achieve the required exactness. Type in the value and then click outside the box to "enter" the value. Set the final temperature value to 500 by dragging the red slider on the temperature control, as shown above.

Running the Simulation

- You are now ready to see what happens when you heat the gas inside a container according to the values you set!

» Check that the initial values displayed in your work window match those shown in Figure 1 above.

Make sure that the HAND tool  is selected in the Work Window Tools palette. The HAND tool is used for normal operation of the simulation (i.e., nondesign work). If you close the tool palette, the cursor automatically defaults to the HAND, ready for you to use the simulation.

» Click the GO tool  to heat the gas. When you click this tool, the simulation begins.

» When the final temperature reaches 500K, click STOP  to halt the simulation.

» To run the process again, click RESET  to return the simulation to its initial conditions. Then click GO.

» To heat the gas in a controlled way, click SINGLE STEP  a number of times and observe the changes at each step. This tool advances the animation one calculation step each time you click it.

If you haven't already saved your lab file, you should do so now. (First click RESET.) Choose the "Save As" option from the File menu and give this lab a name such as *My Presentation* (or an 8-letter name with the .lab extension in Windows).

Section 2: Displaying Outputs

In this section, you build on the lab you made in Section 1, learning about different ways to display your simulation results.

» Open the *Student Edition Chemistry Explorer* by double-clicking its icon. In the dialog box that opens, choose your file to open it.

In Section 1 you already saw one way of displaying the changes: the Model Window meters showed that the gas expanded while heating. Now let's see how you can use *Explorer* to display your results in other ways and to investigate the relationship between the results.

There are a number of ways you may want to display results. For example, you can have a simple numerical display to see the current values, or a line graph which relates the outputs to each other and enables you to follow the trend of changes during the simulation. Let's show the outputs using both these methods.

» Before continuing, ensure that RESET is selected in the Control Tools palette.

Creating a Numeric Display

❶ Open the Work Window Tools palette if it is currently closed. Select the numerical output icon.

❷ When you move the cursor into the work window and click, the cursor changes into a rectangle. Click and drag to draw a rectangle in the work window:

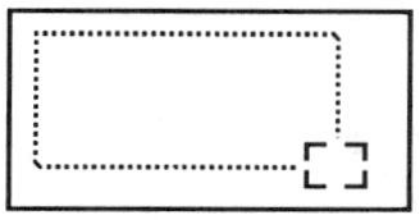

After creating the box, *Explorer* opens a Numeric Display dialog box, where you can enter an expression, whose value you want to display in the work window.

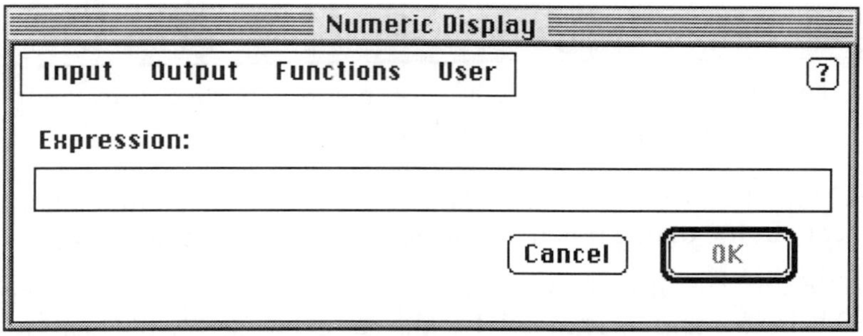

- For the given problem, you will display the current volume and temperature of the gas.

❸ Click the Output pop-up menu and drag down to the Current Conditions (Given Units) submenu, and then across to the Temperature output.

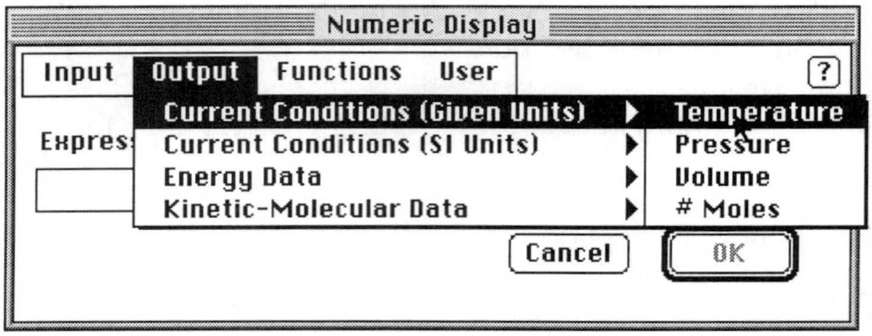

Explorer automatically enters a system code for the Temperature (CURRENT_T) into the Expression field of the dialog box. You do not need to remember this code, since it is entered for you.

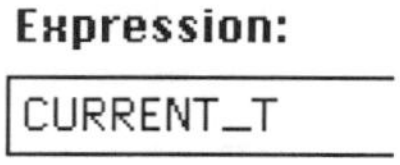

❹ Click OK to accept the display definition.

❺ You can use the ARROW tool to resize and place the display.

❻ Use the TEXT tool to label your Numeric Display. (Remember that after typing the text, you can place and resize the field with the ARROW tool.) Don't forget to mark the units as well.

Repeat these steps for the volume output.

Creating a Line Graph

» Let's see how the *Explorer* line graph can display what happens to the system as the gas is heated.

❶ Click the LINE GRAPH tool.  Click and drag to define the graph the same way you did the numeric displays.

The Line Graph dialog box automatically opens.

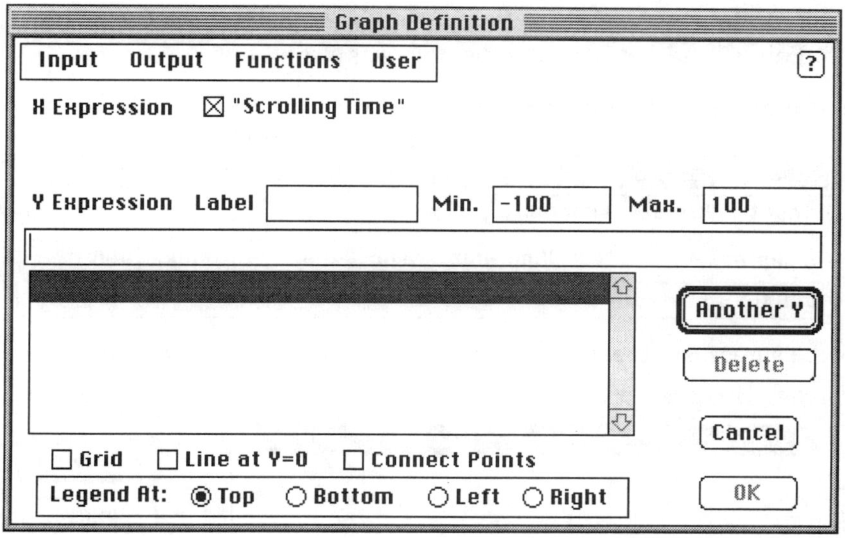

The X expression is defined by default as "Scrolling Time." Scrolling time means that the x-axis scale scrolls automatically to ensure that the current value is always in view. The units are in simulation calculation steps (or cycles).

❷ In our case, click the Scrolling Time check box to deactivate it. Fields (text entry lines) appear for you to define the X expression and the range limits.

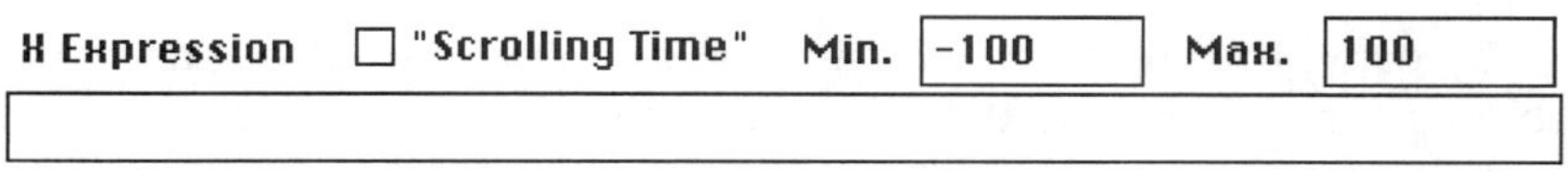

❸ Click in the X expression field to move the cursor. As you did for the first numeric display you created, enter the output variable for current temperature in the field.

❹ The temperature range has been set to 298–500K. Set the minimum X limit to 200 and set the maximum X limit to 600 to ensure that the plot fits horizontally onto the graph.

❺ Now click in the Y expression field to move the cursor. Choose the output variable for Current Volume (Given Units) from the Output pop-up menu.

- To set the Y range, you can use trial and error, or analysis. For example, you know that the value starts at 10L and increases, so you can estimate that the volume will fall in the general range of 10–20 liters.

❻ Set the Y range from 10 to 20.

❼ From the options at the bottom of the dialog box, click the "Connect Points" check box to plot the points as a solid line.

Your dialog box should appear as shown:

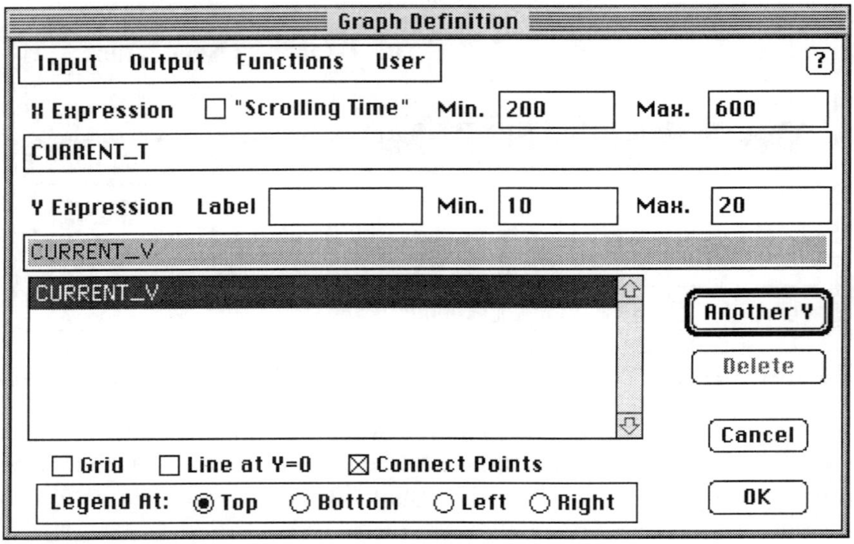

❽ Click OK to accept the graph you have defined.

❾ Use the TEXT tool to label the axes.

You may often find it useful to place the numeric display adjacent to the graph axis that represents the same variable. In this way you obtain an exact instantaneous reading in addition to the graph's depiction of the values' changes over time. In this case, the displays can also share labels. A typical arrangement of this type is shown below:

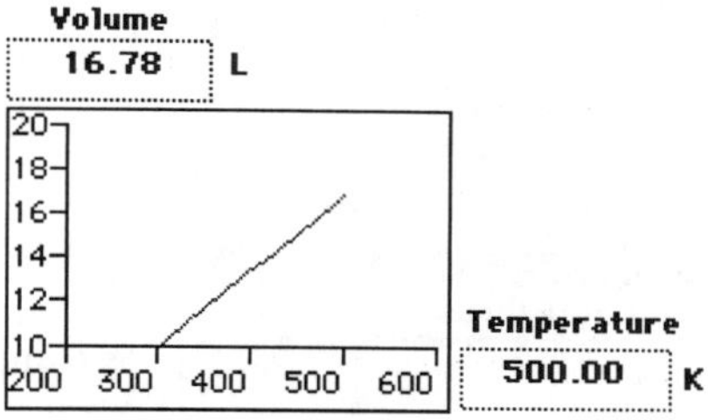

Creating a Bar Chart

The given problem also asks you to describe the energy changes for this process. Dealing with these parameters, we usually speak about heat and work *gained* or *lost* by the system, and whether the internal energy has increased or decreased. How these parameters change depends on the process involved. We suggest displaying the energy outputs as a bar chart:

❶ Click the BAR CHART tool  and draw the bar chart in the same way you did the numeric display.

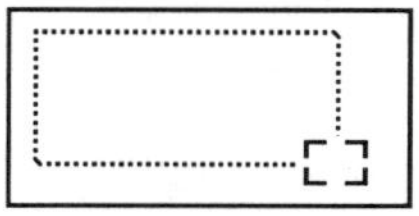

The Bar Chart dialog box opens.

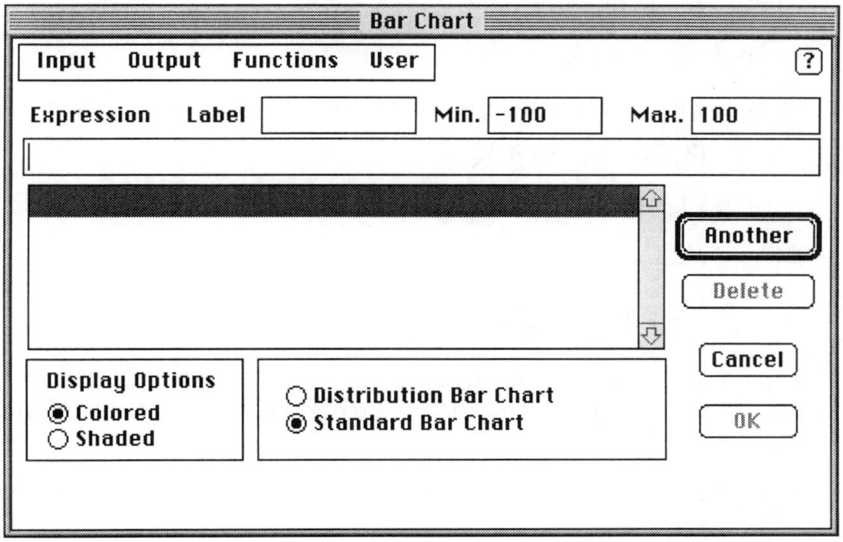

❷ From the Output - Energy Data submenu, choose the Heat output:

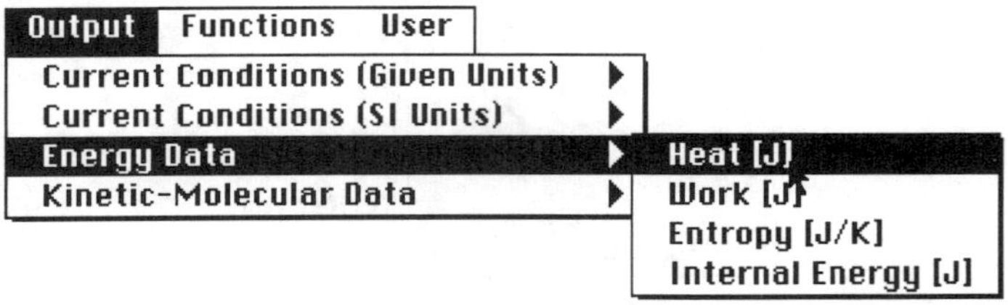

❸ In the Label field, type "ΔH" for the bar on the chart legend if you are working on a Macintosh (for Δ, hold down the Option key and type j) or "delta H" if you are working in Windows.

❹ Click the "Another" button to add a second bar to the chart. For this bar, you need to choose the Work output from the Energy Data submenu. Since the expanding gas does work on its surroundings, you actually want the *negative* of this output value. (When the heat is calculated in the constant pressure mode, ΔH corresponds to the enthalpy change.)

Click in the Expression field. Enter a minus sign (-) from the keyboard, and then choose the Output – Energy Data – Work parameter. Label this bar "W."

❺ Again click the "Another" button. The third bar on the graph should be ΔE; however, the *Gas Laws* model does not have a standard output for this variable. You can display the value for ΔE by creating an expression which equals ΔE: ΔH + W.

❻ Click in the Expression field. Choose the Heat output again.

❼ From the keyboard, enter a plus sign (+).

❽ From the keyboard, enter a left parenthesis and a minus sign (-).

❾ Choose the Work output again.

❿ From the keyboard, enter a right parenthesis.

The field should look like this:

```
CURRENT_HEAT+(-CURRENT_WORK)
```

⓫ Label this expression ΔE if you are working on a Macintosh (for Δ, hold down the Option key and type j) or delta E if you are working in Windows.

⓬ Enter -5000 for the minimum value and 5000 for the maximum value.

Your dialog box should look like this:

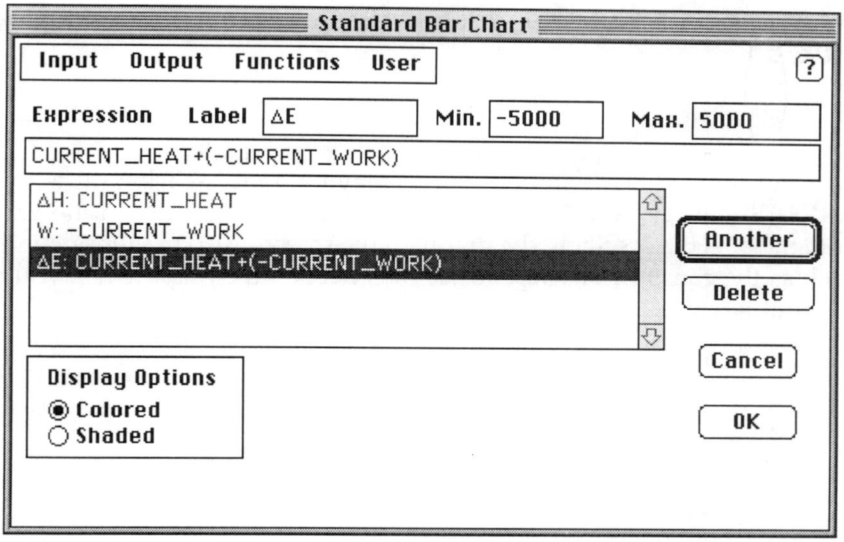

⓭ Click OK to accept the bar chart definition.

⓮ Use the ARROW tool to resize and place the bar chart.

Click GO to see what you can learn from the displays during the simulation.

Below is shown a typical chart (shown after starting the simulation):

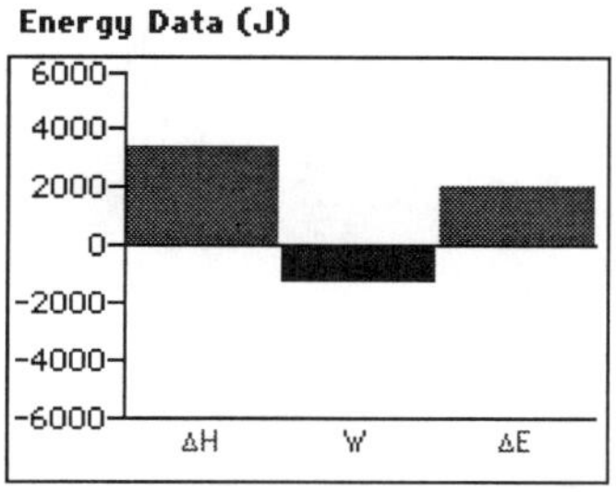

A numeric display box can only give a single, exact output. The bar chart and the pie chart can display more than one output on the same display. The line graph (described above) can display multiple outputs plotted against an expression of your choice. In many cases you will want to use these displays to give more useful, comparative information on the screen.

Can you answer part (a) of the problem?

What changes will occur if the gas is heated from 298K (room temperature) to 500K?

» Save your work.

You can review the definition of a display by double-clicking it with the ARROW tool. Try altering parameters, such as the ranges, in the graph; note the effects of resizing the graph on the screen. Also try altering inputs such as the constant pressure to see the effect on the graph when you run the simulation.

Section 3: Creating Buttons

A very useful item in an *Explorer* work window is a button. Buttons can be used for performing operations related to the specific model, or for general operations like recording data to the spreadsheet or resetting the lab file to the original default values.

» Open your lab file.

Creating a Button

- In studying the energy changes that the gas undergoes during heating, it may be useful to display or hide the energy and/or work arrows in the Model Window.

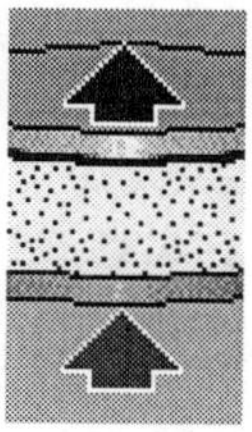

❶ Open the Work Window Tools palette if it is closed. Select the BUTTON tool  in the palette.

❷ Click and drag with the cursor to create a button. (If you are working in Windows, make the button adequately large to type the name on it.)

❸ The cursor blinks on the button, waiting for you to type in a name. Type "Show Arrows." Click anywhere outside of the button to set the text.

❹ With the ARROW tool, double-click the new button to access its dialog box.

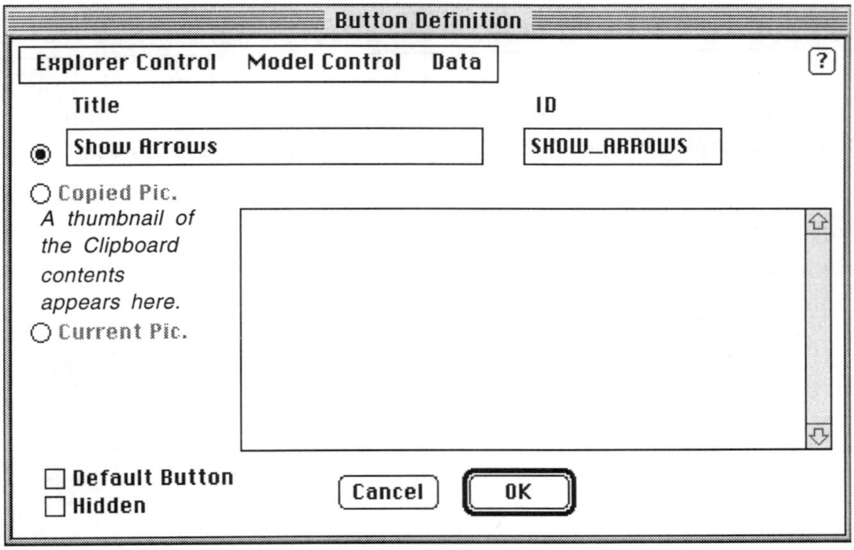

The name that you typed on the button is automatically entered in the "Title" field. If you want to change the name, you can type a new name in this field, or you can change it directly on the button by highlighting it with the TEXT tool and typing a new name. The name is automatically updated in both places.

Explorer automatically assigns an ID to the button. This ID is for internal purposes, and you will not need to use it in the *Student Edition* of *Explorer*.

Across the top of the dialog box are three pop-up menus: Explorer Control, Model Control, and Data. These menus contain lists of operations — called reactions — that are performed when you click the button. You can choose one or more of these reactions for the given button. (These reactions are summarized in Appendix 1.)

- We want this button to turn on the heat and work arrows.

❺ Click the Model Control menu and drag down to the Show Heat Arrow reaction (located in the Show/Hide Energy Arrows submenu.)

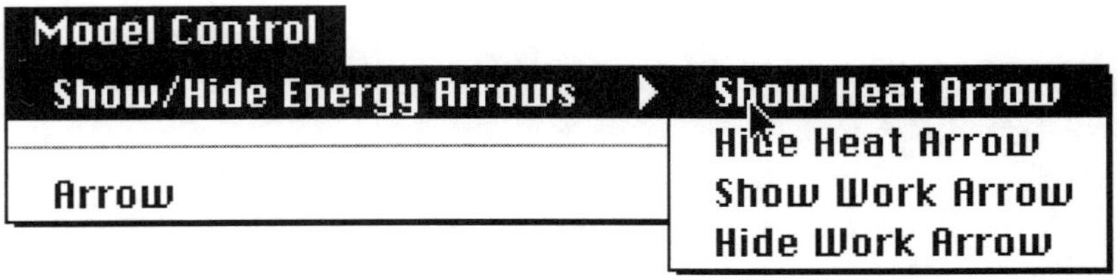

Explorer lists the reaction:

```
Do Model Reaction:  Show Heat Arrow
```

❻ Choose the Show Work Arrow reaction to add it to the button too.

```
Do Model Reaction:  Show Heat Arrow
Do Model Reaction:  Show Work Arrow
```

❼ Click OK to accept the button definition.

» Create a second button with the corresponding hide arrow reactions.

```
Do Model Reaction:  Hide Heat Arrow
Do Model Reaction:  Hide Work Arrow
```

Using the Button

Now let's see how the button works.

» Choose the HAND tool to move from editing to operation mode.

» Click the "Show Arrows" button.

» Click GO to start the simulation. Note the two arrows: the red arrow shows the direction of transfer of heat; the blue arrow shows if work is done on (arrow in) or by (arrow out) the gas.

» Click the RESET tool.

» Click the "Hide Arrows" button.

» Click GO to repeat the simulation. Note that now the arrows do not appear.

The reactions available for defining a button are summarized in Appendix 1.

Save your file.

Section 4: The Spreadsheet

Important aspects of scientific analysis are numerical measurement and the manipulation of measurements made in order to illustrate general conclusions. *Explorer* comes with a built-in Excel™ compatible spreadsheet which enables you to perform numerical analysis and manipulation. Among the functions of the *Explorer* spreadsheet are:

Recording simulation data

Inputting numerical values of expressions to the spreadsheet

Performing operations on the numerical values

Generating graphs from the numerical data

The *Explorer* spreadsheet may be operated in one of two modes:

As a "summary" spreadsheet where you input a result to the spreadsheet and repeat this process for a set of "runs" of an experiment.

As a "dynamic" spreadsheet where data continuously inputs to the spreadsheet during one experiment.

Open your file.

Generating a Summary Spreadsheet

Suppose you would like to find out whether the **work** done by your current gas can be increased. What experiments should you do and how will you summarize all the data you obtain?

You know that $W = -P \Delta V$. Does this mean that increasing P only (keeping all other parameters constant) will decrease the work? Does the temperature range affect work?

Let's see how you can take sets of results and input these to the spreadsheet.

- We will consider the effect of:

  Varying the pressure **P** on the work

  Varying the final temperature **T** on the work

In both of these cases, we will keep all the other input parameters constant, at the values shown in Figure 1 (page 17).

•» For these simulation experiments as for experiments in real life, you will perform a "run," you will then record the values you measure at the end of the run,

» via a button you create,

•» then change an input parameter and repeat the run.

Pressure vs. Work

To record the values at the end of the run, you need to build a button:

❶ Click and drag the BUTTON icon to create a button.

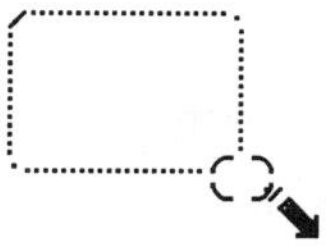

❷ Type "Sample to Spreadsheet" as the title for the button, then click outside the button to set the text.

❸ Use the ARROW tool to double-click the new button to access the Button Definition dialog box.

❹ In the dialog box click the Data menu and drag to the Write Into... reaction:

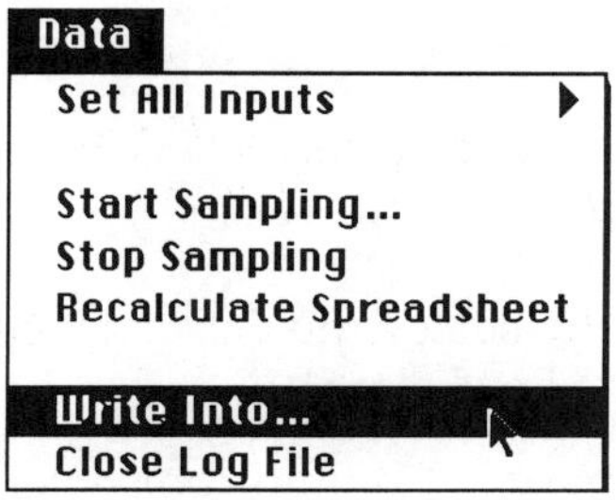

A dialog box opens:

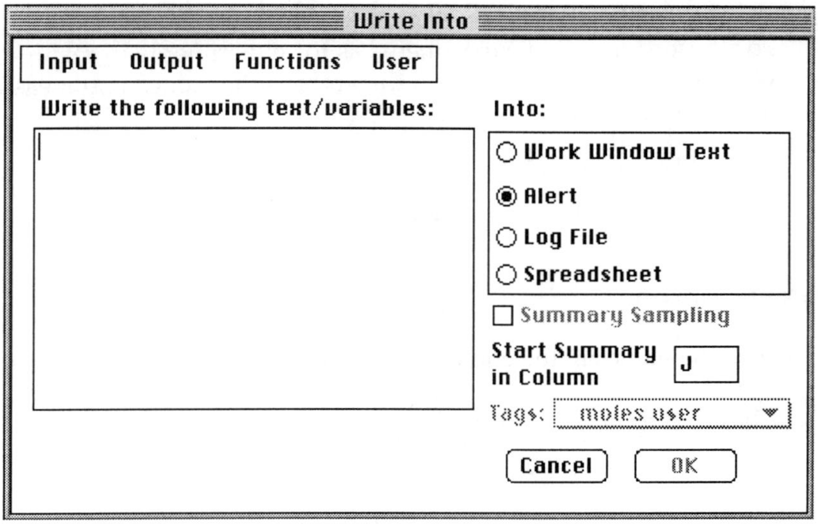

The Write Into dialog box allows you to write text and/or values of variable expressions into one of four places:

A Work Window Text box (not operable in the *Student Edition*)

An Alert Window

A Log File

The Spreadsheet

We want to write into the spreadsheet (every time the button is clicked) the values of the pressure, ΔV, and work.

❺ In the "Into:" field select the Spreadsheet, and then check the Summary Sampling box below it.

❻ In the "Start Summary in Column" box enter **A.** (You can choose any column.)

Summary sampling saves the data in the spreadsheet from run to run. (If summary sampling is not selected, data in the spreadsheet is erased every time RESET then GO are clicked.)

❼ Place the cursor in the "Write the following text/variables:" field to activate it. From the Input menu, select the Initial Pressure input from the Initial Condition Values submenu.

The code {IN$INIT_P} is written into the field. The braces indicate that *Explorer* will write the numerical value of this parameter. If the braces are missing, *Explorer* writes the parameter code as simple text.

8. To sample the ΔV, you must subtract the current volume from the initial volume. Choose the output parameter for Current Volume. After the code appears, place the cursor *inside* the right-most brace:

Write the following text/variables:

```
{IN$INIT_P}{OUT$CURRENT_V}
```

9. Type a minus sign from the keyboard.

10. Choose the input parameter Initial Volume.

The numerical values of expressions *inside braces* are entered into consecutive columns of the spreadsheet, starting in the column designated in the dialog box.

11. Move the cursor to the right of the last brace. Choose the output parameter "Work."

Your Write Into dialog box should appear as below:

Write the following text/variables:

```
{IN$INIT_P}{OUT$CURRENT_V-
IN$INIT_V}{OUT$CURRENT_WORK}
```

12. Click OK to accept the dialog box and return to the Button Definition box. Click OK to accept your button definition.

    To see the results of creating your button, open the spreadsheet from the Spreadsheet menu:

» Choose the HAND tool to move from editing to operation mode.

» Click RESET, then run a simulation using GO.

» Click STOP when the heating has finished, then click your "Sample to Spreadsheet" button.

You should see the results of your run, similar to that shown here, displayed in row 4 of columns A, B and C of the spreadsheet.

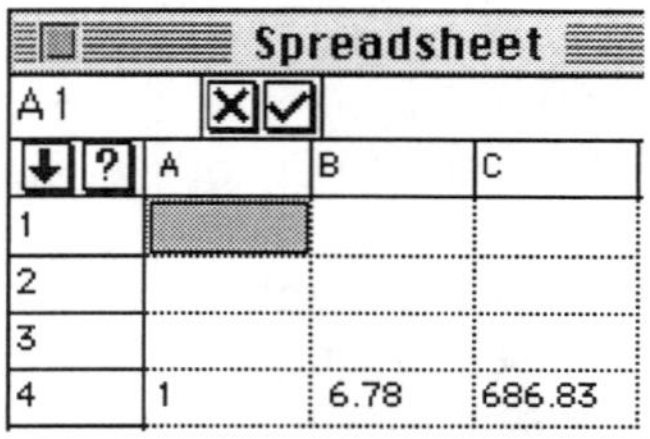

Spreadsheet

A1

	A	B	C
1			
2			
3			
4	1	6.78	686.83

» Let's title the columns.

❶ Click cell A3 to select it.

❷ Type P (atm) and press the Enter key on the keyboard or click the check mark in the edit box.

❸ Select cell B3 and type ΔV if you are working on a Macintosh (for Δ, hold down the Option key and type j) or delta V if you are working in Windows. Press the Enter key.

❹ Select cell C3 and type Work (J). Press the Enter key.

» Erase your trial results in A4, B4, and C4 by selecting these cells and choosing the "Cut" option in the Edit menu.

You are now in a position to use your button and the spreadsheet to record data from simulation runs.

» Click RESET, then ensure that all inputs are as shown in Figure 1.

» Click GO. Click STOP when the process finishes.

» Click the "Sample to Spreadsheet" button.

» Repeat the steps above for pressures 1.5, 2, 2.5, 3 atm. Remember to click RESET *before* changing the pressure value, unless you want to repeat a simulation with the present value.

Your spreadsheet should now display all the run results, similar to those shown here:

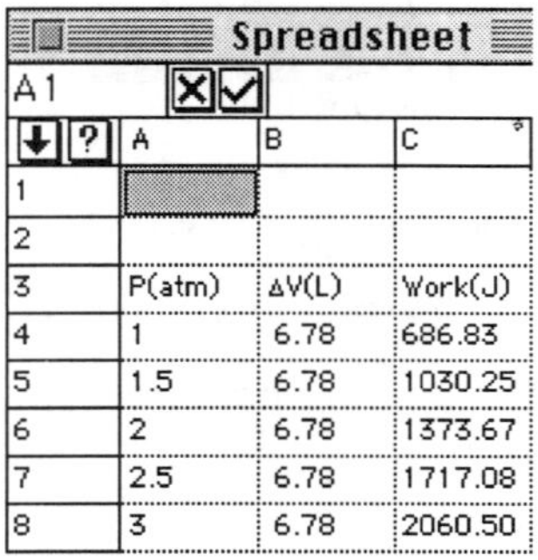

Spreadsheet

A1

	A	B	C
1			
2			
3	P(atm)	ΔV(L)	Work(J)
4	1	6.78	686.83
5	1.5	6.78	1030.25
6	2	6.78	1373.67
7	2.5	6.78	1717.08
8	3	6.78	2060.50

What are your conclusions? In order to explain what happens, investigate the changes by calculating the ratios of the P values between successive runs and the work values from successive runs.

❶ Highlight cell D5.

❷ In the edit box at the top of the spreadsheet,

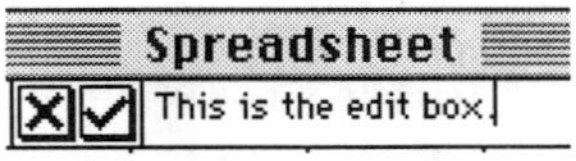

type an equal sign (=). This sign indicates that *Explorer* will calculate the numeric value of the expression in the edit box. Without the equal sign, *Explorer* reads the contents of the edit box as simple text.

❸ Click cell A5. It is entered into the edit box.

❹ From the keyboard, type a slash (/).

❺ Click cell A4.

❻ Click the check box to accept the expression:

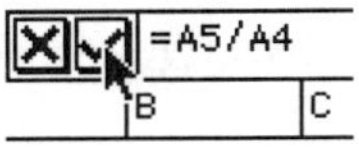

❼ From the Edit menu, choose the Copy option in order to copy the expression. Highlight the next ten rows in column D and the choose Paste option. The expression is pasted into the cells, and the row numbers are automatically updated to the relevant row. For example, the expression in cell D6 reads A6/A5.

Note that if you paste an expression into a cell for which the expression depends on a still-empty cell, *Explorer* displays the code "err." When data is sampled into the referenced cell, then *Explorer* can calculate the value that belongs in this cell.

A	B	C	D
P(atm)	ΔV(L)	Work(J)	
1	6.78	686.83	
1.5	6.78	1030.25	1.50
2	6.78	1373.67	1.33
2.5	6.78	1717.08	1.25
3	6.78	2060.50	1.20
still empty			0.00
still empty			err

Typical results are:

Spreadsheet

A1

	A	B	C	D	E
1					
2					
3	P(atm)	ΔV(L)	Work(J)		
4	1	6.78	686.83		
5	1.5	6.78	1030.25	1.50	1.50
6	2	6.78	1373.67	1.33	1.33
7	2.5	6.78	1717.08	1.25	1.25
8	3	6.78	2060.50	1.20	1.20

Repeat these steps to calculate the value of C5/C4 in cell E5, and paste for additional rows.

From the results, you should see that the ratio of the "P's" (P_2/P_1) is the same as the ratio of the "Works" (W_2/W_1).

» This is a convenient place to save your lab.

Final Temperature vs. Work

To record values of final temperature vs. work (all other parameters being held constant), first click RESET and then set P=1 atm.

» In the spreadsheet directly below your Pressure vs. ΔV results, type the following labels into columns A and B respectively: Temperature (K), Work. Do not leave a blank row, or *Explorer* will sample into this row.

If the column is not wide enough to see the entire label, move the cursor onto the border between column headings and drag the border to the desired size:

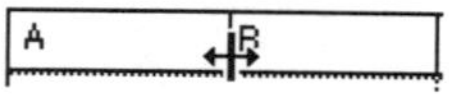

Change the "Sampling to Spreadsheet" button:

❶ With the ARROW tool, double-click the button to open the button definition dialog box.

❷ Double-click the reaction to access the Write Into dialog box:

Write Into Spreadsheet: {IN$INIT_P}{... (Summary)

❸ Delete the contents of the Write the Following... field by highlighting all the text and selecting the Cut option from the Edit menu.

❹ Choose the input parameter Final Condition Values: Temperature.

❺ Add the output parameter Energy Data: Work. The field should look like this:

```
{IN$FIN_T}{OUT$CURRENT_WORK}
```

❻ Click OK to accept the Write Into... reaction. Then click OK in the Button Definition dialog box to accept your new definition of the button and return to your work window.

Now try your new button:

» Click RESET.

» Set the final temperature to 400K.

» Click GO. Click STOP when the process finishes.

» Click the "Sample to Spreadsheet" button.

» Repeat the steps above for final temperature values of 400, 500, 600, and 700K. (Click RESET before you change the temperature.)

Your spreadsheet should now display all the run results, similar to those shown here:

Temperature(K)	Work
400.00	346.82
500.00	686.83
600.00	1026.85
700.00	1366.87

You should see that as the final temperature increases, the work done by the gas also increases.

Saving Your Spreadsheet

You can save any *Explorer* spreadsheet to a special spreadsheet file. This special file is useful, for example, when you want to save the results and then use them at a later date.

Save the spreadsheet you have created:

1. Make sure the spreadsheet window is open and active. (Click in the window to activate it.)
2. In the File menu, select Save Spreadsheet As...

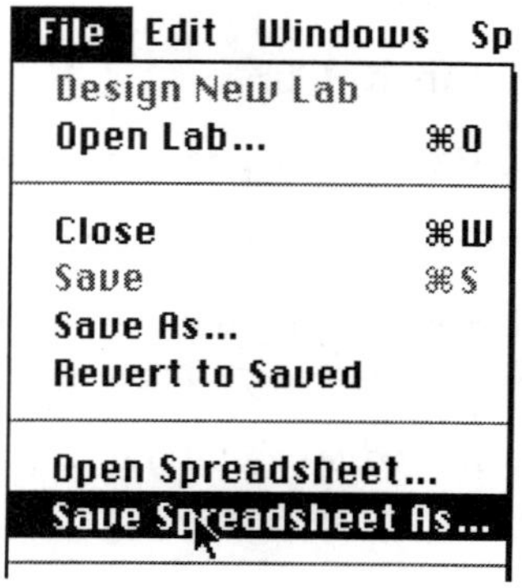

3. Enter a name, such as *Changing Work* , and click Save to accept.

To open a previously saved spreadsheet, choose the Open Spreadsheet option from the File menu. Note that this option is only available when the Spreadsheet window is active.

Generating a Dynamic Spreadsheet

WARNING: In this section the spreadsheet you generated in the Summary Spreadsheet section is overwritten! Make sure your Summary Spreadsheet is saved as explained above.

The spreadsheets you have produced to now – Summary Spreadsheets – have had results entered by you clicking a button, *after* a run. In the spreadsheets we consider now – Dynamic Spreadsheets – results are entered automatically and continuously *during* a run.

We will generate a dynamic spreadsheet of the changes of energy during the heating of the gas.

1. Highlight the labels and data in the spreadsheet and choose Cut from the Edit menu.

❷ In the Spreadsheet menu select Sampling... to open the Spreadsheet Sampling dialog box.

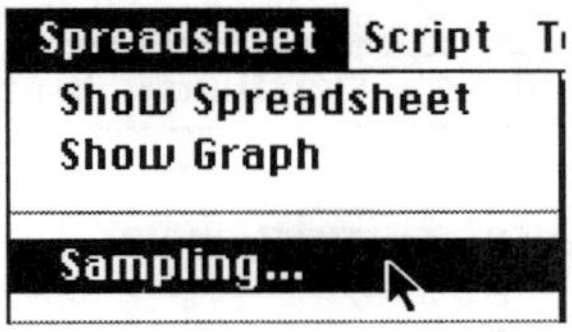

A dialog box opens:

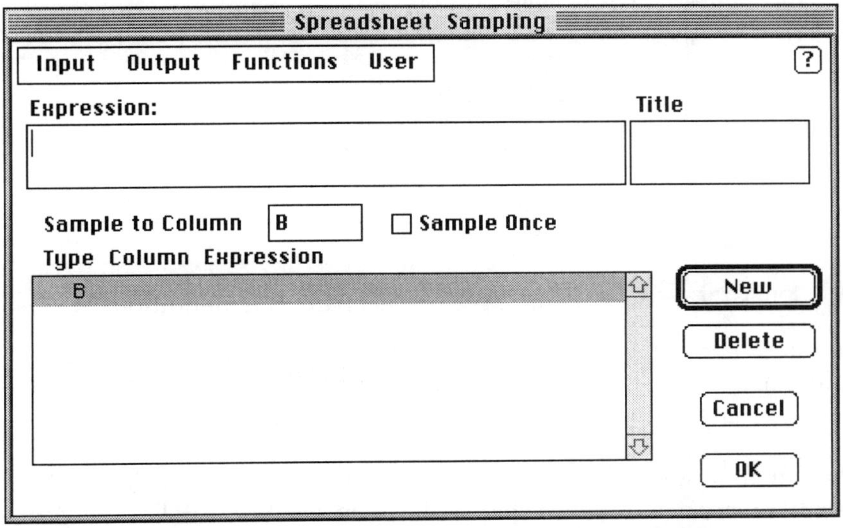

❸ In the Output menu choose Current Volume, to sample the current volume to column B. Title the column V (L).

❹ Click New to set column C.

❺ In the Output menu choose Current Temperature, to sample the temperature to column C. Title the column T (K).

❻ Click New to set column D and in the Output menu choose Energy Data: Heat to sample heat to column D. Title the column Heat (J).

❼ Click New to set column E and in the Output menu choose Energy Data: Work to sample work to column E. (Remember to type a minus sign in front of the parameter!) Title the column Work (J).

❽ Click New to set column F and build the expression "Heat minus Work" to input to column F. Title the column ΔE (deltaE in Windows).

The completed dialog box should look like this:

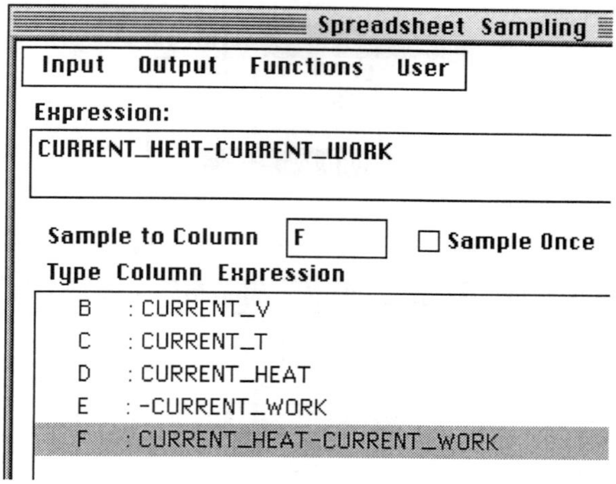

❾ Click OK to accept your Spreadsheet Sampling definitions.

The steps above tell *Explorer* what is sampled to the spreadsheet. The values are automatically sampled to the spreadsheet when you click GO, and sampling continues until the simulation stops.

The dynamic spreadsheet automatically formats numeric values to 8 decimal places. In order to change this, select columns B–F:

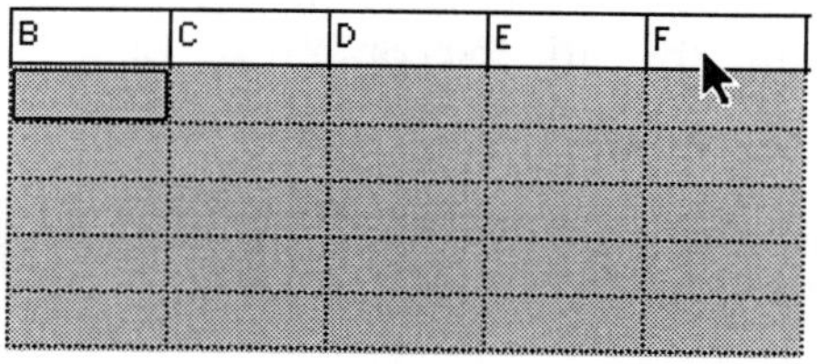

❼ In the Edit menu, choose the Format option and set the columns to Decimal. Enter 2 in the Decimal Positions:

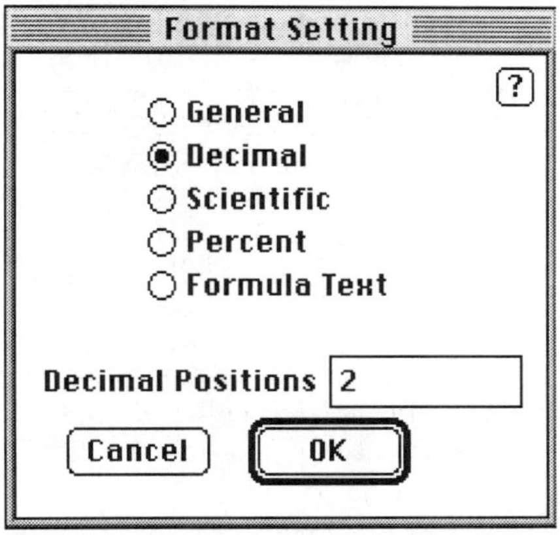

To see the results, open the Spreadsheet. (Remember: You should have deleted your summary results.)

» Click RESET. Set the final temperature value back to 500. Click GO. When the simulation halts, click STOP.

Your dynamic spreadsheet results should appear similar to the example shown here:

V (L)	T (K)	Heat (J)	Work (J)	ΔE
10.00	298.00	0.00	-0.00	0.00
10.68	318.20	171.71	-68.68	103.03
11.36	338.40	343.42	-137.37	206.05
12.03	358.60	515.12	-206.05	309.08
12.71	378.80	686.83	-274.73	412.10
13.39	399.00	858.54	-343.42	515.12
14.07	419.20	1030.25	-412.10	618.15
14.74	439.40	1201.96	-480.78	721.18
15.42	459.60	1373.67	-549.47	824.20
16.10	479.80	1545.38	-618.15	927.23
16.78	500.00	1717.08	-686.83	1030.25
16.78	500.00	1717.08	-686.83	1030.25

Graphing the Spreadsheet Results

You can display your spreadsheet results graphically:

Volume vs. Work

❶ Click and drag the mouse across the cells containing the data you want to graph. You can select an entire column by clicking the letter at the top of the column. You can select nonadjacent columns by holding down the Command key (Macintosh) or the CTRL key (Windows) while you select columns. In this case, select columns B and E.

V (L)	T (K)	Heat (J)	Work (J)	ΔE
10.00	298.00	0.00	-0.00	0.00
10.68	318.20	171.71	-68.68	103.03
11.36	338.40	343.42	-137.37	206.05
12.03	358.60	515.12	-206.05	309.08
12.71	378.80	686.83	-274.73	412.10
13.39	399.00	858.54	-343.42	515.12
14.07	419.20	1030.25	-412.10	618.15
14.74	439.40	1201.96	-480.78	721.18
15.42	459.60	1373.67	-549.47	824.20
16.10	479.80	1545.38	-618.15	927.23
16.78	500.00	1717.08	-686.83	1030.25
16.78	500.00	1717.08	-686.83	1030.25

❷ In the Spreadsheet menu select New Line Graph...

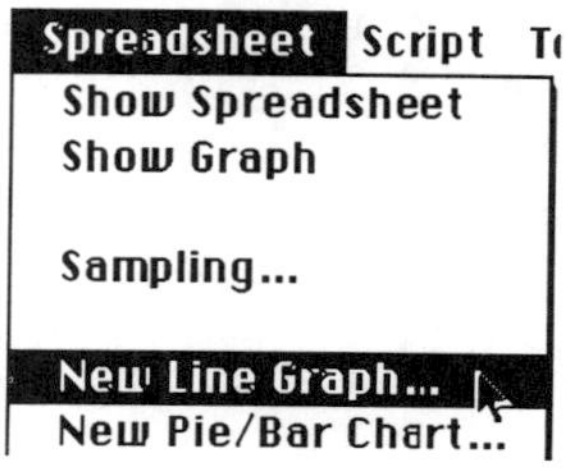

❸ Complete the Assign Ranges dialog box by accepting the default values:

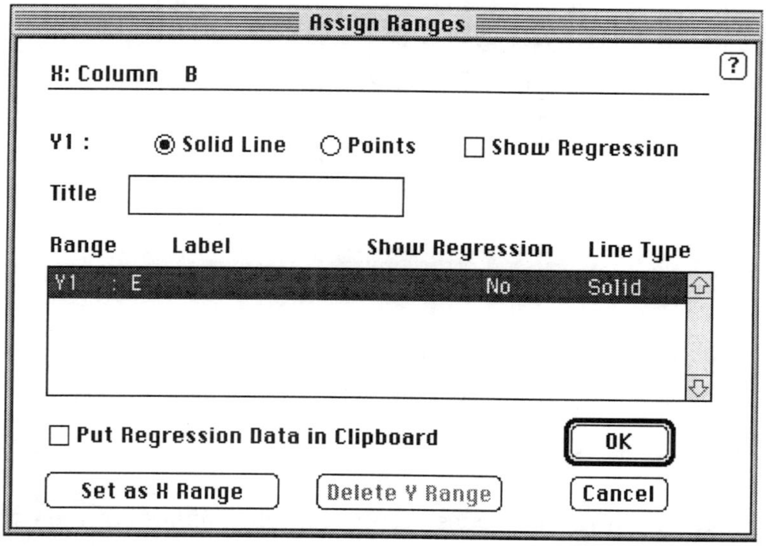

Note that *Explorer* assumes that the left column, column B, will be the x (horizontal) axis of the graph. (You can override this by selecting the Y1: E field and then clicking "Set as X Range.")

➍ Click OK to accept the Assign Ranges dialog box. The Spreadsheet Line Graph dialog box opens automatically for you to define and label the axes.

In this particular case, accept the default to plot all data on both axes. There are cases where you will want to extend the axes beyond the sampled values (notably to set the origin point as 0,0 or to show the y-intercept). You can do so by deactivating the check box and then entering values in the fields that appear:

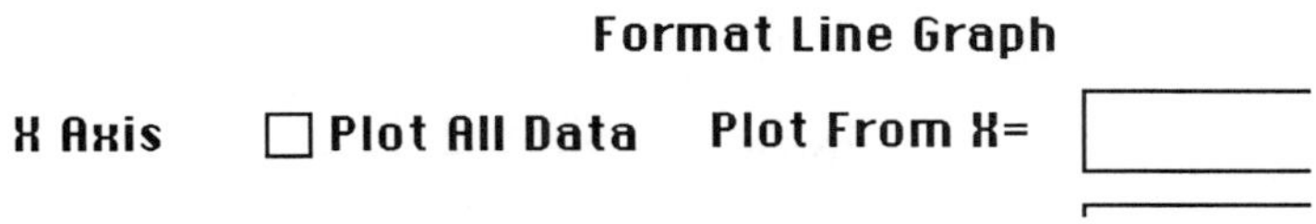

➎ Complete the title and axis label fields. Click OK to accept the dialog box, and generate a graph.

Your graph should appear similar to that shown here:

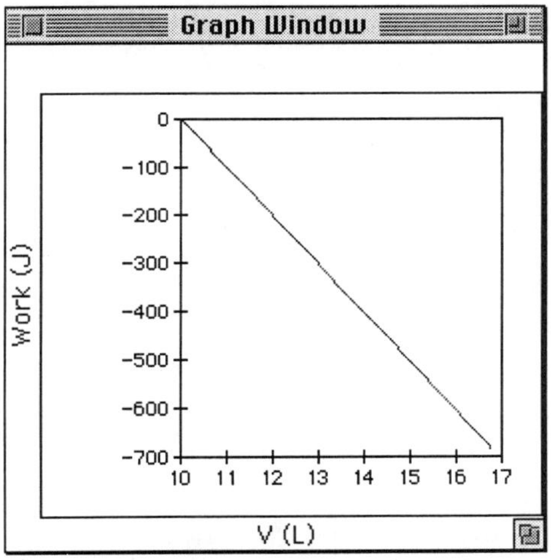

You can also graph all energy data on one graph. Select columns C, D, E, and F and then choose the New Line Graph option from the Spreadsheet menu again. Repeat the steps given above. A typical graph is shown here:

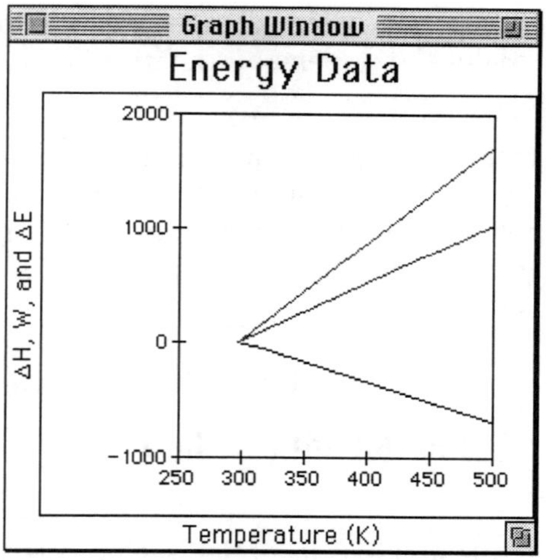

If you save your lab, you will save the last definitions of the spreadsheet, its graph, and the changes you made to the Sampling to Spreadsheet button. Save the lab.

Section 5: Finishing Touches

Your lab now contains all the basic elements necessary to run different simulation setups and to analyze the results in order to solve the initial problem. Before you do so, you may want to do some fine tuning in order to present the material in a logical yet aesthetic way.

We present below some of the *Explorer* features that enable you to personalize your lab, adjusting the layout and graphic presentation to your own tastes.

Work Window Titles

The work window where you have prepared the inputs and outputs for the simulation remains untitled. You can give it a name.

» Choose the "Work Windows Setup" option from the Windows menu.

» Highlight the current name of your work window (probably Untitled Work Window 1). In the name field at the top of the dialog box, type in a new name.

» Click OK to accept the change.

Fonts & Text Styles

You can define the font and point size for every text field on the screen.

» With the ARROW tool, select the desired text field.

» Choose a font from the Font submenu of the Text menu.

» Choose a point size from the Size submenu of the Text menu.

» Choose a style (bold, italic, etc.) from the Style submenu of the Text menu.

If some of your text "escapes" from the text field and you can no longer see it, use the ARROW tool to resize the field.

Color Coordination

Color can play a significant role in your screen layout. You may want to color coordinate all input and output displays related to a particular variable. For example, you can assign a color to each variable V, T, and P, and color-coordinate each one's numeric display with its plot on the graph.

» To color an item in the work window, select it with the ARROW tool. Choose a color from the Colors menu.

» To color an expression on a graph or chart, double-click with the ARROW tool on the display to open its dialog box. Highlight the desired expression from the list of expressions in the display and choose a color from the Colors menu.

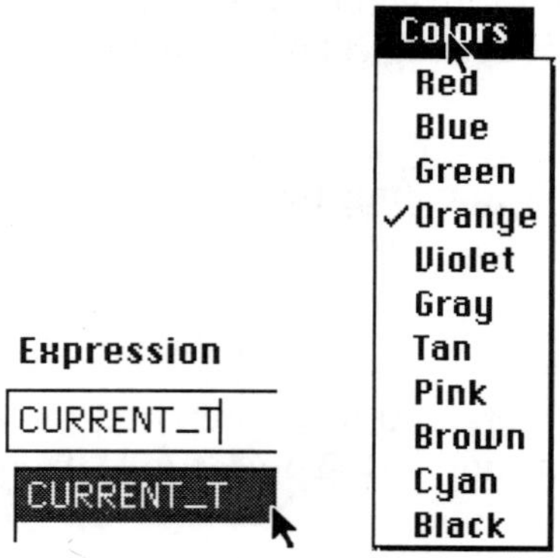

Section 6: Extension — Effect of Number of Moles

The original problem was:

> *A gas is confined to a cylinder with a moveable piston under constant atmospheric pressure. What changes will occur if the gas is heated from 298K to 500K?*

- How will your answers change if the gas has the same initial and final temperatures, and the same initial volume, but the number of moles of gas change?

The problem now involves number of moles, so you need to base your work on the equation PV = nRT.

» Open the Gas Laws Template window from the Windows menu.

» Change the Set Variables option to P, V, T, n:

Two new parameters appear: the initial number of moles input value box and the Initial Condition Selection pop-up menu.

» This pop-up menu indicates which of the variables (P, V, T, n) will be the third one with user-defined initial values (in addition to the "Variable Parameter Selection" currently set to Temperature and the number of moles). Make sure the menu is set to Volume.

Since you are now selecting the number of moles, the *Gas Laws* model will automatically calculate the initial pressure. The Initial Pressure input box in the work window is no longer relevant.

» Copy the # of moles input value box and its label and paste them into your work window so that your window appears similar to the one shown:

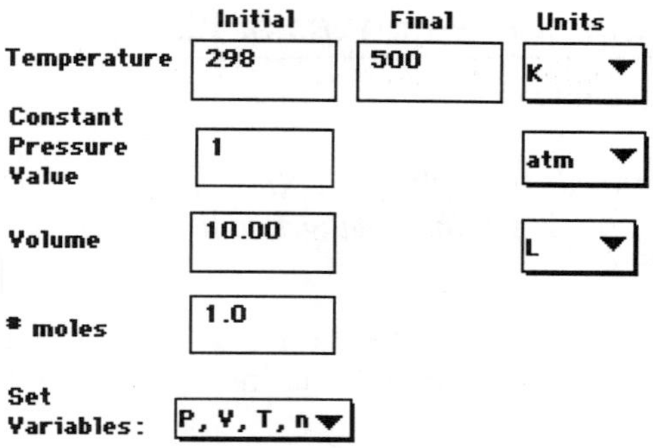

» Reselect the HAND tool to move to operation mode.

» Set the # of moles (n) to 1.

» Click GO to run the simulation.

» Click RESET. Change the # of moles to 2 and run the simulation.

» Click RESET. Change the # of moles to 3 and run the simulation.

Note the effect of changing n on the V vs. T graph, the Energy bar chart, and the spreadsheet data. You should find that the graph appears the same, but that the energy changes shown in the bar chart and spreadsheet increase as the number of moles increases.

Chapter 3: Lab Templates

This chapter contains write-ups that accompany the on–disk project templates, intended to help you get started with your own lab design. The templates are designed by *Chemistry Explorer* model: *Gas Laws, Chemical Equilibrium, the Atom, Chemical Kinetics,* or *Electrochemistry.*

There are four types of templates for each model:

- General template. This template (on disk only) contains a work window with all available model input parameters already displayed and labeled for you. Typical buttons and output displays are also included where appropriate.

  You can use this template as the starting point for your own independent projects.

- Manipulation template. This template (on disk and in text) helps you work through a basic science problem using the required model window tools. If the problem requires elements in the work window, the elements have been prepared for you.

- Analyze template. This template (on disk and in text) shows you how to use the *Explorer* design tools and spreadsheet to answer a given science problem. The model window has been preset so that you can focus your work on the output displays and the spreadsheet.

  Each lab file includes a "Sample Work Window" to show you one way of setting up the displays. If you encounter problems building your own work window, you can check how this sample window was built.

- Design template. This template (on disk and in text) challenges you to set up both the model window and a work window in order to answer the given problem.

Each template is designed to stand as an independent unit. The templates provide you not only with the instructions for *how* to accomplish a task, but they also discuss the reasoning behind *why* you need to perform these tasks. As you work through the templates, you will develop the *Explorer* and reasoning skills to enable you to advance to independent project design.

It is highly recommended that you work the tutorial presented in Chapter 2 before beginning work with the lab templates. The tutorial provides six lessons that escort you through the logic and operations of lab design.

Before you begin to work with a specific model, it is recommended that you work the model's First Look lab file, which presents a basic introduction to the model. The appendices provide a detailed reference on the *Explorer* system and on each model.

Manipulating Elements in the *Gas Laws* Model Window

Related Lab File:
Gas Laws Manipulate (Macintosh); \gas_laws\maniplt4.lab (Windows)

Tools and Controls Used:

 ARROW tool

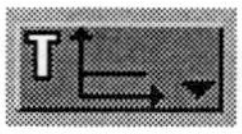

PROCESS TYPE menu

 WORK MODE menu

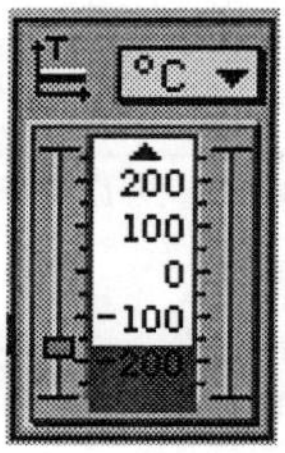

TEMPERATURE control (value and units)

PRESSURE control (value and units)

The Problem: Kinetic-Molecular Theory

A fixed quantity of gas at a constant volume has an initial pressure of 1 atm. How will the pressure change if the temperature is increased from 100 to 600K ?

How does the Kinetic-Molecular theory explain (qualitatively) the effect of a temperature increase at a constant volume?

Initial Setup

You need to define which process and which parameters to set; the question tells you that the process is constant volume, and you are told what temperatures to use.

1. Process Definition

Since the question deals with a constant volume process, use the PROCESS TYPE menu to select the CONSTANT VOLUME process.

2. Variable Parameter Selection

When the process is set to constant volume, you need to choose either setting 1 or 2:

Known Settings	Model Calculates
1. Initial and final temperature, initial pressure	Pressures to final pressure
2. Initial and final pressure, initial temperature	Temperatures to final temperature

Since the question gives you a specific temperature range, choose setting 1 by clicking the activation icon on the TEMPERATURE control.

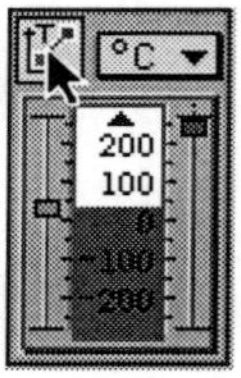

Change the units to those given in the question (K), and set the initial and final temperatures to 100 and 600 respectively:

- Drag the green slider to set the initial temperature to 100K.
- Click the black arrow to change the range to 600–1000. Drag the red slider to set the final temperature to 600. The model "remembers" the first temperature you set, even though you have changed the range. You can check the values by clicking each slider in turn.

3. **Set Variables**

In the template window you can choose one of two sets of variables for working with the amount of gas (i.e., the number of moles):

1. P, V, T. For this default setting, the model calculates the number of moles from the *initial* pressure, volume, and temperature data you input. The number of moles calculated here is kept constant during a simulation; or

2. P, V, T, n. The number of moles becomes an input controlled by the user. In addition, the model uses the initial and final values for the parameter you defined in the "Variable Selection," and the initial value for the parameter you define in the "Initial Condition Selection." The model calculates the fourth parameter in the equation PV=nRT automatically.

For the given problem, choose the Set Variable option P, V, T.

You need to set one more initial value to enable the model to evaluate the equation. Use the PRESSURE control to set the pressure to 1 atm. (Remember to check the units!).

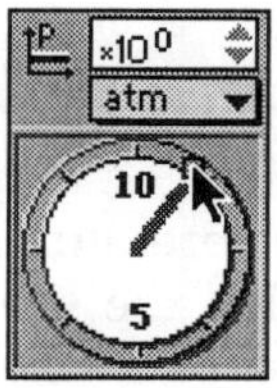

Since the model automatically changes the number of moles to suit the volume you set as the constant volume, set the volume to any reasonable value (e.g., 15 L).

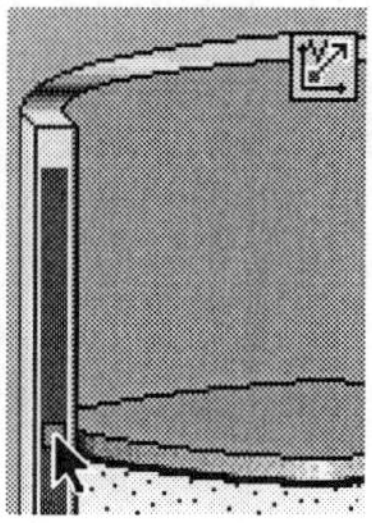

Running the Simulation

Once you have the parameters set as above, click the GO tool and observe the changes in the Model Window. To find the numerical value of the pressure at any time during the simulation, click anywhere inside the PRESSURE control. You can click the STOP tool to stop the simulation and measure pressure or temperature (or volume if desired) in the Model Window.

Explaining the Results: Using the ZOOM MODE

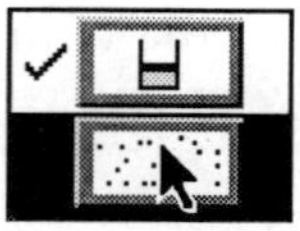

The original problem asks you to explain the above results using the Kinetic-Molecular Theory. The model's zoom view shows a simulation of the behavior of individual molecules. Click the WORK MODE menu and change to zoom mode.

In order to see how the Kinetic-Molecular Theory explains the effects of temperature on pressure, you should run the simulation at various temperatures.

Note that while the simulation runs in zoom mode, the PVT values stay at the values you set, unless you change them, i.e., for the given problem the temperature will not automatically move to the final value.

Run the simulation starting from 100K, the initial temperature, and note the pressure gauge behavior. Click the STOP tool. Increase the temperature to 600K and click GO to watch the changes in the pressure.

Results

The final pressure is 6.00 atm. As the temperature increases, the model shows that the speed of the molecules and the rate of collisions of the molecules with the walls increase. These changes are shown on the bar charts in the work window.

Conclusions and Challenges

There is a linear relationship between pressure and absolute temperature: increasing the absolute temperature by a factor of 6 causes the gas pressure to increase by a factor of 6.

An increase in temperature means an increase in the rms speed, and thus an increase in the average kinetic energy of the molecules. The rate of collisions also increases. Both effects contribute to the increase in pressure.

If you consider a system at constant *pressure*, not constant volume, how does the Kinetic-Molecular Theory explain the effect of volume increase as the temperature increases?

Analyzing Data in *Gas Laws* Work Windows and Spreadsheet

Related Lab Files:
Gas Laws Analyze (Macintosh); \gas_laws\analyze4.lab (Windows)

The Problem: Boyle's Law

Boyle's law states:

"The volume of a fixed quantity of gas maintained at constant temperature is inversely proportional to the pressure."

In verifying the law:

(i) What type of pressure-volume curve is obtained?

(ii) Which outputs should be plotted in order to obtain a linear relationship?

(iii) How does the linear relationship depend on the number of moles present, the temperature, and the gas constant R (R = 0.08206 atm-L/moles-K)?

Initial Setup

Model Window Setup

The Model Window process has already been set to constant temperature (Boyle's law), and arbitrarily the model has been set to accept initial and final pressures, and an initial volume value. The model could alternately have been set to accept initial and final volumes, and the initial pressure.

Since the question is general, the numerical values set are arbitrary.

Creating the Output Displays

Open the Work Window Tools palette from the Windows menu.

Numeric Displays

The given problem asks about the relationship between pressure and volume. In order to see their ranges, create numeric displays of both outputs and run the simulation.

First create a display for the output parameter for the current volume:

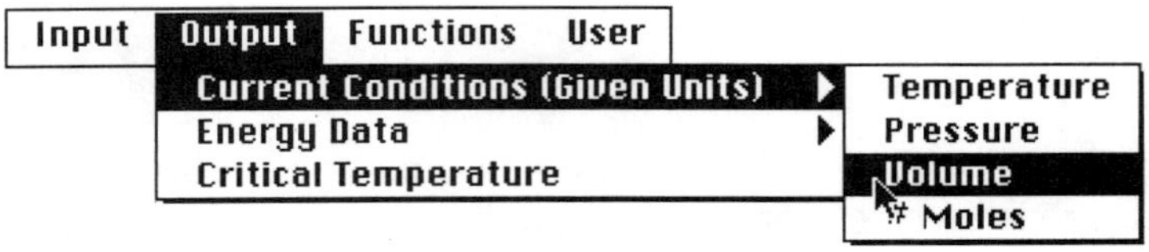

Then create a display for the current pressure. The dialog box should look like this:

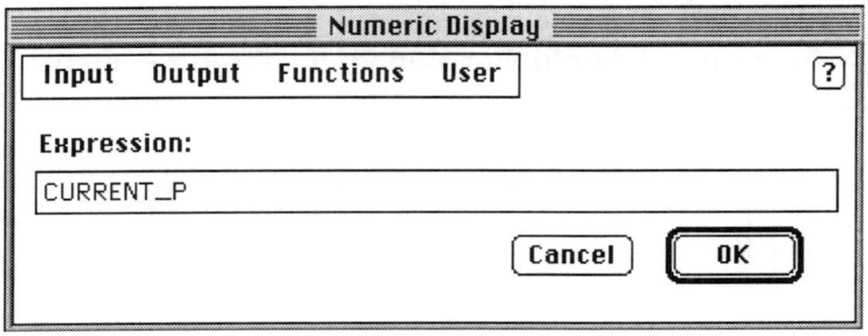

Use the TEXT tool to title the output displays accordingly, including units.

If you are unsure of the expected numerical output values, it helps to format the outputs as "scientific notation" first. Then run the simulation and note the values. Format the displays to accommodate the actual values that you obtained. In this problem, format them to two decimal places.

Line Graph Displays

To find the type of pressure-volume curve, you need to create either a P vs. V or a V vs. P graph. To be consistent with the earlier choice of setting initial and final pressures, it is recommended that you set the pressure as the independent variable (on the x-axis), i.e., you should plot a V vs. P graph. Your dialog box should appear similar to the one shown below:

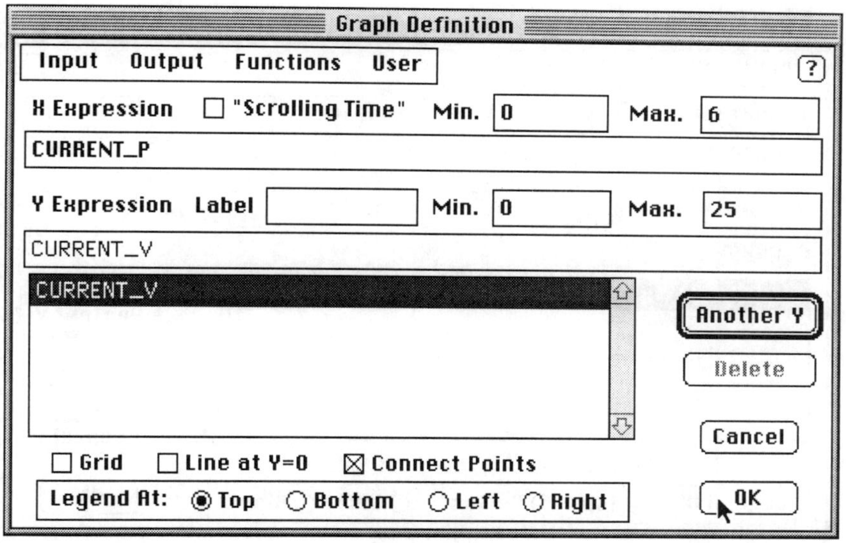

Set the maximum and minimum values in the dialog box according to values you found when you ran the simulation for formatting the numerical displays.

Working Off-Line

The relationship between P and V is PV = K (where K is a constant). To plot this as a linear graph, the expression has to be rearranged to the form y = mx + b. (In this case, b=0.)

PV = K can be rearranged to $V = K\frac{1}{P}$, so that a graph of V vs. $\frac{1}{P}$ should give a straight line through the origin.

Linear Relationship

You can create a graph of V vs. $\frac{1}{P}$ as above. Your dialog box for the new graph should be similar to the example shown below. Note that the 1/ in the Y expression field was typed in directly from the keyboard.

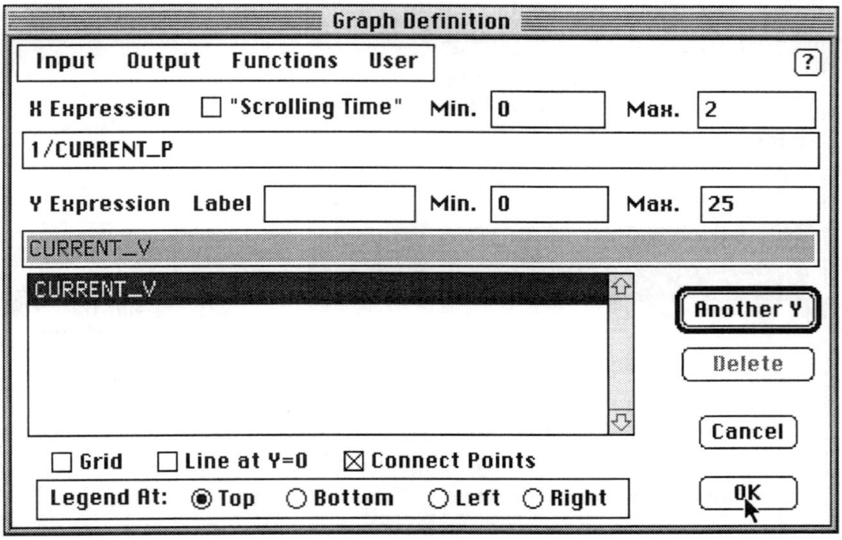

Number of Moles

To find the dependency of the linear relationship and the number of moles, you need to know the number of moles. We suggest using the equation PV=KT, thereby letting the model determine the number of moles, since this equation gives you more freedom in setting the pressure, volume, and temperature. You can create a numeric display showing the number of moles that the model is using:

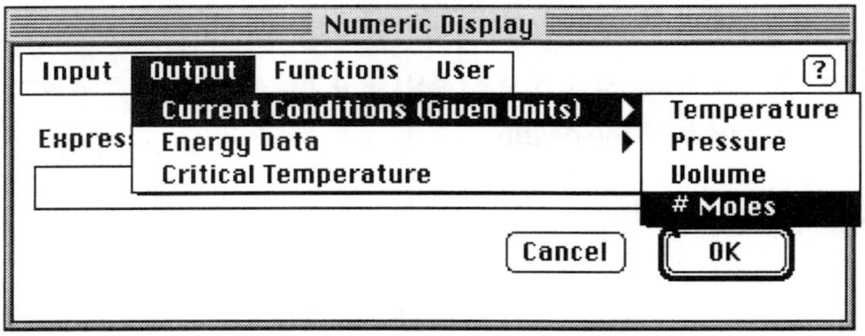

Your work window should appear similar to the following example:

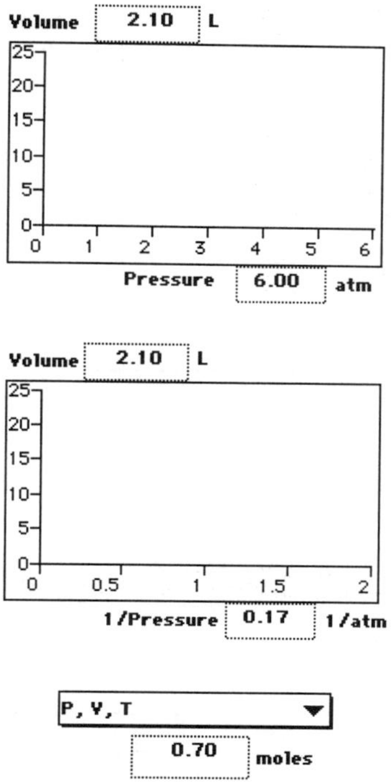

Spreadsheet

To find how the linear relationship depends on the number of moles, it is helpful to run the experiment and sample values of P, V, # of moles, and T to the spreadsheet. Your spreadsheet sampling dialog box should look similar to the example below:

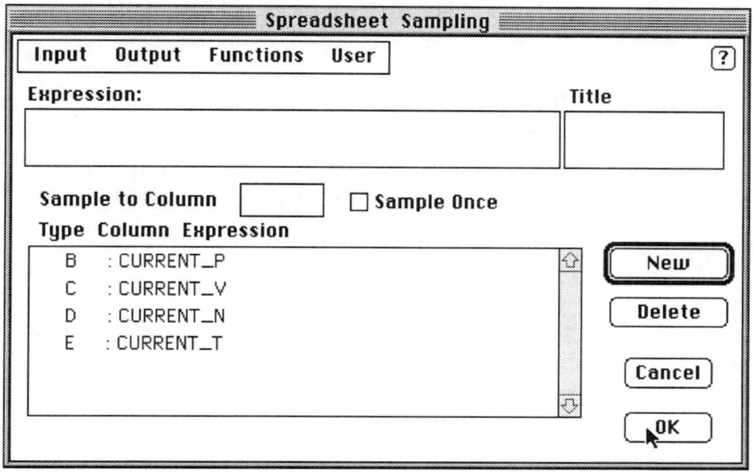

Results, Conclusions and Challenges

Running the simulation shows that the V vs. P graph is curved, and that the V vs. 1/P graph is a straight line which appears to pass through the origin.

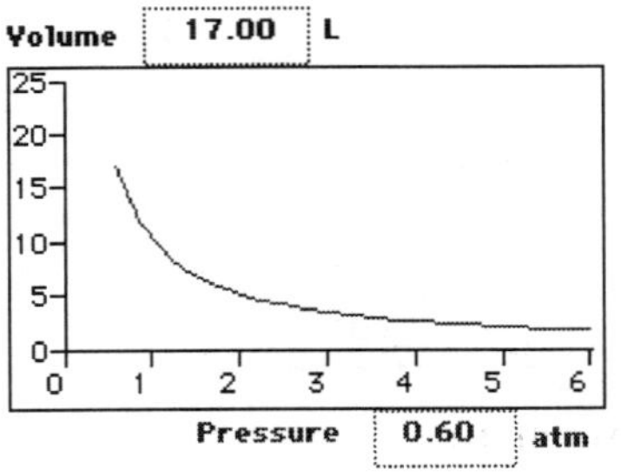

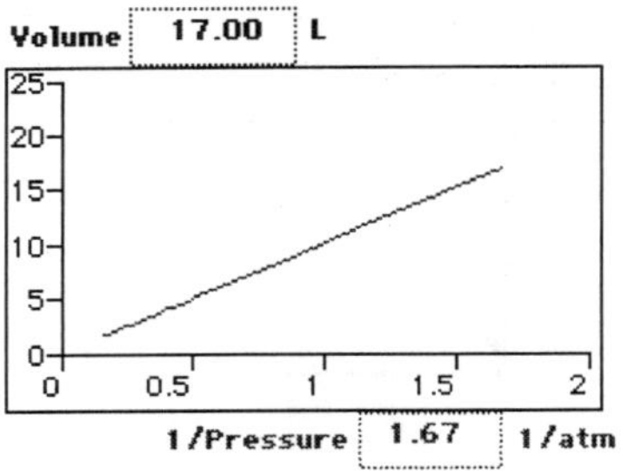

Spreadsheet Calculations

The gas law equation (PV = nRT) implies that the constant K in the Boyle's law equation (PV = K) is equivalent to nRT. If this is the case, then the number of moles n should equal K/RT. The values of K and K/RT can be calculated in the spreadsheet. The steps for calculating K (PV) in column F are:

- Highlight the cell F4.
- In the Formula box at the top, type an equal sign so that *Explorer* will read the expression as a calculated value and not as text:

  =

- Click cell B4 (P) to enter it as a reference cell, type in an asterisk to indicate multiplication, and then click cell C4 (V).

- Click the check box next to the Formula box to accept.

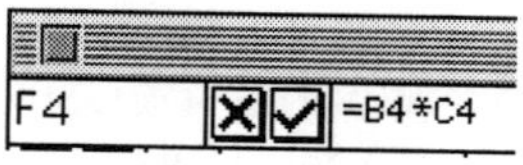

- Copy cell F4. Paste it into cells F5 – F25. The reference to cells B4 and C4 are automatically updated relative to the row number.

- Repeat these steps for cell G4, defining it by the following formula:

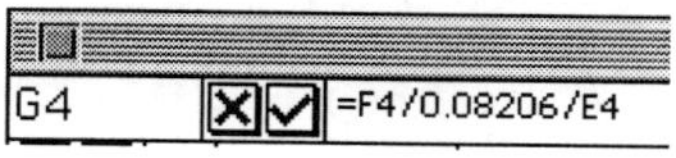

Typical values are shown here.

CURRENT_P	CURRENT_V	CURRENT_N	CURRENT_T	K (=P*V)	K/RT
6	1.7	0.565012	220	10.2	0.564997
5.73	1.7801	0.565012	220	10.2	0.564995
5.46	1.86813	0.565012	220	10.2	0.564996
5.19	1.96532	0.565012	220	10.2	0.564997
4.92	2.07317	0.565012	220	10.2	0.564997

Note that, as expected, column 6 and column 3 have the same values.

How would you expect the results to change if you change the gas from an ideal gas to a real one, such as helium or hydrogen?

Designing a *Gas Laws* Lab

Related Lab Files:
Gas Laws Design (Macintosh); \gas_laws\design4.lab (Windows)

The Problem: Ideal and Real Gases

Devise a method for checking how close to ideality a gas is if values of P, V, n, and T are known for the gas. Justify your method by considering hydrogen, helium and chlorine.

Working Off-Line

An ideal gas, by definition, follows the equation PV = nRT. Thus deviations from the equation will form the basis for measuring how close to ideality a real gas is. There are many ways to measure this deviation numerically; we will use a rearrangement of the basic equation.

The ideal gas equation can be rearranged:

$$\frac{PV}{nRT} = 1$$

and deviations from ideality can be measured by calculating the actual value of $\frac{PV}{nRT}$ and finding how much it differs from 1. (Further consideration from our knowledge of real gases leads us to expect that in practice the independent variables that affect ideality will be the pressure and temperature.) One simplification we make at the outset is to set n = 1, since it is reasonable to assume that deviations will not depend on the *amount* of gas considered. The basic equation is thus:

$$\frac{PV}{RT} = 1$$

Initial Setup

In the template window, use the Set Variables pop-up menu to P, V, T, n. In the value box that appears, set the initial # of moles to 1.

One of the practical problems you need to deal with is that you are considering four parameters (P,V, T, and $\frac{PV}{RT}$), and you need to devise a reasonable way to display variations of the parameters. You can generate three graphs, of $\frac{PV}{RT}$ vs. P, V, and T, and check their behavior.

Work Window Set Up

Open a new work window and the Work Window Tools palette from the Windows menu.

Construct three graphs of $\frac{PV}{RT}$ vs. P, V, and T. Your graph dialog boxes should be similar to the one shown here for $\frac{PV}{RT}$ vs V.

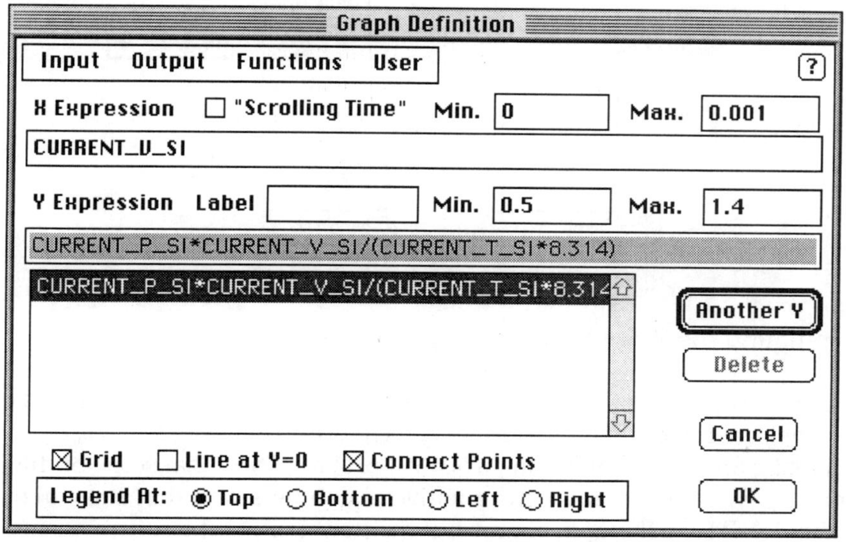

Note that the scales for the vertical axis will be centered around 1 (the ideal value of PV/RT), but that the scale for the horizontal axis will be dependent on the values set in the simulation, and you should be prepared to alter the scales.

For checking purposes it will be useful to have numerical outputs of the four parameters, and the most useful format for checking is scientific notation. You should also have an input for the type of gas, and your work window should appear similar to that shown here:

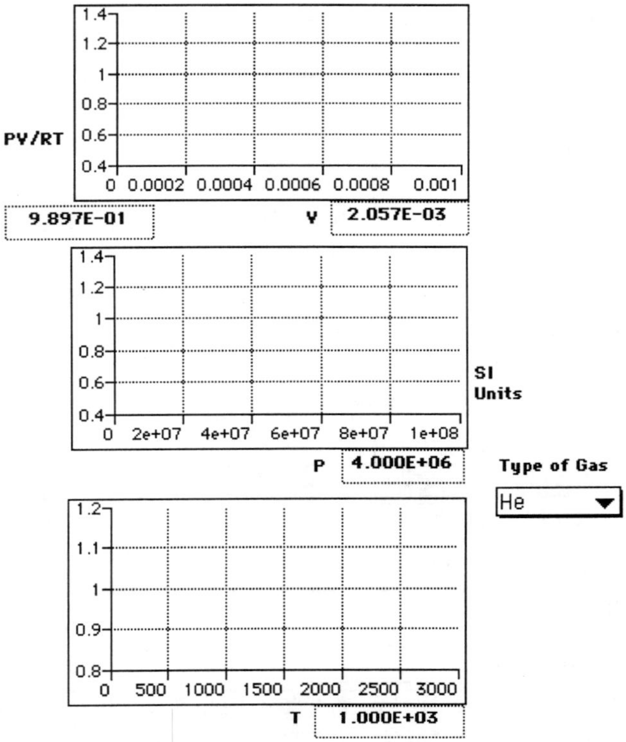

Evaluating the Work Window Display

Try running the simulation under different conditions to see the types of graphs generated. (Remember you can choose constant P, V, or T, or adiabatic for your simulation.) You will probably find that some of the graphs generated are more informative than others, and that a useful one is the PV/RT vs. P graph, running at constant temperature.

Results

Typical results are shown below in SI unit. (The SI units for pressure are Pascal and for volume are cubic meters.)

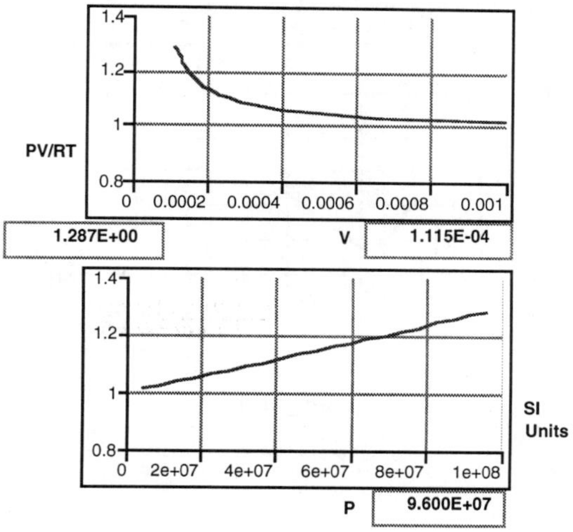

H_2 at 1000K

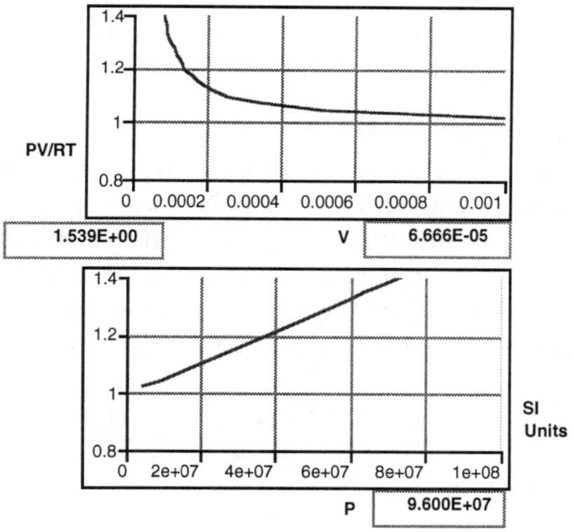

He 500K

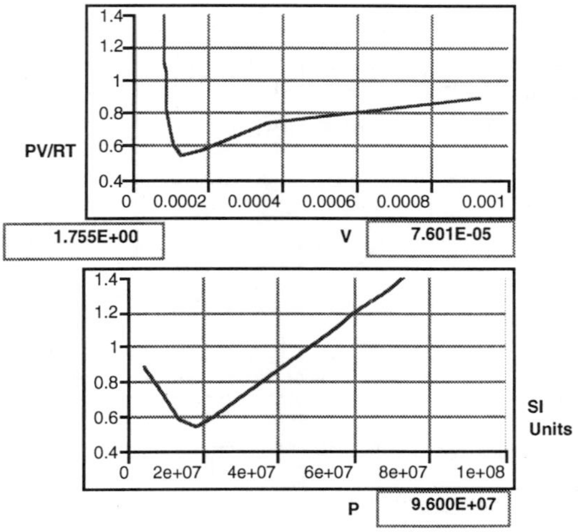

Cl_2 500K

Conclusions

The graphs of PV/RT vs. P and PV/RT vs. V, both at constant T, can be used to display how close a gas is to ideality. The family of graphs of PV/RT vs. P at different temperatures is probably most useful, and you can generate families of graphs using the spreadsheet.

Sample P and PV/RT to the spreadsheet for a particular temperature. Copy the values of PV/RT to an empty column, and repeat the simulation for a different temperature. You can repeat this simulation quickly for three or four temperatures, and then use the spreadsheet graphing facility to display all graphs simultaneously. The graphs for chlorine at 400K, 500K, and 1000K are shown here. Note that as expected, the deviation from ideality decreases as the temperature increases; note also that for all gases, chlorine approaches ideality as the pressure approaches zero.

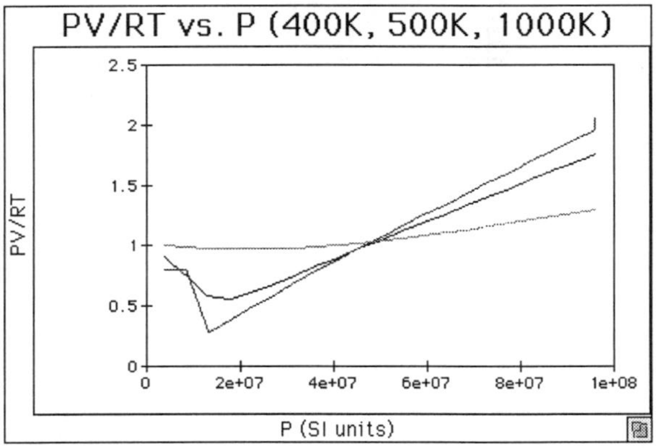

Manipulating Elements in the *Equilibrium* Model Window

Related Lab File:
Equilibrium Manipulate (Macintosh); \equilibr\maniplt3.lab (Windows)

Tools and Controls Used:

ARROW tool

SYSTEM tool

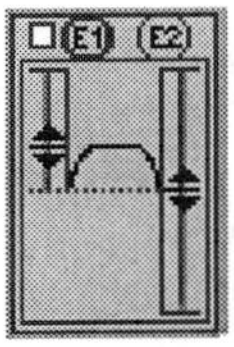

ENERGY control

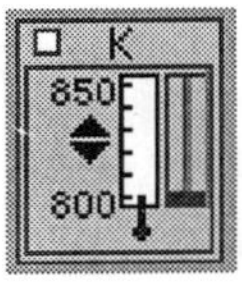

TEMPERATURE control

CONCENTRATION control

The Problem: Le Chatelier's Principle — A System Under a Stress

Le Chatelier's principle states that if one disturbs a system at equilibrium, the equilibrium will shift to minimize the disturbing influence.

Demonstrate this principle for an exothermic reaction of the type

A+B<==>X

Related questions:

- How do the forward and reverse reaction rates change (if at all) with the disturbances?

- How does the equilibrium constant change (if at all) with the disturbances?

Initial Setup

The first step in setting up an equilibrium simulation is to define the reaction. You will use the SYSTEM tool to choose the reaction and to set the stoichiometry. For this problem, use the generic reaction A+B<==>X.

Click the icon representing the desired type of reaction:

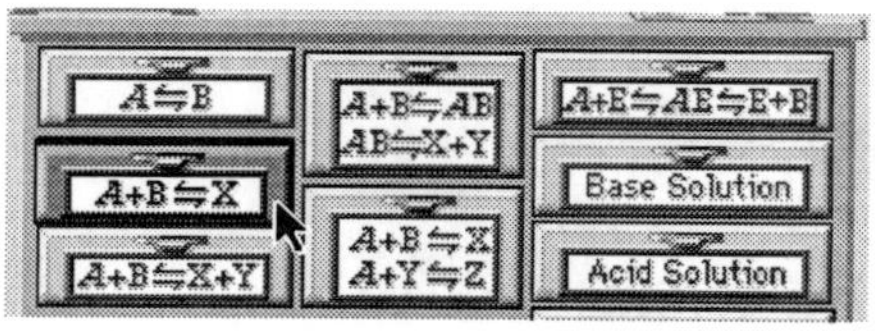

Set the desired stoichiometry for each of the species:

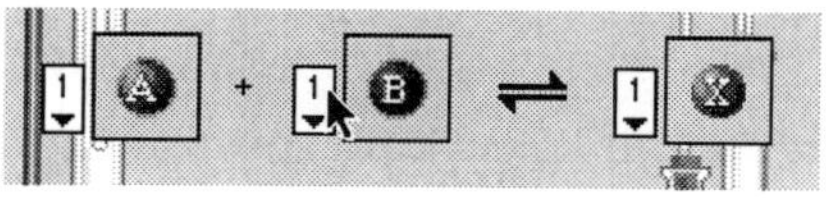

Click the ARROW tool to return to the main view.

Now set the energy parameters of the reaction. Drag the left slider to set the activation energy to 250kJ. Drag the right slider to set the enthalpy to -250kJ. These values represent an exothermic reaction, although the specific values are arbitrary.

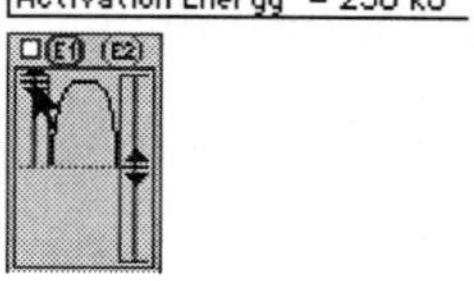

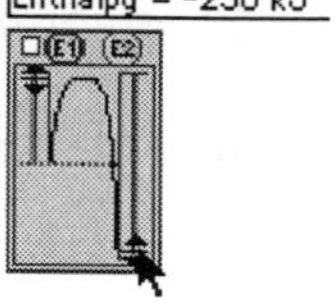

Click the black arrows on the temperature control to set the temperature of the system to 800K:

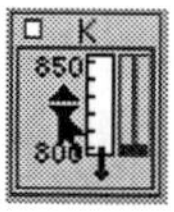

Use the ARROW tool to set the initial concentrations approximately as follows: [A] = 1.00M, [B] = 0.80M, [X] = 0.00M,:

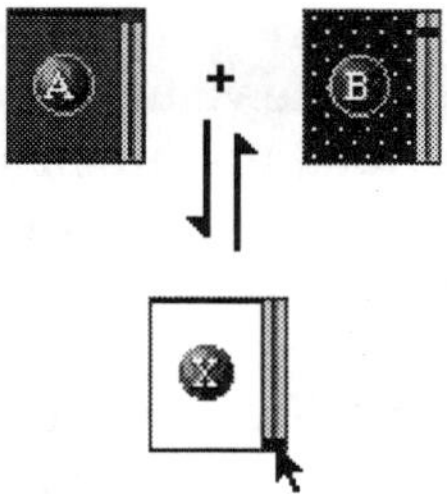

Running the Simulation

Click the GO tool to start the simulation. How can you determine that the reaction has reached equilibrium?

1. Constant concentrations (shown by a constant value of the rate mass-action expression and constant color in the Model Window):

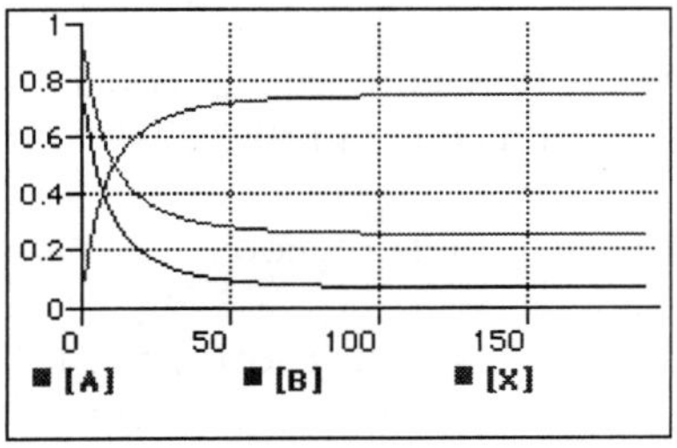

2. Equal rates shown on the pie chart:

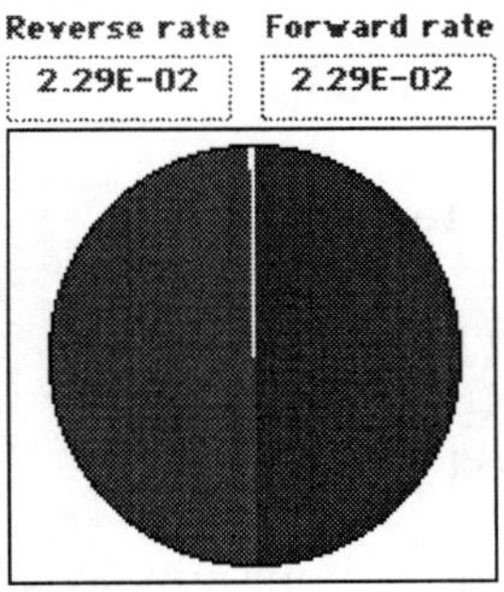

or displayed in the Model Window:

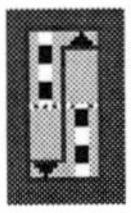

Disturbing the System

While the simulation is still running after equilibrium, use the ARROW tool to increase the temperature. Clicking the black arrows on the temperature control increments the value in steps of 50 degrees. Dragging the red slider increments the temperature in steps of 2 degrees. Since the disturbance creates clearer effects if the changes are greater than 50K, use the black arrows to change the temperature.

Please Note! Since increasing the temperature increases the reaction rate, if you increase the temperature too much you may reach a situation in which the reaction occurs more quickly than the simulation makes calculations, resulting in incorrect results. In this case, *Explorer* alerts you to correct the situation by decreasing the temperature or the concentrations, or by decreasing the time increment, which determines how frequently calculations are made.

- Soon after the change in temperature, what is the effect on the rates?
- Does the system reach another equilibrium?
- If it does, how are the new equilibrium concentrations different from the previous equilibrium concentrations?
- Does the value of the equilibrium constant change?

With the simulation still running, introduce the following disturbances. After each one, wait for the system to return to equilibrium. Answer the above questions for each of the disturbances:

— decrease in temperature

— increase in reactants

— decrease in reactants

— increase in products

— decrease in products

At any time you can return to your initial values by clicking RESET.

Results

For an exothermic reaction of the type A+B<==>X, some of the results are summarized below:

After any disturbance the system returns to a state of equilibrium, which may have different concentrations and rates.

For example, the effects of increasing temperature are: the reverse rate increases, and the forward rate decreases until they equalize again; the product concentrations decrease, and the reactant concentrations increase until a new equilibrium level is reached, for which they remain constant; and the value of the new equilibrium constant is less than the original value.

Challenge

Set an endothermic reaction and compare the changes and their effects on the system.

Analyzing Data in *Equilibrium* Work Windows and Spreadsheet

Related Lab Files:
Equilibrium Analyze (Macintosh); \equilibr\analyze3.lab (Windows)

The Problem: The Equilibrium Constant

Describe the variation in the equilibrium constant as a function of temperature. For the given reaction, how does the temperature affect the equilibrium constant? Find the enthalpy change , ΔH, for the given reaction.

Related questions:

- What will the equilibrium constant be if the temperature is increased by a factor of 2?
- How will the activation energy change (if at all) under the above conditions?

Initial Setup

Model Window Manipulation The Model Window has been preset with the appropriate values for the reaction parameters, initial concentrations, and temperature for the reaction A+B<==>X.

Based on the Arrhenius equation for k, the rate constant, $k = Ae^{-Ea/RT}$. Since the ΔH of a reaction is the sum of Ea for both directions (forward and reverse reactions), the equation can be extended to $K = Ae^{-\Delta H/RT}$, where K is the equilibrium constant. Based on the last equation, the enthalpy change of any given reaction can be calculated from the slope of the graph of lnK vs. 1/T. *Explorer* enables you to create this graph for a series of temperature values.

To address the problem, run the simulation for at least five different temperatures (recommended in an ascending order). For each temperature, click a button to sample the equilibrium constant, K, and the current temperature into the spreadsheet. Then create the spreadsheet graph that illustrates the problem.

Creating the Output Displays

To answer the given question, you need to see numerical values once the system has reached equilibrium.

Open the Work Window Tools palette from the Windows menu.

Numeric Displays

Create the numeric displays for the species concentrations and reaction rates. Open a numeric display box, and in the Numeric Display dialog box select the output variable Current Concentration for Molecule 1.

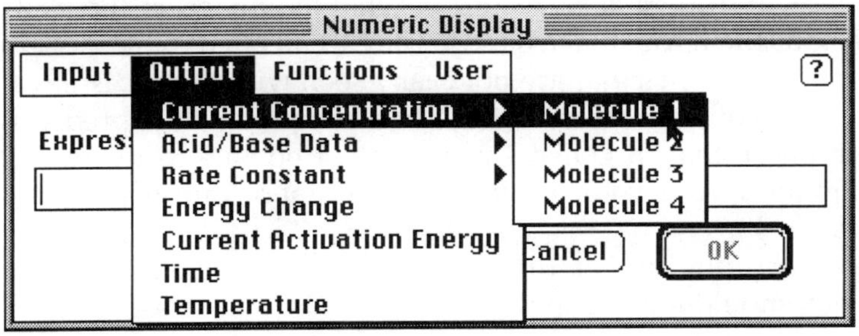

Repeat these steps for the other molecules.

There is no output for the reaction rate, so in order to create a numeric display of the current value, you must enter into the dialog box the *expression* that represents this value, i.e., the rate constant times the concentration (for this specific reaction). The dialog box should look like this:

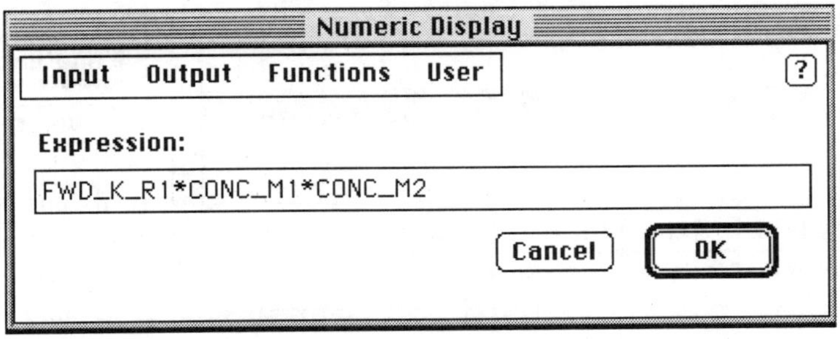

The expression for the reverse rate = reverse rate constant*products concentration.

To measure the concentrations and rates to a reasonable accuracy set the displays to scientific notation with 2 decimal places.

1.00E+00   8.00E-01   6.00E-02

Line Graph Displays

The numeric outputs give accurate values of the reaction at an instant in time, but it is difficult to see trends of behavior (except very simple ones) from them, and there is no record of what happened in the past. Graphs with time as the horizontal axis show any trends that exist, including long-term trends and ones that are not easily seen with other displays. Graphs give a picture of the results to date, often allowing you to predict what *will* happen. (Note that in general you will not be able to read results from graphs to the accuracy you achieve in the numeric outputs.)

Create a concentration vs. time graph for all the displayed species. The dialog box should look as shown:

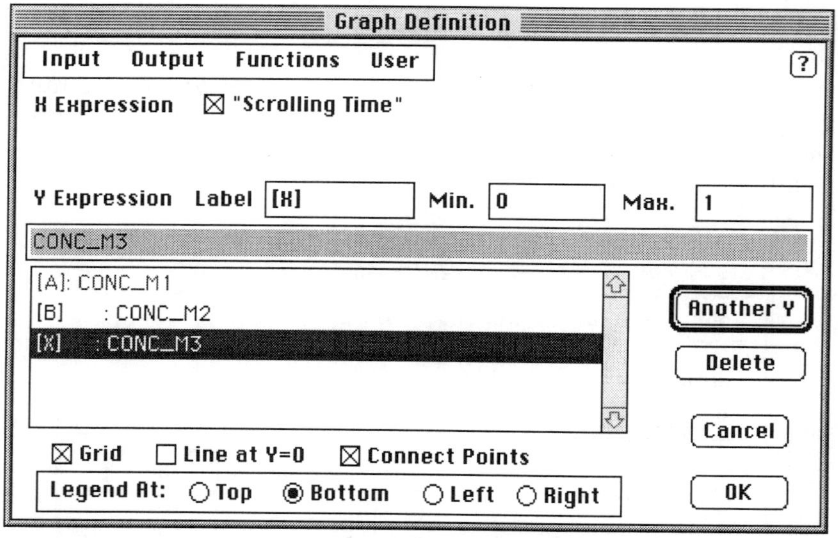

Pie Chart Displays

A pie chart is a visual representation of the ratio of two or more parts relative to a whole. For the given problem, you want to see how the rates relate to each other and when they are equal, so the pie chart is a useful display.

Create a pie chart of the rates. Remember that you must build the expression from other variables. The dialog box should look like this:

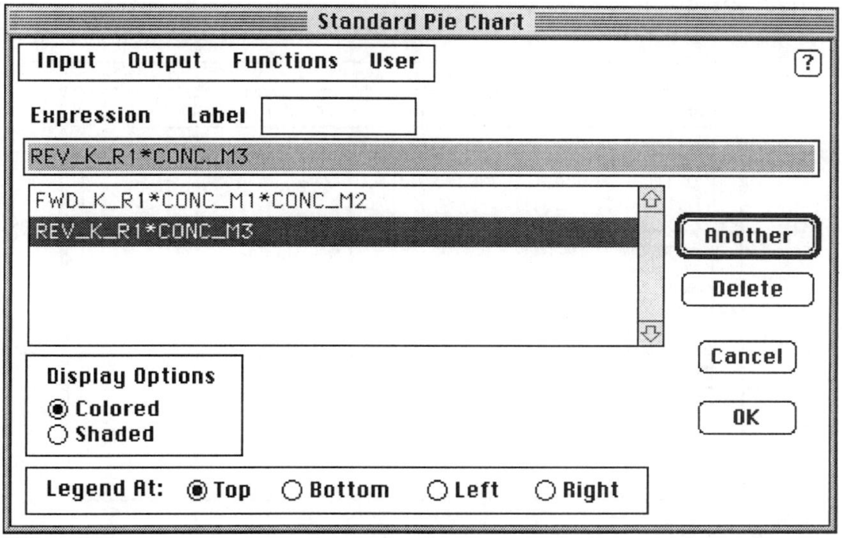

Spreadsheet

The summary spreadsheet allows you to save values from many different experimental runs to one spreadsheet. You can then use the spreadsheet to analyze the results and/or to generate graphs, helping you "see what's been going on."

Summary Spreadsheet

For the given problem you will generate a summary spreadsheet recording the temperature and its corresponding equilibrium constant.

Create a button and label it "Sample to Spreadsheet." Define the button as shown here:

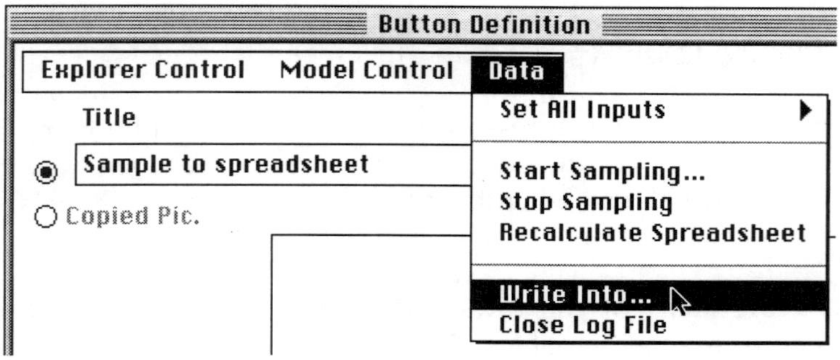

and then fill out the text/variables:

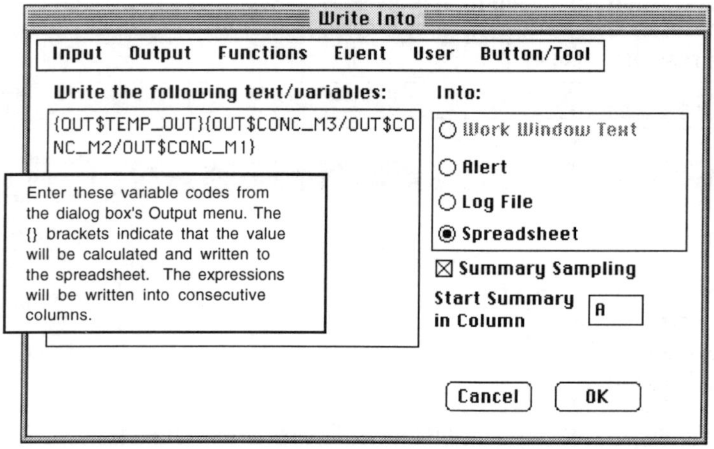

To sample to the spreadsheet, run the simulation and at equilibrium, click the "Sample to Spreadsheet" button. Click RESET, change the temperature, and then GO. After at least five trials, open the spreadsheet.

Note that since you are going to create a graph from the sampled values, it is helpful if you sample them in ascending or descending order. If you sample them in random order, you can afterward select the columns in the spreadsheet and use the "Sort" option to reorder them. For this problem, start with a value of 700 and sample in ascending order.

Spreadsheet Calculations

In order to answer the given problem, you need to calculate the enthalpy change. You must rearrange the Arrhenius equation so that you can calculate the Ea from the slope:

$$\ln K = -\Delta H/RT + \ln A \, .$$

You have two columns of data in the spreadsheet, T and K. You can let the spreadsheet calculate automatically for you the values of 1/T and lnK:

- Highlight the cell C4.
- In the Formula box at the top, type an equal sign so that *Explorer* will read the expression as a calculated value and not as text:

  =

- From the keyboard, type the notation for 1/:

  =1/

- Click cell A4 to enter it as a reference cell in the formula:

  =1/A4

- Click the check box next to the Formula box to accept.

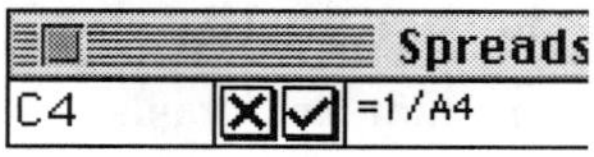

- Copy cell C4. Paste it into cells C5 – C20. The reference to A4 is automatically updated relative to the row number.
- Repeat these steps for cell D4, defining it by the following formula:

  =log(B4)

Note that you can use the "Mathematical Functions" sub-menu in the Edit menu to enter the log. The parentheses are automatically entered. Place the cursor between them and click cell B4.

Spreadsheet Graph

From the above equation, you can see that creating a graph of lnK vs. 1/T will result in a straight line, whose slope is $-\Delta H/R$ and whose Y intersection is lnA.

- Select columns C and D and choose "New Line Graph" from the Spreadsheet menu.

Complete the dialog box as shown. The left column, 1/T, is automatically defined as the X axis.

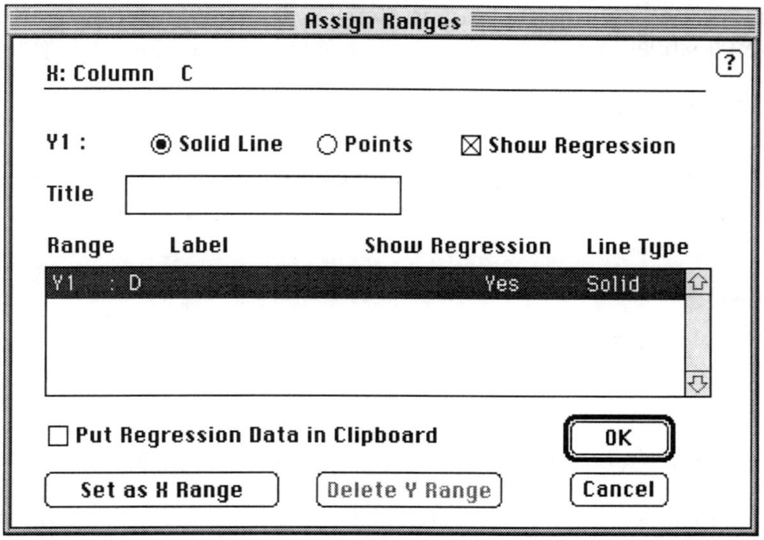

In the second dialog box that opens, title the graph and label the axes:

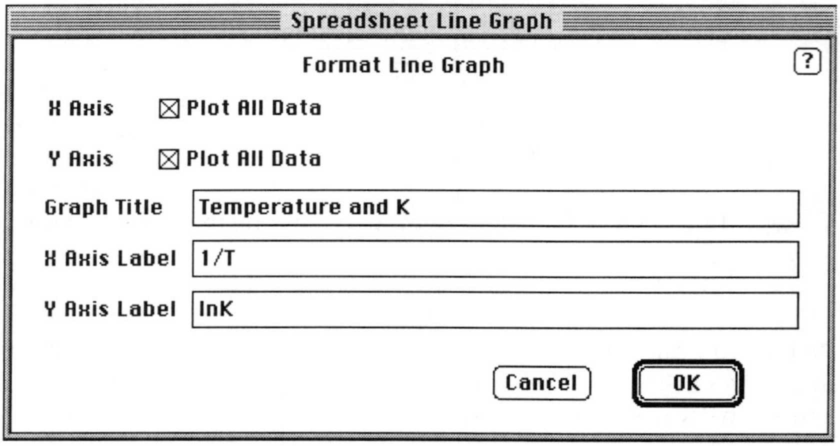

The resulting graph should look something like this:

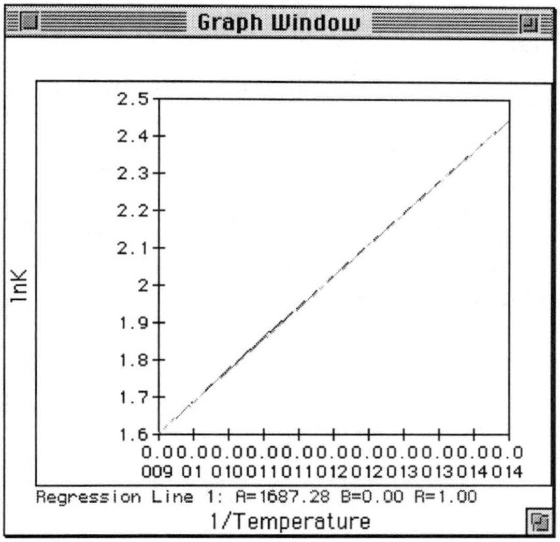

The regression line gives the best fit to a straight line of the plotted points. This line helps compensate if you sampled the data before the reaction was fully at equilibrium. Likewise, it gives three important data: A = the slope, B = the y-intercept, and R = the correlation coefficient. For a perfect fit, R = 1.

We see that the slope is approximately 1690. Since the slope equals $-\Delta H/R$, we have to calculate: $\Delta H = 1690*(-R) = 1690*-8.31J/K\text{-}mol = -14kJ$.

What is the Arrhenius factor for this reaction?

Challenges

In what direction should you change the temperature in order to decrease the enthalpy (e.g., to half its value)?

Which theory explains the dependence between the activation energy and the temperature? How?

Designing an *Equilibrium* Lab

Related Lab File:
Equilibrium Design (Macintosh); \equilibr\design3.lab (Windows)

The Problem: Reaction Kinetics (Rate and Order)

The reactions between nitrogen dioxide, nitrine oxide, and oxygen proceed as follows: $2NO_2(g)$ ------> $2NO(g) + O_2(g)$. At 300°C, the parameters of the reaction are:

Forward Arrhenius factor: 0.00

Reverse Arrhenius factor: 0.53

Activation Energy: 111.00 kJ/mol

Reaction Enthalpy: -113.04 kJ/mol

Explore the reaction:

a. How is the rate of disappearance of NO_2 related to the rate of appearance of each product?

b. Is the reaction first or second order in NO_2?

c. What is the half-life for this reaction?

Answering the Question

The rate of appearance or disappearance is expressed as the change in concentration of a species over a certain period of time. Plotting in the same graph the concentrations of all species that participate in the reaction will display the relationship between them.

To test whether the reaction is first or second order, construct plots of ln $[NO_2]$, and $1/[NO_2]$ against time. The linearity of the graphs will help you answer the question. Using the *Equilibrium* model, you may simulate a certain reaction and observe *all* species' concentration changes together. Use the spreadsheet to derive the above plots and any expression for the variation in concentrations of any species in order to check how it changes with time. At the same time, you can find the rate constant and the half-life of the given reaction.

Setting Up the Model Window

First you must use the SYSTEM tool to select the reaction type and to balance the stoichiometry. For the *Student Edition Chemistry Explorer*, you will use generic species and a reaction type of two reactants and one product in order to simulate the reaction $2NO_2(g)$ ------> $2NO(g) + O_2(g)$:

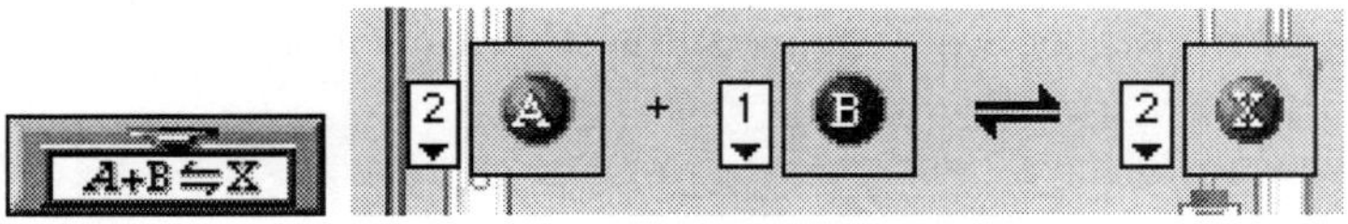

Note that when you choose a system in the *Chemical Equilibrium* model, the reaction can go in either direction, determined by the Arrhenius factors that you set independently.

The problem states the values for the activation energy (111 kJ/mol) and the reaction enthalpy (-113.04 kJ/mol). Use the energy control to set these two values. Note that the energy control is not exact enough to set these specific values. Since you are using generic species to represent $2NO_2$ and 2NO, you can round off the energy and enthalpy values without affecting the results of the problem.

Use the temperature control to set the temperature to 573K.

Since the reaction deals with the decomposition of NO_2, it is the only species that must initially be present in the vessel. Since NO_2 is represented here by X, you must give this species an initial concentration. While you can use any concentration that you wish in order to answer the problem, using the maximum value of 1.00 M will display the phenomenon most fully. The other two species are not present in the vessel at the outset of the reaction, and therefore they have an initial concentration value of 0.

In order to set the Arrhenius factors to the proper values, it is recommended that you copy these inputs from the template window into a new work window. Although you can set these values in the Model Window by dragging the end of the arrows showing the direction of reaction, it is hard to achieve the required precision. The input value box in the work window allows you to set exact values. By setting the forward Arrhenius factor to 0, you simulate the decomposition of NO_2 (X).

Run the simulation to see how the reaction looks. If it runs too quickly or too slowly at this temperature, reset the simulation and change the time increment input to control it. A smaller time increment slows the reaction because the model makes more frequent calculations; a faster time increment speeds up the reaction, but sometimes to the point that equilibrium is achieved almost immediately. It is recommended that you start with a time increment value of 1.

Building a Work Window

Open the Work Window Tools palette from the Windows menu.

Inputs

As recommended above, you may want the input boxes for the Arrhenius factors in your work window for better control. Likewise you may wish to copy the concentration value boxes from the template into your work window. Finally, copy the time increment value box. Make sure that all the input boxes are labeled.

Outputs

To display the changes in concentrations over time, create a graph showing the current output concentration of molecules 1–3:

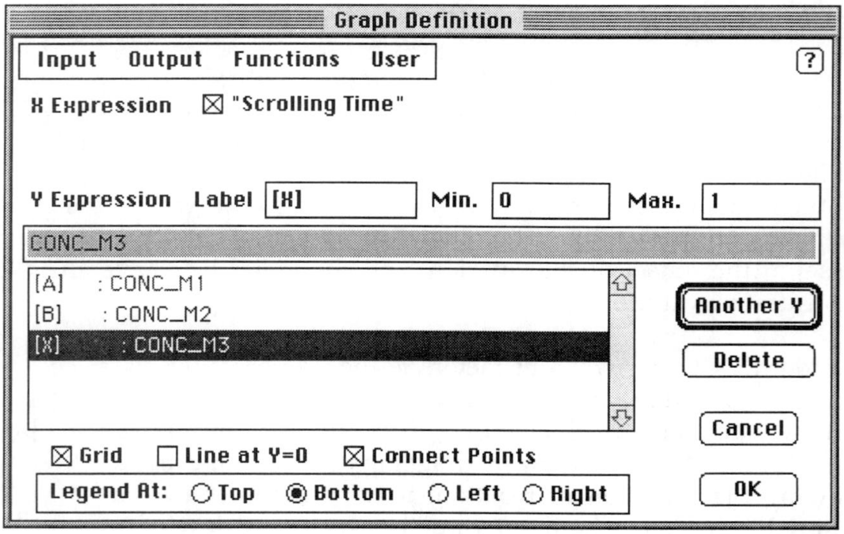

You can color the concentration plots to match the background color of the concentration controls in the Model Window. Highlight each Y expression in the dialog box and choose the appropriate color from the Colors menu.

Next create a numeric display for each of the concentration values. Label the displays and coordinate their colors with the corresponding plots on the graph. Format the displays in scientific notation to two decimal places.

If you are not sure what scale to assign to the graph, you can create the numeric displays first, and then run the simulation to find the appropriate range.

Sampling to the Spreadsheet

In order to answer the question, plot graphs of ln [NO_2] and 1/[NO_2] vs. time to test the linearity. If ln [NO_2] vs. time is linear, the reaction is first order in NO_2 and if 1/[NO_2] vs. time is linear, the reaction is second order in NO_2.

You can sample time and the concentration of NO_2, and then calculate ln [NO_2] and 1/[NO_2] directly in the spreadsheet as you have done in other problems.

Alternately, you can define the expressions directly into the spreadsheet, so that the calculated values are sampled directly into the spreadsheet:

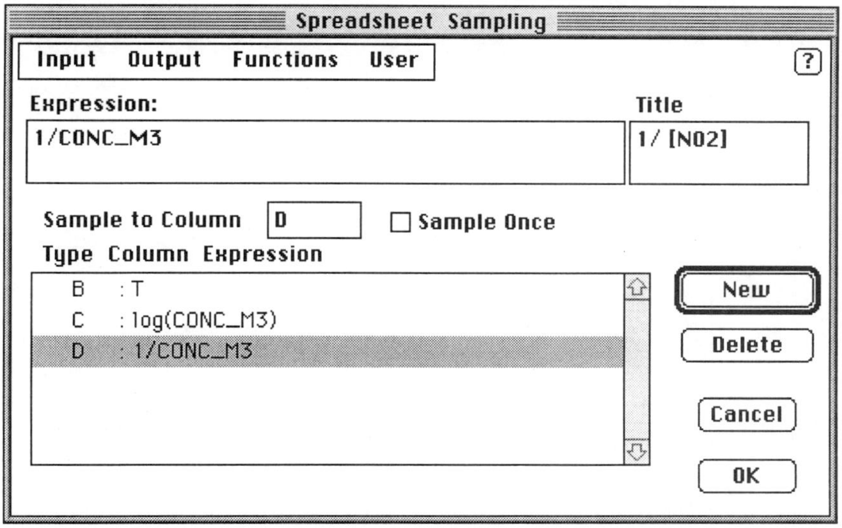

Generating Graphs from the Summary Spreadsheet

Create a spreadsheet graph of ln [NO_2] vs. time. Use the regression values to check for linearity.

Create a new spreadsheet graph of $1/[NO_2]$ vs. time. Select nonadjacent columns in the spreadsheet by holding down the Command key (Macintosh) or the CTRL key (Windows) while you click the column titles.

Use the spreadsheet to find the half-life by locating the time it took for the concentration of NO_2 to reach half its initial concentration.

Conclusions and Additional Challenges

The plot of $1/[NO_2]$ vs. time is linear; therefore, the reaction is second order in NO_2.

How will your answers change if you vary the initial conditions, e.g., the initial concentration of NO_2? Or the temperature? Can you make some general predictions?

Manipulating Elements in the *Atom* Model Window

Related Lab File:
Atom Manipulate (Macintosh); \atom\maniplt1.lab (Windows)

Note to Macintosh users: If your computer has more than 4 Mb of RAM, we recommend that you increase the memory allotment for *Explorer* in order to get the best results with this template. To increase the allotment, before launching the *Student Edition Chemistry Explorer*, select its icon on the desktop. Choose the "Get Info" option from the File menu. Change the preferred size for Memory Requirement to 2500 (or whatever your computer will support).

Tools and Controls Used:

 BOHR ATOM icon

 EXPERIMENT (Absorption or Emission) control

 SCALE control

LAMP WAVELENGTH control

 CHARGE DISTRIBUTION icon

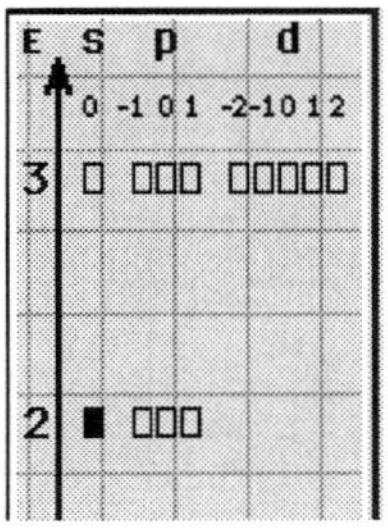

ORBITAL ENERGY LEVEL control

 2-D/3-D VIEW control

The Problem: First Excited State of the Hydrogen Atom

Considering the hydrogen atom, show how the Bohr model for n = 2 compares to the 2s quantum mechanical model.

Working Off-Line

The Bohr model predicted that the electron circulates in a circular orbit about the nucleus. The quantum mechanical model assumes that the position of the electron is given by a probability distribution giving the probability of the electron being at a particular position relative to the nucleus. We can use the model to compare the results in the two cases.

Initial Setup

Bohr Atom

Use the ARROW tool to select the Intermediate Period and ensure that the BOHR ATOM button is pressed (and not the "Classical Dilemma" on the left).

You want to be able to show the simulation running with n = 2, and without transitions to other levels occurring during the simulation. One method of doing this is to set the simulation to absorption mode, while illuminating the atom with wavelengths greater than those generating transitions.

Use the ARROW tool to select the *absorption experiment*:

Use the ARROW tool to select the second energy level:

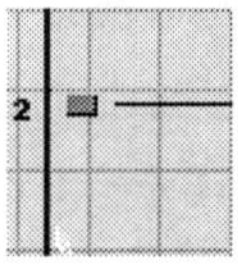

Use the ARROW tool to select the optimum scaling of the orbital display:

The screen should look like this:

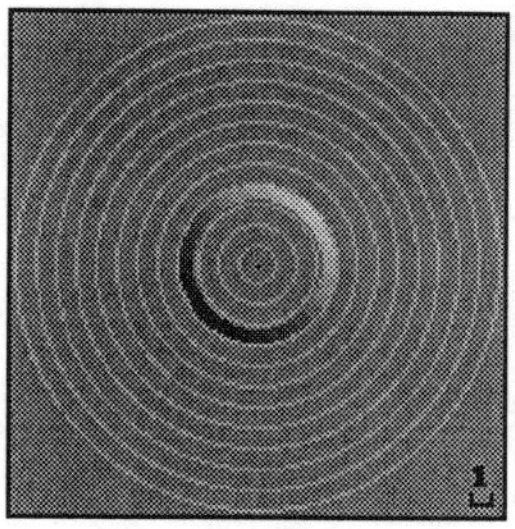

Use the ARROW tool to set the illumination wavelength range so that the electron will stay at the second level. (If necessary, use trial and error to find a correct wavelength range.)

Click GO to run the simulation and check to see that transitions do not occur.

Bohr Results

With the simulation running, click with the ARROW tool on the orbiting electron to find the radius of the orbit. You should find that r = 4.0 (Bohr radii).

Quantum Mechanical Model

In order to compare the Bohr atom to the quantum-mechanical model, RESET the simulation and then click the MODERN PERIOD icon. Ensure that the CHARGE DISTRIBUTION icon is pressed (and not the "Electron Configuration").

Set the 2s state:

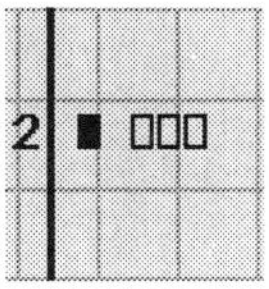

Select the optimum scaling factor and set the simulation to display a 2D view:

Use the ARROW tool to select a direction along which to plot normalized values of $|\psi|^2$.

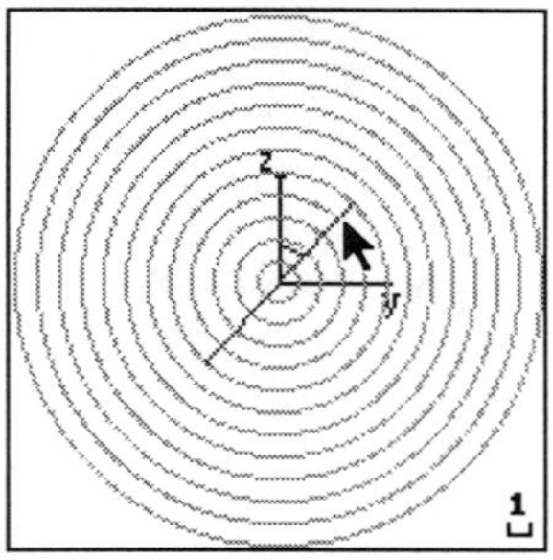

Run the simulation.

When the simulation has finished, it draws a graph in the Model Window giving the normalized value of $|\psi|^2$ along the direction you chose.

Quantum Mechanical Results

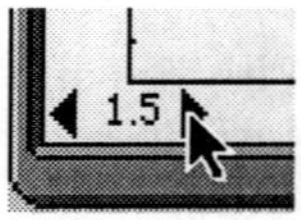

The graph shows there is a local maximum for $|\psi|^2$ at a distance of approximately 4 Bohr radii from the center. You can see the results on the graph more clearly if you change the initial radius to 1.5.

Conclusions and Challenges

The Bohr model shows that for n = 2, the electron orbit radius is 4 Bohr radii. The Bohr model has the electron moving in a circle centered on the nucleus, and so is circularly symmetric. The quantum mechanical model shows that for the 2s state the $|\psi|^2$ distribution is spherically symmetric, with a local maximum for $|\psi|^2$ at 4 Bohr radii.

How do the results for the 2p state differ from those for the 2s state? How are the results similar?

Analyzing Data in *the Atom* Work Windows and Spreadsheet

Related Lab Files:
Atom Analyze (Macintosh); \atom\analyze1.lab (Windows)

Note to Macintosh users: If your computer has more than 4 Mb of RAM, we recommend that you increase the memory allotment for *Explorer* in order to get the best results with this template. To increase the allotment, before launching the *Student Edition Chemistry Explorer*, select its icon on the desktop. Choose the "Get Info" option from the File menu. Change the preferred size for Memory Requirement to 2500 (or whatever your computer will support).

The Problem: The Classical Dilemma

Rutherford's results showing a planetary atom created a dilemma, since classical science predicted that the orbiting electron would spiral into the nucleus, and that at the same time its energy would decrease and that it would radiate energy. Assuming that the classical science predictions held,

a. How would the electron energy vary with its distance from the nucleus?

b. How would the radiated energy wavelength change as the electron approached the nucleus?

Initial Setup

Model Window The Model Window has been set to show the "classical dilemma" simulation, and the electron energy has been set to -2.0eV.

Creating the Output Displays

The questions involve electron energy, electron distance from the nucleus, and wavelength radiated. All three parameters are available as "Intermediate Period" outputs.

You have a number of ways to display the data, but since both parts of the question request comparisons of two parameters, the best choice is probably line graphs. Here we describe how to produce two suitable line graphs.

Open the Work Window Tools palette from the Windows menu.

Line Graph Displays

Create a line graph of electron energy vs. electron orbit radius. You need to choose the scales for both parameters so that the graph displays the required information, and you may need to change the scales to optimize your graph, using trial and error. A typical dialog box is shown below:

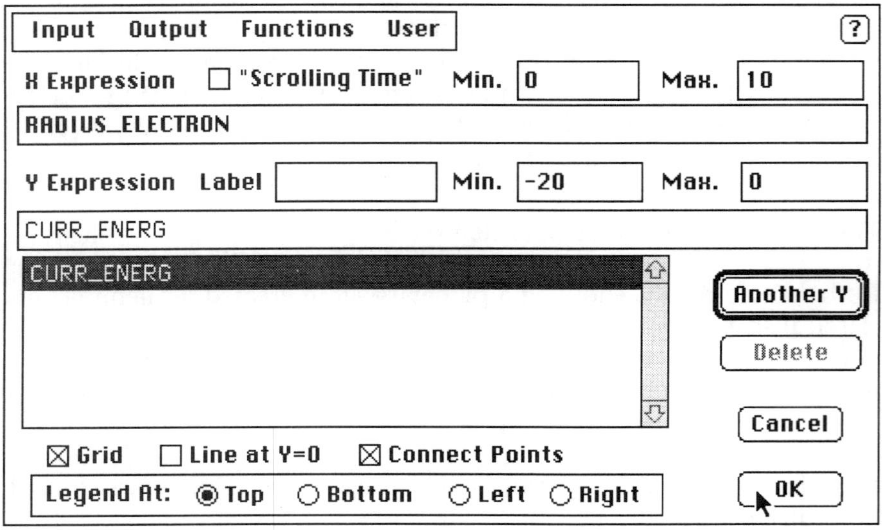

Create a graph for wavelength vs. radius; your dialog box for this graph should be similar to the example below:

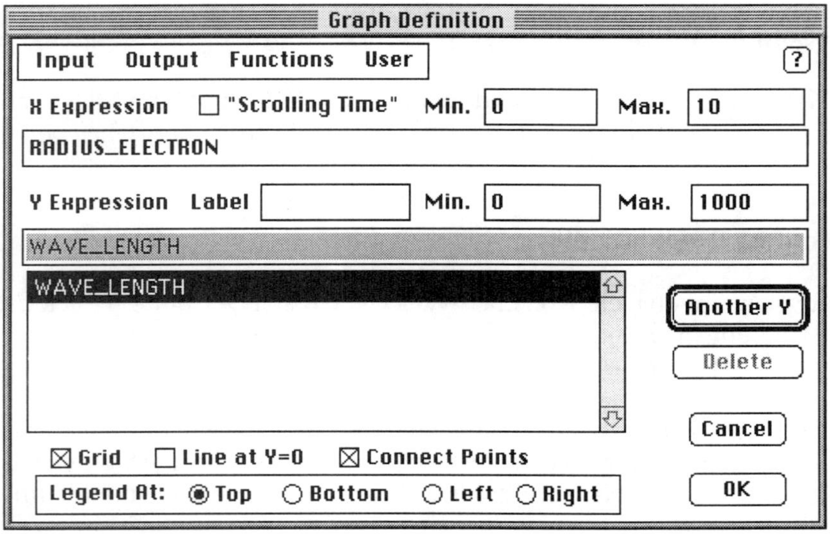

Use the TEXT tool to label both graphs.

Running the simulation will generate graphs of the relationships. Typical graphs will appear as here:

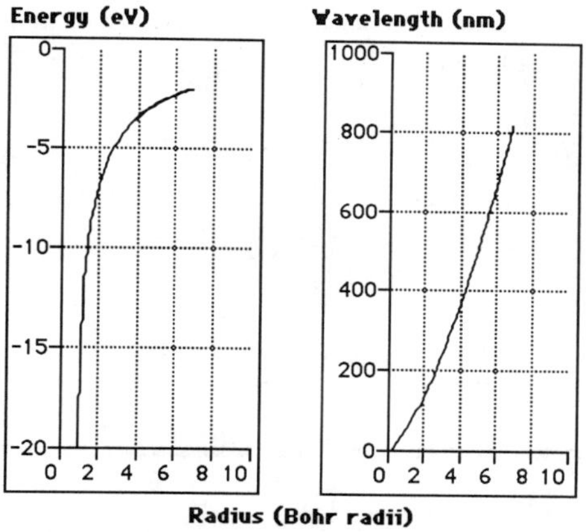

Results

In one sense the graphs *are* the results, since they display the required relations. In order to analyze the results to give general relationships, it is convenient to record the results of a run to the spreadsheet, and then analyze the spreadsheet results graphically.

To sample the results to the spreadsheet, use "Sampling..." in the Spreadsheet menu to open a definition dialog box. Choose the Orbital Radius output for the column B expression. Title the column Radius.

Click the "New" button and define the wavelength and the energy to columns C and D respectively.

Your dialog box will be similar to that shown here:

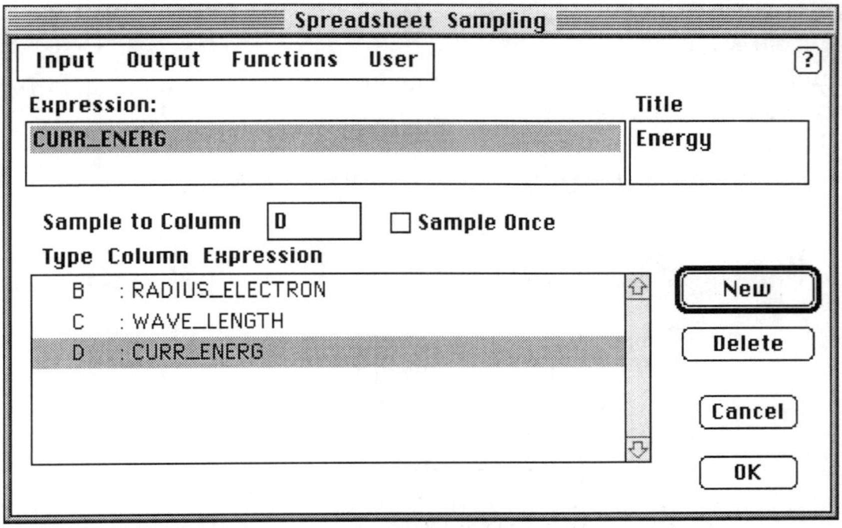

A typical run should generate results similar to these shown:

Radius	Wavelength	Energy
6.85278	818.094	-1.98614
6.8314	814.267	-1.99236
6.80988	810.423	-1.99865
6.78823	806.561	-2.00503
6.76644	802.68	-2.01149
6.7445	798.781	-2.01803
6.72243	794.862	-2.02465
6.70021	790.924	-2.03137

The raw results may be graphed in the spreadsheet graph, giving graphs similar to those you have generated in the work window. In addition, you can generate derived results in the spreadsheet, and verify hypotheses. For example, you may hypothesize that E (energy) = k/r, where r is the radius.

To verify this result, you can use the spreadsheet to calculate 1/r. Highlight cell E4. In the edit box at the top of the spreadsheet, type:

= 1/

Then click cell B4 to enter the radius value. Click the check to accept the expression.

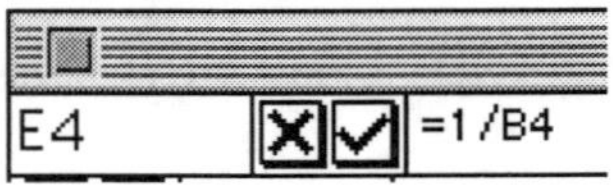

Select cell E4 and choose the Copy option from the Edit menu. Highlight at least 120 rows in column E and choose the Paste option. The expression is entered into each cell, and the reference to cell B4 is updated to the relevant row.

To verify your hypothesis, graph column E vs. column D; if the relationship is correct, the graph should be a straight line. Highlight the two columns and choose the New Line Graph option from the Spreadsheet menu. Column D is automatically defined as the x-axis:

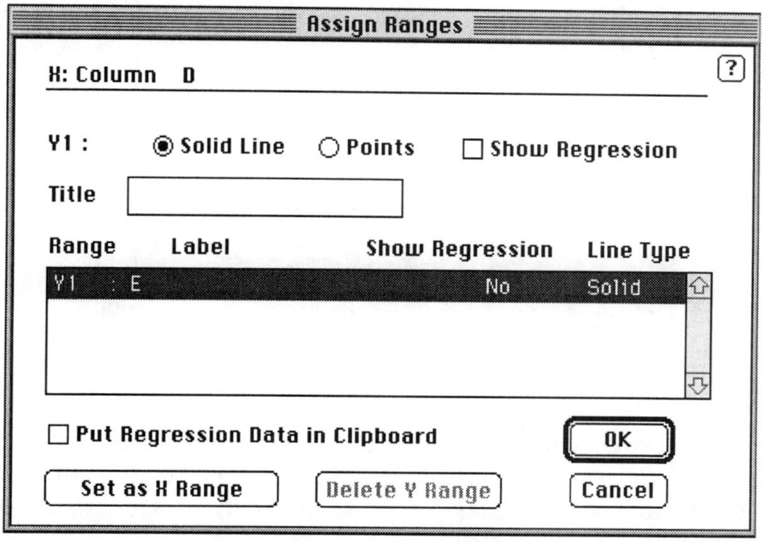

If you look at the values in the spreadsheet, you see that the final values of the energy are of a much larger order of magnitude, and therefore it is recommended not to include them on the graph. You can limit the data plotted by deactivating the Plot All Data check box for the X axis (energy) and entering your own ranges:

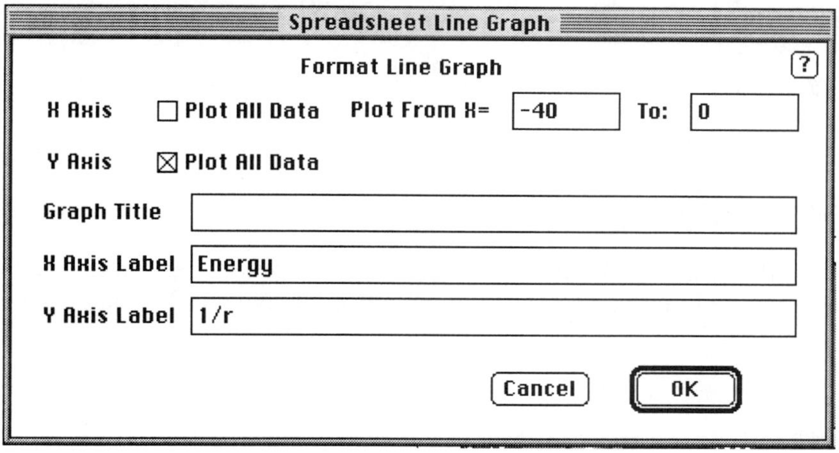

The graph should look like this:

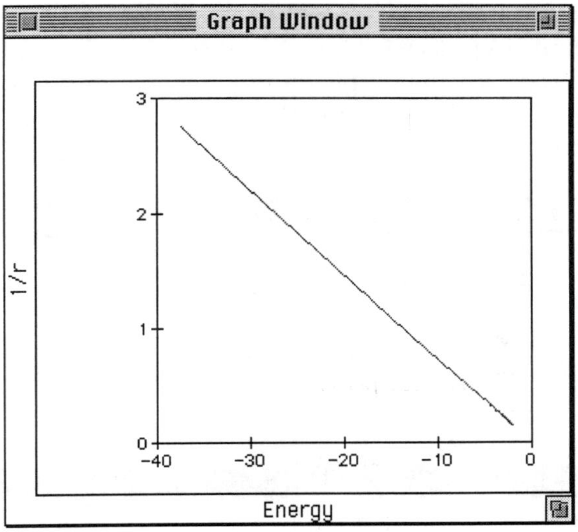

Conclusions and Challenges

From your results you should see that the electron energy is inversely proportional to the distance from the nucleus. The relationship between wavelength and radius may be more difficult to see quantitatively, although the graph shows that it is not a linear one.

One general method to find how two parameters y, x are related is to assume that they obey an equation of the form:

$y = A\,x^n$.

Taking logs of both sides of this equation gives

$\log y = \log A + n \log x$, which is in the form of a straight line equation:

$Y = B + n\,X$

Thus plotting log y vs. log x will generate a straight line.

Using this method in the spreadsheet (the spreadsheet mathematical functions include logarithms) shows that the wavelength is proportional to $(\text{radius})^{2/3}$.

Designing an *Atom* Lab

Related Lab File:
Atom Design (Macintosh); \atom\design1.lab (Windows)

Note to Macintosh users: If your computer has more than 4 Mb of RAM, we recommend that you increase the memory allotment for *Explorer* in order to get the best results with this template. To increase the allotment, before launching the *Student Edition Chemistry Explorer*, select its icon on the desktop. Choose the "Get Info" option from the File menu. Change the preferred size for Memory Requirement to 2500 (or whatever your computer will support).

The Problem: Dual Nature of the Electron

Sometime after Bohr's initial work, Louis de Broglie proposed that any mass m moving with velocity v has associated with it a wavelength given by $\lambda = \frac{h}{mv}$.

Bohr initially assumed that the electron orbited in a circle about the nucleus, being held there by electrostatic attraction. Based on this assumption, the speed of an electron in an orbit of radius r is given by $v = \frac{15.91}{\sqrt{r}}$. (All units are in SI units.)

Bohr's "semi-classical" proposal, which explained the spectrum of the hydrogen atom, implied that electrons moved in quantized circular orbits with a radius r given by $r = 5.3 \bullet 10^{-11} \bullet n^2$ (r is in meters.)

Use the above relationships to show how, for the different Bohr orbits, the wavelengths predicted by de Broglie relate to the circumference of the orbit.

Working Off-Line

The question asks us to find values of λ, and corresponding values of the orbits circumference. To find λ we need to find the values of h and m, and the values of v for different values of n.

h and m are constants; from tables $h = 6.626{\bullet}10^{-34}$ SI units and $m = 9.109{\bullet}10^{-31}$ kg. Using these values we can generate outputs for λ in a work window; we can also generate an output for the circumference and compare the outputs.

Building a Work Window

Open the Work Window Tools palette from the Windows menu.

Outputs

The circumference of an orbit is given by $C = 2\pi r$, and the model outputs the orbit radius in terms of Bohr radii (Bohr's predicted radius when $n = 1$). To convert the radius to SI units you need to multiply by $5.3 \cdot 10^{-11}$.

Create a numeric display for the circumference. The expression should be similar to that shown here:

```
2*pi*5.3e-11*RADIUS_ELECTRON
```

You also need to create a display for the wavelength according to the de Broglie equation, and you can do this in one step by substituting for v in the equation to give:

$$\lambda = \frac{h}{m\,(15.91/\sqrt{r})}$$

The corresponding numeric display expression is:

```
6.626e-34/(9.109e-31*(15.91/sqrt(5.3e-11*RADIUS_ELECTR ON)))
```

(Note that in the example above, the field is too small to show everything entered; for cases like this use the → key on your keyboard to enter and see the remaining entry.)

You now have outputs for the circumference and the wavelength, and you will need to compare them. A simple way of comparing is to make a third numeric display of circumference/wavelength. The numeric display expression should be similar to:

```
(2*pi*5.3e-11*RADIUS_ELECTRON)/(6.626e-34/(9.109e-31*(
15.91/sqrt(5.3e-11*RADIUS_ELECTRON))))
```

(If you check carefully, you will see there are sets of parentheses added so that the expression is correct.)

Your final work window should be similar to the example here:

Circumference (m) 1.6317E-08

Wavelength (m) 2.3299E-09

Circumference/Wavelength 7.0033E+00

Note that we have formatted all values in scientific notation, since the values for wavelength and circumference are extremely small.

Running the Experiment

In the Intermediate Period Model Window you need to set up a method for changing the value of n so you have control over n. One method is to set the experiment to absorption  and set the wavelength of the illuminating light so that the atom cannot make transitions. This method takes trial and error, but if you select n = 7 (the highest value of n allowed by the model) and increase the wavelength, you can find some wavelengths where transitions cannot occur.

Run the experiment, and observe the values of the outputs as you change the value of n in the Model Window.

Observations in the Model Window

Up to now we have concentrated on the numerical results, but the Model Window is also designed to display the electron wavelengths graphically. In the Model Window, click  to show the de Broglie wavelengths schematically.

Conclusions

You should observe that the circumference/wavelength ratio is a simple whole number (to within 0.1%). Since the ratio of the orbit length to the "electron wavelengths" is a simple number, one way of visualizing the electrons in their orbit is to assume that they formed standing waves. These are displayed in the Model Window.

Manipulating Elements in the *Kinetics* Model Window

Related Lab File:
Kinetics Manipulate (Macintosh); \kinetics\maniplt5.lab (Windows)

Tools and Controls Used:

 ARROW tool

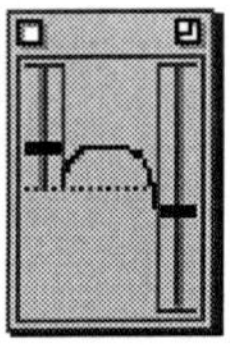 ENERGY control

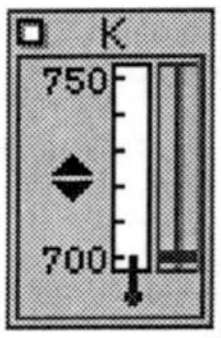

TEMPERATURE control

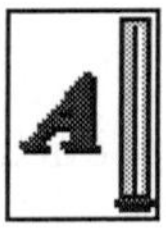

PARTCLES control

The Problem: Energy Profile

For the reaction A + B < = > C + D — and assuming you start with equal concentrations of the four species — what conditions for activation energy and enthalpy are best for maximizing the concentrations of C and D?

How does the temperature affect your answer?

Initial Setup

To start with equal concentrations, you need to set the values of A, B, C, and D to the same value. Drag the slider on the # PARTICLES control to set each species to 25 particles.

Click the GO tool to run the simulation with the given conditions.

Working Off-line

Consideration of the energy of the system tells us that the equilibrium position will depend on the value of ΔH, i.e., for positive values of ΔH the position will favor the reactants A,B; for negative values of ΔH the position will shift towards the products C,D. For a given ΔH, the activation energy E_a should not affect the final concentrations.

If the reaction A + B -> C + D gives out heat, i.e., $\Delta H < 0$, then Le Chatelier's principle tells us that the system will compensate for a reduction in temperature by shifting more to the right. If the reaction A + B -> C + D absorbs heat, i.e., $\Delta H > 0$, then Le Chatelier's principle tells us that the system will compensate for a reduction in temperature by shifting more to the left.

Running the Lab

Try setting extreme values for ΔH, E_a, and temperature in order to see the effects quickly. Whenever possible, you should wait until the system appears to have stabilized reasonably well before drawing conclusions. Your goal is to change all parameters that affect the reaction in such a way that the product concentrations are maximized.

Conclusions

The reaction final state is affected by the enthalpy of the system (other factors, such as entropy change, remaining the same). Decreasing it ($\Delta H<0$) increases the product concentrations. The activation energy affects the rates of the reaction and not the final state — decreasing E_a increases the reaction rates. The temperature affects both the rates of reaction and the products' concentrations. As the temperature increases, the rates of the forward and reverse reactions increase. In the case of an exothermic reaction ($\Delta H<0$), reducing the temperature increases the product concentrations.

The best conditions for maximizing C and D concentrations (theoretically) will be minimum Ea, minimum ΔH (in this case $\Delta H = -250$ kJ/mol), and T = 5K.

Challenges

Set the optimal conditions for energy profile and temperature. Now change the initial concentrations to see the effect.

How do the particle size and mass affect the rate?

Analyzing Data *Kinetics* Work Windows and Spreadsheet

Related Lab Files:
Kinetics Analyze (Macintosh); \kinetics\analyze5.lab (Windows)

The Problem: Effusion of Gases

Demonstrate Graham's Law of diffusion by finding the times taken for molecules with different masses to diffuse into a vacuum. Graham's Law is usually presented as diffusion through a small orifice, but it also applies for diffusion through a large orifice, simulated here by removing a wall.

Initial Setup

Model Window The Model Window has been set up to have 50 particles initially in the left side of the container, and a vacuum in the right side. (In the template window, you can set the container to two regions.)

Working Off-line

Graham's law of diffusion may be stated in the form $r = \frac{K}{\sqrt{M}}$, where r is the rate of diffusion of a gas with molecular mass M, and K is a constant. To verify the law, you must measure rates of diffusion and masses of the diffusing molecules.

One method for measuring the rate of diffusion is to find the time it takes for a fixed percentage of the molecules to diffuse into the vacuum. In this case, you will measure the time for 20% of the molecules to diffuse, i.e., if you start with 50 molecules on the left, you need to find the time for 10 molecules to diffuse to the right.

Note that the rate of diffusion is the number of molecules passing in a given time, whereas you are measuring the inverse of this — the time t taken for a given number to pass. You can rewrite the equation in the form:

$$\frac{1}{t} = \frac{K'}{\sqrt{M}}, \text{ or } t = K' \sqrt{M}$$

so that you can verify the law by plotting t vs. $\sqrt{M}$, which should generate a straight line through the origin.

Creating the Output Displays

Open the Work Window Tools palette from the Windows menu.

Numeric Displays

Create a numeric display for the time:

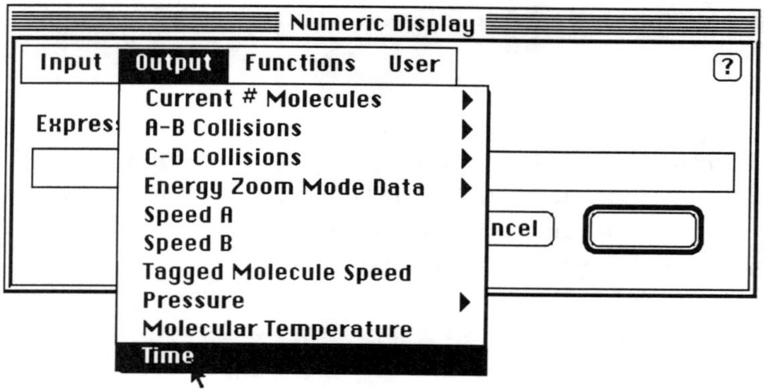

and for the number of molecules that are in the right side:

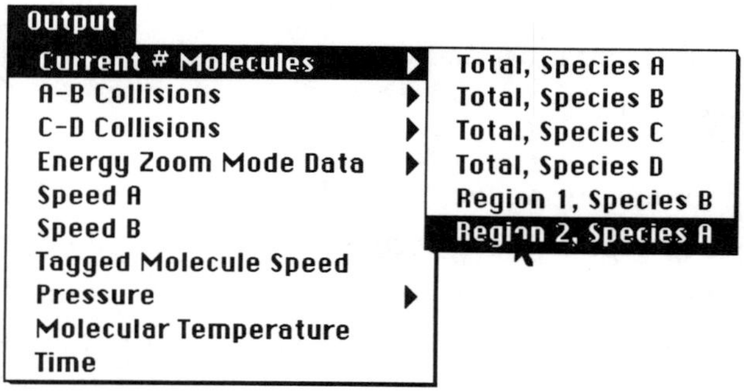

The dialog box for the latter should look like this:

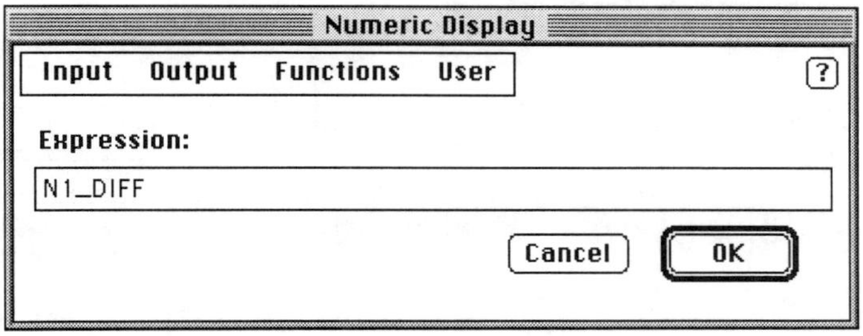

Since you may be unsure of the general value for time, or even the order of magnitude involved, it is recommended that you format the time display in scientific notation to two decimal places.

The current time *display* is very important in this type of experiment. Therefore, it is useful for you to be able to control the increments by which time is determined and the calculations are made. Copy the Time Increment input value box from the Template window into your work window.

Use the TEXT tool to label the elements in your work window.

Line Graph Displays

Although for this problem you need to find the time for a specific occurrence, it may help you to have a graph of the number of molecules in the right region vs. time. Then, if you miss the exact time for the 10th molecule to diffuse, you may be able to reconstruct it from the graph. Your graph dialog box will be similar to the one shown here:

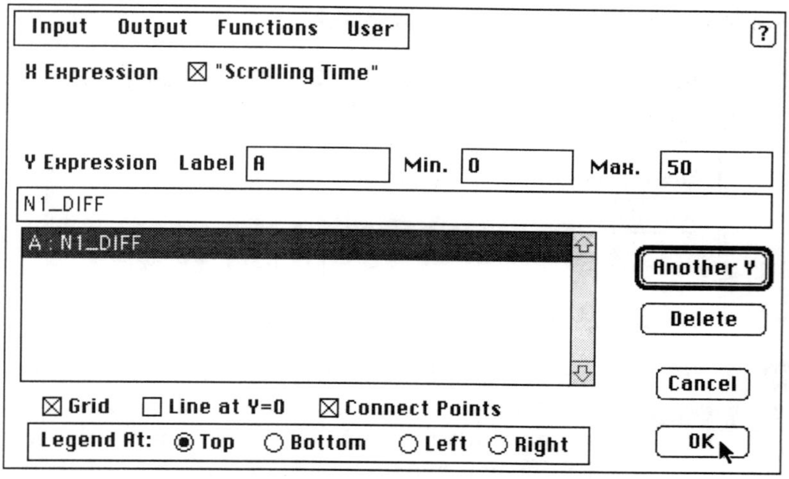

Remember to label your graph.

Building a Button

You need to remove the separating wall at the same time you start the simulation, so that the simulation time reflects accurately the diffusion time. A convenient way of performing both actions at once is to build both reactions into one button. Your button dialog box should appear similar to the one shown here:

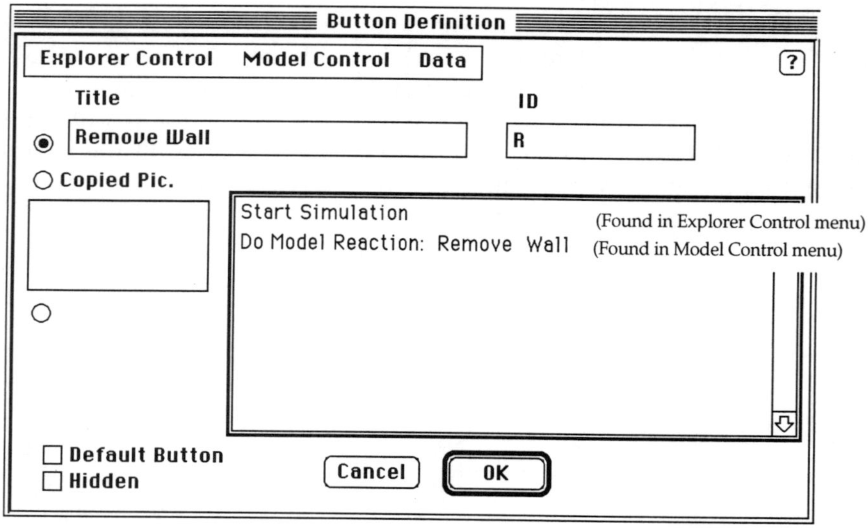

Note that the reaction order *is* important! (Try reversing them to see what happens.)

Trying Out Your Simulation

Before finalizing your lab, try it out. Make sure that the lab is initially in RESET mode, then click your "Remove Wall" button.

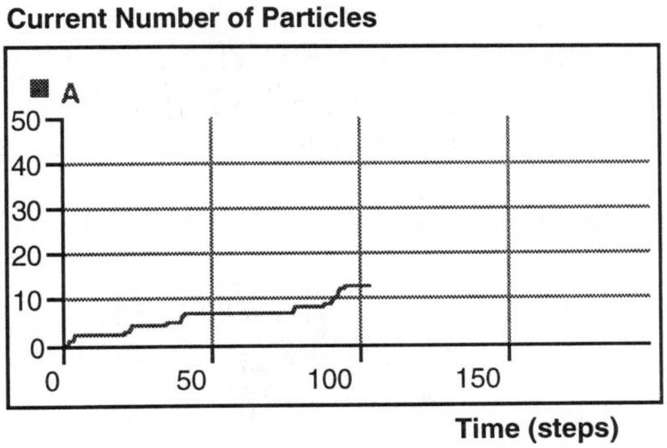

Your output graph should appear similar to the one above. (In this case the mass was set to 50g, the temperature to 850K, and the Time Increment Δt to 0.1 sec.) Note that from the nature of the model there are large statistical variations in the results of the *Kinetics* model, so the results you obtain will vary from run to run. For example, the graph shows that 10 molecules diffused after 90 calculation steps; the next run took 118 steps for 10 molecules to diffuse.

Taking Results

Once the simulation is working as above, you are in a position to take results, initially in a form similar to the two-column table shown here. If you lower the temperature before beginning the trials, you will have better control over "capturing" the tenth molecule. We recommend a temperature of approximately 300K.

Mass of molecules	Time for 20% to diffuse

Use the *Explorer* spreadsheet to record your raw results. Click the "Remove Wall" button. When nearly 10 molecules have passed to the right side, use the SINGLE STEP tool until the tenth molecule just crosses over. Manually enter the molecule mass and the time into the spreadsheet.

Click RESET. Use the SPECIES tool to change the molecule mass and repeat the experiment. It is recommended that you use mass values in ascending order to produce a smooth spreadsheet graph.

Your spreadsheet should appear similar to the one shown here:

Mass	Time
2.5	55.00
10	133.00
40	223.00
75	248.00
100	329.00

Conclusions

Use the spreadsheet to generate a third column corresponding to $\sqrt{M}$:

Copy cell C2 and paste it into the next four rows in column C. The reference to cell A2 is automatically updated to the relevant row number.

Spreadsheet

C2 =SQRT(A2)

	A	B	C
1	Mass	Time	SQRT(Mass)
2	2.5	55.00	1.5811

Your final spreadsheet will be similar to the one shown here:

A	B	C
Mass	Time	SQRT(Mass)
2.5	55.00	1.5811
10	133.00	3.1623
40	223.00	6.3246
75	248.00	8.6603
100	329.00	10.0000

In order to verify the law, select columns B and column C and generate a graph with column C as the x axis. Plot the column B values as points and show the regression line:

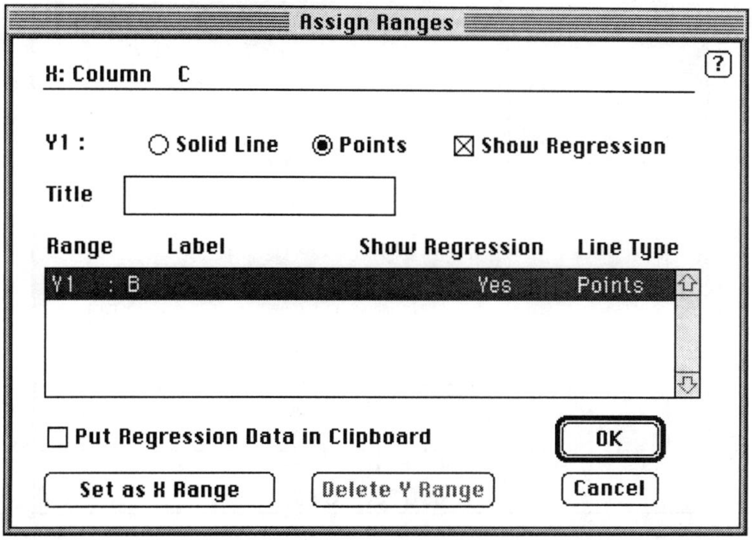

In the next dialog box that opens automatically, set the axes scales to show the origin and label the graph:

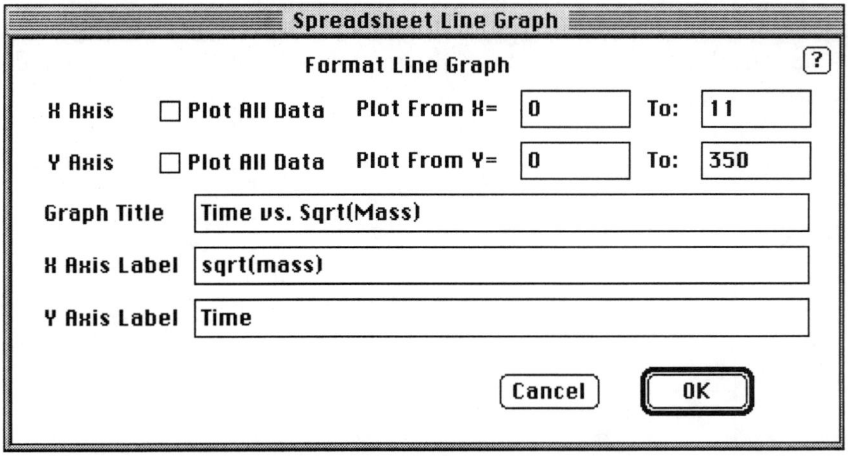

Your graph should appear similar to the one shown here:

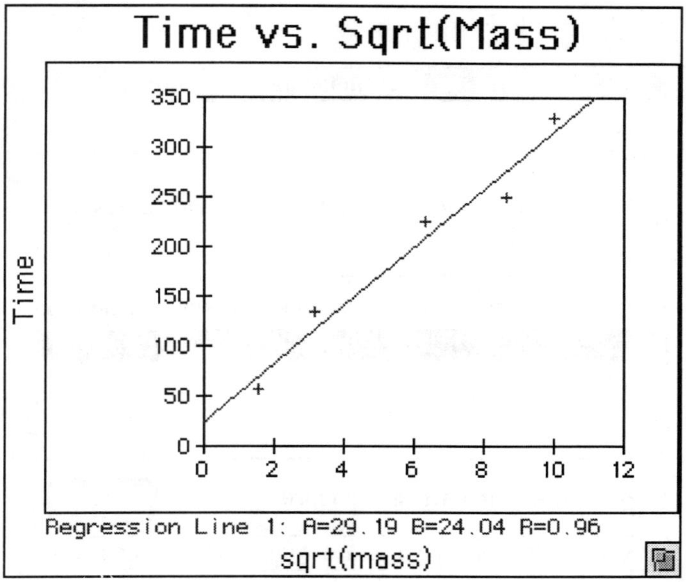

Theoretically the graph should pass through the origin (in the regression line data, B — the y-intercept — should equal 0) and R, the regression coefficient, should be very close to 1.

Note that in most cases the points do not fall directly on the line because of the statistical nature of the results; the model deals with relatively small numbers of particles.

Designing a *Kinetics* Lab

Related Lab Files:
Kinetics Design (Macintosh); \kinetics\design5.lab (Windows)

The Problem: Contradicting a Law of Nature?

At the end of the nineteenth century, James Clerk Maxwell proposed a method for changing the temperature of a gas without expending energy. A variation on his proposal is as follows:

Suppose that a gas container has a shutter, which can be opened or closed to let gas molecules escape. Open the shutter to let slow molecules out, but close it when fast molecules approach. The temperature of the gas will thus steadily increase.

Can this system work?

Approaching the Problem

The *Chemical Kinetics* model has a moveable shutter on the right outer wall, so you can set up the system to let molecules out of the container. The model also has an output for the temperature of the molecules in the container, and you can use this to check if the proposal works.

Model Window Setup

To simplify the setup, use only one type of molecule and start with a fairly low number, say ten. Set a low initial temperature, approximately 300K, so that you can see the rise if the proposal works!

Use the SPECIES tool to set the molecules to a size where they are easily visible, approximately 5 units. Ensure that the aperture in the wall is closed.

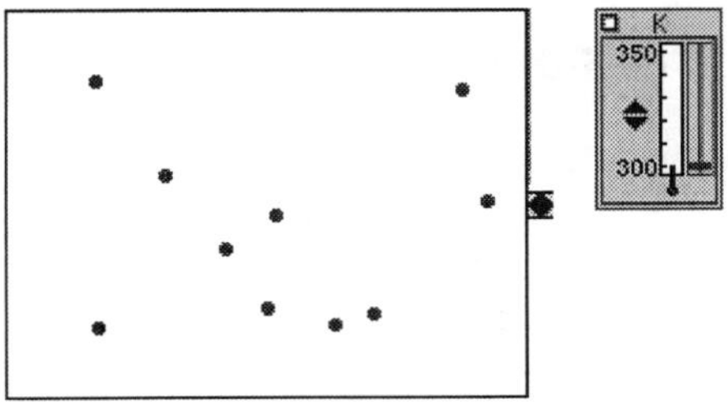

Initial Work Window Setup

Open the Work Window Tools palette from the Windows menu.

The outputs needed for evaluating the proposal are current number of molecules and current temperature. It will be helpful to copy the time increment into your work window, so you can speed up or slow down the simulation.

A typical work window is shown here:

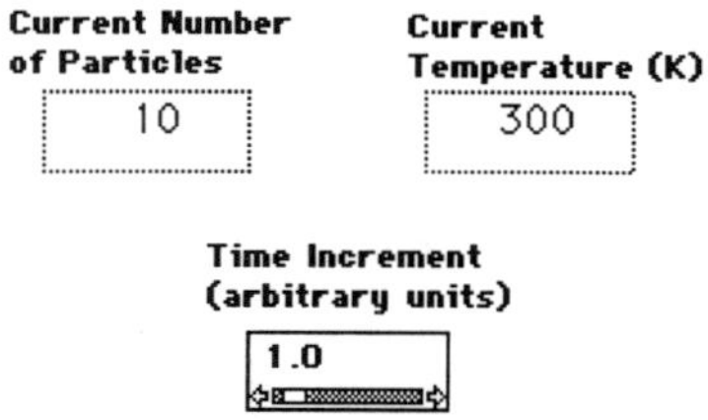

Running the Simulation

Run the simulation, then open the aperture to a fixed width. Once you have seen the effect, reset the simulation and run it with the aperture closed. Note that selecting the handles of the aperture stops the simulation, allowing you to open or close the aperture. Deselecting the aperture handles allows the simulation to continue.

Although it is somewhat easier to let the fast molecules out and to leave the slow molecules inside, with a little practice you can learn to let only the slow molecules out.

Results

After some practice, it is relatively easy to consistently increase the temperature, so that starting with 10 particles at 300K you can finish with 5 particles at 450K or more.

Conclusions and Challenges

Maxwell's proposition appears to be correct, but this is a simulation. In practice, gas molecules are moving at orders of magnitude faster than the "molecules" on the screen. Even if this problem could be overcome, there is a more fundamental objection — in order to "see" real molecules you have to input energy to the molecules, and this will automatically change the molecular speed. (In the simulation you are not inputting energy to the "molecules" in order to see them.)

Can a similar system be designed to separate gas molecules of different diameters? This system can be simulated by setting up two types of molecules with different diameters, say 20 and 2, setting the aperture diameter to be smaller than 20, then running. Note that this system does not need the input of energy to view the molecules, since the aperture should automatically perform the sorting.

Manipulating Elements in the *Electrochemistry* Model Window

Related Lab File:
Electrochemistry Manipulate (Macintosh); \elecchem\maniplt2.lab (Windows)

Tools and Controls Used:

ARROW tool

ELECTROLYTE control (TYPE and CONCENTRATION)

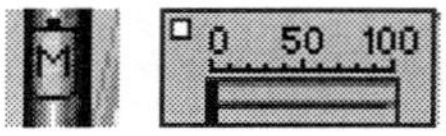

MASS control

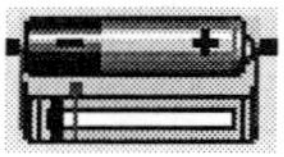

BATTERY control

ELECTRODE control

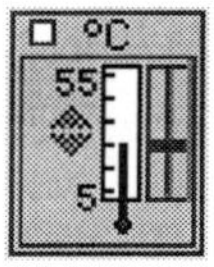

TEMPERATURE control

The Problem: Electroplating with Cd

One important application of an electrolytic cell is producing metals from their solutions. Find the conditions necessary to produce the metal Cd through electrolysis of $CdCl_2$.

Related questions:

How will the results change if you use:

i. an active electrode

ii. molten $CdCl_2$

Initial Setup

Your first step must be to set up the electrolytic cell.

Use the CONCENTRATION control in the Model Window to set the initial concentration of the $Cd^{2+}_{(aq)}$ ions to 0.1M. Although you can choose any initial concentrations, starting with a maximum value for this lab will produce the maximum mass of Cd.

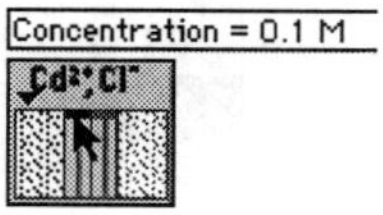

Note that the [Cl^-] is automatically set to twice the [$Cd^{2+}_{(aq)}$]. Recall that the solution is $CdCl_2$.

Next use the BATTERY control in the Model Window to set the desired voltage that will run this electrolysis. Click the red marker in the battery control to find the **emf**. A possible setting can be: 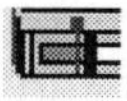 The desired voltage must be greater than the emf value.

Does the value of emf depend on concentration? In order to answer, change the concentration of $Cd^{2+}_{(aq)}$ and click the emf marker to read its value in the information window. Did it change after you changed the concentration? For the sake of this problem, set the concentration of $Cd^{2+}_{(aq)}$ back to 0.1 M.

What reactions do you expect to occur in this cell? Click the arrows to see:

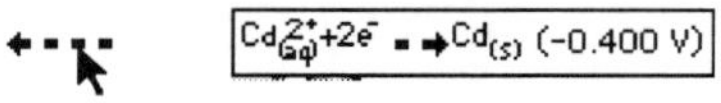

Once you have set up the cell, click GO to run the electrolysis process.

Click the dynamic arrows to answer:

- What is the oxidation reaction? In which electrode does it take place?
- What is the reduction reaction? In which electrode does it take place?

What Cd mass did you get? In how many hours?

Using the Setup

What changes will occur in the cell if the inert anode is replaced by an active electrode?

Click RESET. Click the ELECTRODE control to replace the Pt electrode with a Cd electrode:

What other changes have occurred from the previous setup?

- Notice that the emf has changed automatically. You may change the initial voltage, but it is not necessary.

- What reactions do you expect will occur in this case? Click the arrows to see. Now that the electrode is active, the oxidation reaction is changed and the initial mass of Cd should cover the electrode. Open the mass control and set an initial mass larger than zero.

Run the simulation to see the effect.

- What is the oxidation reaction? In which electrode does it take place?

- What is the reduction reaction? In which electrode does it take place?

What Cd mass did you get? In how many hours?

Expanding the Options

Another way to produce Cd(s) is to use molten salt of Cadmium. Simulate this method to see which gives better results.

Use the ELECTROLYTE control to change the electrolyte to $CdCl_2$(m):

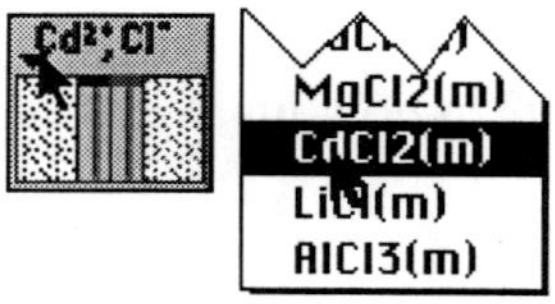

What changes do you see in the cell (emf, possible reactions, etc.)?

What changes do you need to make in the setup in order to run this electrolysis?

Run to verify and see the effect.

If you did not make all the necessary changes, the electrolysis will not run.

Remember that the electrolyte is molten salt and in this state there is no conductivity. Since only molten salt can conduct electricity, you must run the process under a temperature that is higher than the melting point of the given electrolyte!

Use the TEMPERATURE control to set the temperature to about 850°C and run the simulation.

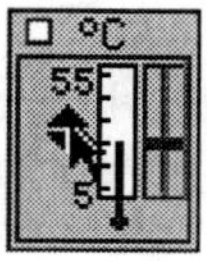

What Cd mass did you get? In how many hours?

Conclusions and Challenges

You have investigated three ways to produce Cd: aqueous electrolyte with inert electrode; aqueous electrolyte with active electrode; and molten salt. For each case, there are different desired initial conditions, different reactions, and different results.

Find the optimum conditions (maximum mass in minimum investment) out of all the above experiments. Remember that voltage, temperature, and time are money!

Analyzing Data in the *Electrochemistry* Work Windows and Spreadsheet

Related Lab File:
Electrochemistry Analyze (Macintosh); \elecchem\analyze2.lab (Windows)

The Problem: Edison's Coulometer

Thomas Edison's invention of the light bulb in 1879 created a wide demand for electricity in the home. Since electricity costs money to generate, there was a problem how to measure the amount of electricity consumed by each household.

Edison invented a coulometer for alternating current. Zinc plated out at the cathode of the coulometer. Every month the cathode was removed and weighed to determine the quantity of electricity used. If the cathode increased in mass by 1.62g and the coulometer drew 0.35% of the current entering the home, how many coulombs of electricity were used in that month?

Initial Setup

Model Window The Model Window has already been set with the desired electrolytic cell. The cell contains 0.05M Zn^{2+} and 0.1M NO^{-3}, the electrode is platinum, and the battery's voltage is 2.9V.

Creating the Output Displays

Open the Work Window Tools palette from the Windows menu.

Numeric Displays

To answer the problem, you need to know when the electrode mass is 1.62g, so you should set up two numeric displays: time and electrode mass.

Create a numeric display field. In the model's output menu, the outputs for the electrolytic cell are grouped by electrode type, so you must know the type of electrode on which the Zn is produced. As you know, the Zn is the product of a reduction reaction which takes place near the cathode. Therefore you should select the output as shown below:

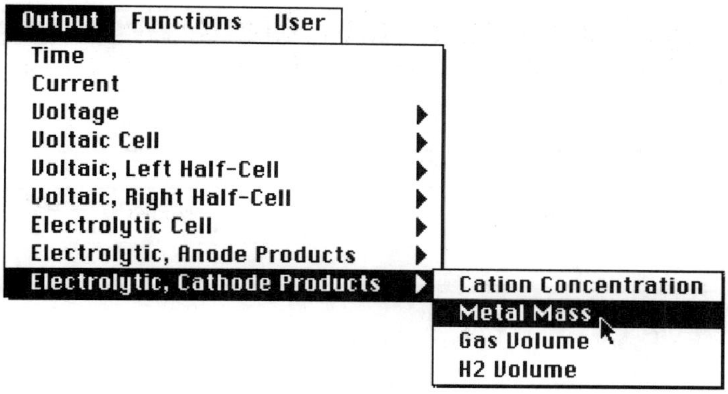

Use the TEXT tool to label your numeric display.

To measure the mass to the desired accuracy, set the display to two decimal places.

In order to answer the question, you need to know how many coulombs were used in a *month*. Create a numeric display for the number of days. Since the time unit is in hours, you must divide by 24:

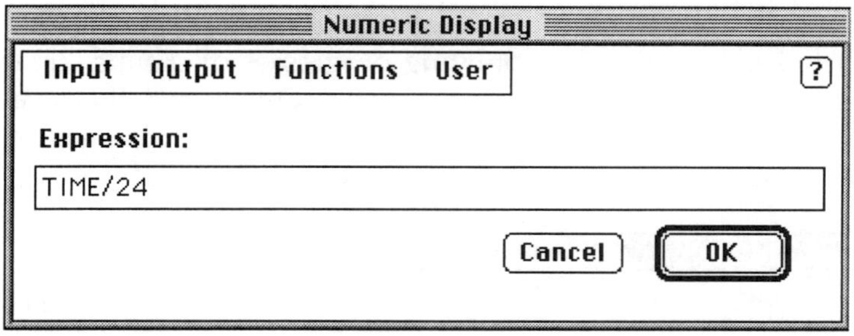

In order to arrive at the answer to the problem you need to calculate the total number of coulombs, but you may also want to follow the current number of coulombs drawn by the coulometer. Create a numeric display field. The current number of coulombs can be defined by the expression:

Coulombs = amperes * seconds

To build this expression from the variables available in the *Electrochemistry* model, you need to multiply the output value for current by the time unit (1 hour) by 3600 to convert the time to seconds. Since the problem states that only 0.35% of the coulombs are drawn by the coulometer, you must then divide this expression by 0.0035:

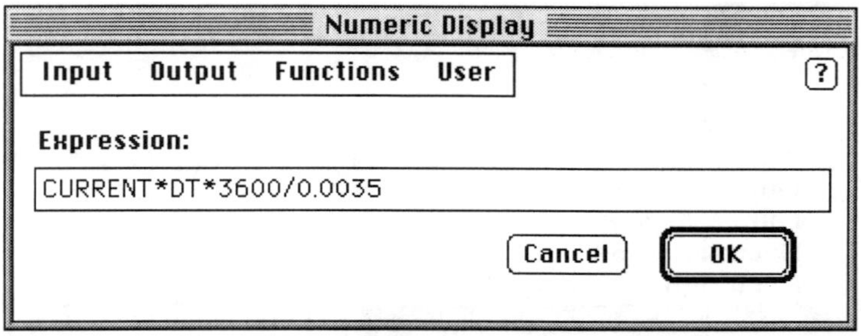

Use the TEXT tool to label your numeric display.

To measure the coulombs to the desired accuracy, set the display to scientific notation with two decimal places.

Line Graph Displays

You have already created a numeric display that shows the electrode mass at a given moment. It may also be of interest to see how it changes over time: Does the mass increase at a constant rate? How quickly does it increase?

In order to see how the electroplating progresses over time, create a line graph of the electrode mass vs. time. Use the same output that you used in creating the numeric display for electrode mass. The dialog should look like this:

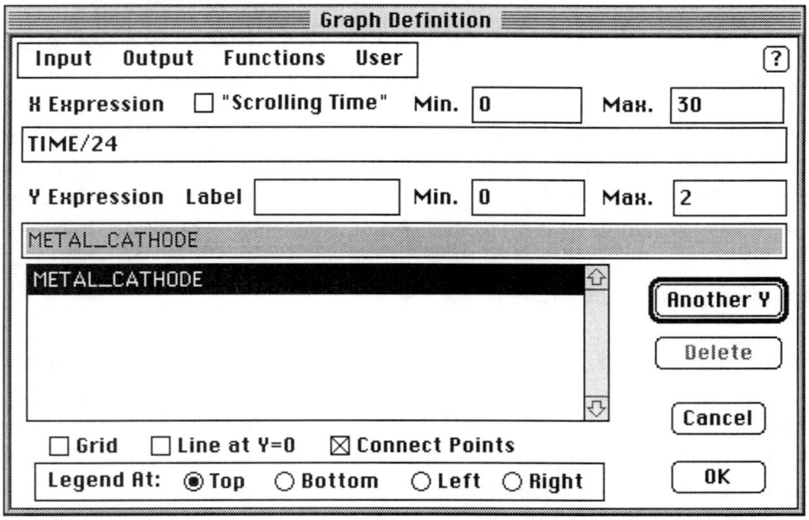

Use the TEXT tool to label the graph appropriately.

Bar Graph Displays

You may also want a graphic representation of the current number of coulombs drawn by the coulometer. Create a bar chart using the same expression that you used for the numeric display:

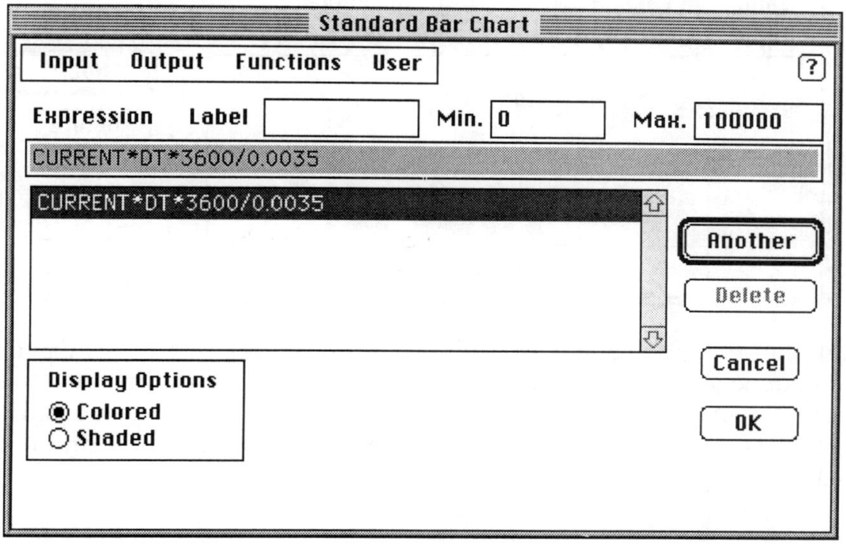

Now that the work window has been completed, run the simulation for preliminary results. To stop the simulation at 30 days, click GO and watch the time display. When it nears 30, click the SINGLE STEP tool. Keep clicking this tool until the number of days equals 30. The mass of Zn produced should be about 1.62 grams.

Using the Spreadsheet

When you run the simulation, the results show the instantaneous data, i.e., current number of coulombs. In order to answer the question, you need to calculate the number of coulombs for the *whole month*. You can use the spreadsheet to summarize the number of coulombs when the mass has reached a value of 1.62g.

Choose the "Sampling" option in the Spreadsheet menu. Use the expressions as given above to define sampling for the electrode mass and the current number of coulombs:

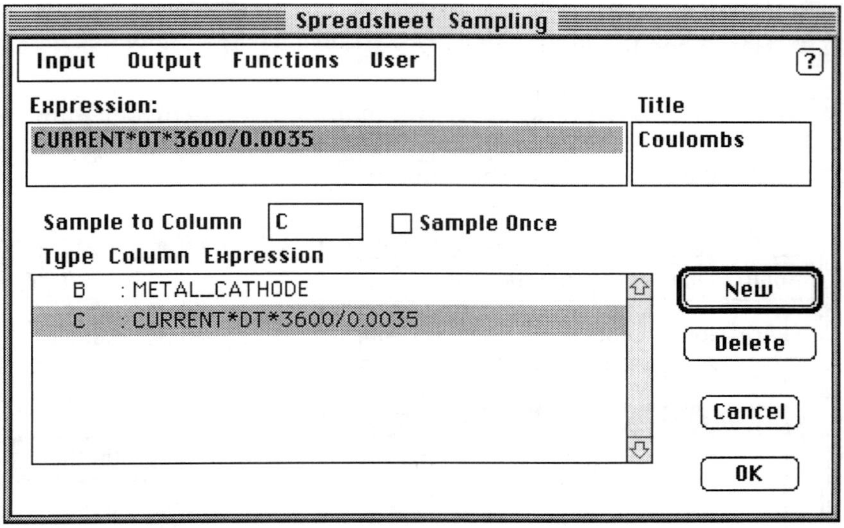

Run the simulation and stop it after the mass has reached 1.62g. Open the spreadsheet. To sum up the total coulombs drawn in one month, highlight an empty cell. In the Formula box at the top, type an equal sign so that *Explorer* will read the expression as a calculated value and not as text:

=

From the "Mathematical Functions" option in the Edit menu, choose sum().

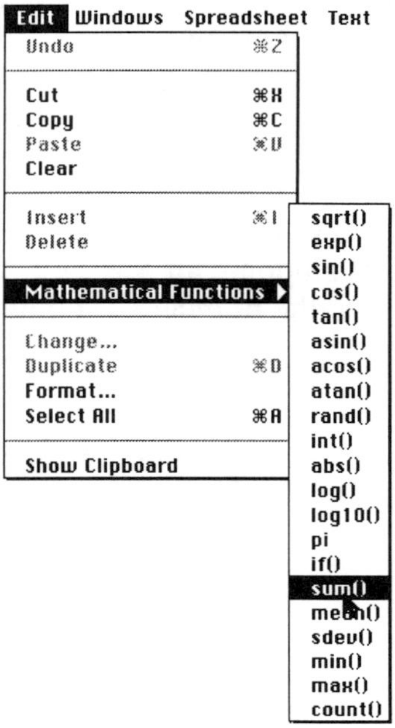

Place the cursor inside the parentheses and select cells C4–C27. Click the check box next to the Formula box to accept.

=SUM(C4*C27)

A	B	C	D
	Mass (g)	Coulombs	1.37119e+06
1	0.09	7.31E+04	
2	0.17	7.16E+04	
3	0.25	7.01E+04	
4	0.34	6.86E+04	
5	0.42	6.71E+04	
6	0.49	6.57E+04	
7	0.57	6.43E+04	
8	0.64	6.28E+04	
9	0.72	6.15E+04	
10	0.79	6.01E+04	
11	0.86	5.87E+04	
12	0.93	5.74E+04	
13	0.99	5.60E+04	
14	1.06	5.47E+04	
15	1.12	5.34E+04	
16	1.18	5.22E+04	
17	1.24	5.09E+04	
18	1.30	4.97E+04	
19	1.36	4.85E+04	
20	1.42	4.73E+04	
21	1.47	4.61E+04	
22	1.52	4.50E+04	
23	1.58	4.38E+04	
24	1.63	4.[illegible]+04	

The result for the given problem is 1.37e6 C used over 30 days.

Challenge:

Can you use the electrolyte NaCl(aq) instead of $ZnCl_2$(aq) to measure the coulombs? Explain. Which other electrolytes can you use? Use the electrolyte control to see the possibilities.

Designing an *Electrochemistry* Lab

Related Lab File:
Electrochemistry Design (Macintosh); \elecchem\design2.lab (Windows)

The Problem: Concentration Cells

Explore a cell containing 0.5M $Zn^{2+}(aq)$ in one compartment and double concentration of $Zn^{2+}(aq)$ in the other if Zn electrodes are used in both.

a. What is the **emf** of this cell?

b. What parameter(s) define(s) the electron flow direction?

c. Under what conditions is the cell **emf** zero?

d. Plot the graph that describes the Nernst equation for this cell. Find E° of this cell and calculate the Faraday's constant from the slope of the graph.

Setting Up the Model Window

Select the voltaic cell: 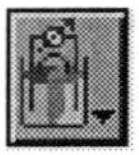

The question deals with a *concentration cell,* which is a voltaic cell that contains the same substances in both the anode and cathode compartment but in different concentrations.

Use the BALANCE tool to set the Zn reactions in both half-cells and balance the cell reaction. Set initial concentrations: 0.5M and 1M.

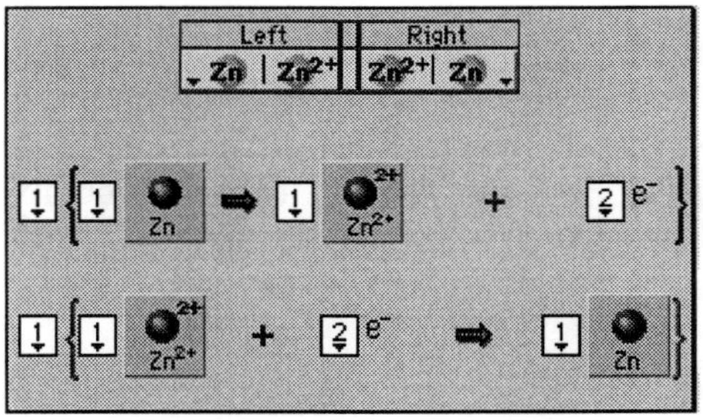

Building a Work Window

Open the Work Window Tools palette from the Windows menu.

Inputs

Most of the changes you will make in the simulation conditions you can make directly in the Model Window. Inputs that you may want to copy into your work window are the time increment and units, and the solution colors. It is recommended that you set the time increment to 2000 hours so that you can see the changes quickly.

Outputs

Since part of the problem is to determine which direction the electrons flow, it may not be immediately clear which half cell contains the reactant and which the product. Create numeric displays of the product and reactant concentrations for both half-cells. Format the displays to two decimal places and label.

Create a numeric display and a bar chart for the voltage. Note that there are two outputs for the cell voltage: EMF (the result of Nernst equation) and Voltmeter Reading (which considers the cell resistance). Remember that the cell emf is the voltage when there is no current in the cell! In order to answer (b) and (c) you may want to use both outputs. You can display them both on one bar chart:

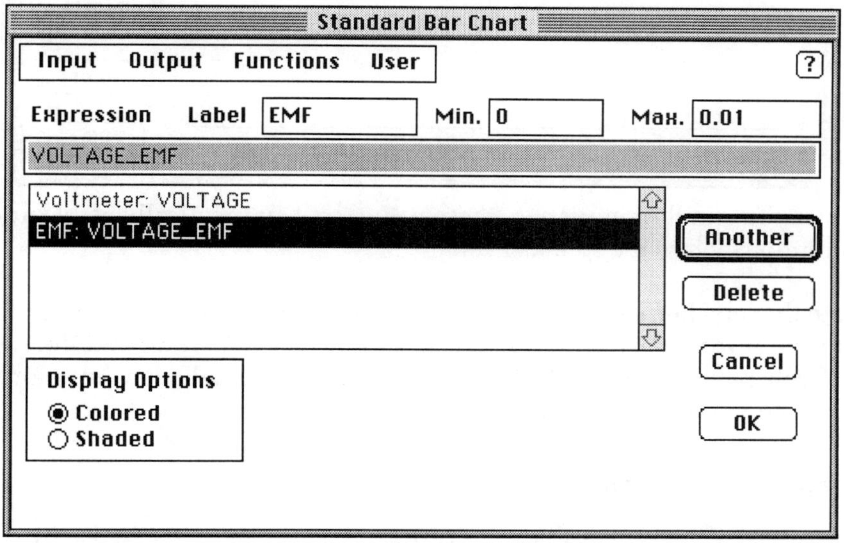

Sampling to the Spreadsheet

In order to plot the Nernst equation, you need to sample the EMF output and logQ during the simulation.

Choose the "Spreadsheet Sampling" option in the Spreadsheet menu. For sampling to column B, choose the Voltage: EMF output. Label it.

For sampling to column C, you must build the expression for logQ. In the *Electrochemistry* model, the products and reactants for each half-cell are defined by the concentrations you set in the half-cells. For example, if you set the higher concentration in the left half-cell, divide the left half-cell product concentration by the right half-cell reactant concentration, and take the log of this expression.

The Sampling dialog box should look like this:

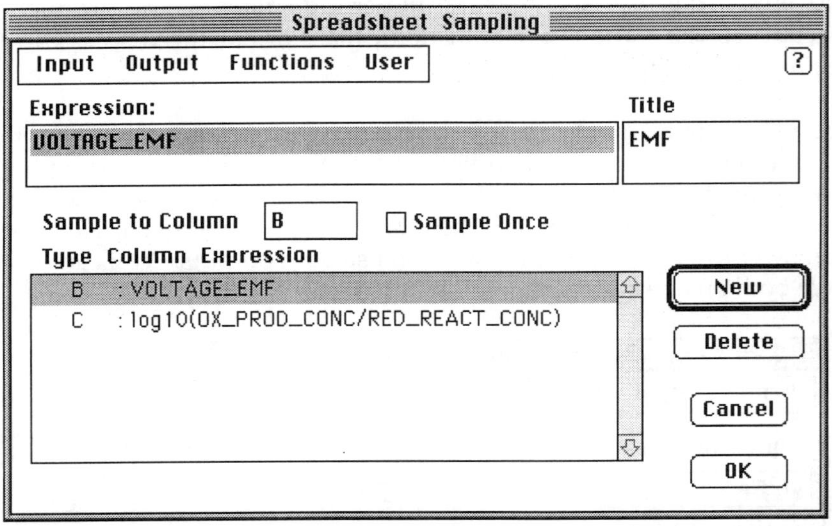

When you run the simulation, these values will automatically be recorded continuously into the spreadsheet.

Note that an alternate way to achieve the same results would be to sample into the spreadsheet the left half-cell product concentration and the right half-cell reactant concentration. Then you could calculate in another spreadsheet column the log of their ratio.

Answering the Question

To answer the questions: (a) recall that the cell emf is measured when there is no current (i.e., in RESET mode); (b) by changing concentrations while the simulation is running, show that in a concentration cell the electron direction is defined by the relative concentrations in the two half-cells; (c) set equal concentrations (try several values) to show that in this case the voltage is zero; (d) calculate E° and F (Faraday constant) from the graph based on the Nernst equation (see below).

Generating Graphs from the Spreadsheet

Create a graph of EMF vs. logQ. Show the regression data and put it into the clipboard. The spreadsheet graph dialog box should look like this:

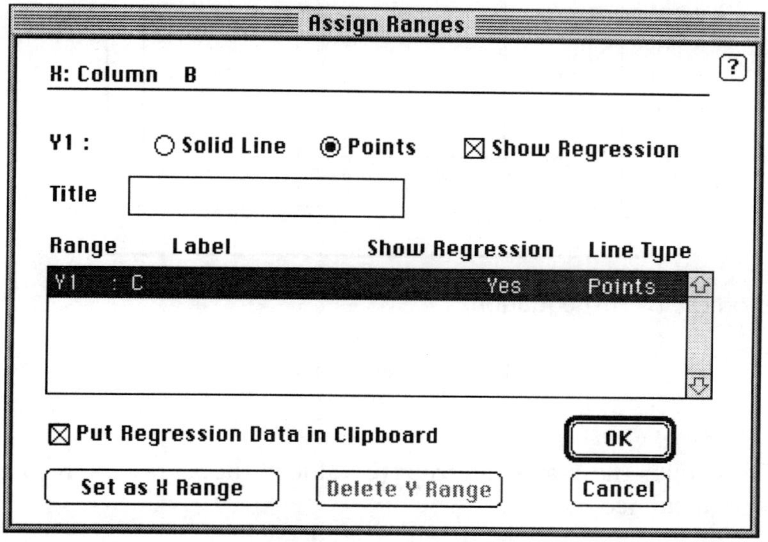

Your graphic should look similar to this:

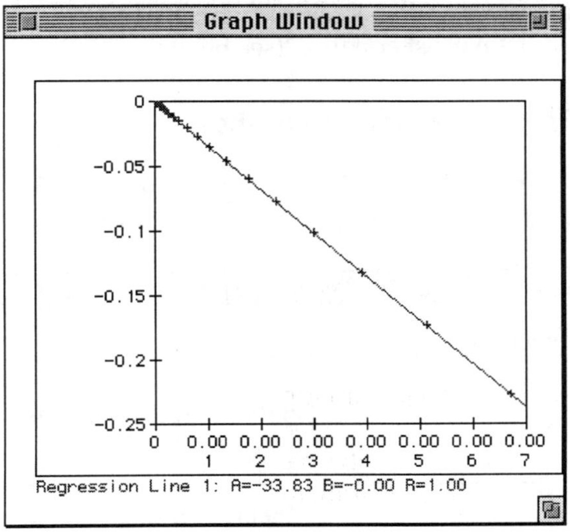

Based on the Nernst equation:

$$E = E° - \left(\frac{2.3RT}{nF}\right) \times \log Q$$

E° of this cell is the Y-axis intercept (**b** value in the regression results) and F can be calculated by:

F = slope x n / 2.3RT (slope is **a** value in the regression results)

You can highlight a cell in the spreadsheet and paste these regression data from the clipboard into the spreadsheet.

Conclusions and Additional Challenges

In concentration cells, the cell data such as cell voltage or electron flow depend on the relative concentrations of both half-cells (and not on the reactions' potentials).

How will your answers change if you vary the cell reactions within the concentration cell, e.g., use Cu electrodes in both half-cells? Can you make some general predictions for answers?

Appendix 1:
Explorer System Reference

1. Explorer Control Tools

The *Explorer* control tools are the same for every *Explorer* series product.

RESET Tool

The RESET tool returns the simulation setup to its initial values so you can repeat the simulation or set up a new one.

If you changed any values in an input box or with a model tool while the GO or STOP tool was active, the simulation will not "remember" these changes. You must click RESET before changing values in order to have them take effect for the next simulation.

GO Tool

This tool starts the simulation with the current values.

STOP Tool

This tool stops the simulation at its present state.

SINGLE STEP Tool

This tool advances the simulation one calculation step and then stops each time you click the tool. It is particularly useful for watching a simulation "in slow motion."

HELP Tool

Click this tool and then one of the following elements on the screen to get a help window: input boxes, pop-up menus, model tools, dialog boxes.

2. Work Window Tools Palette

The Work Window Tools palette contains nine tools.

The first tool, the HAND, is used for normal operation of an *Explorer* lab, such as changing input values and clicking buttons.

The other eight tools are for use in designing or editing lab files. The following pages show how to use each tool.

Hand Tool

The HAND tool is for working within the Work window. It has no design functions.

1) Drag the scroll box to change a value

2) Type in a value
(click outside the box to enter)

3) Click the scroll arrows to change a value

4) Choose an option value

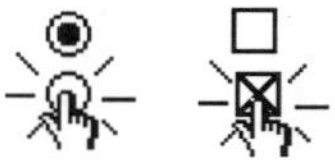

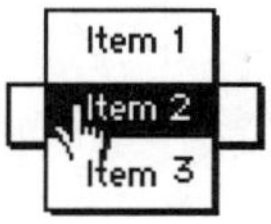

5) Click a button

Arrow Tool

The ARROW tool is for selecting and modifying elements in the Work window.

1) Select an item

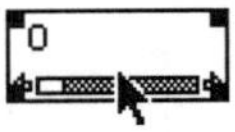

2) Double-click on an item to open its dialog box

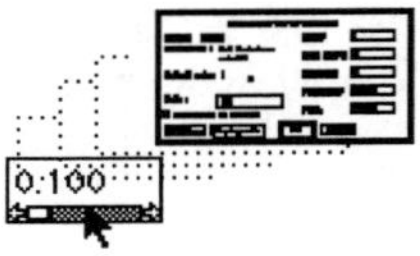

3) Drag an item

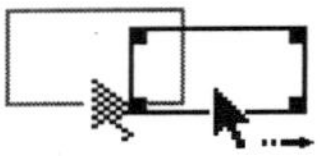

4) Resize an item

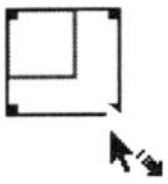

Text Tool

The TEXT tool is for adding text in the Work window.

1) Define the text field

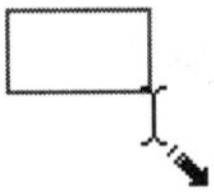

2) Type in the text

3) Click outside the box to finish

You can then select the text with the ARROW tool and use the "Font," "Size," and "Style" options in the Text menu to change the look of the text.

 Numeric Display

1) Define the Numeric Display box:

2) Click the mouse to open this dialog box:

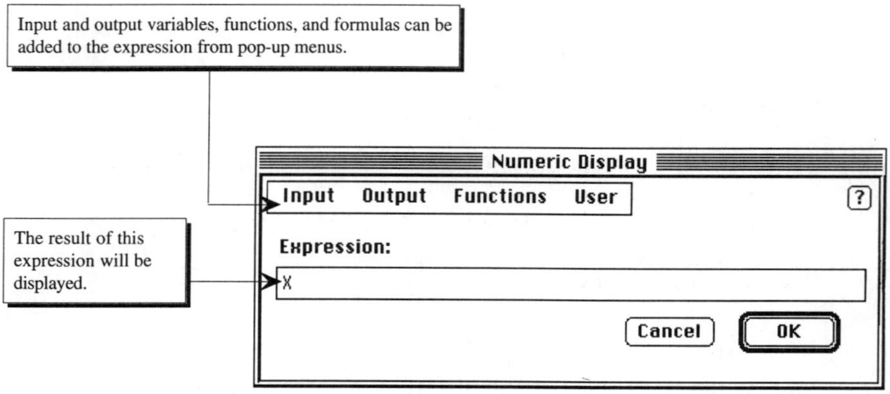

3) Use the ARROW tool to determine the size, shape, and location of the box, and to adjust its format using the "Format" command in the Edit menu.

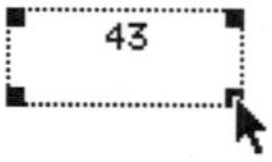

Result:

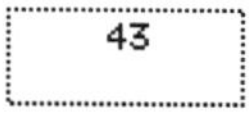

Bar Chart

1) Define the Bar Chart box:

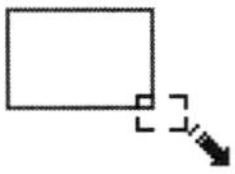

2) Release the mouse to open this dialog box:

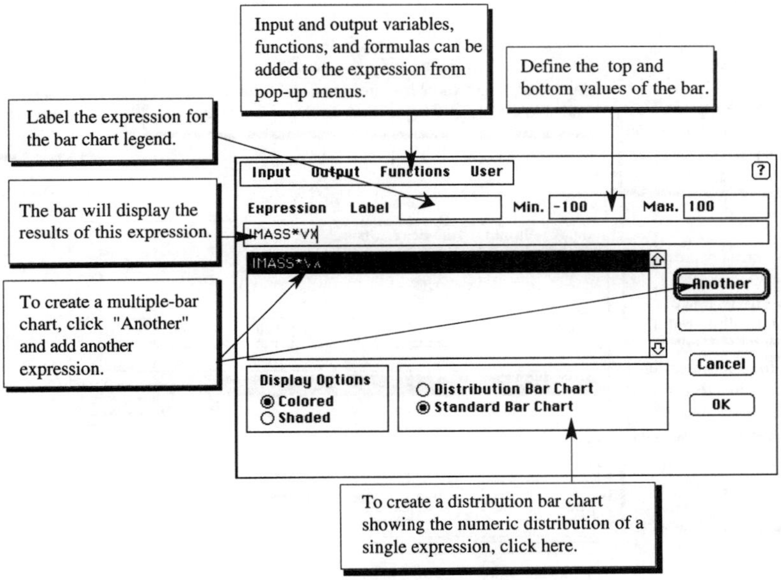

Result:

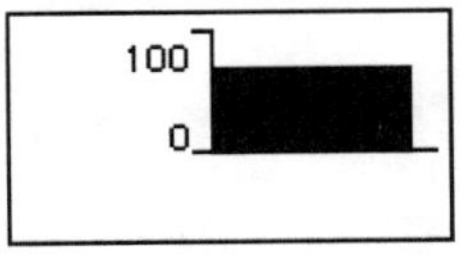

Line Graph

1) Define the Line Graph box:

2) Release the mouse button to open this dialog box:

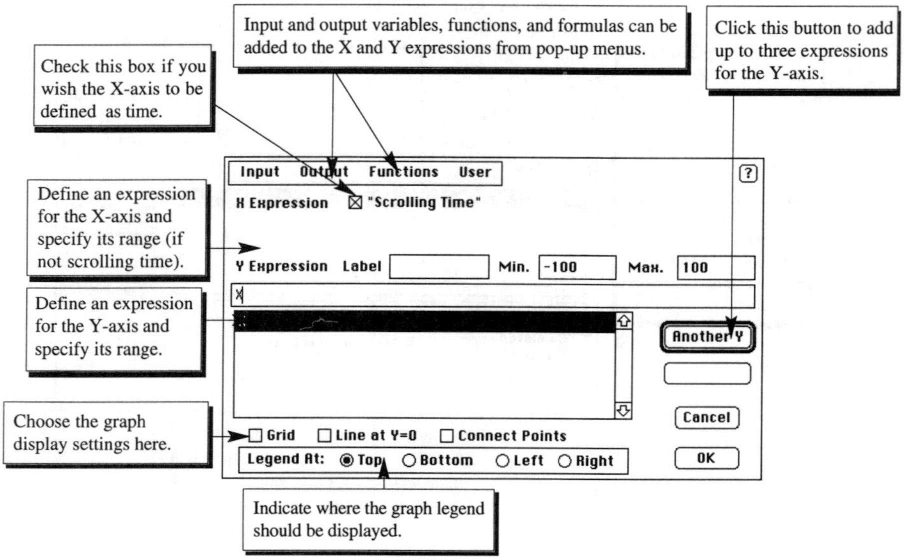

Result:

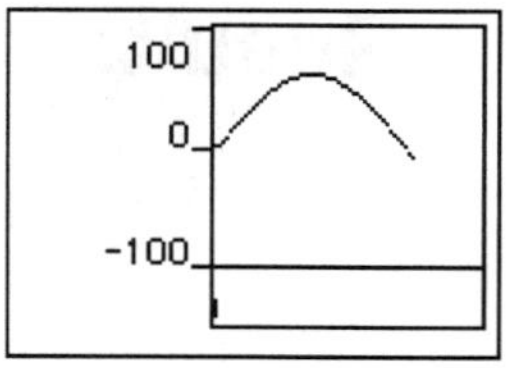

Pie Chart

1) Define the Pie Chart box:

2) Release the mouse button to open this dialog box.
Click on the "Distribution Pie Chart" radio button. The following dialog box opens:

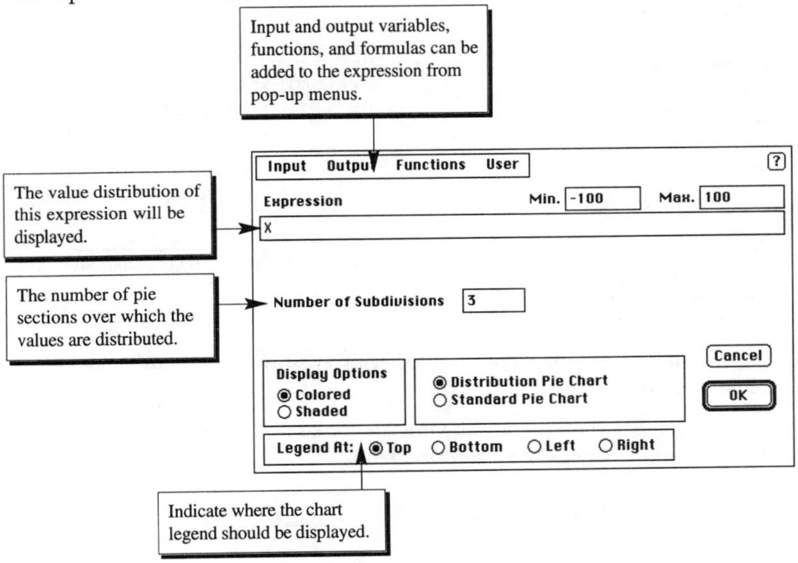

Result:

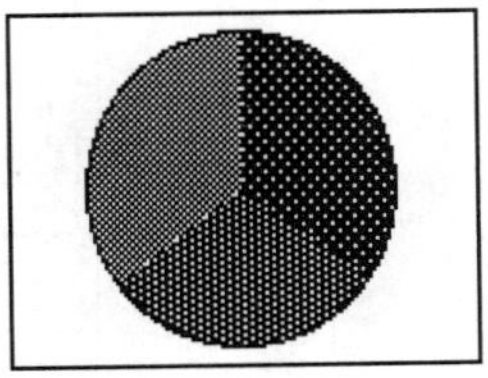

Creating a Button

1) Define the button using the BUTTON tool:

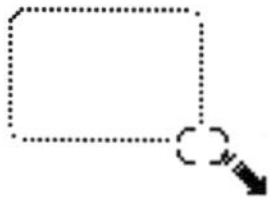

2) Type in the button name, then click the cursor outside the Button box:

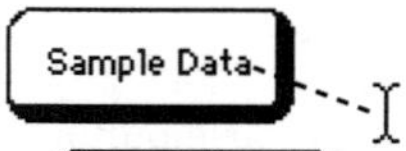

3) Select the ARROW tool and double-click the new button to open the button dialog box:

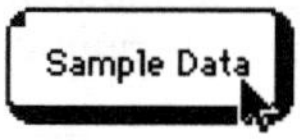

4) In the dialog box, choose a reaction clicking on one of the menu boxes such as "Data":

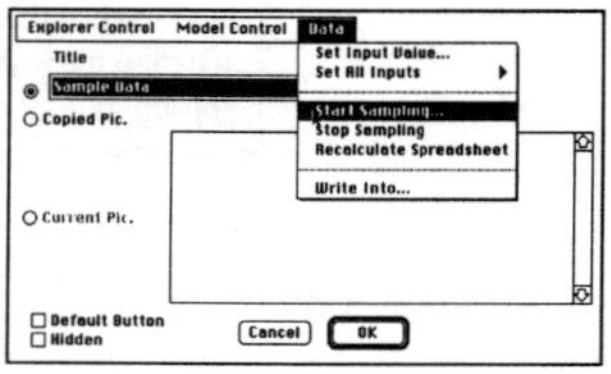

5) A pop-up menu appears. Drag to select a reaction, such as "Start Sampling."

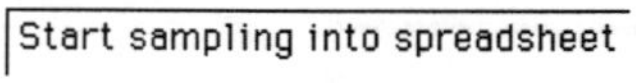

6) The reaction text is displayed in the Script window. Select additional reactions for this button, if desired.

7) You can paste a picture onto the button, make it invisible, or make it a default button. Check the appropriate box.

8) Close the dialog box by clicking on OK:

Result:

The HAND tool now operates your new button.

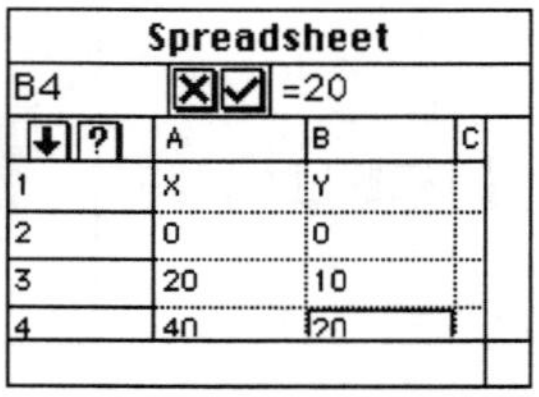

More Button Information

Button Reactions

Explorer Control reactions perform operations related to the general *Explorer* system:

Reset Simulation: This reaction returns the current simulation to the last initial values that were in effect. It has the same effect as clicking the RESET tool.

Start Simulation: This reaction starts the simulation from the current values. If the simulation is in reset mode when this reaction is invoked, the simulation starts from the initial input values. If the simulation is in stop mode when this reaction is invoked, the simulation continues from the current output values in the model window. It has the same effect as clicking the GO tool.

Stop Simulation: This reaction stops the simulation in the model window. It has the same effect as clicking the STOP tool.

Single Step: This reaction enables you to advance the simulation one calculation step at a time. This reaction is useful for putting the animation into "slow motion." It has the same effect as clicking the SINGLE STEP tool.

Beep: This reaction emits a beep when the button is clicked.

Exit Explorer: This reaction allows you to exit *Explorer* to the Finder or Program Manager. You are prompted to save the lab before exiting.

Model Control reactions perform operations related to the specific model, and they vary from one model to another. Some models have only a few model reactions, while others have numerous reactions.

Data reactions perform operations related to the handling of data during the simulation.

Set All Inputs to Default Values: This reaction resets all input variable values to the values that were in effect when the lab was opened. This reaction is useful for creating an "Initial Conditions" button that returns the lab to its startup conditions. Note: For such a button, it is recommended that you place a Reset Simulation reaction before and after the Assign All Inputs to Default Values reaction.

Set All Inputs to Current Values: This reaction updates the values of the input variables according to the current values in the model window. This reaction is useful if you want first to play with input values until you get them "right," and then use these values for setting up your simulation.

Start Sampling to Spreadsheet: This reaction begins the data sampling into the spreadsheet. In the *Student Edition Chemistry Explorer*, the spreadsheet has been prescripted to start sampling each cycle whenever the GO tool is clicked. You can overwrite this using the Stop Sampling to Spreadsheet reaction in combination with this reaction, e.g., to change the sampling rate.

Stop Sampling to Spreadsheet: This reaction stops the sampling of expression values to the spreadsheet.

Recalculate Spreadsheet: This reaction forces the spreadsheet to recalculate all formula values. It has the same effect as choosing the Re-Calculate option from the Spreadsheet menu.

Write Into: This reaction enables you to write a message into an alert, the spreadsheet, or an external log file. Note that you can include variables in the message, and that the value of the variable at the moment the message is written will be inserted into the message. Names of variables must be enclosed in {} in order to be read as a value and not as text. For example:

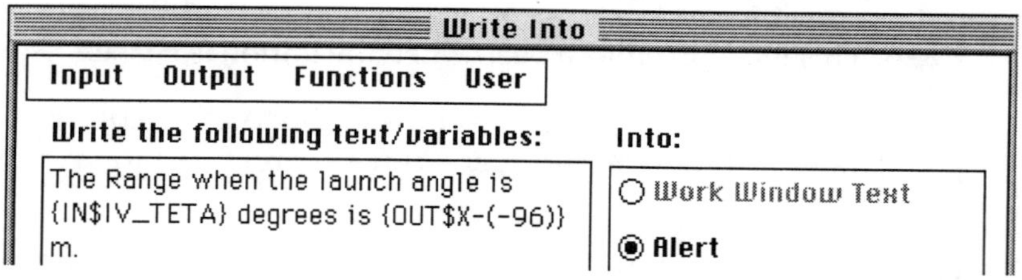

generates an alert when the button is clicked:

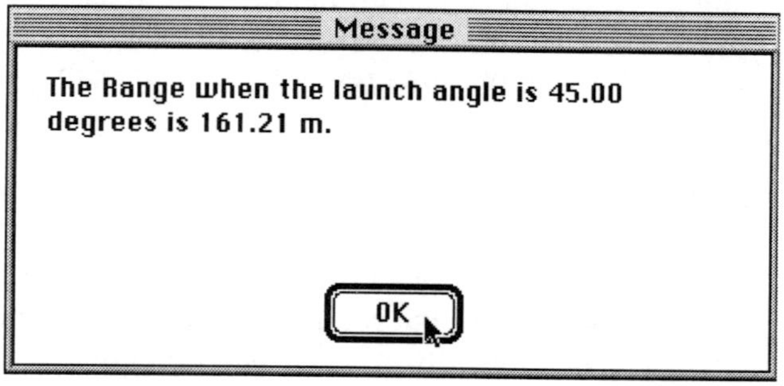

Close Log File: This reaction closes the external log file currently open as a result of a "Write Into... Log File" reaction.

Editing Buttons

You can assign more than one reaction to a button. *Explorer* performs the operations in the order that they are listed. Often the order is insignificant, but sometimes the order can affect the result. Consider this example of reactions listed in a button called START:

```
Reset Simulation
Start Simulation
```

This will reset the input values to their original values, then start the simulation. If the order is reversed, this will not be the result! (Try it!)

You can double-click with the ARROW tool to access the button's dialog box. Then you can add new reactions or delete existing ones by selecting them and choosing Cut from the Edit menu.

You can paste a picture onto a button instead of giving it a name. For example, you may want to make a button that looks like this:

Create the desired picture in a graphics program. Copy it to the Clipboard. Inside the button dialog box, click Copied Pic. A thumbnail of the Clipboard contents appears under the radio button to show you the picture that will be displayed on the button.

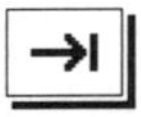

Input Tool

The INPUT tool enables you to place input value boxes in the Work window.

1) Place the tool in an empty space in the window and click. A pop-up menu opens.

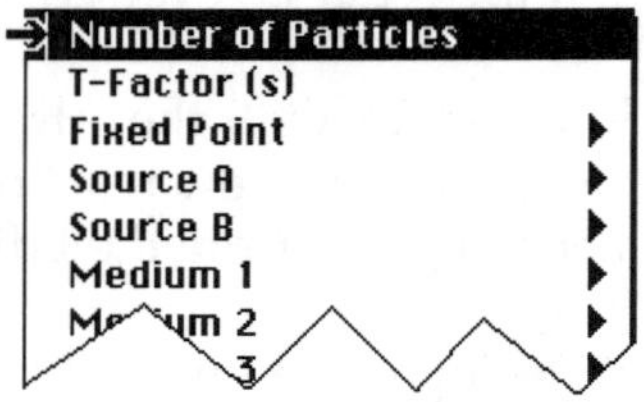

2) Drag to the desired variable and release. An input box is placed in the window.

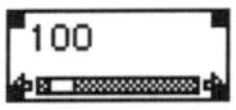

3) Double-click on the box to open the input variable's dialog box. You can define its attributes here.

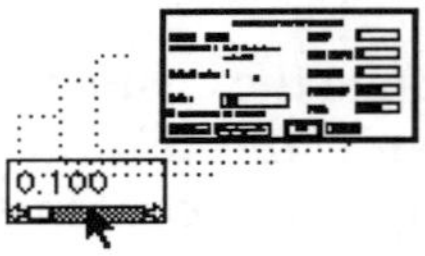

4) You can select the input box and choose the "Format" option from the Edit menu to set the number of decimal places.

3. The Spreadsheet

Sampling Data to the Spreadsheet

During an experiment run, you may output data automatically to a spreadsheet. This process is called "sampling." Steps for sampling data to the spreadsheet are outlined on the following pages.

Once you've defined the sampling for the spreadsheet, it normally starts and stops sampling automatically when the experiment starts and stops. You can, however, create a button to stop sampling.

Spreadsheet Summary Sampling

Sometimes it is useful to collect data from a series of simulation trials into one spreadsheet for comparison and analysis. Steps for summary sampling to the spreadsheet are outlined on the following pages.

Spreadsheet Graphing

Once data has been recorded in a spreadsheet, it can be graphed in a special Spreadsheet Graph window. Steps for creating a spreadsheet graph are outlined on the following pages.

Spreadsheet Calculations

You can also use the spreadsheet to analyze the output data after the simulation run is completed. You may add columns to perform specific calculations on data collected from sampling. Alternatively, you may use the spreadsheet without sampling data at all, by manually entering and analyzing experimental observations. In this way, it can be used exactly like any other spreadsheet program. For more details on the spreadsheet, refer to More Spreadsheet Information (page 147).

Spreadsheet Sampling

Choose the "Sampling" option from
from the Spreadsheet menu.

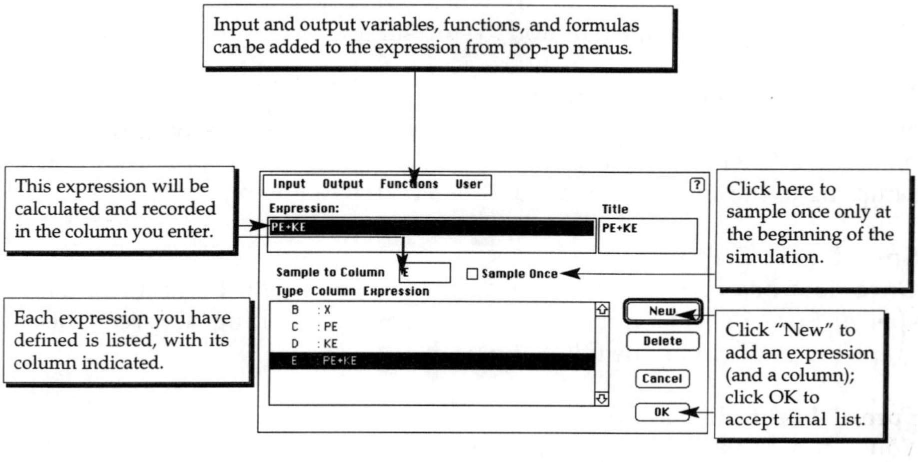

Result:

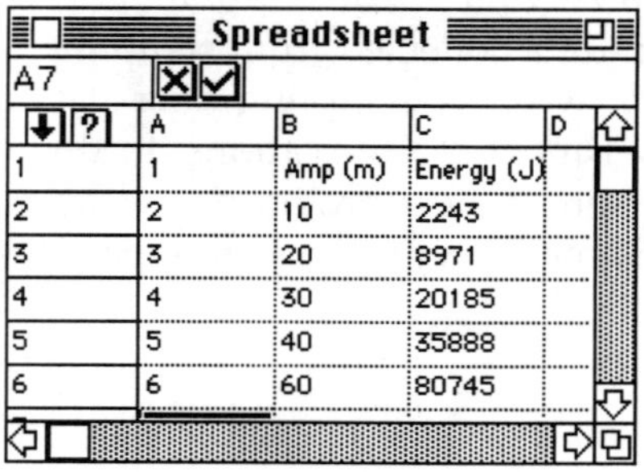

	A	B	C	D
1	1	Amp (m)	Energy (J)	
2	2	10	2243	
3	3	20	8971	
4	4	30	20185	
5	5	40	35888	
6	6	60	80745	

Spreadsheet Summary Sampling

1) Create and label a new button.

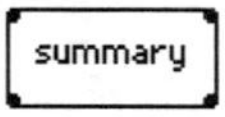

2) Double-click the button with the ARROW to open a dialog box:

Pull down the Data Reactions menu.

Select "Write Into."

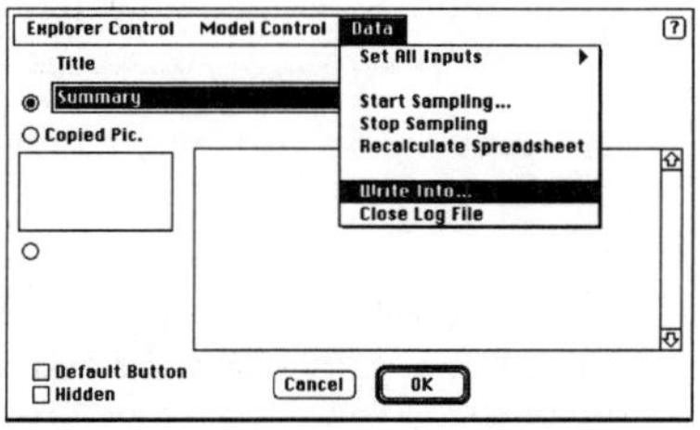

3) The "Write Into" dialog box appears. Select "Spreadsheet" as destination, and check "Summary Sampling." Choose variables from pop-up menus. Each of the variables is assigned to a column according to the order in which you list them in the dialog box. You may also enter text. Set the column into which the data is sampled.

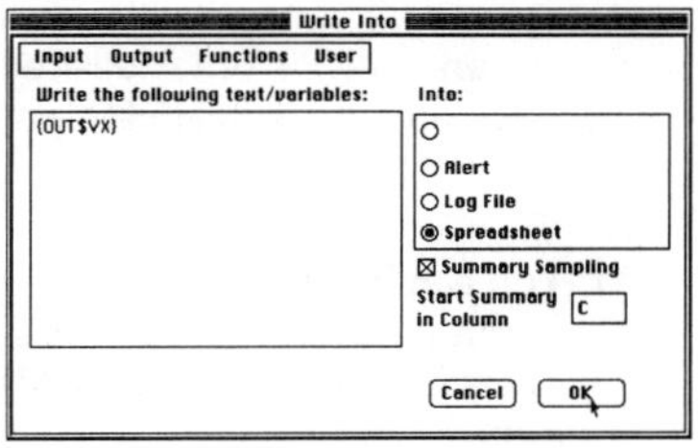

Result:

Every time you press the button, values of the designated variables are entered into the spreadsheet.

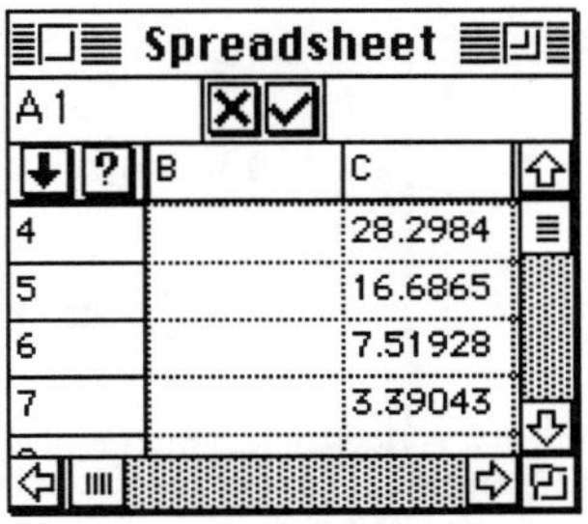

Spreadsheet Graphing

1) Select the columns or portions of columns that contain the data you wish to plot. If the columns are not adjacent, hold down the CMD key (Macintosh) or the CTRL key (Windows) while you select each column.

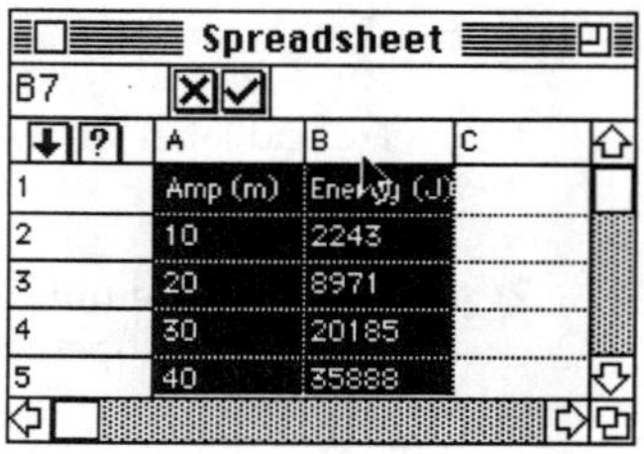

2) From the Spreadsheet menu, choose "New Line Graph."

3) The left-most column you selected is automatically defined as the x-range of the graph. To change this, highlight the desired column in the list and click on the "Set as X-Range" button. Set the other desired options. Click on OK to create the graph.

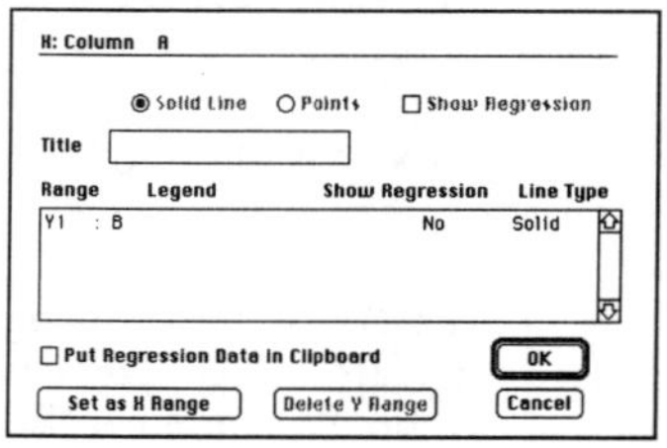

4) You can label the axes and title the graph by choosing the "Axis and Titles" option in the Spreadsheet menu.

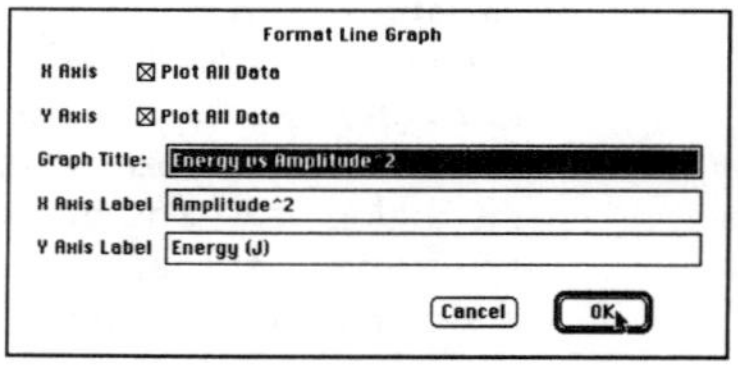

Result:

Note that various other graph formats and statistical analyses are available for selection in the Spreadsheet menu or the dialog box.

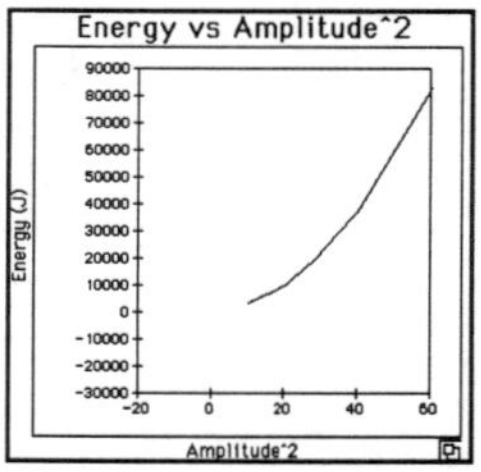

The graph is displayed in the Spreadsheet Graph window

More Spreadsheet Information

Spreadsheet Cell Contents

Spreadsheet cells contain one of three types of information:

- a number, either entered directly or sampled from the simulation;
- a formula; or
- a text label.

When the cell contains a simple number, the edit box contains the unformatted number (to 8 decimal places). The cell itself displays the formatted number, with the number of decimal places set by the user.

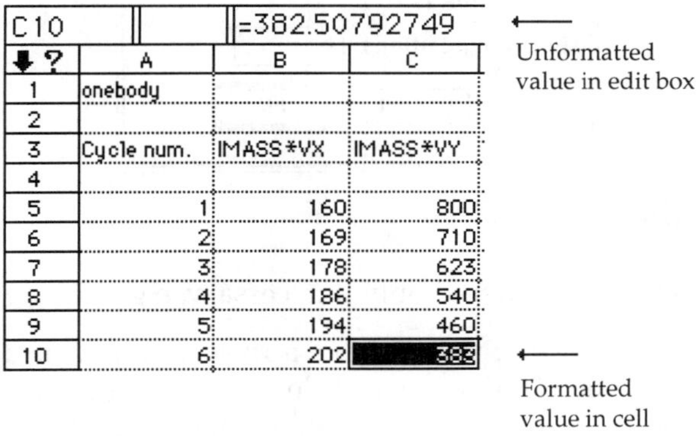

The contents of the selected cell are displayed in the edit box at the top of the spreadsheet. When the cell contains a formula, the expression is displayed in the edit box. The value of the expression is displayed in the cell.

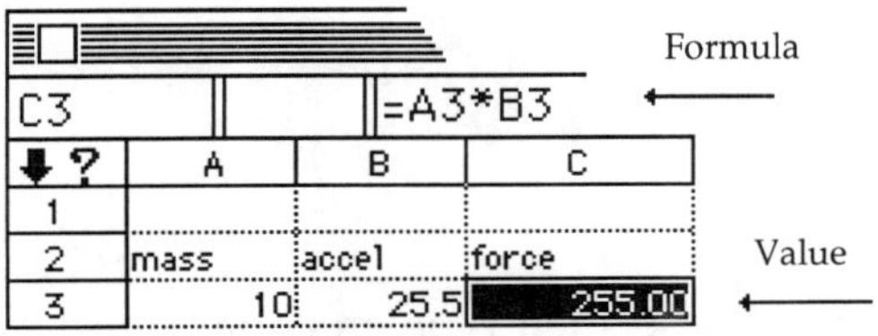

Entering Text or Numbers

To enter text (such as a label) or a number in a cell:

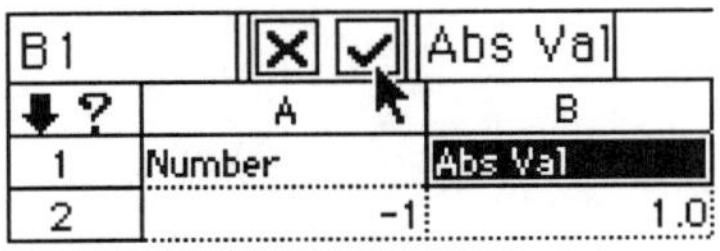

❶ Select the cell in the spreadsheet.

❷ Type the text or number. The edit box is activated and what you type is automatically entered into it.

❸ Click ☑ to enter the value, or ☒ to cancel.

Formatting Numbers in the Spreadsheet

The Format option in the Options menu lets you define the format of numbers written in the spreadsheet cells.

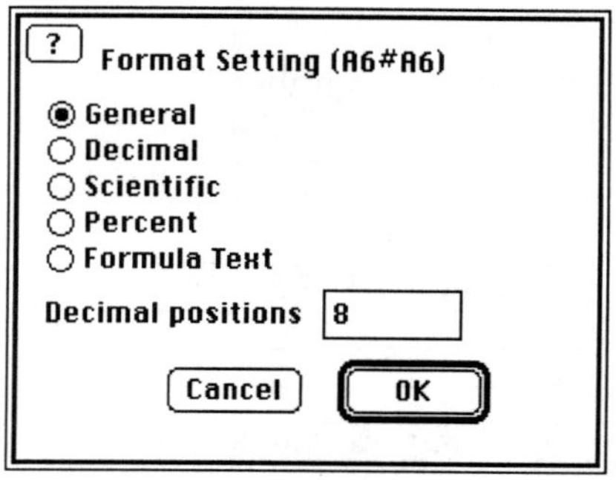

The available formats are:

- General: Decimals entered only if necessary, e.g., 17.
- Decimal: All numeric values have a fixed number of decimal positions, e.g., 17.000. The number of decimal positions depends on the number you enter in the Format options dialog box.
- Scientific: Numbers are written in scientific notation, e.g., 1.700E+001.
- Percent: Numbers are written as percentages, e.g., 17%.
- Formula Text: The cell's formula is written as text and not as a calculated value.

Entering a Formula

To enter a formula:

❶ Select the cell and then type an equal sign (=). The edit box is activated and what you type is entered into it.

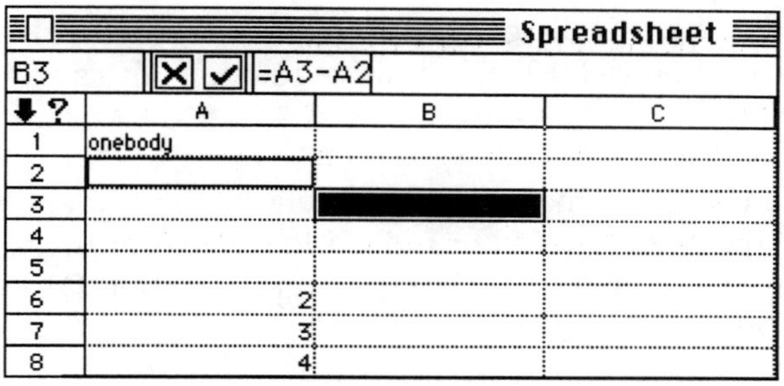

❷ Type in the formula. The formula *must begin with an equal (=) sign*. Otherwise, *Explorer* reads the contents of the edit box as text.

❸ To enter a reference to another cell into the formula, click the actual cell while still in formula writing mode. Its address is automatically entered into the formula. You can also select a range of contiguous cells—their addresses are automatically written into the formula (using the convention A1#A10, where # means "through").

When you click the check box, *Explorer* checks the formula to make sure the syntax is correct. If *Explorer* detects an error, a dialog box appears giving you the choice to convert the formula into text or to try again. The vertical cursor indicates the position of the error.

☞ **NOTE**: If a formula cannot be calculated immediately because it contains a reference to a cell that is currently empty, the cell is marked with the symbol "Err" until a calculation can be made.

Selecting Multiple Cells

Selecting multiple cells in a spreadsheet is similar to selecting elements in other Macintosh/Windows windows.

- To select an entire column, click the letter label in the column bar.
- To select an entire row, click the number label in the row bar.
- To select a range of cells, click the first cell and drag the mouse to the last cell; or, click the first corner of the range, and then hold down the Shift key and click the cell constituting the opposite corner. The entire cell range is selected.

Relative vs. Absolute Reference

A reference to another cell is automatically a *relative* reference. If the cell for which the formula is being written is copied or moved to another address, the cell reference is updated to the new location. For example, suppose cell A6 contains the formula A1+5. If cell A6 is copied into cell

D10, cell D10 automatically contains the formula D5+5. A1 is 5 cells before A6; D5 is 5 cells before D10. The relative relationship is maintained in the formula.

You can also choose to make a cell reference an *absolute* reference. In this case, even if the cell is copied or moved to another address, the cell reference remains the same. In the above example, the cell D10 would contain the original formula A1+5.

- To enter an absolute column reference, add a dollar sign ($) before the letter label.
- To enter an absolute row reference, add a dollar sign before the number label of the reference cell address.

In the example above, you would type A1+5.

Absolute Reference Option

You can also define a single cell or a range of cells as an absolute reference by selecting the desired cells and then choosing the Absolute Reference option from the Spreadsheet menu. Whenever you enter a reference to one of these cells into a formula for another cell, it is automatically entered as an absolute reference.

The contents of cells that have been defined as absolute references appear in italics.

Copying a Formula over a Range

When you paste a formula into a selected range of cells, *Explorer* works out the relative references for the entire range. For example, to produce a series of integers from 1 to 10 in column A:

❶ Enter the formula =1 in cell A1.

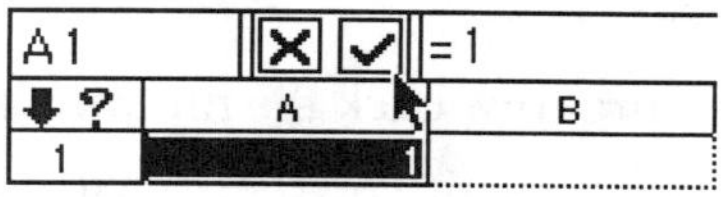

❷ Enter the formula =A1+1 in cell A2.

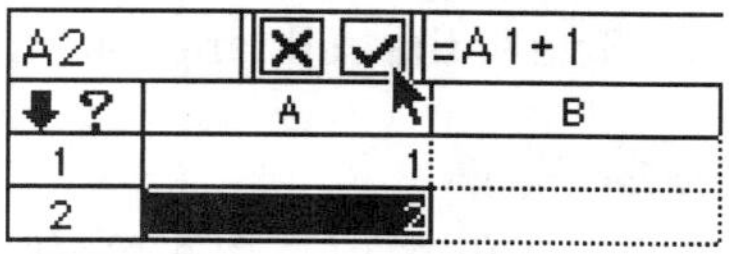

❸ Select cell A2 and choose Copy from the Edit menu.

❹ Select the range of cells A3#A10 and choose Paste from the Edit menu.

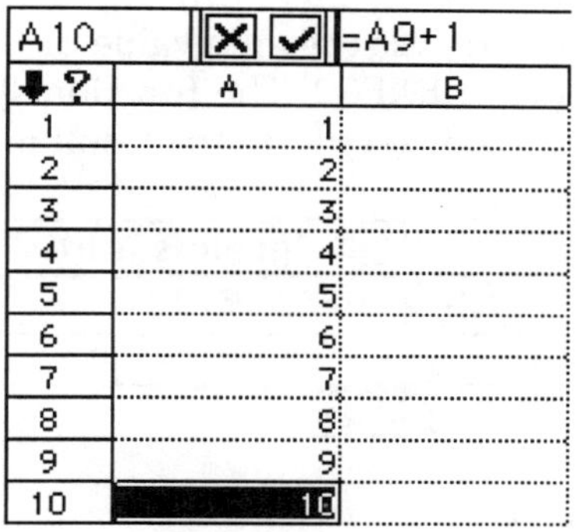

Explorer automatically completes the series, using relative references.

A range is punctuated with a number sign (#), e.g., B1#C8 represents cells B1 through B8 *and* C1 through C8.

Functions

You can type a mathematical function directly, or you can choose a function from the list called up by the Functions option on the Edit menu.

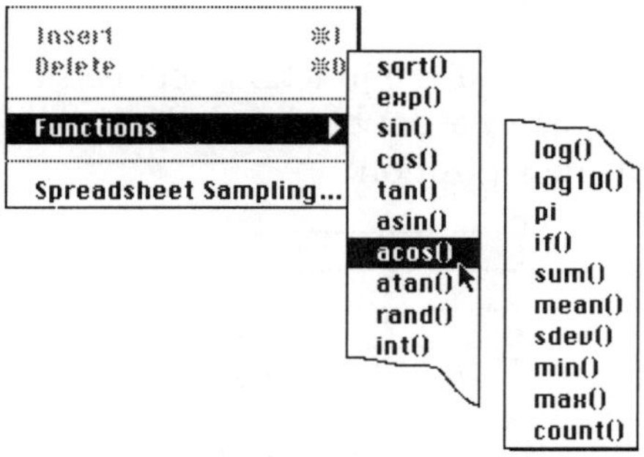

When you select a function from the list, the vertical cursor automatically appears inside the parentheses, waiting for you to enter the value or cell address the function will use as an argument.

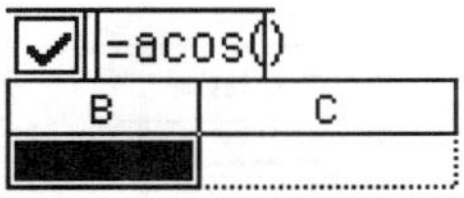

☞ **NOTE:** The Functions list shown here contains additional functions not found in the pop-up function menus in the various dialog boxes; these extra functions are ones that can be performed only on a range: sum, mean, sdev (standard deviation), min, max, and count (the number of nonzero values in a range).

The IF Function

IF is a function that lets you enter simple IF/THEN statements containing three arguments: IF(EXPR; EXPR1; EXPR2), where *Explorer* calculates EXPR. If EXPR is true, *Explorer* uses the value of EXPR1. If EXPR is false, *Explorer* takes the value of EXPR2. The function shown below computes the absolute value of cell B8 and places the result in cell B1.

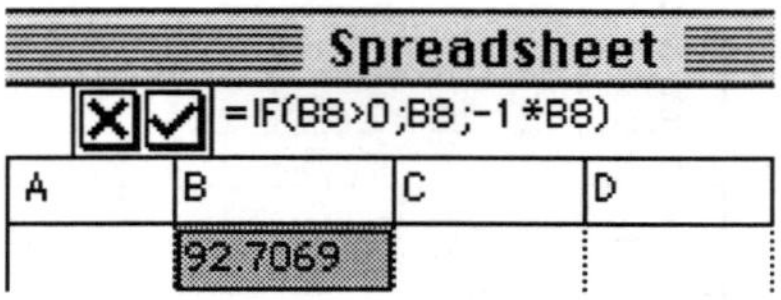

Modifying Spreadsheets

You can copy, cut, paste, and clear cells, rows, or columns using the appropriate options in the Edit menu. You can also change the width of columns and insert and delete columns.

TIP: Cutting or clearing a spreadsheet cell that contains a formula deletes the entire formula definition, not just the current value. You can set the value of such a cell equal to zero by emptying the cells referenced in the formula.

Changing Spreadsheet Column Width

There are two ways to change the width of a column.

❶ Place the ARROW at the junction between two column labels; the tool changes to a double-headed arrow. Drag the column division to its new position.

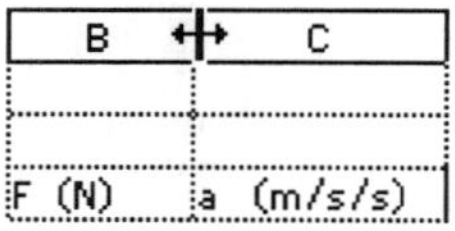

❷ Alternatively, you can click a column label to select a column. Choose Column Width from the Spreadsheet menu and enter the column width in characters in the dialog box.

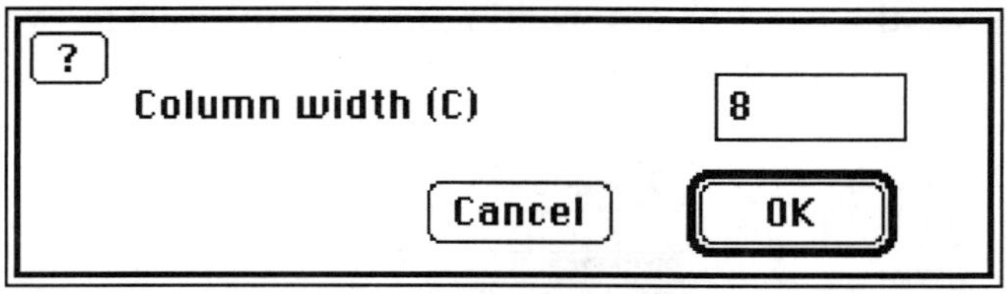

Inserting and Deleting Columns

You can insert or delete columns or rows using the appropriate option from the Edit menu. All rows or columns subsequent to the inserted or deleted one are renumbered accordingly. Any references to a deleted row or column change to question marks.

Converting to Values

If a cell's content is defined as a formula containing a cell reference, you can convert the current value to a constant numeric value that does not depend on the value in the referenced cell.

1. Select the desired cell or a range of cells.
2. Choose the Convert to Values option on the Spreadsheet menu.

The cell contents are automatically converted to a constant numerical value.

Spreadsheet Graphing

Once you have sampled or entered data into the spreadsheet, you can use the data to generate line graphs, bar charts, and pie charts. The graphs and charts you generate from data recorded in the spreadsheet are different from the dynamic graphs and charts displayed in Work windows. A fundamental difference is that you have direct access to the raw data the spreadsheet graph is based on. This means, for example, that you can manipulate the data (e.g., square all the values in a column) and see immediate results in the graph.

Here are some of the situations in which the spreadsheet graph may be useful:

- When you want to show a graph that summarizes data from several different trials.
- When you want to display your data points with a regression line—to see how closely the data points fit along a straight line—with associated statistics.
- When you want to show a bar chart or pie chart based on a frequency table, along with the results of a chi-square test.

Setting up a Spreadsheet Graph

Generating any kind of graph (line graph, bar chart, or pie chart) from spreadsheet data is normally a three-step process.

1. Enter the data to the spreadsheet, typically by running a simulation with predefined sampling. You can also enter the data manually as explained above.

❷ Highlight the ranges (generally columns) containing the data you want to graph.

❸ Choose the option representing type of graph you want to make ("New Line Graph," "New Pie/Bar Chart"), then select options in the dialog box for that graph type.

Detailed instructions are given in the sections that follow.

Spreadsheet Line Graph

In order to generate a line graph, follow these steps:

❶ Select two or more columns in the spreadsheet. To select non-adjacent columns, hold down the CMD key (Macintosh) or the CTRL key (Windows) while you select each column.

❷ From the Spreadsheet menu, choose the "New Line Graph" option. A graph definition dialog box opens.

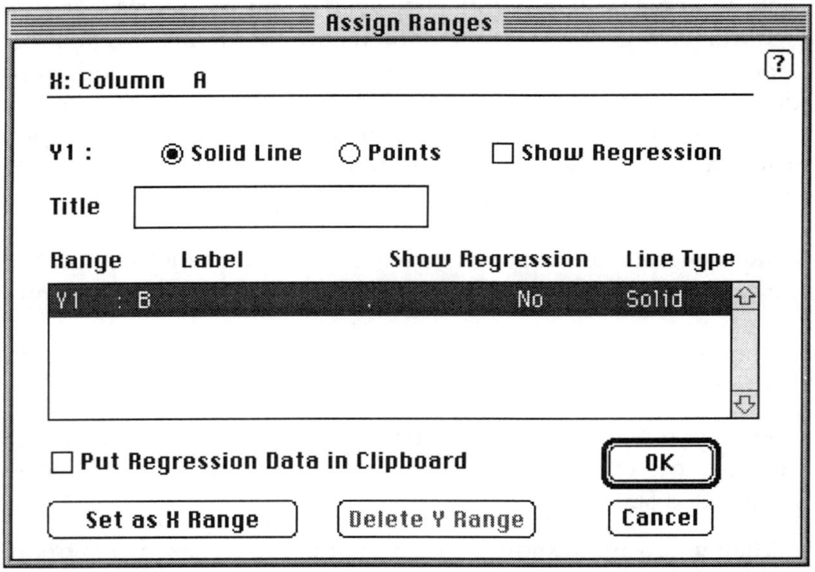

❸ The left-most column that you selected is automatically defined as the X-range of the graph. To change the range, select the desired column from the list of Y-ranges and click the "Set as X-Range" button.

❹ You can set a number of options for each Y-range. Select the desired range in the list.

- The default for each range's plot is a solid line connecting the plot points. Click the "Points" radio button to plot the discrete points without connecting them. This option is useful if the data was not sampled in ascending or descending order.

- Activate the "Show Regression" check box to show the regression line for the given Y-range. A regression line is the best straight line fit of the plotted points, used to smooth out experimental variations. The line graph will also display the slope (A), the Y-intercept (B), and the correlation coefficient (R). For a perfect fit, R=1.

  You can activate the "Put Regression Data in Clipboard" check box to write the regression values to the clipboard. While it remains on the clipboard, you can select an empty cell in the spreadsheet and paste the data into it.

- Enter a title for the given Y-range. The title will appear on the graph legend.

- If you have selected more columns than you need, you can delete the superfluous Y-range(s). Select the desired range from the list and click the "Delete Y-Range" button.

❺ When you OK the "Assign Ranges" dialog box, the "Format Line Graph" dialog box automatically opens. You can enter this dialog box directly by choosing the "Axis & Title" option from the Spreadsheet menu.

The default enables you to use the data in the spreadsheet to set the minimum and maximum values for the axes. By plotting all the data in the selected columns, the graph "autoscales" itself to the sampled data values.

If you want to change the values for one of the axes, deactivate its "Plot All Data" check box. For example, you may want to extend one of the axes to start from point 0 instead of from the minimum value sampled.

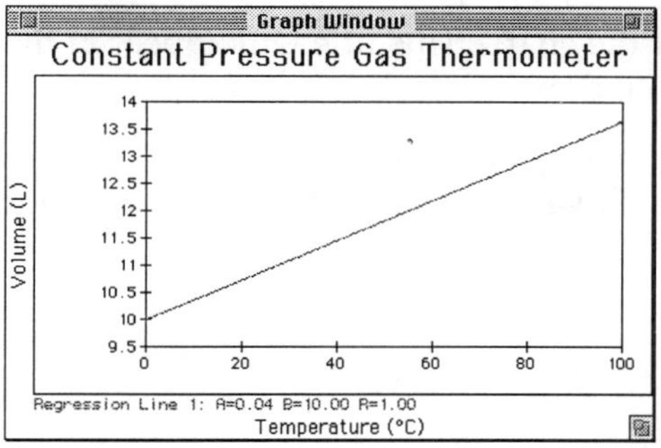

Plot All Data, Resulting in Autoscaling

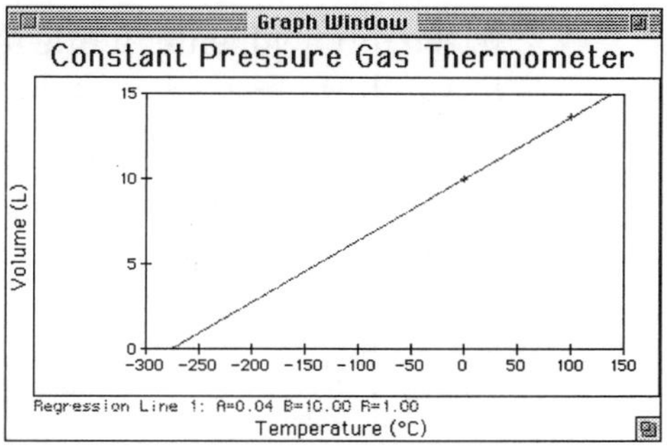

Plot with User-Defined Axis Limits, Shows Absolute Zero

☞ **NOTE**: Sometimes the plotted values are so close together that autoscale "zooms in" on the graph. In this case, slight variations in plotted values may appear more significant than in actual fact.

❻ You can enter a title for the entire graph, as well as labels for the X- and Y-axes.

❼ When you click OK in this dialog box, the graph is automatically generated. If you enter more data into the plotted columns, the spreadsheet graph updates automatically to plot the new data.

Creating a Distribution Bar Chart or Pie Chart

To create a bar chart or pie chart, follow these steps:

❶ Select one or more columns or a range of values in the spreadsheet.

❷ From the Spreadsheet menu, choose the "New Pie/Bar Chart" option. In the dialog box that opens, click the "Bar" or "Pie" radio button to set the form of the chart.

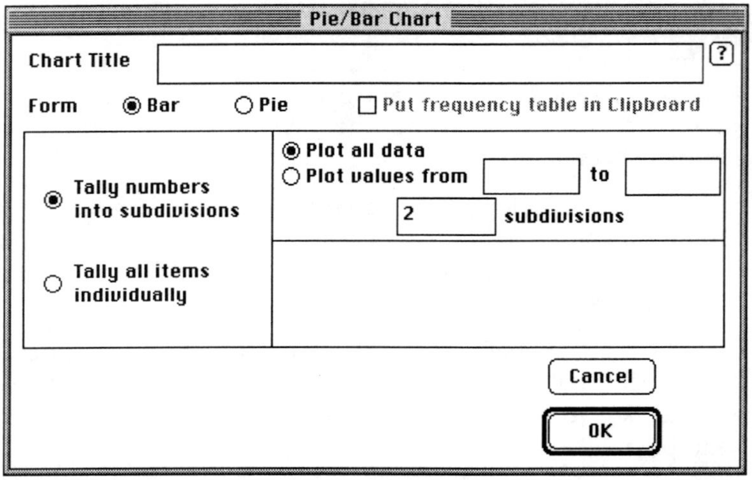

❸ In the top field, enter a title for the chart. This title will appear at the top of the Spreadsheet Graph window.

❹ You must choose between displaying the values in a fixed number of subdivisions and displaying each value as its own subdivision. For example:

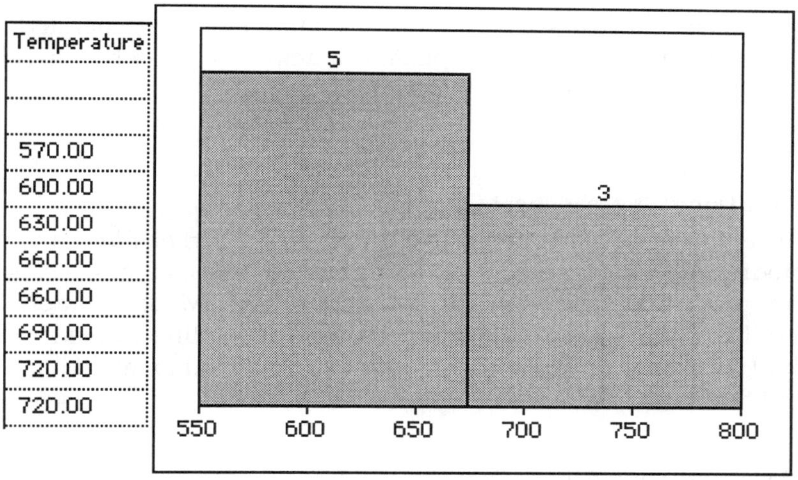

Tally numbers into 2 subdivisions

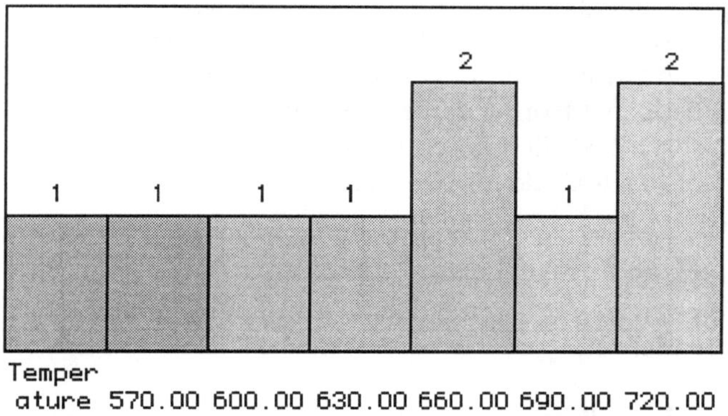

Tally all items individually

❺ If you have chosen to display the values according to number of subdivisions, you can set the following options:

- Plot all of the values selected in the spreadsheet, or
- Plot only values that fall into the minimum and maximum values that you set. This option is useful for eliminating extreme values from the plot.
- Define the number of subdivisions you want to display on the chart. *Explorer* calculates the values represented by each subdivision based on the total range of values.

❻ If you have chosen to display all values individually, you indicate whether the data is raw data or data for a frequency table.

7. When you click OK, the chart is automatically generated. If you enter more data into the plotted columns, the pie or bar chart updates automatically to plot the new data.

Modifying a Spreadsheet Graph or Chart

Once you have generated a line graph, bar chart, or pie chart, you can modify any aspect of the plot: data graphed, titles, ranges, etc. Select the new data that you want to plot, choose the "Modify Graph/Modify Pie/Bar Chart" option (the option changes depending on the current type of plot), change any other definitions that you want to modify (e.g., titles), and then click OK.

Managing Spreadsheet Data

Explorer offers you several ways of managing simulation data stored in a spreadsheet. For example, you can save data sampled to the current spreadsheet in a spreadsheet file, then open a new spreadsheet and sample additional data. You can also merge data into the currently open spreadsheet from a spreadsheet you have previously saved. You can also import a text file into your spreadsheet or export the contents of a spreadsheet to a text file for editing.

Merging Spreadsheets

You can merge a spreadsheet you have previously saved with the active spreadsheet.

1. Place the insertion point on the open spreadsheet at the point where you want the upper left corner of the merged spreadsheet to appear.
2. Choose the Merge option from the File menu and select the spreadsheet that you want to merge into the present one.

Extracting a Spreadsheet Range

You can also take a section of the active spreadsheet and write it to a new spreadsheet:

1. Select the range that you want to extract.
2. Choose the Extract option from the spreadsheet File menu.
3. In the dialog box, type the name of the file to which you want to write the selected range. The data remains in the active spreadsheet and is also copied to the new file.

Exporting Spreadsheet Data

To export data from a spreadsheet to a text file on disk:

1. With the spreadsheet open, select the column from which you want to export data.
2. Select Export from the File menu. A dialog box opens.
3. Check Write Column Titles if you want column titles to be included in the exported data.
4. Type the name of the new file you want to export data to and click Save.

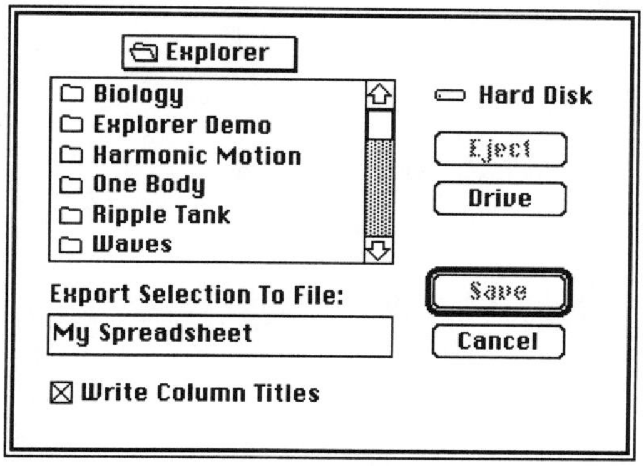

Appendix 2:
The *Chemistry Explorer* Models

1. General Information

The information provided in this appendix is intended for your reference during lab design. For a general introduction to each of the chemistry models, refer to the "First Look" lab files provided on disk.

Model Window

Each model has a unique model window appropriate to the topic, which displays the simulation animation. The specific animation displayed depends on the initial values and parameter settings defined for the given simulation.

Tools

Each model has a set of model tools along the left side of the model window. The tools are for setting simulation values, for introducing and removing elements from the model window, and for taking measurements.

Model Parameters: Inputs, Outputs, and Reactions

Each model has numerous parameters: input parameters, which are set by the user; output parameters, which are calculated by the mathematical model; and reactions, which can be used for creating buttons to control aspects of the simulation.

Note that for the *Student Edition*, access to the model parameters has been limited to about 10 inputs and outputs each. Therefore, you will not be able to recreate some of the simulations included in the *Run Time Explorer* packages.

Inputs

You can control up to approximately a dozen variables in each model. Often the input variable values will remain constant for the duration of an experiment unless you change one with a model window tool during the simulation.

Other "input variables" are simply initial conditions for the experiment and will typically change as the experiment is run.

If an input variable is changed while the simulation is initialized (using the RESET tool), the simulation "remembers" the new value of the input variable the next time the simulation is started. If the input variable is changed with a model window tool without resetting the simulation, the initial values of the variable are not "remembered" for the next experiment run.

Outputs

You can display, graph, or record to a spreadsheet the current value of variables called "output variables" generated by the model. You can display, graph, or record to a spreadsheet most quantities of interest.

Reactions

Each model includes special reactions that can be invoked by creating a button.

2. The Atom

The model describes the historical process scientists have used to build our present ideas of the structure of the atom. The model is divided into three periods, the Early period (... – 1913), the Intermediate period (1913 – 1925), and the Modern period (1925 – ...).

1. Early Period

This period is based on simulating Rutherford's classical bombardment experiment. The model allows the user to simulate the effects that *would* have happened if the experiment had been available at earlier times in the period. You can simulate firing alpha particles or protons (discovered in the twentieth century) at a Dalton rigid body atom (proposed in the early nineteenth century) or a Thomson plum pudding atom (proposed in the late nineteenth century).

The results of bombarding the different types of atoms are shown both schematically and on a distribution histogram. It was as a result of seeing the distribution that Rutherford was able to say that he "knew how the atom was built," and that in particular the Dalton and Thomson atoms were incorrect.

2. Intermediate Period

This period is based on analysis of simulated spectra. The model allows the user to simulate one of two situations:

- Before Bohr

One of the problems created by the Rutherford solar system model (a central nucleus surrounded by planetary electrons) was that classical physics predicted that electrons rotating around a nucleus would radiate energy (including ultraviolet energy) continuously, eventually collapsing into the nucleus. The consequent spectrum would be continuous. You can simulate this happening and observe the resulting spectrum.

- Bohr Atoms

Bohr proposed that spectral lines are caused by transitions between different stationary electron states within the atom. You can observe both emission and absorption spectra from different transitions for the hydrogen atom. You can vary parameters for the starting and final transition levels, and the range of wavelengths to be observed. In addition, the charge on the nucleus can also be changed to simulate Bohr "hydrogen-like" atoms.

3. Modern Period

There are two sections for this period.

- Charge Distribution Diagrams

This section is based on graphical simulations of Schrödinger's equation, which predicts the charge probability $|\psi|^2$ (pronounced Psi square) of the electron cloud surrounding the nucleus. $|\psi|^2$ corresponds to the probability of the electron being at a particular point at a particular time. You can generate two– or three–dimensional pictures of the charge distribution for different electron configurations of the hydrogen atom. In addition to the pictures, graphs and histograms of the charge distribution vs. r or theta can be produced.

- Electron Configuration

This section enables you to simulate atoms up to atomic number 48, fill **s**, **p** and **d** electron shells for the atom, and check to see if the filled shells correspond to the ground state of the specified atom.

Early Period

Dalton's Rigid Body Atom

Thomson's Plum Pudding

Rutherford's Planetary Model

The distribution histogram is available for the three Early Period experiments: Dalton, Thomson, and Rutherford. The bars show the distribution that would likely have resulted when the historical model was simulated. The blue outline indicates the correct statistical result of the Rutherford model.

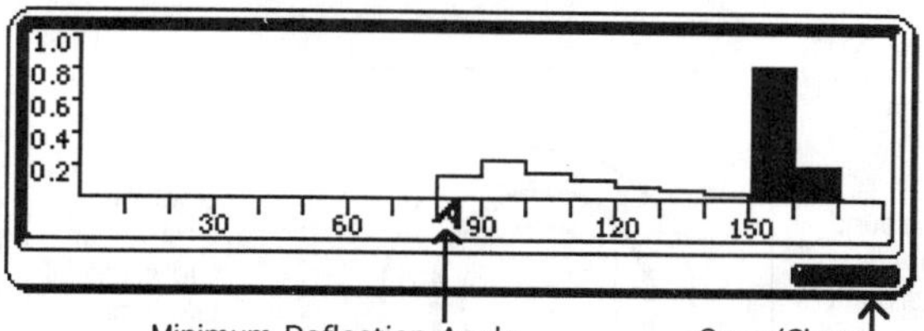

Minimum Deflection Angle (Determined by Beam Width)

Open/Close Histogram Display

The source for particles to be used in simulating a bombardment experiment.

Displays # of particles in source

Click to define the particles as protons (top button) or alpha particles (bottom button).

Drag the slide to set the enegy of the bombarding particles (MeV).

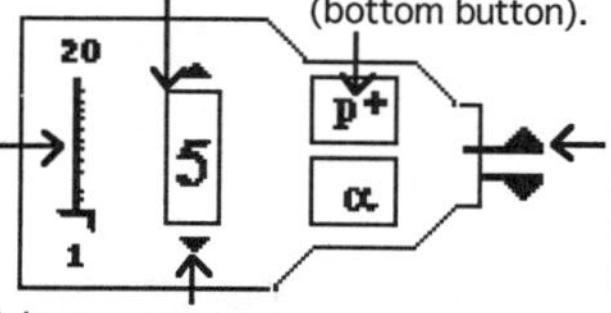

Drag up or down to change the beam width (nm).

Click on arrows to change the initial # of bombarding particles.

Zoom View: A magnified view of the foil, with an even more magnified view of a single atom

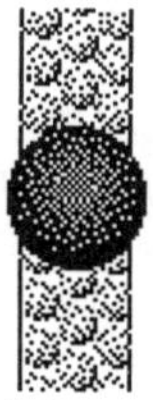

Dalton Zoom View

Thomson Zoom View

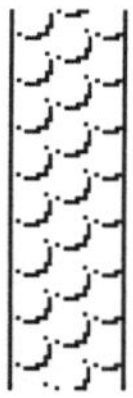

Rutherford Zoom View

Intermediate Period — Before Bohr

Before Bohr: Classical Dilemma

Bohr Atoms

Simulates a "classical" electron rotating around the nucleus. The background color indicates the wavelength the electron would emit as it rotates.

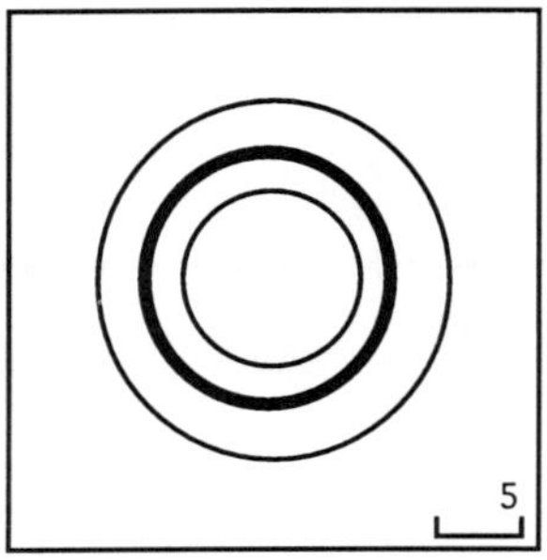

Shows the selected electron energy for the classical dilemma simulation.

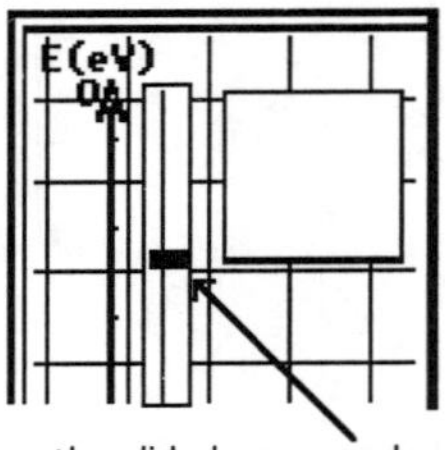

Drag the slide bar up or down to set the initial energy in eV.

Click the arrows to change the minimum wavelength.

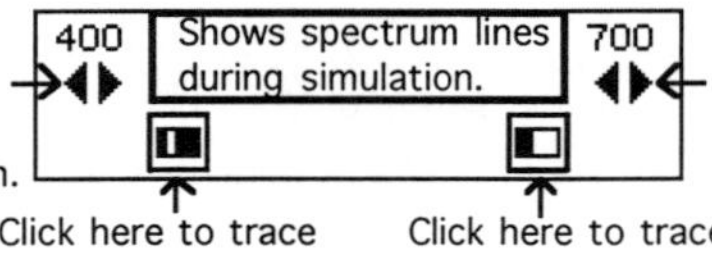

Click the arrows to change the maximum wavelength.

Click here to trace only the current lines.

Click here to trace all lines as they are produced.

Intermediate Period — Bohr

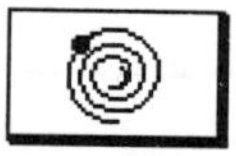

Before Bohr: Classical Dilemma

Bohr Atoms

These controls enable you to determine if an absorption or emission experiment is performed, and the wavelength limits to be used. If the experiment is set to absorption, you can also set the lamp's minimum and maximum wavelength.

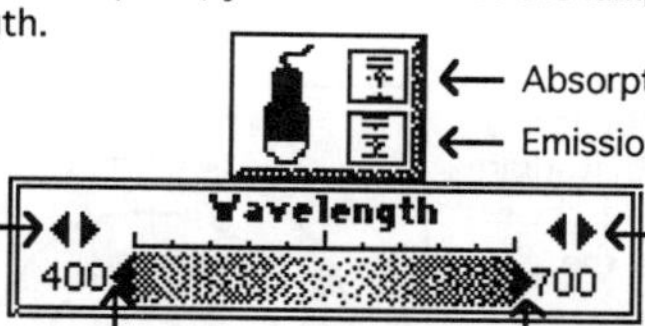

Absorption (Light On)

Emission (Light Off)

Click the arrows to set the minimum wavelength of the spectrum.

Click the arrows to set the maximum wavelength of the spectrum.

Drag the arrow to set the lamp's minimum & maximum wavelength.

Simulates a "Bohr" electron rotating about the nucleus. The background color indicates the wavelength the electron would emit as it undergoes a transition.

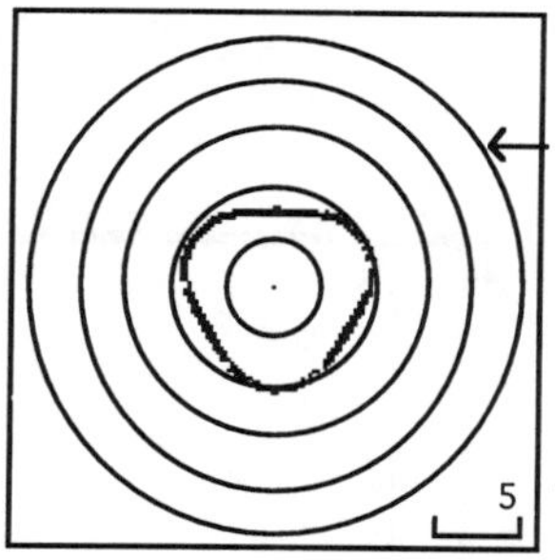

The grid control turns on or off a grid

The scale in Bohr radii, a_o

Shows the energy levels Bohr proposed, and transitions between these levels.

Click on a box to set the Initial Quantum Number $n_{initial}$.

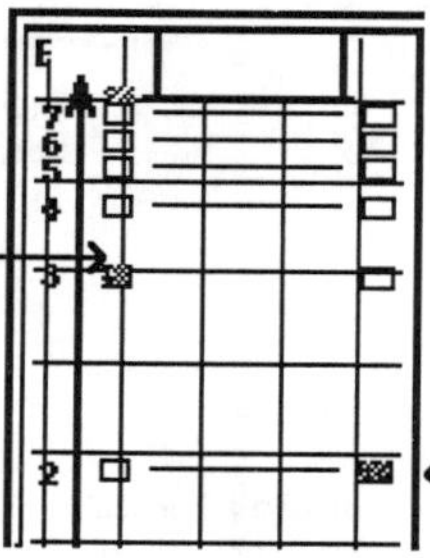

During the simulation, the level changes are drawn.

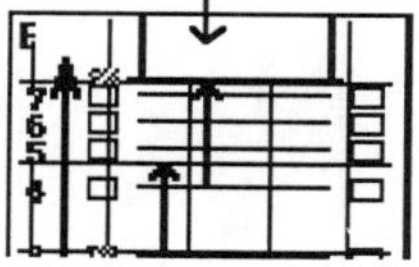

Click on a box to set the Final Orbital Level n_{final} (only available if the light is turned off).

Modern Period — Charge Distribution

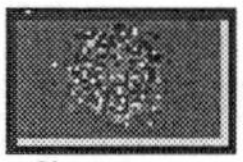

Charge Distribution

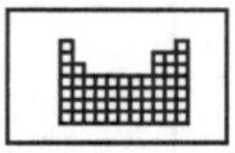

Electron Configuration

Shows a dynamic simulation of the electron density surrounding the hydrogen atom nucleus, generated from Schrödinger's equation.

If θ (theta) is constant, drag the red circle to set the radius.

If radius is constant, drag the line to set θ (theta).

(Toggle the Constant Radius/ θ in the Charge Distribution Graph.)

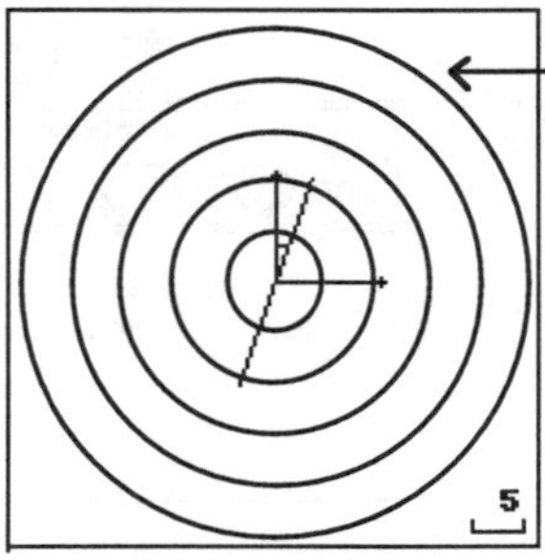

The grid control turns grid on or off

Toggles between views

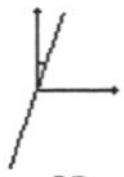

2D

3D

The scale in Bohr radii, a_o

Shows a graph of the normalized electron density surrounding the nucleus, vs. r or θ (theta).

The normalized electron density value

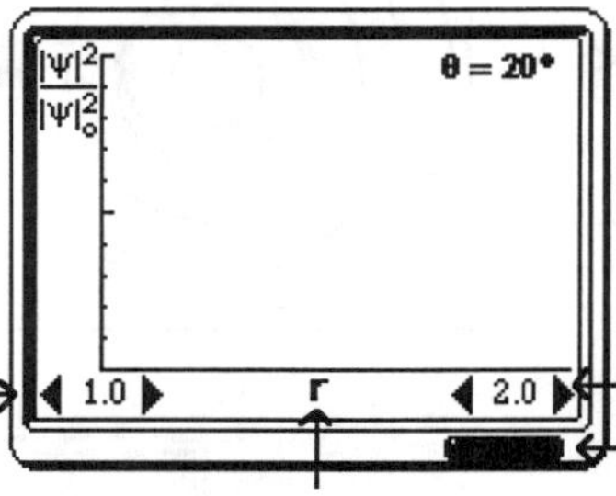

The constant value

Click on arrows to set the maximum value of variable radius/ θ .

Click on arrows to set the minimum value of variable radius/ θ .

Click here to toggle between variable radius and variable θ.

Click the switch to toggle the graph on/off.

Shows orbital energy levels in hydrogen and hydrogen-like ions (only one electron).

Click on one of the boxes to choose a particular electron orbital (from 1s to five types of 3d orbital). As the box is clicked, the corresponding values for n, l and m are set.

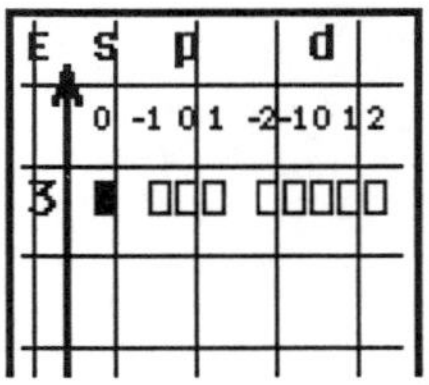

Modern Period — Electron Configuration

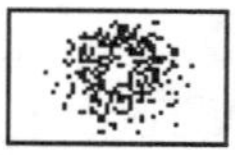

Charge Distribution

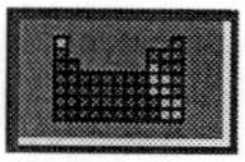

Electron Configuration

Electrons can be inserted into the orbitals shown. A filled orbital must have paired electrons - one spin up, one spin down. The model does not accept an invalid filling.

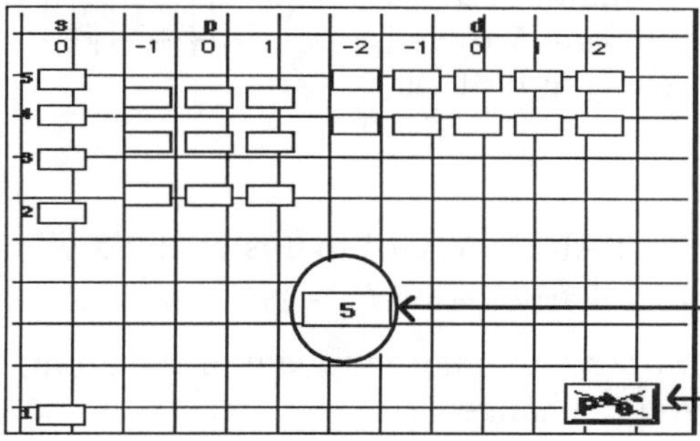

Nuclear Charge (Atomic Number)

Click to clear the diagram

Spin up and spin down electrons.
Click and drag to the orbital.

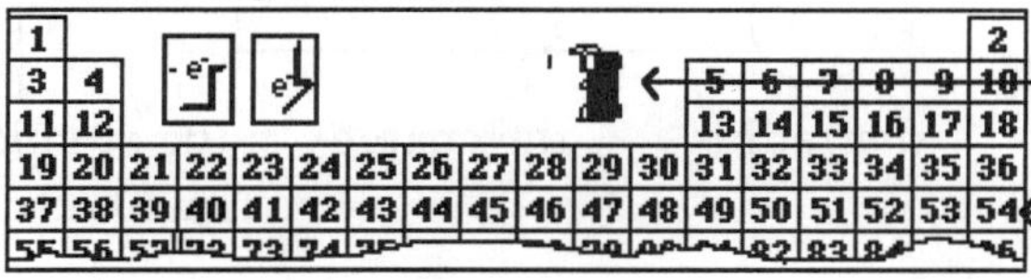

Clear electrons from the orbital by dragging them to the trash.

Click on the desired atomic number.

The periodic table enables you to set the atomic number (up to 48), and then to use the spin up or spin down electron icons to fill s, p and d electron shells in the orbital diagram.
The fill can also be performed automatically to generate the ground state of the specified atom using the model reaction "Fill Configuration."

Model Window Tools

ARROW Tool

The ARROW tool has many functions in the model window:

- Clicking icons to change periods and views

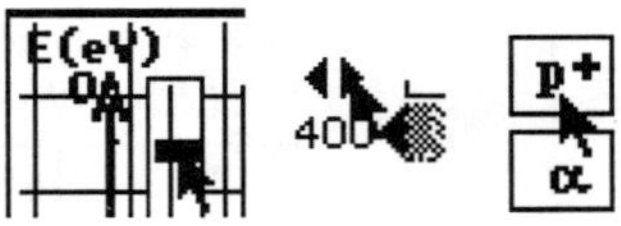

- Changing parameters by
  dragging slide bars
  clicking arrows
  clicking buttons

EARLY ATOM Tool

The EARLY ATOM tool is enabled when model is set to Early Period (... – 1913).

It opens an enlarged view of the atom, enabling you to change

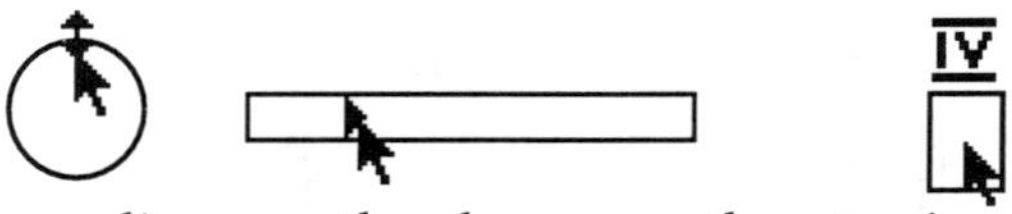

the radius | the charge | the atomic group number of the atom

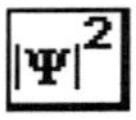

WAVE FUNCTION DETECTOR Tool

The WAVE FUNCTION DETECTOR tool is enabled when the model is set to Modern Period (1925 – ...), Charge Distribution. The tool acts as a probe in the orbital, giving outputs when the two-dimensional (2D) view of the atom is selected.

Measurement outputs are r, θ theta, and electron density. These measurements are automatically written to the spreadsheet.

The tool is only active when the simulation is running.

MODEL WINDOW HELP Tool

The MODEL WINDOW HELP tool is for exploring the many features in the model window. Click it on an item to get a help window. Note that this tool is independent of the *Explorer* help tool that works outside the model window.

Early Period Input Variables

VARIABLE	DESCRIPTION
Period Definition: Early Period	Sets the Early Period model to be simulated in the model window. The options are: **Dalton** rigid body (1803); **Thomson** plum pudding (1897); **Rutherford** planetary (1909) models. Can also be set in the model window by clicking the icons:  • symbol: **ATOM_EARLY**
Source Data: Particle Energy	The energy in electron volts of the bombarding particles. Range: 0–20 MeV. Can also be set on the source in the model window: 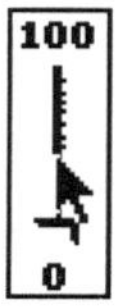 • symbol: **ENERGY**
Source Data: Beam Width	The width of the bombarding beam. Range: 5–100 nm. Can also be set in the model window:  • symbol: **BEAM**

Early Period Output Variables

VARIABLE	DESCRIPTION
Early Period: Deflection Angle	The current value of the angle of deflection in degrees. Use this output to build distribution histograms of the deflected particles. • symbol: **ANGLE_HIST**
Early Period: Force	The average force, in Newtons, acting on the bombarding particle while it is being deflected. This output is **not active** if the model is set to Dalton. • symbol: **FORCE**

Early Period Reactions

REACTION	DESCRIPTION
Show/Hide Theoretical Distribution	These reactions show and hide respectively the theoretical distribution (given by Rutherford's model) outlines in the distribution histogram.
Clear Experimental Distribution	This reaction clears the bars showing the probable distribution of the historical model in the distribution histogram.

Intermediate Period Input Variables

VARIABLE	DESCRIPTION
Period Definition: Intermediate Period	Sets the Intermediate period model to be simulated in the model window. The options are: **Before Bohr** giving a classical dilemma, or **Bohr**'s (Atoms) permitted orbits solution. The two models produce different spectra when the simulation runs. Can also be set in the model window by clicking the icons: 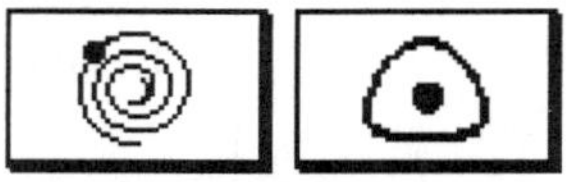 • symbol: **ATOM_PLAN**
Before Bohr: Initial Energy	Applies if the model is set to Before Bohr. Sets the energy of the atom before running the classical dilemma simulation. Range: –14–0 eV. Can also be set in the model window: 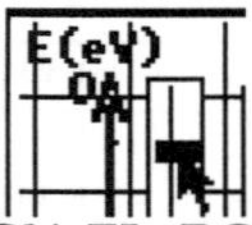 • symbol: **ENERGY_EL_BOHR**

Intermediate Period Input Variables, Cont.

VARIABLE	DESCRIPTION
Bohr: Experiment (Absorption or Emission)	Applies if the model is set to Bohr. Sets the experiment in the model window to Absorption or Emission. **Absorption:** Light on; the view shows a continuous lamp emission, and an absorption spectrum is produced when the simulation runs. **Emission:** Light off; the emission spectrum is produced when the simulation runs. Can also be set in the model window by clicking the "light switches." 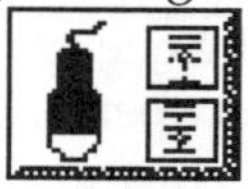 Absorption (Light On) Emission (Light Off) • symbol: **MODE_BOHR_SPONT**
Bohr: Standing Wave	Applies if the model is set to Bohr. Toggles the electron displayed in the Bohr zoom between a standing wave (determined by the n quantum number) and a circular orbit. Can also be set in the model window:  • symbol: **STAND_WAVE**
Spectrum Data: Spectrum Trace	Set to **Yes** to trace all the lines as they are produced in the spectrum, and to show all the lines when the simulation has stopped. Set to **No** to display the current spectrum lines. Can also be set in the model window:  • symbol: **TRACE_SPECTR**

Intermediate Period Output Variables

VARIABLE	DESCRIPTION
Intermediate Period: Orbit Radius	The Bohr radius in nanometers corresponding to the electron level; also the radius of the electron orbit at the start of a "Before Bohr" (classical dilemma) simulation. • symbol: **RADIUS_ELECTRON**
Intermediate Period: Wavelength	The wavelength (in nanometers) output when a transition between electron levels is made. • symbol: **WAVE_LENGTH**
Intermediate Period: Δ E (Absorption or Emission)	The energy output in eV when a transition between energy levels occurs; positive for absorption, negative for emission. • symbol: **ENERGY_ABS_EMISS**
Intermediate Period: Current Electron Energy	The energy in eV corresponding to the electron level. This value is always negative since the energy of the electron at infinity is assumed to be zero. • symbol: **CURR_ENERG**
Current n Quantum Number	The current value of the n quantum number. • symbol: **CURR_LEVEL**

Modern Period Input Variables

VARIABLE	DESCRIPTION
Period Definition: Modern Period	Sets the model window display. The options are: **Charge Distribution**, which displays a modern electron density distribution and electron configuration table, or **Electron Configuration**, which displays the latter and a modified periodic table and enables you to fill atom shells. Can also be set in the model window by clicking the icons: 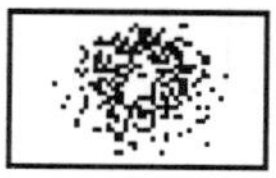 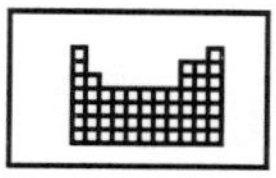 • symbol: **ATOM_MODERN**
Screen Setup: # Dots	The total number of dots that will be generated in the model window. When this number is reached, the simulation stops. Minimum value: 100 dots • symbol: **NUMBER_POINT_MODERN**
Distribution Display: Distribution Display Definition	Applies if the model is set to Charge Distribution. Sets the charge distribution display to constant theta mode or constant radius mode. **Constant theta:** the graph of electron density vs. radius is plotted for a value of theta equal to "Initial Theta." **Constant radius:** the graph of electron density vs. theta is plotted for a value of r equal to "Initial Radius." • symbol: **MODERN_GR_GIST**
Distribution Display: Draw Histogram	Applies if the model is set to Charge Distribution. Choose **Yes** to enable histogram calculation and plotting inside the work window. You can then create a distribution bar chart with the BAR CHART tool, plotting the "Modern: Distribution" output variable. Choose **No** to disable the histogram calculation and plotting. • symbol: **MODERN_GIST**

Modern Period Input Variables, Cont.

VARIABLE	DESCRIPTION
Distribution Display: Draw Graph	Applies if the model is set to Charge Distribution. Choose **Yes** to open the graph in the model window. You can then create a graph in the work window. Use the GRAPH tool to plot outputs "Modern Period: Current Electron Density" vs. "Modern Period: Current Radius/Theta (Graph Data)." Choose **No** to close the graph. Can also be set in the model window.  • symbol: **MODERN_GR**

Modern Period Output Variables

VARIABLE	DESCRIPTION
Modern: Current Radius (Graph Data)	If the "Distribution Display Definition" option input has been set to Constant Theta, this output gives the radius value scanned by the model after concluding the "# Dots" (input). The model scans from the "Initial Radius (R)" input value to the "Final Radius (R)" input value (both found in the Distribution Display input menu.) If the "Distribution Display Definition" option input has been set to Constant Radius, this output is the value of "Initial Radius (R)." The values are output after the model has concluded the "# Dots" given. • symbol: **RAD_GR_MODERN**
Modern: Current Theta (Graph Data)	If the "Distribution Display Definition" option input has been set to Constant Radius, this output gives the theta value scanned by the model after concluding the "# Dots" (input). The model scans from the "Initial Angle (Theta)" input value to the "Final Angle (Theta)" input value (both found in the Distribution Display input menu.) If the "Distribution Display Definition" option input has been set to Constant Theta, this output is the value of "Initial Angle (Theta)." The values are output after the model has concluded the "# Dots" given. • symbol: **TETA_GR_MODERN**
Modern: Current Psi Square (Graph Data)	The unnormalized value of electron density, Psi square. The value is updated as the model scans through r or theta. • symbol: **PSI2_GR_MODERN**
Modern: Radius/Theta (Histogram Data)	This output is used to prepare a distribution histogram (with the BAR CHART tool) of the charge distribution vs. r or theta, similar to that shown in the model window. Outputs only exist if the "Draw Histogram" input has been set to Yes. • symbol: **GIST_MODERN**

Modern Period Output Variables, Cont.

VARIABLE	DESCRIPTION
Modern: # Points Displayed	The number of points that have been displayed. The final output value is dependent on the values you have entered for the "Screen Setup: # Dots" and the "Screen Setup: Dot Increment" input options. • symbol: **CURR_DRAW_POINT**
Current n Quantum Number	The current value of the n quantum number. • symbol: **CURR_LEVEL**

Modern Period Reactions

REACTION	DESCRIPTION
Electron Configuration: Clear All Electrons	Clears all the electrons entered in the periodic table.
Electron Configuration: Clear All	Clears both the electrons and the atomic number in the periodic table.
Electron Configuration: Fill Configuration	Fills the electron shells according to the entered atomic number.

3. Chemical Equilibrium

The *Equilibrium* model simulates changes in the concentration of molecules in homogeneous chemical equilibrium reactions. The reactions take place in a container of fixed volume. The concentration changes are described as a function of such variables as temperature, initial concentration and activation energy. The type of reaction — simple, concurrent, or consecutive — can also be set.

The model also simulates acid-base titration, enabling you to control the drop volume and concentration of the contents of the buret, and to choose from a variety of indicators to follow pH changes in the titrations.

Note: For the *Student Edition*, the list of species available in the *Chemical Equilibrium* model has been limited to the generic species, A, B, X and Y.

Equilibrium Mode

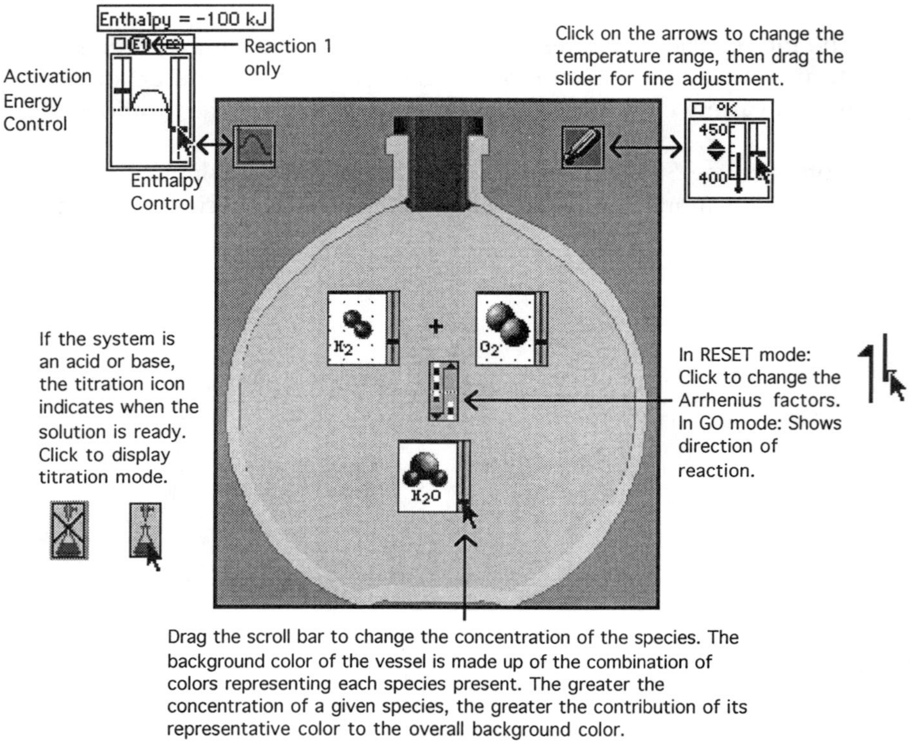

Titration Mode

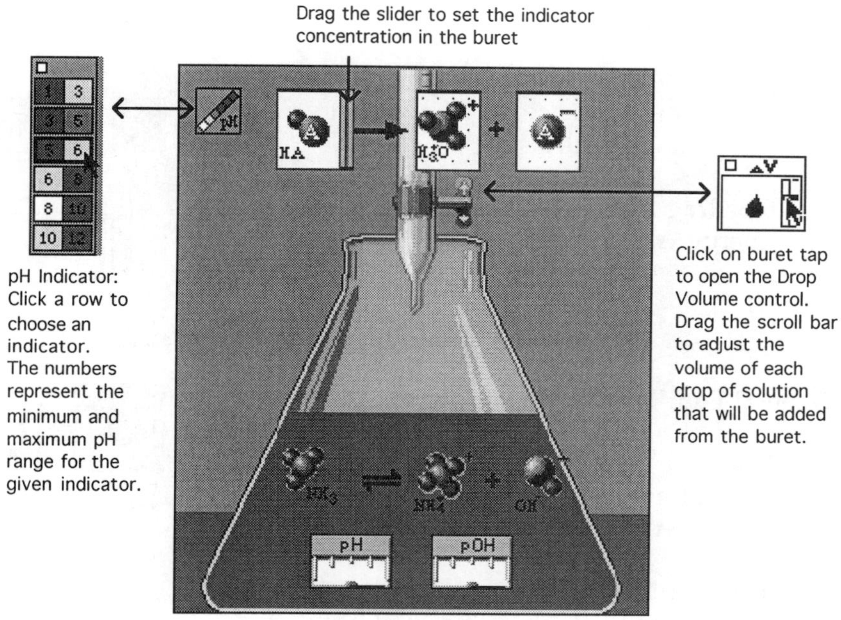

The Model Tools

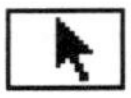 ARROW Tool

You can use the ARROW tool:

General:

- To open and close the energy and temperature controls

- To drag sliders to change the activation energy, ΔH, and temperature

General Equilibrium Mode:

- To drag sliders to change the initial concentrations of the reactants and products, and the Arrhenius factors

- To switch to titration mode (only when the base or acid solution is at equilibrium and the X disappears)

Titration Mode:

- To click the tap to open the drop volume control for the species being added to the flask

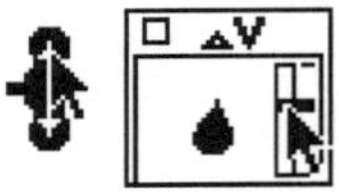

- To open the pH indicator selector, to choose an indicator, and to set the concentration in the buret

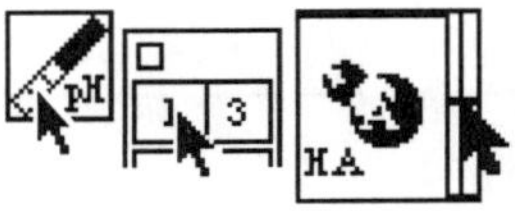

SYSTEM Tool

The SYSTEM tool enables you to define the chemical equation for the reaction in the vessel. Note that for the *Student Edition*, you are limited to the generic species A, B, X, and Y.

Click the "drawer" icon for the desired system in the bottom table. A chemical equation appears.

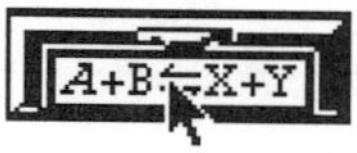

Click the coefficient in front of the species to define the number of molecules (i.e., stoichiometry).

Chemical Equilibrium Input Variables

VARIABLE	DESCRIPTION
Init. Concentration 1–4	The initial concentration of the given molecular species, measured in mole/l. • symbol: **INIT_M1,...,INIT_M4**
Reaction #1,Parameters: Forward Arrhenius Factor	The value of the Arrhenius factor in the forward reaction. • symbol: **A_FORWARD**
Reaction #1,Parameters: Reverse Arrhenius Factor	The value of the Arrhenius factor in the reverse reaction. • symbol: **A_REVERSE**
Catalyst With/Without	Defines whether or not a catalyst is added to the reaction. • symbol: **CATALYST**

Chemical Equilibrium **Input Variables, Cont.**

VARIABLE	DESCRIPTION
Catalyst Concentration	The concentration of the catalyst (where present) in mole/l. • symbol: **CONC_CATALYST**
Base: Base-Dissociation Constant	The equilibrium constant for the general weak base. Note: If you set the base to NH_3, set this value to 1.8e-5. • symbol: **BASE_DISS_CONST**
Acid: Acid-Dissociation Constant	The equilibrium constant for the general weak acid. Note: If you set the acid to acetic acid, set this value to 1.8e-5. • symbol: **ACID_DISS_CONST**
Concentration Changes: V-P Multiplier	The multiplication factor by which the current volume or pressure in the container is increased or the division factor by which it is decreased when one of the following four reactions is invoked: Decrease Volume, Increase Volume, Decrease Pressure, Increase Pressure
Time Increment ΔT	The time between "snapshots" of the simulation. If Δt=1.0, each step represents one second of real time. **Note:** The recommended starting Δt is 0.05, but many simulations will require a bigger value. *Explorer* notifies you when the Δt is too large for the model to remain stable between calculation steps. • symbol: **DT**

Chemical Equilibrium **Output Variables**

VARIABLE	DESCRIPTION
Current Concentration; Acid/Base Concentration	The current concentration of molecules of the given species: molecules 1-4, H30+, OH-, base, or acid. • symbol: **CONC_M**
Forward Reaction Rate Constant	The rate constant Kf for the forward reaction, as defined by $$K_f = Ae^{-E_A/RT}$$ Units of Kf depend on the rate expression: Rate of forward reaction = Kf[reactants] General units of Kf are $l^x / mol^x / s$ where x=(no. of particles in rate expression)-1. • symbol: **FWD_K_R**
Reverse Reaction Rate Constant	The rate constant Kr for the reverse reaction, as defined by $$K_r = Ae^{-(E_A+\Delta H)/RT}$$ Units of Kr depend on the rate expression: Rate of forward reaction = Kr[reactants] General units of Kr are $l^x / mol^x / s$ where x=(no. of particles in rate expression)-1. • symbol: **BWD_K_R**
Energy Change	The current energy change taking place between the reaction vessel and the surroundings, in kJ/mole. • symbol: **ENERGY**
Current Activation Energy	The current activation energy taking into account the effect of a catalyst. • symbol: **CURRENT_ACT_EN**
Time	The time in simulated seconds since the simulation was activated. • symbol: **T**

Chemical Equilibrium **Output Variables, Cont.**

VARIABLE	DESCRIPTION
Temperature	The current temperature in the vessel. • symbol: **TEMP_OUT**

Chemical Equilibrium **Model Reactions**

REACTION	DESCRIPTION
Increase/Decrease Volume	These two reactions change the total volume in the container. "Decrease Volume" divides the current total volume by the value given in the "Multiplier" input. "Increase Volume" multiplies the current total volume by the value given in the "Multiplier" input.
Increase/Decrease Pressure	These two reactions change the total pressure in the container. "Decrease Pressure" divides the current total pressure by the value given in the "Multiplier" input. "Increase Pressure" multiplies the current total pressure by the value given in the "Multiplier" input.

4. Chemical Kinetics

The *Chemical Kinetics* model simulates the relationships among concentration, temperature, Ea, and time. The model window shows up to four different species of "molecules" — A, B, C, and D — inside a two-dimensional container. The molecules may react according to the reaction $A + B \rightleftharpoons C + D$, or the reaction may be forbidden.

Kinetics Model Window

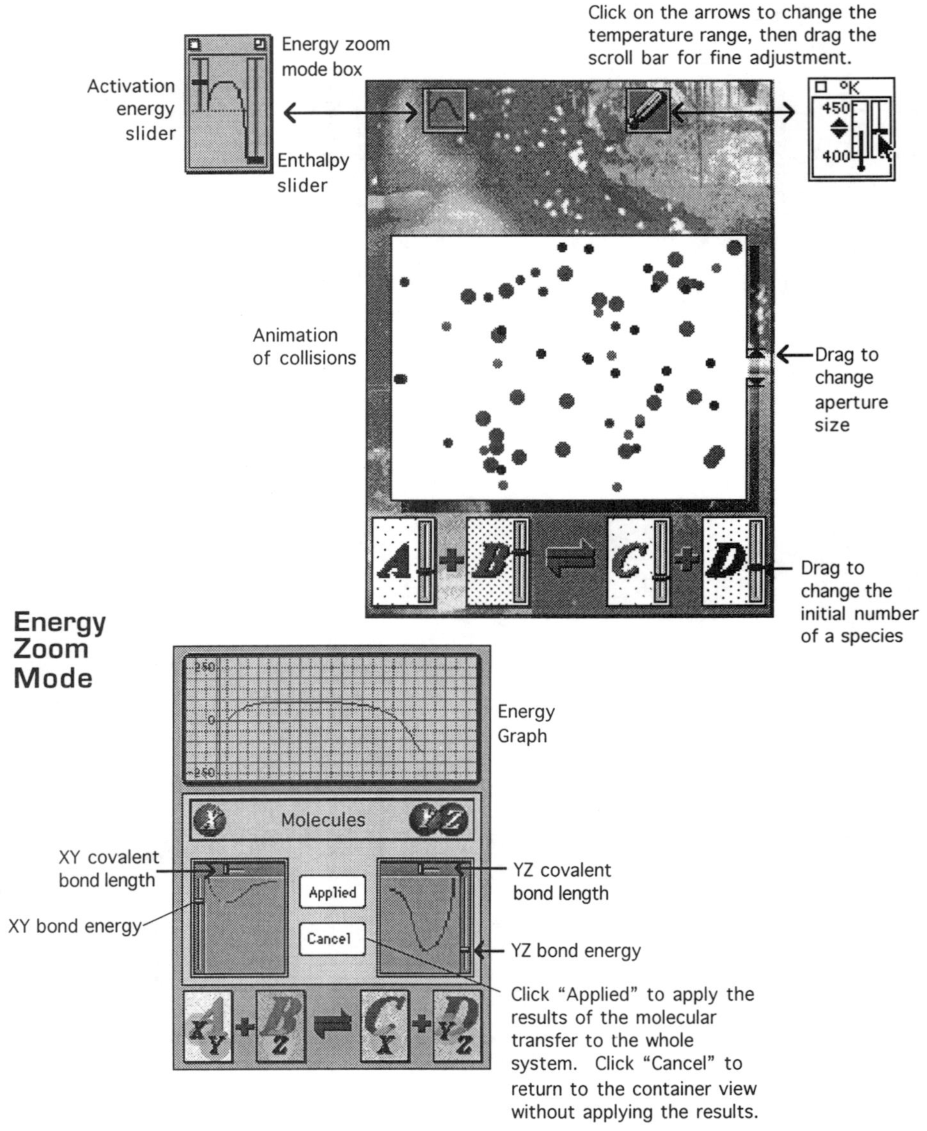

Model Tools

ARROW Tool

The ARROW tool has a number of functions:

1. Open and close the controls

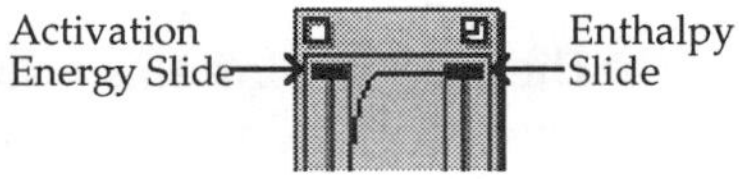

2. Set values in the energy control

3. Change to energy zoom mode

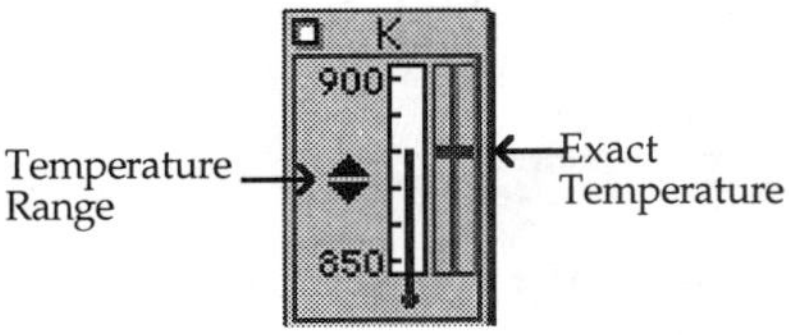

4. Set values in the temperature control

5. Set the number of molecules of each species

6. Set the size of the opening in the external wall of the container

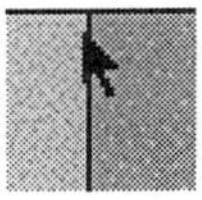

7. Remove a wall when simulation running with 2 regions

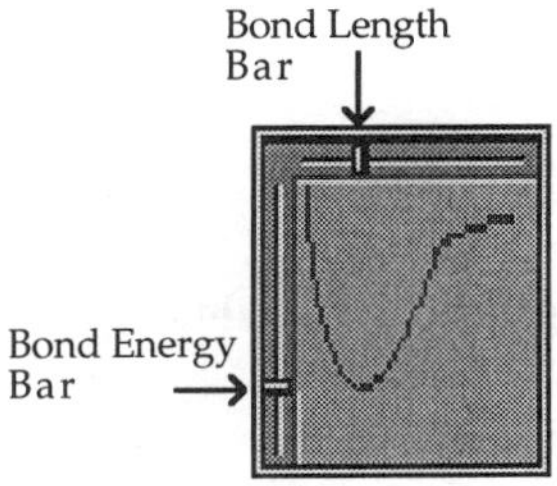

8. Change values in energy zoom mode (XY+Z->X+YZ reaction); accept or cancel the new values

SPECIES Tool

The SPECIES tool enables you to set the mass and the diameter of the species.

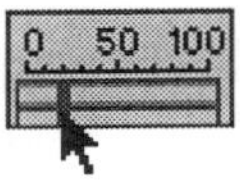

1. Drag the mass control slide bar to adjust the masses of the initial species A, B, C. The model automatically calculates the mass of species D as a result of the other three masses.

2. Drag the diameter arrow to adjust the diameter of each species.

Special Features

• Two Regions

When you set the "# of Regions" option to 2 Regions, the container is automatically divided, and an internal wall is present. One region contains A molecules and the other contains B molecules.

You can remove the wall during the simulation, but the container remains divided into regions, enabling you to track the number of B molecules that diffuse into Region 1 and the number of A molecules that diffuse into Region 2. You can also get pressure readings for each region, whether the wall is present or not.

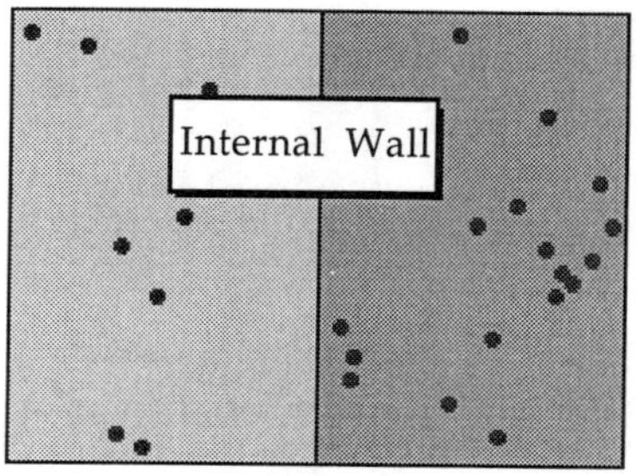

• Tagged Molecules

When you set the "Tagged Molecule?" option to yes, the model arbitrarily selects a molecule. You can then display an output of the molecule's speed and track its wall collisions and collisions with other molecules. The tagged molecule appears with its velocity vector.

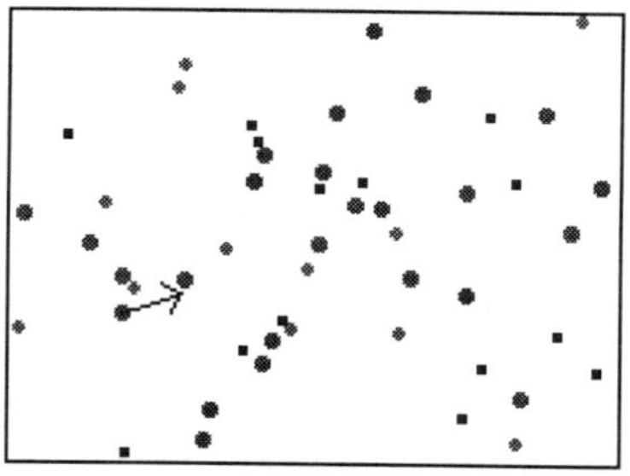

- **Maxwell-Boltzmann Speed Distribution**

When you set the "Speed Distribution: Display?" option to yes, the model makes Maxwell-Boltzmann calculations. You can create a histogram of the speeds of the A or the B molecules. Use the BAR CHART tool and click the Distribution Bar Chart radio button. Enter the Speed A or Speed B output for the expression.

- **Constant # Molecules**

Under normal simulation conditions, the molecules can escape through the aperture in the container, and the number of molecules in the container gradually decreases.

When you set the "Constant # Molecules?" option to yes, the molecules that escape the container are replaced in order to maintain a constant number inside the container. This enables you to simulate a very large container in which the number of molecules remains essentially constant. Note that this option does not affect the value of the "Molecule ->" output, which counts the number of molecules that escape whether or not they return to the container.

Chemical Kinetics **Input Variables**

VARIABLE	DESCRIPTION
Reaction Parameters: Reactions?	Defines whether the species are allowed to react or not. If the reaction is allowed, there will be a chemical reaction when appropriate collisions occur. If the reaction is forbidden, there will be physical changes at collision (velocity changes), but no change of species. In this case, the number of a specific species will not change during the simulation. • symbol: **REACTIONS**
Container Parameters: # Regions	Determines if the container is one or two regions. If it is two regions, an internal wall automatically separates the regions at simulation start. One region has A molecules, the second region has B molecules. The wall can be removed while the simulation is running. The number of molecules and the pressure in each region can be displayed separately. • symbol: **WALL**
Speed Distribution Diagram: Display?	Determines if the model makes Maxwell-Boltzmann calculations and displays the speed distribution diagram. **Yes**: Maxwell-Boltzmann calculations are made during the simulation, and a histogram of the speed distribution displays. **No**: Maxwell-Boltzmann calculations are not made. **Note:** The simulation runs faster when the calculations are **not** made. • symbol: **MAXWELL**
General Data: Time Increment (Δt)	Determines the time interval between model calculations and "snapshots" of the model window animation. A lower value gives more calculations and therefore more accurate results; a higher value speeds up the accumulation of collision data.

Chemical Kinetics **Input Variables, Cont.**

VARIABLE	DESCRIPTION
General Data: Tagged Molecule?	Determines whether the model tags an arbitrarily chosen molecule. If set to Yes, the tagged molecule's velocity vector is automatically displayed in the model window. The molecule's speed can be displayed and its collisions can be tracked. • symbol: **FIRST_MOLEC**
General Data: Constant # Molecules?	Determines whether the molecules that escape from the container are returned in order to keep a constant number of molecules in the container. • symbol: **MOLEC_OUT**

Chemical Kinetics Output Variables

VARIABLE	DESCRIPTION
Current # of Molecules: Species A–D	The total number of molecules of the given species. • symbol: **N**
Current # of Molecules: Region 1, Species B	The number of molecules of species B that have diffused into the region initially occupied solely by species A. **Note:** This output becomes active when there are two regions and the internal wall is removed. • symbol: **N2_DIFF**
Current # of Molecules: Region 2, Species A	The number of molecules of species A that have diffused into the region initially occupied solely by species B. **Note:** This output becomes active when there are two regions and the internal wall is removed. • symbol: **N1_DIFF**
A–B Collisions: Total #	The total number of A–B collisions in the last simulation cycle. • symbol: **FORWARD_COLL**
A–B Collisions: Effective #	The number of A–B collisions that produced the reaction A + B —> C + D in the last simulation cycle. • symbol: **FORWARD_REAC**
C–D Collisions: Total #	The total number of C–D collisions in the last simulation cycle. • symbol: **REVERSE_COLL**
C–D Collisions: Effective #	The number of C–D collisions that produced the reaction A + B <— C + D in the last simulation cycle.
Energy Zoom Mode Data: Activation Energy	The activation energy of the reaction XY + Z —> X + YZ, measured in kJ/mol. • symbol: **ENERGY_A_XYZ**

Chemical Kinetics **Output Variables, Cont.**

VARIABLE	DESCRIPTION
Energy Zoom Mode Data: Enthalpy	In the reaction XY + Z —> X + YZ, the difference in energies of the XY bond and the YZ bond; measured in kJ/mol. • symbol: **ENTHALPY_R_XYZ**
Energy Zoom Mode Data: Potential Energy	The current value of the potential energy of the system XY —> YZ. • symbol: **U**
Energy Zoom Mode Data: X–Y Distance	The current distance between X and Y as the reaction X + YZ —> XY + Z proceeds. The distance is measured in arbitrary units. • symbol: **X_U**
Speed A	The distributed speed of the species A molecules when Maxwell-Boltzmann calculations are being made. This output gives the speed of each and every molecule, and therefore has no significance as a numeric output. These outputs are commonly displayed on a histogram. **Note:** To create the histogram, use the BAR CHART tool and click the Distribution Bar Chart radio button. Enter the speed A output for the expression. • symbol: **VELOCITY_1**
Speed B	The distributed speed of the species B molecules when Maxwell-Boltzmann calculations are being made. This output gives the speed of each and every molecule, and therefore has no significance as a numeric output. These outputs are commonly displayed on a histogram. **Note:** To create the histogram, use the BAR CHART tool and click the Distribution Bar Chart radio button. Enter the speed B output for the expression. • symbol: **VELOCITY_2**

Chemical Kinetics **Output Variables, Cont.**

VARIABLE	DESCRIPTION
Tagged Molecule Speed	The current speed of a specific molecule chosen at random by the model at the beginning of the simulation. This output is valid if the "Tagged Molecule?" option is set to Yes. • symbol: **VELOCITY_FIRST_MOL**
Pressure	For one region, the current value of the pressure on the external walls, created by the molecules colliding with the walls. This output is given in arbitrary units. • symbol: **PRESS**
Pressure, Region 1	The current value of the pressure on the walls of region 1, created by the molecules colliding with the walls. This output remains active even if the internal wall has been removed. This output is given in arbitrary units. • symbol: **PRESS_1**
Pressure, Region 2	The current value of the pressure on the walls of region 2, created by the molecules colliding with the walls. This output remains active even if the internal wall has been removed. This output is given in arbitrary units. • symbol: **PRESS_2**
Molecular Temperature	The current average temperature of the molecular energy, in degrees Kelvin. • symbol: **MOLEC_TEMP**
Time	The current time during the simulation. • symbol: **TIME**

Chemical Kinetics **Model Reactions**

REACTION	DESCRIPTION
Remove Wall	This reaction removes the separating wall inside the container, allowing the two separated species to mix.

5. Electrochemistry

The *Electrochemistry* model simulates the relationship between electrical and chemical energy by simulating an electrolytic device or a voltaic cell. You can control the initial conditions such as the electrolyte type, ion concentration, and electrode (metal strip). The simulation outputs show results such as the process time, the amount of metal accumulated near the cathode, the current, and the voltage.

Voltaic Cell Mode

The voltmeter indicates the voltage produced by the cell.

The bulb lights up when there is sufficient voltage.

The **temperature control icon**
Click on the black arrows in the temperature control to adjust the general range of the temperature value. Slide the red bar to set the exact value.

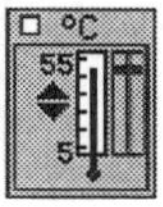

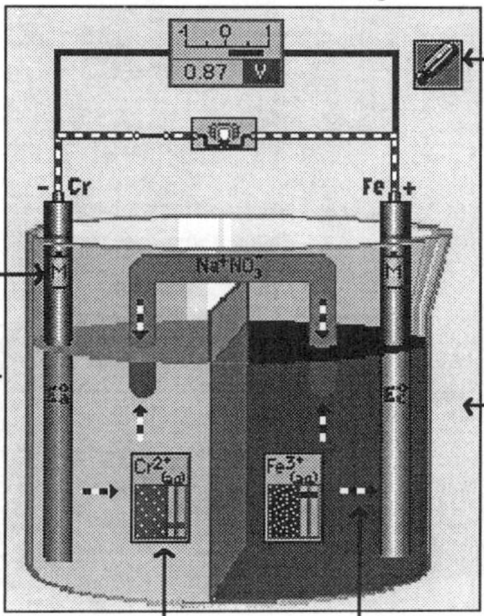

The **mass control icon** (Available for electrodes that participate in the reaction)
Slide the blue bar to set the electrode mass. The color of the electrode changes to indicate its relative mass.

The **potential control icon**
Displays the E° of the current half-reaction.
For a generic reaction, slide the blue bar to change the potential of the half-cell.

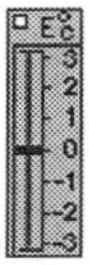

The **concentration control**
Slide a bar to set the concentrations. The density of the dots and the color of the solution indicate the relative concentrations.

Click on a **reaction arrow** to see the reaction taking place.

$Fe^{3+}(aq) + 3\bar{e} \rightarrow Fe(s)$

Electrolytic Cell Mode

The **battery** supplies voltage to the cell. Slide the blue bar to set the voltage. The red marker indicates the minimum emf.

The **temperature control icon** Click on the black arrows in the temperature control to adjust the general range of the temperature value. Slide the red bar to the exact value.

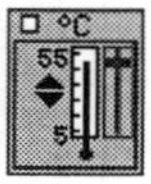

The **ammeter** indicates the current in the cell during the simulation.

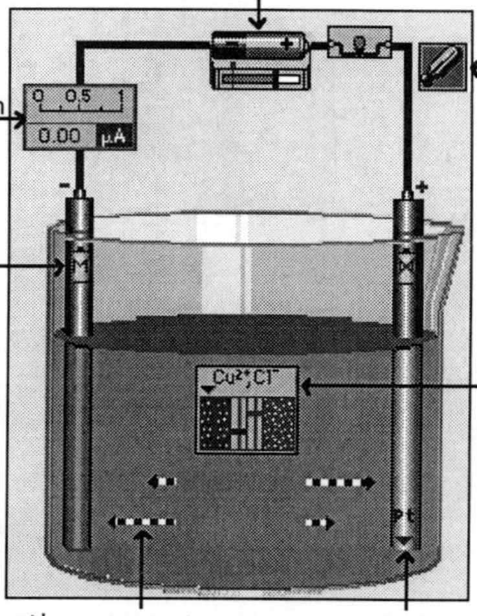

The **mass control icon** (Available for electrodes that participate in the reaction) Slide the blue bar to set the electrode mass. The color of the electrode changes to indicate its relative mass.

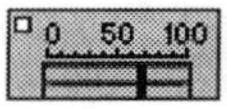

The **electrolyte control** Click on the title bar to choose an electrolyte from the pop-up menu. Slide a bar to set the concentrations. The density of the dots and the color of the electrolyte indicate the relative concentrations.

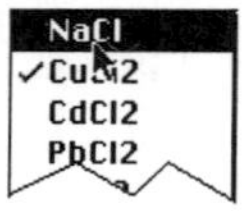

Click on a **reaction arrow** to see the reaction taking place.

$Cu^{2+}(aq) + 2\bar{e} \rightarrow Cu(s)$ (0.337V)

The **electrode type icon** (Available for some electrolytes) Click on the icon and choose a metal type for the electrode.

Model Tools

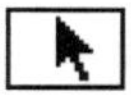 ARROW Tool

The ARROW tool enables you:

- To change cell type (from the pop-up menu)

- To open & close the temperature and the potential controls

 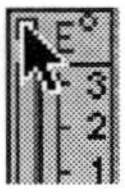

- To change values by dragging the slide bar in different controls, e.g., concentrations, potential, etc.

- To set the electrolyte type

- To set the current electrode type

- To change the battery voltage

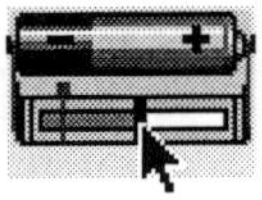

- To click the dynamic arrows to observe the reactions while the simulation is running.

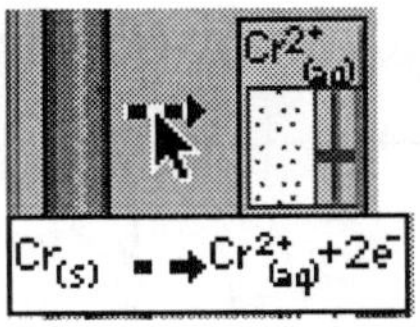

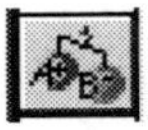 BALANCE Tool

The BALANCE tool defines the half-reactions in the voltaic cell and balances the stoichiometry of the reactions. The tool is disabled in the electrolytic cell mode.

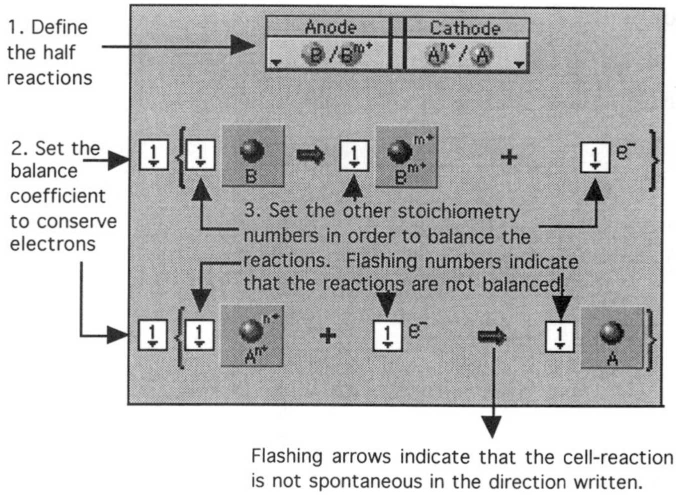

Electrochemistry Input Variables

VARIABLE	DESCRIPTION
Electrolytic Cell: Battery Voltage	The voltage of the battery connected to the electrolytic cell, over a range of 0–10 V. • symbol: **BATTERY_VOLTAGE**
Voltaic Cell: External Resistance	The resistance of the bulb or other device attached to the cell, measured in ohms. • symbol: **EXT_RESIST**
Left & Right Half-Cell: Reaction Potential	The potential of the given half-reaction, measured relative to the standard hydrogen electrode, over a range of -3.1 to 3.1 V. • symbol: **OXIDATION_POTENTIAL** **REDUCTION_POTENTIAL**
Time Data: Time Increment & Unit	The time increment determines the time between "snapshots" of the simulation animation. If $\Delta t=1.0$, each calculation step represents one time unit. The time units may be defined as seconds, minutes, or hours. • symbol: **DT & TIME_UNIT**
Colors	The color assigned to the given substance in the model window. • symbol: **COL_XXX**

Electrochemistry Output Variables

VARIABLE	DESCRIPTION
Time	The current number of time units that the simulation has been in GO mode. The units — seconds, minutes, hours — are determined by the Time Unit input. • symbol: **TIME**
Current	The present value of the current in the cell, measured in amperes. • symbol: **CURRENT**
Voltage: Voltmeter Reading	The voltmeter display, measured in volts. • symbol: **VOLTAGE**
Voltage: EMF	The current cell EMF, as calculated from the Nernst equation, measured in volts. • symbol: **VOLTAGE_EMF**
Voltaic Cell: Free Energy	The current free energy of the redox reaction in the voltaic cell, given in J/mol. • symbol: **FREE_ENERGY**
Voltaic Cell: Resistance	The current total resistance (the internal resistance of the cell plus the external resistance of the bulb) given in Ω. • symbol: **RESISTANCE**

Electrochemistry **Output Variables, Cont.**

VARIABLE	DESCRIPTION
Left & Right Half-Cell: Reactant Concentration	The current ion concentration of the reactant in the given half-cell solution, measured in mols. • symbol: **OX/RED_REACT_CONC**
Left & Right Half-Cell: Product Concentration	The current ion concentration of the product in the given half-cell solution, measured in mols. • symbol: **OX/RED_PROD_CONC**
Left & Right Half-Cell: Electrode Mass	The mass of the metal accumulated on the electrode of the given half-cell, measured in grams. • symbol: **METAL_ANODE/CATHODE_GALV**
Left & Right Half-Cell: Gas Volume	The current volume of gas produced at the electrode of the given half-cell, measured in liters. • symbol: **GAS_ANODE/CATHODE_GALV**
Electrolytic Cell: Minimum EMF	The minimum EMF, measured in volts. • symbol: **MIN_EMF**

Electrochemistry **Output Variables, Cont.**

VARIABLE	DESCRIPTION
Electrolytic Cell: Anion/Cation Concentration	The current concentration of the anion/cation in the cell solution. • symbol: **ANION_CONC/CATION_CONC**
Electrolytic Cell: Anode/Cathode Metal Mass	The mass of the metal accumulated on the electrode of given half-cell, measured in grams. • symbol: **METAL_ANODE/METAL_CATHODE**
Electrolytic Cell: Anode/Cathode Gas Volume	The current volume of gas produced at the electrode of the given half-cell, measured in liters. • symbol: **GAS_ANODE/GAS_CATHODE**
Electrolytic Cell: Anode/Cathode O2/H2 Volume	The current volume of oxygen/hydrogen produced at the anode/cathode, measured in liters. • symbol: **O2_ANODE/H2_CATHODE**

Electrochemistry **Model Reactions**

REACTION	DESCRIPTION
Hide/Show Control	These reactions show and hide respectively the given control. Controls that can be displayed or hidden include: • General: Ammeter Minimum EMF Cell Type Menu Temperature Control • Electrolytic Cell Mode: Electrolyte Concentration • Half-Cell Mode: Anode/Cathode Concentrations

6. Gas Laws

The *Gas Laws* model simulates the behavior of a gas enclosed in a cylinder with a moveable piston. The gas can be set to be an ideal gas, one of a number of real gases, or a generic real gas. The process may be constant volume, constant pressure, constant temperature, or adiabatic. The model makes measurements of variables such as pressure, temperature, volume, work done, and heat out during the simulation.

Model Window View

Gas Laws Model Window

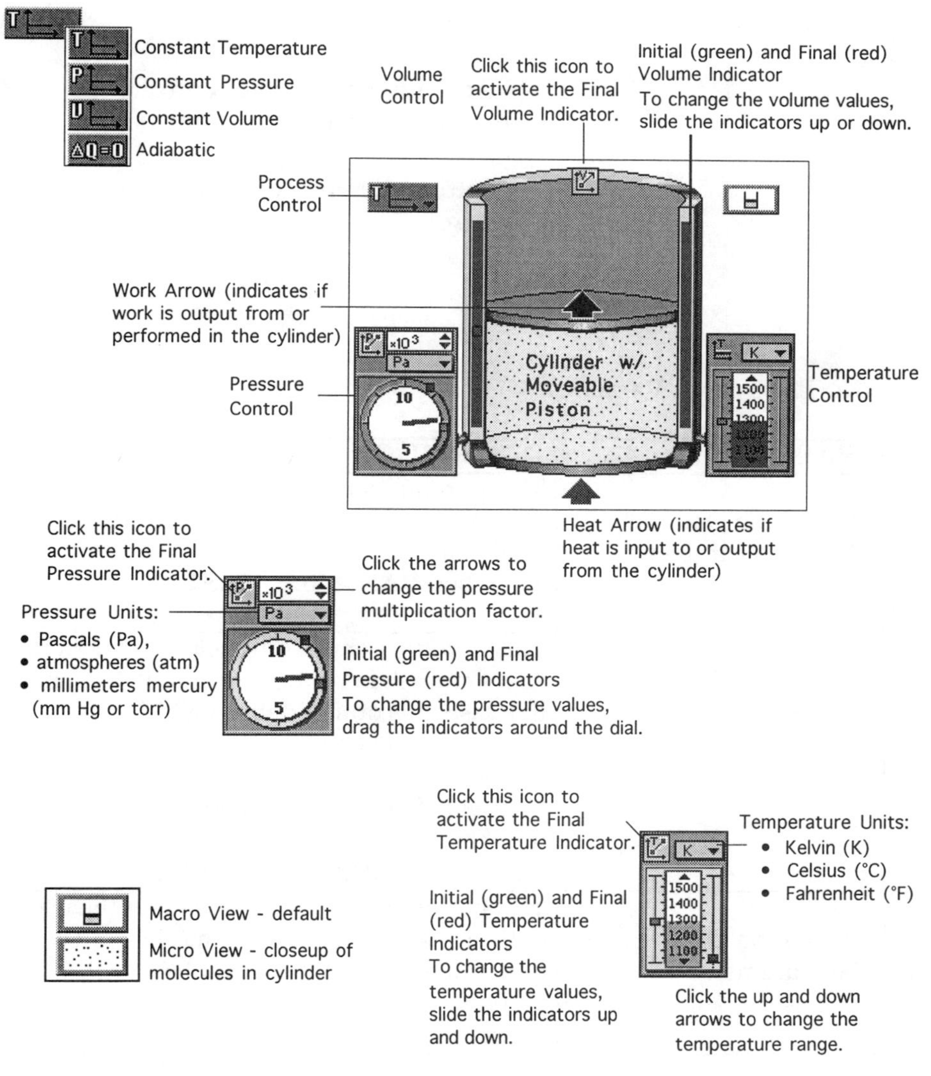

Model Tools

 ARROW Tool

The ARROW tool is used for setting options and values directly in the model window:

- Setting Process Type and Zoom Mode 

- Activating a variable control and setting its initial and/or final values 
- Setting the units on a variable control 

Gas Laws Input Variables

VARIABLE	DESCRIPTION
Definition Data: Variable Parameter Selection	This pop-up menu defines which of the variables will have user-defined initial and final values. • symbol: **VARIABLE_FIRST**
Definition Data: Initial Condition Selection	This pop-up menu is only relevant when the "Set Parameters" option is set to P, V, T, n. In this case, this option defines which of the variables will be the third one with user-defined initial values (in addition to the "Variable Parameter Selection" and the number of moles). • symbol: **VARIABLE_SECOND**
Definition Data: Set Parameters	This pop-up menu defines whether the user defines just P, V, and T (the model determines the number of moles), or P, V, T, and n. • symbol: **MOLES**
Initial Condition Values: Temperature	Sets the initial temperature inside the cylinder. The temperature units are set with the "Temperature" input parameter in the Units pop-up menu. • symbol: **INIT_T**

Gas Laws **Input Variables, Cont.**

VARIABLE	DESCRIPTION
Initial Condition Values: Pressure	Sets the initial pressure. The pressure units are set with the "Pressure" input parameter in the Units pop-up menu. • symbol: **INIT_P**
Initial Condition Values: Volume	Sets the initial volume. The volume units are set with the "Volume" input parameter in the Units pop-up menu. • symbol: **INIT_V**
Initial Condition Values: # Moles	Sets the initial number of moles; relevant only if the Set Variables option is set to "P, V, T, n." • symbol: **INIT_N**
Final Condition Values: Temperature	Sets the final value only if "Temperature" is selected in the "Variable Parameter Selection" option menu. In this case the model does not consider the values set in the other two "Final Condition Values" input boxes. • symbol: **FIN_T**
Final Condition Values: Pressure	Sets the final value only if "Pressure" is selected in the "Variable Parameter Selection" option menu. In this case the model does not consider the values set in the other two "Final Condition Values" input boxes. • symbol: **FIN_P**
Final Condition Values: Volume	Sets the final value only if "Volume" is selected in the "Variable Parameter Selection" option menu. In this case the model does not consider the values set in the other two "Final Condition Values" input boxes. • symbol: **FIN_V**
Gas Data: Gas Type	The "Gas Type" pop-up menu enables you to define the gas for the simulation: Ideal Gas Generic Real Gas — enables you to set the other four gas data inputs for any gas. Real Gases — a list of real gases. • symbol: **GAS_TYPE**

VARIABLE	DESCRIPTION
Gas Data: Molar Mass	The program automatically sets the molar mass according to gas chosen in the "Gas Type" input variable. • symbol: **MOL_MASS**
Gas Data: Molar Heat	The program enables you to set the molar heat capacity (C_V) as $\frac{3R}{2}$, $\frac{5R}{2}$, or $\frac{7R}{2}$ according to gas chosen in the "Gas Type" input variable. • symbol: **HEAT_CAPACITY**
Gas Data: van der Waals Constants a and b	The program automatically sets the constants according to gas chosen in the "Gas Type" input variable. If type is set to generic, the user can set these values. • symbol: **VAN_DER_WAALS_A/B**
Units: Temperature	Defines the temperature units as K, °C or °F. • symbol: **UNIT_T**
Units: Pressure	Defines the pressure units as Pa, atm, or mm Hg (torr) • symbol: **UNIT_P**
Units: Volume	Defines the volume units as m^3 or L. • symbol: **UNIT_V**

Gas Laws **Output Variables**

VARIABLE	DESCRIPTION
Current Conditions (Given Units): Temperature	The current value for the temperature variable, given in the units defined by the user (with the "Units" inputs or directly in the model window). • symbol: **CURRENT_T**
Current Conditions (Given Units): Pressure	The current value for the pressure variable, given in the units defined by the user (with the "Units" inputs or directly in the model window). • symbol: **CURRENT_P**
Current Conditions (Given Units): Volume	The current value for the volume variable, given in the units defined by the user (with the "Units" inputs or directly in the model window). • symbol: **CURRENT_V**
Current Conditions (Given Units): # Moles	The current value for the number of moles variable. • symbol: **CURRENT_N**
Current Conditions (SI Units): Temperature	The current value for the temperature variable; given in SI units. • symbol: **CURRENT_T_SI**
Current Conditions (SI Units): Pressure	The current value for the pressure variable; given in SI units. • symbol: **CURRENT_P_SI**
Current Conditions (SI Units): Volume	The current value for the volume variable; given in SI units. • symbol: **CURRENT_V_SI**
Energy Data: Heat [J]	The current value of heat input or heat output; measured in joules. • symbol: **CURRENT_HEAT**

Gas Laws **Output Variables, Cont.**

VARIABLE	DESCRIPTION
Energy Data: Work [J]	The current value of work done on, or work done by, the gas; measured in joules. • symbol: **CURRENT_WORK**
Energy Data: Entropy [J/K]	The current value inside the simulated container for entropy; measured in joules/kelvin. • symbol: **CURRENT_ENTROPY**
Energy Data: Internal Energy [J]	The current value inside the simulated container for internal energy; measured in joules. • symbol: **CURRENT_INTER_ENERGY**
Kinetic-Molecular Data: Average Kinetic Energy	The current average value of the kinetic energy of the molecules in the gas; measured in joules. • symbol: **AV_KIN_ENERGY**
Kinetic-Molecular Data: rms Speed	The current rms speed of the molecules in the gas; measured in m/s. • symbol: **RMS_SPEED**

Gas Laws **Model Reactions**

REACTION	DESCRIPTION
Show/Hide Energy Arrows	These reactions show and hide respectively the heat or work arrows in the model window. The red arrow indicates the direction of the heat flow. The blue arrow indicates the source of work (environment or gas).

VOLCANIC ASH

David H. Sharp and L. M. Simmons, Jr., editors

LOS ALAMOS SERIES IN BASIC AND APPLIED SCIENCES

DETONATION
Wildon Fickett & William C. Davis

NUMERICAL MODELING OF DETONATIONS
Charles L. Mader

THE THEORY OF ATOMIC STRUCTURE AND SPECTRA
Robert D. Cowan

VOLCANIC ASH

Grant Heiken
Kenneth Wohletz

LOS ALAMOS NATIONAL LABORATORY

UNIVERSITY OF CALIFORNIA PRESS
Berkeley · Los Angeles · London

UNIVERSITY OF CALIFORNIA PRESS
Berkeley • Los Angeles • London

University of California Press
Berkeley and Los Angeles, California

University of California Press
London, England

ISBN 0520-05241-2
Printed in the United States of America

1 2 3 4 5 6 7 8 9

CONTENTS

PREFACE

The characterization of volcanic ash and its relationship to eruption phenomena have been of interest to scientists since the time of Aristotle. Interest in this subject was renewed during the mid-1800s when European naturalists began studying volcanic fields in the Mediterranean, South America, and Indonesia. With the invention of the scanning electron microscope (SEM), a full range of pyroclast sizes could be investigated and chemical changes on grain surfaces measured. Used in conjunction with optical microscopy and the electron microprobe, the SEM has allowed us to begin systematic characterization of volcanic ash.

The purpose of this book is to provide a reference and guide to the characteristics of volcanic ash. We began assembling data for the book after the eruption of Mount St. Helens, when we received many inquiries regarding the physical and chemical properties of volcanic ash. Questions came not only from geologists but from people in the fields of environmental science, public health, engineering, aeronautics, agriculture, and the atmospheric sciences. The book is designed to be used by that broad spectrum of scientists and engineers.

We began the book on our own time but eventually were supported by research funds from Los Alamos National Laboratory. For their encouragement throughout the project, we thank John Whetten, Wes Myers, John Birely, David Sharp, and L. M. Simmons of Los Alamos. Technical support in the laboratory was provided by Roland Hagen and David Mann.

We collected most of the samples ourselves, but we also received many samples and much help from the volcanological community. Those who contributed include: William Rose, Jr., Michigan Technological University; Stephen Self, University of Texas/Arlington; Michael Sheridan, Arizona State University; Richard Fisher, University of California/Santa Barbara; Richard Hay, University of Illinois; Johan Varekamp, Wesleyan University; Colin Wilson, University of Auckland; Steve Carey, University of Rhode Island; Rodey Batiza, Washington University; J. V. Wright, British Petroleum Co.; Arturo Alcarez, Philippine Commission on Volcanology; Gudrun Larson, Nordic Volcanological Institute; Barry Dawson, University of Sheffield; R. Funiciello, University of Rome; David McKay (who was one of the very early collaborators on this project), Johnson Space Center; Leopold Gélinas, École Polytechnique, Montreal; John Stix, University of Toronto; the Portuguese Geological Survey; The Smithsonian Institution/ Natural History Museum; the late Sigurdur Thorarinsson; and the late Roald Fryxell.

Many people participated in construction of the book. Typing of the various drafts was done by Barbara Hahn and Marcia Jones. Line illustrations were drawn by John Tubb and Janet Repa. The many photos were reduced in the photo lab under supervision of Dan Morse. Book design was by Lisa Mirski of the University of California Press. Technical reviews were provided by Richard Fisher and Stephen Self. Editing this book was a difficult job that required strict attention to conformity among the various sections and steadfast patience; this job was undertaken by Jody Heiken. Her persistence in demanding grammatical, technical, and visual consistency greatly improved this book. E. Katherine Valdez took the challenging responsibility of composing, layout, and continual design revisions required for the complex presentation of this book. She not only devoted every spare moment but also worked into evenings and over weekends to satisfy the time constraints for completion. We greatly appreciate Kathy's effort and professional job.

We especially acknowledge Richard Fisher and Michael Sheridan, who have continually offered scientific support and enthusiasm. And finally: without the understanding and good sense of our wives, Jody Heiken and Ann Wohletz, this book would not have been possible.

This book is a compilation of a large amount of volcanic ash data that is, we hope, presented in a usable form. Along with the data, we have offered both conservative and liberal interpretations in an attempt to fit observations into both classical and contemporary volcanological thought.

Ash particles are like snowflakes in that no two are alike. They present an infinite variety of morphological, chemical, and physical characteristics. To the novice, systematic classification may seem an insurmountable task, but with persistence this practice yields increasing organization and understanding of this new aspect of volcanology. Although this work is in its infancy, it is becoming increasingly apparent that, with careful progress and the incorporation of techniques and logic from other related sciences, the unifying potential of ash analysis to many aspects of the geologic, planetary, and atmospheric sciences can be great. In one sense, each ash grain is like a field area to be systematically mapped. On a statistical basis, the distinguishing characteristics of ash sample sets, consisting of thousands of grains, define a general relationship that is readily incorporated into the macroscopic world. Because each ash grain yields a multitude of data, much of which has not yet been adequately described or measured, ash deposits are a new world to be studied, utilized, and interwoven into geological practice—sufficient work for generations of scientists. This book represents our present view of what should prove to be only the initial stages in the study of volcanic ash.

Houses buried in volcanic ash—Sakurazima, Japan, 1914.

INTRODUCTION

Explosive volcanic activity has always captured the public's attention. Indirectly, there is interest in such events as the last days of Pompeii and Herculaneum, the eruption of Krakatau, and the spectacular eruptive activity on Io, one of Jupiter's moons, as seen by the Voyager Spacecraft. More recently, villages and towns in many parts of the Earth have been affected directly. The 1980 eruptions of Mount St. Helens, Washington, paralyzed much of the Pacific Northwest with products of explosive activity—volcanic ash. Volcanic ash was also responsible for many deaths and the evacuation of thousands after the eruptions of El Chichón, in the State of Chiapas, Mexico. The eruption plume from El Chichón has also been responsible for spectacular twilight effects seen over much of the Earth during 1982 and may yet have an effect on climate. Around the house, volcanic ash is used in the form of mechanic's soap, tooth powder, and household cleansers, and as aggregates for cement blocks and insulation.

Volcanic ash deposits are found throughout the entire geologic record. Pyroclastic rocks are present as layers in metamorphic rocks over a billion years old, interbedded with sedimentary rocks deposited throughout Paleozoic and Mesozoic time, and in recent deposits only a few years old. Many of these deposits have served as the source or host of a variety of mineral resources such as uranium, lithium, copper, and diamonds. Reserves of zeolite minerals (molecular sieves) are most often found in altered volcanic ash deposits. Volcanic ash is also an important constituent of soil in many areas of the world; because it often determines the fertility of soils, it is critical for development of agriculture. Blocks of lithified volcanic ash are cut and used for construction of houses and public buildings throughout all parts of the world where there are volcanoes.

This book is intended to serve as a reference for those interested in the identification of volcanic ash and its physical properties. For the geologist and volcanologist, we have attempted to provide examples of ash from all eruption types and from a variety of eruption sequences. These examples may be used in interpretation of explosive eruption phenomena and physical characteristics of the magmas. Ash characteristics, along with chemical characterization, may be used in tephrochronology, the study and correlation of ash layers in the geologic record. Tephrochronologists determine the ages of volcanic eruptions and locate source vents for ash layers; these data are especially important for the appraisal of volcanic hazards (Westgate and Gold, 1974). We have provided a number of examples of relict volcanic ash textures in soils and meta-tuffs to help identify ancient eruption sequences.

Examples of volcanic ash collected in eruption plumes may be of interest to those in the atmospheric and environmental sciences. There is increasing research on the atmospheric effects of volcanic ash since the eruption of Mount St. Helens. Characterization of ashes from ash fall deposits and eruption plumes could also be of interest to engineers in the aerospace industry; in 1982, three jet aircraft were forced down because they had incurred engine damage after flying into eruption plumes.

This book is devoted to descriptions of volcanic ash from different eruption types, sequences of ash layers, and weathered and metamorphosed ash. Divisions of eruption and, therefore, ash types are based upon the mechanism of ash formation (discussed later in this chapter) and the chemical composition of the ash. The samples upon which this book is based were collected by the authors or thoughtfully contributed by members of the volcanological community; in our opinion, the samples typify each major type of explosive eruption and its ash deposits. Each example is presented independently; plates and text are on facing pages so that this book can be used as a reference text for laboratory application. Appropriate references follow each section. Because the terminology used here may be unfamiliar to many readers, an illustrated glossary is included. If used as a text, this book should be used in conjunction with *Pyroclastic Rocks* by Fisher and Schmincke (1984). The Fisher-Schmincke book is a thorough study of the genesis and field occurrence of volcanic ash deposits.

REFERENCES:

Fisher, R. and Schmincke, H. (1984). *Pyroclastic rocks*, Springer-Verlag, Berlin, 472 pp.

Westgate, J. A. and Gold, C. M., 1974. *World bibliography and index of Quaternary tephrochronology*. Univ. of Alberta, Edmonton, 528 pp.

Glossary

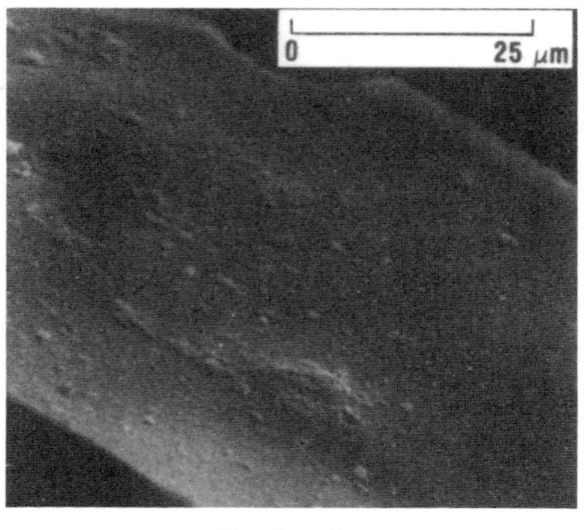

Adhering dust

Aggregate

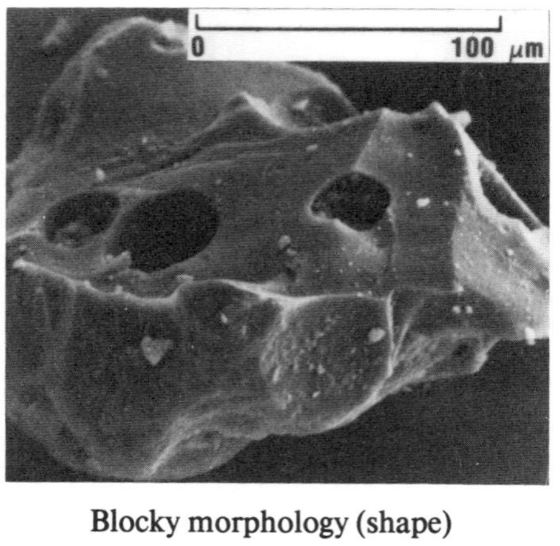

Blocky morphology (shape)

Achnelith
: Term used by Walker and Croasdale (1972) to describe droplets from Strombolian lava fountaining. It is from the Greek *achne*—spray.

Accretionary lapilli
: More or less spherical masses of cemented ash ranging in diameter from a few millimeters to several centimeters. Sometimes referred to as "volcanic hailstones," "pisolites," "volcanic mud-balls," or "chalazoidites" (in New Zealand).

Adhering dust
: Fine-grained volcanic dust sticking to the surfaces of coarser ash grains.

Aerosol
: A suspension of colloidal particles in a gas. The particles may be fine ash, sublimates, ice and water, or acid droplets present in a dilute volcanic plume.

Agglutinates
: Vitric pyroclasts consisting of a cluster of smaller pyroclasts that are annealed or "welded" to one another.

Aggregate
: A mass of pyroclasts bonded by electrostatic forces, sublimates, or alteration products such as carbonate, zeolites, or clay.

Aphyric (non-porphyritic)
: An igneous texture in which phenocrysts are absent in a fine-grained or glassy matrix.

Ash
: Pyroclasts 2 mm or less in diameter. (Fisher, 1961; Schmid, 1981).

Ash fall
: A rain of airborne volcanic ash falling from an eruption cloud and the resulting deposit. An ash fall deposit is generally characterized by mantle bedding of well-to-moderately sorted ash. Syn: "ash fallout."

Ash flow
: Synonym for "pyroclastic flow" but referring to pyroclastic flows with pyroclasts under 2 mm in diameter.

An
: Abbreviation for anorthite ($CaAl_2Si_2O_8$). Plagioclase compositions are expressed as the percentage of anorthite in solid solution with albite ($NaAlSi_3O_8$).

Base surge
: A ring-shaped cloud that moves horizontally outward from the base of a vertical explosion column, initially at hurricane velocities, as a turbulent density current. Base surge deposits are thinly bedded and consist of dune-like beds, inversely graded flat-lying beds, and massive beds showing little or no internal grading. Base surges were first described in association with nuclear weapons tests in the Pacific and then in the deposits of maar volcanoes.

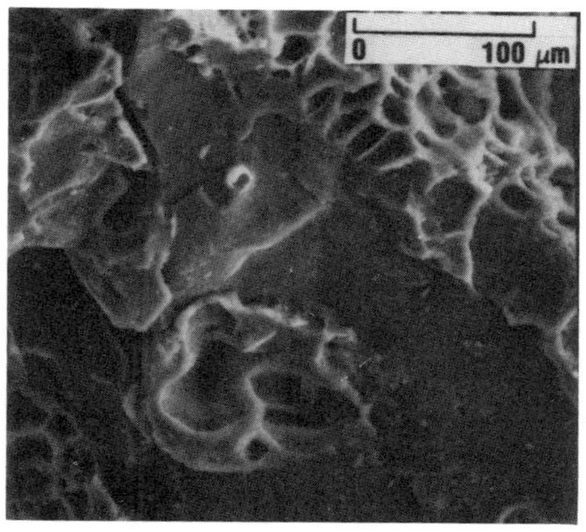

Bubble-wall texture

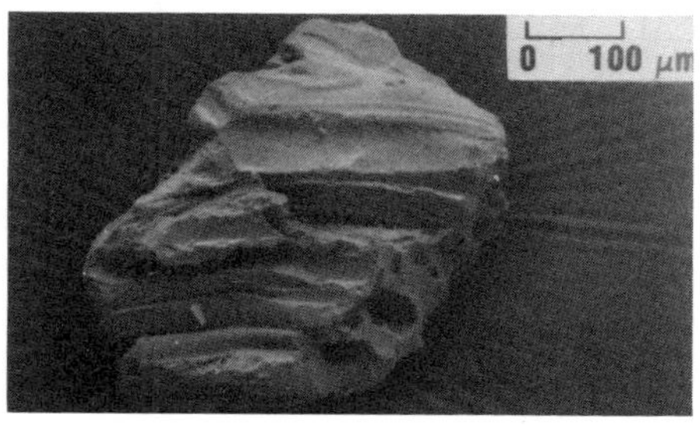

Chipped edges

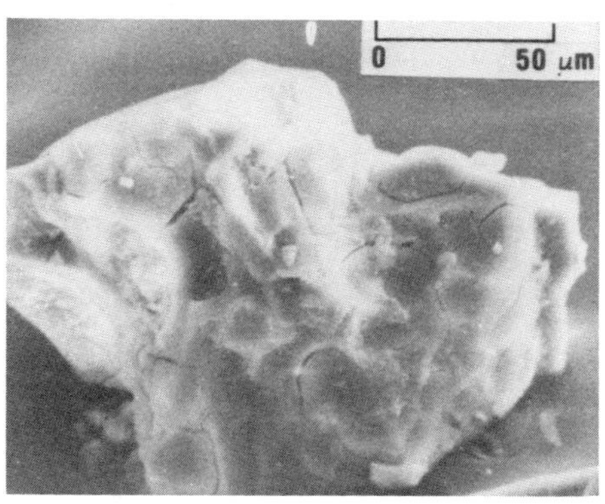

Cracks

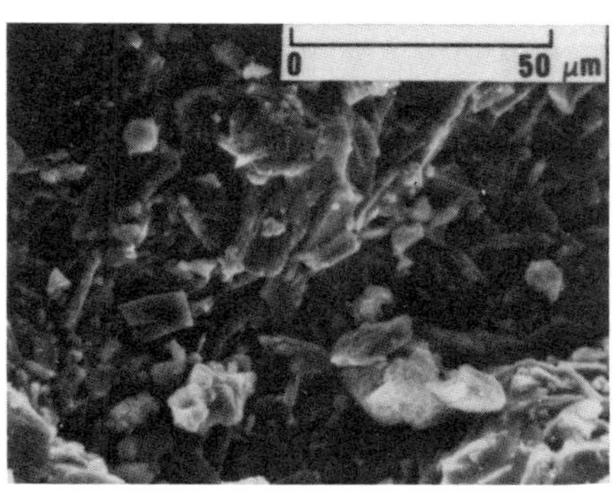

Diktytaxitic

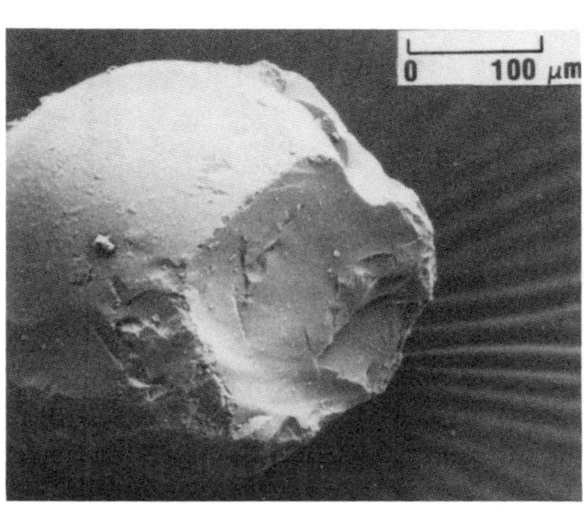

Dish-shaped fracture

Block
A pyroclast larger than 64 mm that was solid when ejected (AGI). It may have been juvenile or torn from older rocks in the vent or crater walls.

Blocky morphology (shape)
Overall grain shape characterized by planar or curviplanar surfaces that intersect nearly at right angles. Equant, blocky grains have a nearly isometric form that may grade into a tabular or prismatic form. Typical of phreatomagmatic ash grains.

Bomb
A pyroclast larger than 64 mm that was viscous (partly or completely fluid) when ejected (AGI).

Botyryoidal
Having the form of a bunch of grapes (AGI). Clusters of glass or minerals consisting of spherical and/or ovoid shapes.

Breccia
A fragmental rock made of highly angular coarse fragments with or without a finer grained matrix.

Bubble-wall texture
Texture shown by phenocrysts with a thin coating of vesicular glass (Fisher, 1963).

Chipped edges
A textural feature of pyroclasts in which small, concave fractures (dish-shaped, v-shaped, or uneven) are concentrated along thin, angular edges of a pyroclast. This feature is present when angular grains are rounded because of transport abrasion.

Cinder
Synonym of "scoria." Used in ancient times when volcanoes were thought to contain fire from combustibles within the Earth.

Cracks
Cracks in volcanic ash grains form by mechanical (fragmentation and percussion) and chemical (alteration) processes. The latter, illustrated here, are tens of micrometers long and several micrometers wide, project through a glass skin several micrometers thick, and result from volume changes during hydration.

Diatreme
A general term for a volcanic vent or pipe drilled upward through enclosing rocks (usually flat-lying sedimentary rocks) by the explosive energy of gas-charged magmas. The surface expression of a diatreme is a maar crater and, after extensive erosion, is a tuff or tuff-breccia neck.

Diktytaxitic
A textural term for an igneous rock with irregular, angular cavities between groundmass minerals. In many instances, the groundmass minerals have a thin coating of glass.

Dish-shaped fracture
Fracture surfaces characterized by circular, or nearly circular, concave areas that are the result of conchoidal fracture of glassy material. Such fractures are typical of grain modification during transport abrasion.

EDS
Energy dispersive spectral analysis. A method of chemical analysis based upon the energy spectrum of x-rays emitted from a sample illuminated by an electron beam. This method is readily adaptable for use with a scanning electron microscope.

Eruption plume
A convoluted, rolling mass of partly condensed water vapor, dust, and ash that often rises to great heights. Syn: "eruption cloud."

Explosion pit
A bowl-shaped crater surrounded by a low rim of ejecta; generally formed by phreatic or phreatomagmatic explosions in which little or no juvenile material is ejected.

Fluidal texture
Smooth, round surfaces of pyroclasts formed by surface tension within a fluid droplet before it is chilled.

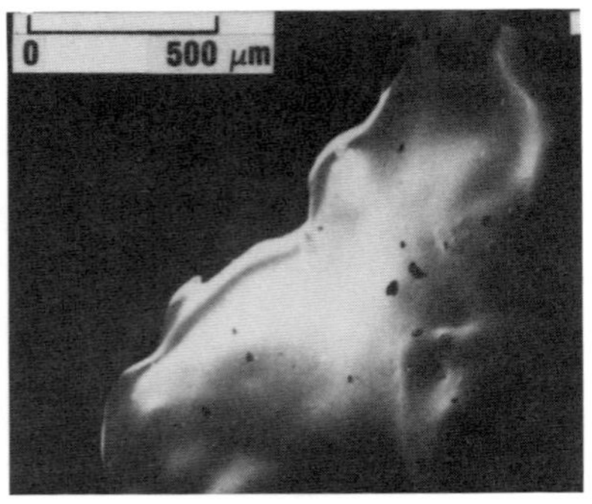

Fluidal texture

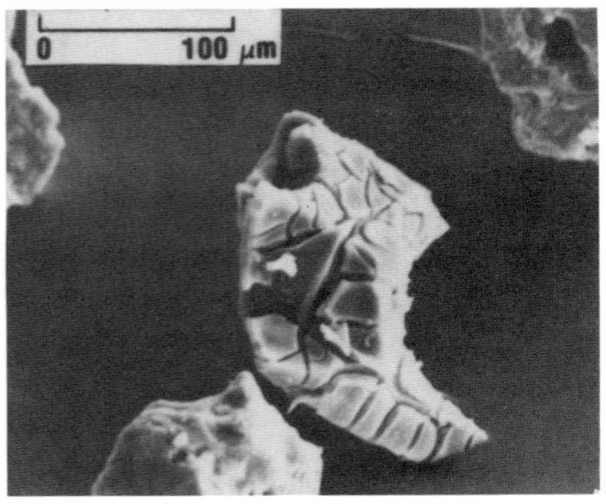

Hydrated surface

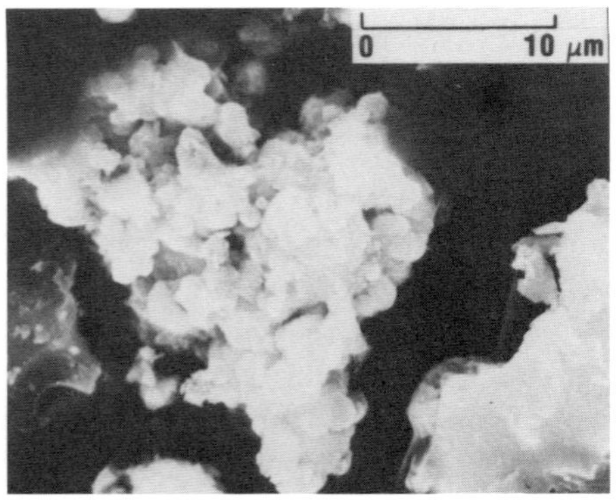

Moss-like morphology

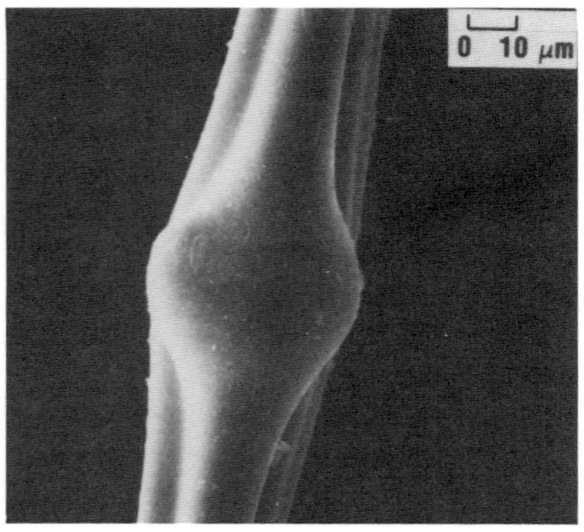

Pele's hair

Fo
: Abbreviation for forsterite (Mg_2SiO_3). Olivine compositions are expressed as the percentage of forsterite in solid solution with fayalite (Fa—Fe_2SiO_3).

Glomeroporphyritic
: Igneous texture denoting clustering of phenocrysts in a finely crystalline groundmass.

Hackly
: Showing jagged points on a fracture surface (AGI).

Hawaiian eruption
: Eruption of low-viscosity gas-charged magmas as lava fountains and thin, rapidly moving lava flows.

Hyaloclastite
: Volcaniclastic rocks generated by non-explosive to mildly explosive granulation of volcanic glass when basaltic magmas are quenched on contact with water. These are common products of deep-sea volcanism (Honnorez and Kirst, 1975) and subglacial eruptions.

Hyalocrystalline
: A textural term applied to igneous porphyritic rocks composed of equal or nearly equal amounts of crystals (phenocrysts) and glassy groundmass.

Hyalotuff
: Pyroclastic rocks generated by phreatomagmatic explosions when volcanoes erupt in shallow water (Honnorez and Kirst, 1975). The term is applied mainly to the tuffs of nearshore maar volcanoes.

Hydrated surface
: A lighter colored, vitreous skin covering all or part of a vitric pyroclast; it is formed by hydration of the outer part of a pyroclast. The surface is often cracked by desiccation of the skin.

Hydromagmatic
: A general term to include all processes, subsurface or surface, involving interaction of magma or magmatic heat with meteoric or connate water in the Earth. Syn: "phreatomagmatic" when describing volcanic eruptions.

Hydrovolcanic
: A term encompassing all volcanic activity that results from the interaction of lava, magmatic heat, or gases with meteoric or connate water at or near the surface of the Earth. This volcanic activity is generally associated with vapor explosions. Syn: "phreatomagmatic."

Ignimbrite
: A rock-unit term applied to tuffs deposited by pyroclastic flows (AGI).

Juvenile pyroclast
: Pyroclast derived directly from magma reaching the surface (AGI).

Lahar
: A debris flow or mudflow of pyroclastic material that is mainly derived from pyroclastic deposits.

Lapilli (s.: lapillus)
: Pyroclasts between 2 and 64 mm (Fisher, 1961).

Lava fountain
: A jet of low-viscosity lava shot into the air by release of volatiles from the melt (AGI).

Lithic pyroclasts
: Pyroclasts consisting of previously formed rocks; fragments of igneous, metamorphic, or sedimentary rock or pieces of earlier lavas in the vent area that are included in a pyroclastic deposit.

Littoral cone
: A tuff cone formed on or adjacent to a lava flow when it enters a body of water. Such cones are the result of steam explosions (AGI). Syn: "rootless cone."

Maar volcano
: A small volcano characterized by a crater that is wide (several hundred meters to several kilometers diameter) relative to its height and having a crater floor that commonly lies below the general level of surrounding topography. Maars form during phreatomagmatic eruptions.

Magma
: Molten rock within a terrestrial planet.

Pele's hair has intrigued geologists and historians for centuries. Typically, it consists of long strands that are circular in cross-section, with diameters from 1 to 300 μm. The thinner varieties may or may not be vesicular and are near-perfect cylinders. Elongate vesicles within the hairs may be as thin as several micrometers in diameter. Larger (>100-μm-diameter) varieties may be smooth or ribbed cylinders; the ribs reflect elongate vesicles near and parallel to the hair surface. Lumps in the hairs are present over phenocrysts or ovoid vesicles. In some cases, elongate glass threads are bent or twisted into knots. Some are attached to ovoid droplets, with shapes akin to a "tail" (Duffield et al., 1977).

Near-surface acceleration of lava in fountains, caused by expansion of gases, drives jets of low-viscosity basaltic lava (10^2 to 10^3 poise), which form Pele's hair. Unstable jets that break up into a spray produce melt droplets, formed by surface tension, which are then chilled to glass while in flight. Many of the streamers of chilled glass (Pele's hair) and smaller droplets are carried off by wind, although some fall back into the fountain to form composite droplets after multiple collisions with rising melt droplets. Many Pele's hairs are broken up by air turbulence and the hair and droplets are frequently broken upon hitting the ground or colliding with larger lapilli and bombs. No one has measured the length of a Pele's hair because they are so delicate. Pele's hair also forms at lava cascades and lava falls, where clots of low-viscosity lava are "spun" by strong air currents passing over the turbulent lava.

Reticulite [the thread lace scoria of Wentworth (1938) or "Limu" of the native Hawaiians] is a remarkable open network of lattice-shaped, triangular basaltic glass rods of extremely low bulk density. Nearly all of the vesicles are coalesced, although some thin glass membranes remain as vesicle walls. Reticulite is extremely delicate and generally is broken into lumps of several centimeters or less in diameter. Reticulite is formed by the extreme attenuation of a low-viscosity liquid during vesiculation. Deposits of reticulite are rarely preserved because they are so easily broken down mechanically and chemically.

Other pyroclast forms include ragged "cinders" or spinose, vesiculated droplets (Walker and Croasdale, 1972). Although the range of shapes is great, most pyroclasts have the smooth skin formed by surface tension within droplets of melt ejected from lava fountains.

Ultramafic Tephra Formed in Lunar Lava Fountains. Fine-grained, homogeneous glass spheres and composite spheres (those with glass mounds adhering to their surfaces) of mafic to ultramafic composition have been found in clastic horizons deposited on lunar lava plains at three Apollo landing sites. It has been proposed that these glass spheres were produced in fountains of ultramafic lava (Heiken et al., 1974) because they are so similar to pyroclasts produced in Hawaiian eruptions. Some differences, including the remarkable sphericity, may be the result of very low viscosities of the erupting lavas (calculated viscosities of 3 to 10 poise at 1250°C). For comparison, Hawaiian lava fountains have viscosities of $\sim 10^2$ to $\sim 10^3$ poise. It was first believed that these glass spheres might have been of impact-melt origin; none, however, contain shock damaged crystals or exhibit the heterogeneity so common to glass particles of impact origin.

Strombolian Eruptions. Strombolian eruptions of slightly more viscous lava ($\sim 10^3$ to $\sim 6 \times 10^3$ poise) produce distinctive tephra that are considerably different than those formed in Hawaiian eruptions. They are characterized by the ejection of partly molten bombs and ash that forms cinder cones and their associated tephra blankets. Ash produced during Strombolian activity consists of a variety of pyroclasts, from irregular sideromelane droplets to blocky tachylite (black to dark brown microcrystalline pyroclasts).

Sideromelane pyroclasts are smooth-skinned, vesicular droplets that form an irregular, partly open framework of vesicle walls. Where smooth "skin" is preserved, it consists of lumpy, botryoidal surfaces. Many grain shapes reflect interior vesicle shapes, which range from spherical to deformed ovoids. Blocky sideromelane fragments are mostly droplet fragments. The gradation from clear sideromelane droplets into microcrystalline tachylite grains reflects the degree of chilling and crystallization of the melt droplet.

Angular, blocky tachylite pyroclasts range in shape from those with irregular, open-vesicle networks to nonvesicular blocky grains. Vesicularities (with vesicle volumes of 10 to 20%) tend to be lower than in the sideromelane pyroclasts. Grain surfaces and vesicle walls are not smooth, but rough; they exhibit both fracture surfaces and diktytaxitic textures.

In both sideromelane and tachylite pyroclasts, melt viscosities are not low enough to allow formation of the smooth-skinned spheres and ovoids that are typical of tephra produced by Hawaiian eruptions. Most of these pyroclasts are rapidly quenched after eruption; few appear to have been plastically deformed upon hitting the ground or have particles welded to grain surfaces (agglutinates).

The variety of pyroclasts indicate the variety of conditions that may exist in a Strombolian eruption. Sideromelane pyroclasts exhibit the most pronounced deformation by surface tension and, thus, the lowest viscosities. They are produced by a higher temperature magma, which may be the case if lava flux from the conduit is high or if the vent is open and some jetting is possible (similar to characteristic Hawaiian activity). If the vent is clogged or lava flux is low, occasional bursts of cooler, less vesicular melt, which pools for a while, and "accidental" pyroclasts derived from slumps into the crater will produce an abundance of the blocky, tachylite pyroclasts.

Ash of Peralkaline Silicic Compositions:

Explosive eruptions of rhyolitic-trachytic magmas produce both pyroclastic flows and ash fall tephra, although the volume of these units, associated with stratovolcanoes and calderas, is small when compared with that of some rhyolitic volcanoes (Schmincke, 1974). These magmas have relatively low viscosities when erupted (e.g., $10^{2.8}$ to 10^6 poise at temperatures of >950°C and >2% water).

Peralkaline pyroclast shapes vary in samples from basal to upper portions of thin pyroclastic flow units. Tricuspate (Y-shaped) or blade-like shards and pumice with elongate (tubular) or spherical vesicles are most common near the base. Within the upper portion of these pyroclastic flows, vitric pyroclasts have spherical shapes; some of the larger ones (generally 0.2 to 3 mm) are hollow. Some spheres up to 5 cm in diameter, described at Mt. Suswa, Kenya, have highly vesicular crystalline cores and thin glass rims. In some deposits, glass spheres form the matrix material of coarser grained pumice and obsidian fragments (Hay and Hildreth, 1976; Schmincke, 1974). Ash fall pyroclastic units of glass spheres of melilitic composition from Kenya have also been described (R. L. Hay, 1976, personal communication).

There is considerable disagreement about the genesis of these spherical pyroclasts. In his comprehensive review of peralkaline silicic welded tuffs, Schmincke (1974) suggests that at least two processes are responsible for their genesis: (1) "pulling together" of hot plastic shards within the pyroclastic cloud before final deposition and (2) expansion of volatiles within shards after deposition. Hay and Hildreth (1976) believe that peralkaline glass spheres are formed when jets of low-viscosity magmas break up into sprays, which is similar to the process of droplet formation in Hawaiian eruptions. Supporting their hypothesis is the presence of spherical pyroclasts of melilitic composition in some Kenyan ash fall units.

All of the above processes may account, to some degree, for the spherical pyroclasts present in tuff deposits of peralkaline silicic composition.

Ash of Andesitic to Rhyolitic Composition:

Petrologic and morphologic characteristics of andesitic to rhyolitic pyroclasts are related to higher viscosities (10^3 to 10^6 poise) and the higher volatile contents of their source magmas. Viscosities of above 10^3 poise do not allow the formation of droplets by surface tension; pyroclast shapes are dependent upon the geometry of vesicles developed within the highly viscous magmas. The resulting tephra consist mostly of varieties of pumice or their broken products and are characteristic of nearly all large-volume pyroclastic deposits. They make up the extensive ash fall and pyroclastic flow deposits produced during highly explosive eruptions and are associated with composite cones, some ash and pumice cones, and large calderas. These pyroclast types represent some of the products of Vulcanian and most products of Peléean and Plinian eruptions.

Andesitic Volcanic Ash. Andesitic ashes consist of complex mixtures of vitric, crystal, and lithic components. Vitric components consist of brownish to colorless, homogeneous to heterogeneous glasses, the latter having been derived from mixed magmas. In equant to elongate pumice fragments the particle shape depends on vesicle density and geometry. Many have parallel, highly elongate "pipe-like" vesicles that are parallel to the long axis of the pyroclast. There are no smooth particle surfaces; exposed broken vesicle walls provide sharp, irregular surfaces. Only the vesicle walls are smooth. Smaller particles, which are bits of vesicle wall, are Y-shaped curved plates or simple needle-like forms.

Phenocrysts broken out during explosive eruptions result in a crystal component in these ashes. Broken, angular mineral fragments are common; the particle shapes are governed by the fracture pattern of the mineral. Most lithic fragments (a wide variety of rock types from vent and crater walls) are equant to elongate and angular to rounded. Surface textures are entirely dependent on the physical properties of the rock type.

Dacitic and Rhyolitic Volcanic Ash. Most dacitic and rhyolitic pyroclasts are pumices or their broken products. Pumice has been the object of scholarly studies for a few thousand years; in the 4th century B.C., Theophrastos called it "worm-eaten wood" and thought it to be derived from sea-foam and fire. A thorough historical background of pumice studies may be found in an article by Perlaki (1966).

Nearly all pumices are angular and highly vesicular. Narrow, elongate pumices usually contain highly elongate vesicles (with aspect ratios of >20:1) that are circular to lens-like in cross section and 5 to 20 μm in diameter. Pumice pyroclasts containing phenocrysts are generally heterogeneous. Some of the highly elongate varieties contain large, ovoid cavities around which thinner, elongate vesicles are bent. Larger, less elongate pumices contain 100- to 500-μm-long ovoid vesicles with aspect ratios of less than 20:1. Another pumice variety is massive with remarkably small vesicles (5 to 15 μm in diameter) and relatively thick (10- to 100-μm) vesicle walls.

Most finer grained silicic tephras consist of comminution products of pumice (shards), individual phenocrysts (crystal fraction), and some lithic fragments (xenoliths). Merrill (1885) described curved or Y-shaped glass plates from a vitric-crystal ash from Nebraska; he assumed that the shards had been formed by the pulverization of pumice. Judd (1888) came to a similar conclusion in his studies of silicic volcanic ash.

Many phenocrysts are coated with vesicular glass rinds; vesicles are flattened ovoids that are parallel or perpendicular to phenocryst surfaces and range from 80 to 300 μm long. This is the "bubble-wall texture" of Fisher (1963) and Meyer (1972). Meyer proposed that the volume of glass clinging to a phenocryst was inversely correlated to the "explosivity" or kinetic energy of an eruption.

A rare and unusual variety of rhyolitic ash was produced during a phase of the eruption at Novarupta (in the Valley of 10,000 Smokes), Alaska in 1912. Perret (1950) describes minute hollow bubbles, thickened on one side and often girdled by a raised ridge or septum. Some are divided by a median wall. The mechanism by which this pyroclast type originates is not known.

Many dacitic to rhyolitic domes and flows may represent the last phases of volatile-depleted activity. There are, however, minor ash deposits associated with these domes. Explosive release of volatiles (gas eruption) often occurs within the dome vents, producing lithic and vitric ashes (Rose, 1972). The coarser tephra include pumiceous bombs and massive blocks. Dacitic ashes from this eruption type consist of equant, hyalocrystalline to holocrystalline, slightly vesicular lithic ash fragments. Most grain surfaces are hackly to conchoidal, depending on the glass component of the dome lavas.

Original pyroclast shapes characteristic of silicic tephra may be drastically modified by heating, plastic deformation, and welding within ash flow tuff units. For a complete synthesis of welding and post eruption modification of pyroclasts, see Ross and Smith (1961).

Formation of Rhyolitic and Dacitic Volcanic Ash. In a early analysis of volcanic ash formation, Verhoogen (1951) addressed the problem of nucleation, bubble growth, and rise of these bubbles within a magma. In that analysis, he assumed that volcanic ash forms when the rising, expanding bubbles coalesce, which causes the lava to lose its cohesion. Later work by McBirney (1963) and others shows that the rate of bubble rise within a magma is far too slow to be responsible for the formation of pyroclasts except in basaltic volcanoes with magmas of relatively low viscosities. On the basis of observations that silicic pyroclastic deposits nearly all contain an abundance of pumice pyroclasts that are composed of multiple, intact vesicles and are not composed entirely of bubble walls (shards), it is most unlikely that ash is formed by bubble coalescence alone. The lack of glass droplets within silicic pyroclastic deposits rules out the jetting of melt and formation of a spray that are characteristic of Hawaiian eruptions.

McBirney and Murase (1970) determined that if vesicularity within a melt is low, ash formation takes place when the gas pressure within the bubbles exceeds the tensile strength of the liquid. When the melt is highly vesicular, the loss of coherence of the melt occurs only when gas pressure exceeds the surface tension of the bubble walls. McBirney and Murase, however, also assume bubble coalescence is a requirement for disruption of the magma.

Variations within the form and petrography of erupted tephra may provide clues to variations of temperature, viscosity, and volatile content within a magma body before its eruption. To illustrate this hypothesis, the following is an example using data from Murase and McBirney (1973) for a rhyolite magma that contains 2 to 3% H_2O. Gas pressure should exceed lithostatic pressures at depths of 1 to 2 km. The magma body (at least the upper portion) begins to vesiculate in place. This process continues until rock units overlying the magma chamber are fractured and an eruption begins. Expansion of gases in the vesiculating magma is concurrent with a temperature drop from 1000 to 750°C and a subsequent increase in viscosity from $\sim 10^5$ to 10^8 poise. The vesiculated mass becomes too viscous to flow rapidly through the vent; therefore, another mechanism is called upon to disrupt the froth. This disruption may be accomplished by the large pressure discontinuity at the magma/atmosphere interface. The pressure change creates a shock wave that moves ahead of the ash out of the vent and an expansion wave that passes down into the magma chamber and allows further vapor expansion and disruption of the froth. This phenomenon is described by Rittmann (1936) and Bennett (1974). A high-pressure, condensed fluid (steam), suddenly released, will undergo an unsteady, simple wave expansion of the type already known in conventional shock-tube flow. In a shock tube, one-dimensional expansion of a compressed driver gas creates and drives a shock wave ahead of the contact surface between the driver and ambient gas. At the same time, an expansion wave moves into the driver gas and allows its decompression. The expansion wave moving into a magma chamber disrupts nearly solid foam particles, which are subsequently accelerated out of the vent (Sandford et al., 1975; Heiken, 1978). Because of the high angularity of pumice fragments and their comminution products, this mechanism for the disruption of semi-molten or nearly solid, highly viscous froth by explosive expansion is an acceptable mechanism for formation of dacitic or rhyolitic volcanic pumice and shards.

The theory proposed by Bennett (1974) is compatible with observed tephra, except in one respect. Bennett proposes that ash is formed by condensation of magma after eruption (expansion of melt). Pyroclasts produced by condensation should be mainly droplets or melt blobs (spheres, etc.). This is not the case for tephra from highly explosive eruptions of silicic magmas.

Whatever hypothesis is used to explain the formation of volcanic ash, as the eruption proceeds, hotter, less-viscous melt moves into the void left by the melt previously erupted, vesiculates, and continues the eruption. If most volatile components are concentrated in the top of a magma chamber, a highly vesicular froth is eventually depleted as the eruption proceeds. In many large-volume pyroclastic deposits this is indeed the case, and less vesicular, higher density pumice is ejected during waning phases of the eruption.

The degree of vesicle elongation within pumice is most likely due to movement of the vesiculating magma and not due to elongation toward the magma/atmosphere interface, as proposed by Perlaki (1966). Viscosities are too high for a bubble to grow in that manner.

Rhyolitic pumices from many eruptions are texturally heterogeneous; they consist of "pockets" of large, ovoid vesicles and thin, highly elongate vesicles. The "pockets" consist of 1- to 2-mm-long, ovoid vesicles in clusters that surround or are adjacent to phenocrysts. These pockets, in turn, are enclosed in vesicular glass that is characterized by highly elongate vesicles having much smaller volumes (Heiken, 1983). Differences in vesicle types can be explained by heterogeneous vesicle development. It is possible for vesicle growth to begin at depths of 2 or 3 km in rhyodacitic or rhyolitic melts with a water content of 3% or more (Sparks, 1978). Early vesicle growth may begin at phenocryst surfaces, when those surfaces act as bubble nuclei. Vesicle pockets that are radial to phenocryst surfaces have been described by Heiken and Eichelberger (1980). The larger pockets with thin internal vesicle walls have a higher bulk viscosity than does the surrounding melt and may resist deformation by flow. As magma rises, vesiculation begins around the pockets. Closer to the surface, later stage vesicles are sheared into highly elongate, tube-like forms as they flow around the competent, stronger framework that is created by pockets of ovoid vesicles. At or near the surface, this heterogeneous, vesicular mass could be fragmented, perhaps by a pressure wave passing down into the vent.

Finer grained tephra (shards) are the products of pumice comminution; it is not yet clear to what degree the finer shards are products of explosive disintegration of the magma froth or the mechanical interaction (grinding) of pumices against each other as they are erupted. Both processes appear to be important.

Ash of Carbonatite Composition: Unique pyroclasts containing a substantial portion of sodium carbonate are erupted from the strato-volcano Oldoinyo Lengai, Tanzania (Dawson et al., 1968). The activity varies from the eruption of minor ash plumes to major Plinian eruptions.

The ash consists of sodium carbonate (trona) with crystals of nepheline, pyroxene, wollastonite, apatite, melanite, and pyrite. It is a crystal ash mostly composed of perfect euhedral to anhedral nepheline crystals. Ash particles are coated with clumps of blade-shaped trona crystals. Ash fall tephra of carbonatite composition may contain vitric spheres in addition to other components (R. L. Hay, personal communication).

VOLCANIC ASH FROM PHREATOMAGMATIC (HYDROVOLCANIC) ERUPTIONS

Phreatomagmatic (sometimes called *hydrovolcanic*) eruptions are caused by the explosive generation of steam when rising magma contacts bodies of water (such as the sea, lakes, and marshes), ground water, snow, or ice. Although the dominant factor in phreatomagmatic eruptions is the process of steam explosion, magma composition and physical properties are also important factors determining the mechanism of fragmentation during interaction of magma with external water.

In contrast with magmatic fragmentation, which has been studied from both theoretical approaches and experimental evidence, phreatomagmatic fragmentation has been investigated largely through field observation. *Hyaloclastite* is the term suggested by Honnorez and Kirst (1975) for glass found with pillow basalts that were produced by nonexplosive quenching and fracturing of basaltic glass; *hyalotuff* is used to identify rocks formed by explosive fragmentation of glass during phreatomagmatic eruptions in shallow water. Hyaloclastite may be formed at water depths of over 500 m, where hydrostatic pressure is high enough to inhibit vesiculation in basaltic magma (McBirney, 1963). Consideration of the explosive mechanism and the presence of blocky, equant glass shapes has resulted in the general conclusion that phreatomagmatic ash is formed by thermal contraction and shattering of glass (von Waltershausen, 1853; Peacock, 1926; Fisher and Waters, 1969; Bonatti, 1970; Heiken, 1971). This conclusion is supported by Carlisle (1963), who observed nonvesicular teardrop- and spindle-shaped hyaloclasts in hyaloclastite deposits and in experimentally produced curved splinters and pointed chips of sideromelane. Also, Honnorez and Kirst (1975) observed blocky grain shapes in blast furnace slags that were quenched by water. Recent experimental studies, however, indicate a much more complex fragmentation mechanism for phreatomagmatic ash formation. These experiments have studied the explosive interaction of molten materials with water: a phenomenon commonly termed *fuel-coolant interaction* (FCI-Buchannan, 1974; Peckover et al., 1973; Nelson et al., 1981).

The explosive contact of water with melt begins with the formation and collapse of steam films on the melt surface. This process is cyclic on a micro- or millisecond time scale; it results in the generation of fluid instabilities at the contact, water jet penetration of the melt, and stress waves propagating into the melt. These mechanisms fragment the melt, thereby increasing surface area and heat-transfer rates. The effects of superheating and detonation waves are critical to the vaporization of water. Superheating is a process involving nonequilibrium heat transfer and homogeneous nucleation that causes instantaneous vaporization. It is still poorly understood, but it causes explosive efficiencies several orders of magnitude higher than in normal boiling processes. Detonation waves are shock waves that propagate through the melt to cause fragmentation and mixing with water by fluid instabilities and vapor film collapse. These waves are sensitive to the system size and permit interaction of large volumes of melt and water within a short time.

Ash morphology has been used to elucidate the mechanism of phreatomagmatic eruption; Walker and Croasdale (1972), Heiken (1972; 1974), and Honnorez and Kirst (1975) used optical microscopy and scanning electron microscopy (SEM) to characterize hydrovolcanic ash produced in Surtseyan eruptions and submarine extrusion of pillow basalts. Equant, blocky shapes and curviplanar surfaces with a paucity of vesicles resulted when fragmentation was due to contact of magma with meteoric water. Grain surfaces are mostly planar, quite often meeting at right angles; these flat fracture surfaces are characteristic of hyaloclastic pyroclasts. Vesicularities range from low (mostly less than 30% of volume) to nonvesicular; vesicles are usually spherical and widely spaced.

Particle size as well as shape has also been studied in assessing the formation of volcanic ash. Numerous published size analyses of pyroclasts have shown the relationship of particle size to mechanism of transport and deposition. Walker (1971) demonstrated the size characteristics of flow and fallout tephra by plotting the sorting coefficient versus median diameter, and later, Walker (1973) showed that phreatomagmatic pyroclasts have a much higher degree of fragmentation than do magmatic pyroclasts. These studies and others suggest the important relation of pyroclast size to eruptive mechanism. Size distributions of phreatomagmatic pyroclasts are strongly controlled by eruption energy that, in turn, determines dispersal mechanisms and the resulting deposit types.

Comparison of natural samples and ash debris from experiments modeling hydrovolcanic explosions allowed Wohletz (1983) to develop several theoretical models of hydrovolcanic ash formation. Although there are physical limitations when comparing man-made metallic melts with magma, experiments suggest that the fragmentation mechanisms are basically the same for a wide range of melt compositions. Experimentally produced ash particle shapes are blocky and equant, spherical and drop-like mossy aggregates and the plate-like shapes of broken bubble walls. These shapes, as well as their sizes, show strong resemblance to those of ash produced by hydrovolcanic eruptions. Because rapid vaporization of water is the driving mechanism of hydrovolcanic explosions, studying the pyroclastic material generated by these explosions provides information on the efficiency of the heat-transfer process. Heat transfer is dominantly conductive and requires large surface areas to reach explosive rates. Experimental debris show decreasing grain size with increasing explosive efficiency. The maximum size of experimental explosion debris is 2 to 3 mm; nonexplosive debris are coarser grained. The finest debris are <1 μm. Similarly, explosive hydrovolcanic debris have median grain sizes less than 1 or 2 mm, which is also the median size of many magmatic tephra deposits.

Fragment shapes produced during melt/water interactions reflect fragmentation by brittle failure during passage of stress waves and viscous melt/water mixing by fluid instabilities. The most efficient explosive interactions appear to result in the latter of these two fragmentation mechanisms. Hence, the production of fine-grained debris with large surface areas and fluid shapes is predicted for highly explosive eruptions of fluid basalt. However, melt viscosity and strength strongly encourages development of instabilities. Therefore, for melts of intermediate and silicic composition, brittle fragmentation dominates.

An important consideration is that both magmatic and hydrovolcanic fragmentation mechanisms may operate during an eruption. This situation is illustrated by Self and Sparks (1978) for phreatoplinian silicic eruptions in which magma is initially disrupted by exsolution and expansion of magmatic volatiles, producing relatively coarse-grained pyroclasts; further fragmentation (fine-grained pyroclasts) is caused by explosive interaction with water (Tazieff, 1968). Theoretical consideration of experimental fragmentation mechanisms suggests that stress waves produced by high-pressure vaporization of water at the magma/water interface may, in part, induce vesiculation in the melt.

VOLCANIC ASH FROM LITTORAL CONES

Volcanic ash similar to that of maar volcanoes is characteristic of littoral cones and is formed by steam explosions that are generated when lava flows into a body of water (Fisher, 1968). Evidence at Puu Hou, Hawaii, suggests that explosions occurred when steam was trapped beneath crusted lava flow surfaces. The varied mixture of pyroclasts includes: (1) crystalline basalt fragments and olivine phenocrysts, (2) scalloped sideromelane fragments with abundant vesicles, (3) stretched sideromelane glass with small vesicles, elongate shapes, and fluidal edges (frozen droplets), and (4) irregular or equant sideromelane fragments with few or no vesicles (Fisher, 1968). This mixture is easily characterized by the variety of pyroclast types: blocky and pyramidal sideromelane fragments, rough-surfaced fragments of tachylite, and sideromelane flow crust.

Well-developed vesicularity near lava flow surfaces accounts for the oddly shaped hyaloclastite components of these tephra. When the lava was chilled and the resulting quench was shattered by thermal stresses, irregular pyramidal- and crescent-shaped fragments were formed. The steam explosion also carries out droplets of unquenched lava and bits of solid flow crust. Blocky hyaloclastites also occur as matrix material in pillow lavas and in lava deltas when lava flows into rivers, lakes, and glaciers (Fuller, 1931).

VOLCANIC ASH FROM PHREATIC ERUPTIONS

Active hydrothermal systems often erupt considerable volumes of tephra at stratovolcanoes, domes, or hot springs; for example, the 5×10^6 m^3 of "mud" erupted from Lake City Hot Springs, California in 1951 (White, 1955). Among those who study such eruptions, there is often considerable controversy over the presence or absence of pyroclasts derived from juvenile magma; therefore, the identification of juvenile pyroclasts is important.

Tephra from active hydrothermal (steam) eruptions consist mainly of hydrothermally altered lithic and mineral fragments, commonly in a clay matrix. Included are subangular lithic ash (or lava) fragments in various states of alteration; they range from fragments with a trace of hematite stain to those that are mostly a mixture of clays, zeolites, and pyrite. Only relict textures remain to identify pyroclast types. Particle surfaces are often covered with aggregates of zeolite crystals or clay. Basaltic glass fragments may be rimmed with palagonite, a mixture of smectite clays and iron oxides. Ash particles are similar to the rock units exposed in vent walls and are broken from them by escaping steam; they are erupted to form a local low-volume pyroclastic deposit.

REFERENCES:

Bennett, F. D., 1974. On volcanic ash formation. Amer. Jour. Sci., **274**: 648-661.

Bonatti, E., 1970. Deep sea volcanism. Naturwissenschaften, **57**: 379-384.

Buchanan, D. J., 1974. A model for fuel-coolant interactions. J. Phys. D: Appl. Phys., **7**: 1441-1457.

Carlisle, D., 1963. Pillow breccias and their aquagene tuffs in Quadra Island, British Columbia. Jour. Geol., **71**: 48-71.

Dawson, J. B., Bowden, P., and Clark, G. D., 1968. Activity of the carbonatite volcano, Oldoinyo Lengai, 1966. Geol. Rundschau, **57**: 865-879.

Duffield, W. A., Gibson, Jr., E. K., and Heiken, G. H., 1977. Some characteristics of Pele's hair. U.S. Geol. Surv. Jour. Res., **5**: 93-101.

Fisher, R. V., 1963. Bubble-wall texture and its significance. Jour. Sed. Pet., **33**: 224-235.

Fisher, R. V., 1968. Puu Hou littoral cones Hawaii. Geol. Rundschau, **57**: 837-864.

Fisher, R. V. and Waters, A. C., 1969. Bed forms in base surge deposits: Lunar implications. Sci., **165**: 1349-1352.

Fuller, R. E., 1931. The aqueous chilling of basaltic lava on the Columbia River Plateau. Amer. Jour. Sci., **21**: 281-300.

Hay, R. L. and Hildreth, W., 1976. Globule ignimbrites in Kenya. Geol. Soc. Amer. Abs. with Prog., **8**: 589.

Heiken, G. H., 1971. Tuff rings: examples from the Fort Rock-Christmas Lake Valley, south-central Oregon. J. Geophys. Res., **76**: 5615-5626.

Heiken, G., 1972. Morphology and petrography of volcanic ashes. Geol. Soc. Amer. Bull., **83**: 1961-1988.

Heiken, G., 1974. *An atlas of volcanic ash.* Smithsonian Contr. Earth Sci., **12**: 101 pp.

Heiken, G., 1978. Plinian-type eruptions in the Medicine Lake Highland, California and the nature of the underlying magma. J. Volcanol. Geotherm. Res., **4**: 375-402.

Heiken, G., 1983. Textural analysis of tephra from a rhyodacitic eruption sequence, Thira (Santorini), Greece. *In*: Proc., Workshop on characterization and quantification of surface features on clastic and pyroclastic particles (in press).

Heiken, G. and Eichelberger, J., 1980. Eruptions at Chaos Crags, Lassen Volcanic National Park, California. J. Volcanol. Geotherm. Res., **7**: 443-481.

Heiken, G., McKay, D. S., and Brown, R. W., 1974. Lunar deposits of possible pyroclastic origin. Geochim. Cosmochim. Acta, **38**: 1703-1718.

Honnorez, J. and Kirst, P., 1975. Submarine basaltic volcanism: morphometric parameters for discriminating hyaloclastites from hyalotuffs. Bull. Volcanol., **32**: 441-465.

Judd, J. W., 1888. On the volcanic phenomenon of the eruption, and on the nature and distribution of the ejected materials. *In*: G. J. Symons (Ed.), *The Eruption of Krakatoa and Subsequent Phenomena*, London, Trubner and Co., 1-56.

McBirney, A. R., 1963. Deep-sea volcanism. Bull. Volcanol., **26**: 453-469.

McBirney, A. R. and Murase, T., 1970. Factors governing the formation of pyroclastic rocks. Bull. Volcanol., **34**: 372-384.

Merrill, G. P., 1885. On deposits of volcanic dust and sand in southwestern Nebraska. U.S. Nat'l. Mus., Proc., **8**: 99-100.

Meyer, J. D., 1972. Glass crust on intratelluric phenocrysts in volcanic ash as a measure of eruptive violence. Bull. Volcanol., **35**: 358-368.

Murase, T. and McBirney, A. R., 1973. Properties of some common igneous rocks and their melts at high temperatures. Geol. Soc. Amer. Bull., **84**: 3563-3592.

Nelson, L. S., Buxton, L. D., and Planner, H. N., 1980. Steam explosion triggering phenomena, part 2: Corium-A and Corium-E and oxides of iron and cobolt studied with a floodable arc-melting apparatus. Sandia National Laboratories, SAND77-0998, NUREG/CR-0122, 138 pp.

Peacock, M. A., 1926. The basic tuffs. *In*: G. W. Tyrell and M. A. Peacock (Eds.), The petrology of Iceland, Royal Soc. Edinburgh Trans., **45**: 51-76.

Peckover, R. S., Buchanan, D. J., and Ashby, D. E. T. F., 1973. Fuel-coolant interactions in submarine volcanism. Nature, **245**: 307-308.

Perlaki, E., 1966. Pumice and scoria: their notion, criteria, structure and genesis. Acta Geol. Hung., **10**: 13-29.

Perret, F., 1950. Volcanic ejectamenta. *In*: *Volcanic Observations* (F. Perrett, Ed.), Carnegie Inst. Pub. **549**: 115-125.

Rittmann, A., 1936. Vulkane und ihre Tätigkeit. Ferdinand Enke Verlag, Stuttgart, 188 pp.

Rose, Jr., W. I., 1972. Santiaguito volcanic dome, Guatemala. Geol. Soc. Amer. Bull., **83**: 1413-1434.

Ross, C. S. and Smith, R. L., 1961. Ash-flow tuffs: their origin, geologic relations and identification. U.S. Geol. Surv. Prof. Paper **366**: 81 pp.

Sandford, II, M. T., Jones, E. M., and McGetchin, T. R., 1975. Hydrodynamics of caldera-forming eruptions. Geol. Soc. Amer. Abs. with Prog., **7**: 1257-1258.

Schmincke, H. U., 1974. Volcanological aspects of peralkaline silicic welded ash-flow tuffs. Bull. Volcanol., **38**: 594-636.

Self, S. and Sparks, R. S. J., 1978. Characteristics of widespread pyroclastic deposits formed by the interaction of silicic magma and water. Bull. Volcanol., **41-3**: 196-212.

Sparks, R. S. J., 1978. The dynamics of bubble formation and growth in magmas: a review and analysis. Jour. Volcanol. Geotherm. Res., **3**: 1-37.

Tazieff, H., 1968. Sur le mécanisme des eruptions basaltiques sous-marines a faibles profundeurs et la genese d'hyaloclastites associeés. Geol. Rund., **57**: 955-966.

Verhoogen, J., 1951. Mechanisms of ash formation. Amer. Jour. Sci., **249**: 29-739.

von Waltershausen, W. S., 1853. Über die vulkanischen gesteine in Sizilien und Island und ihre submarine umbildung. Gottingen.

Walker, G. P. L., 1971. Grain-size characteristics of pyroclastic deposits. J. Geol., **79**: 696-714.

Walker, G. P. L., 1973. Explosive volcanic eruptions—a new classification scheme. Sond. Geol. Rund., **62**: 431-446.

Walker, G. P. L. and Croasdale, R., 1972. Characteristics of some basaltic pyroclastics. Bull. Volcanol., **35**: 303-317.

Wentworth, C. K., 1938. Ash formations of the Island of Hawaii. Haw. Volc. Obs. 3rd Spec. Rpt., 183 pp.

Wentworth, C. K. and MacDonald, G. A., 1953. Structures and forms of basaltic rocks in Hawaii. U.S. Geol. Surv. Bull., **994**: 98 p.

White, D., 1955. Violent mud-volcano eruption of Lake City Hot Springs, northeastern California. Geol. Soc. Amer. Bull., **66**: 1109-1130.

Wohletz, K. H., 1983. Mechanisms of hydrovolcanic pyroclast formation: size, scanning electron microscopy, and experimental results. Jour. Volcanol. Geotherm. Res., **17**: 31-63.

Classification of Pyroclastic Rocks

Classification is a difficult task, especially because the categories are not precise and usually grade one into another. Classification schemes of volcanic ash are based upon grain size, composition of pyroclasts and magma, genesis of the pyroclasts, and mechanism of deposition. The most commonly used of these schemes is grain size and is similar to that used in sedimentology to classify clastic sediments and sedimentary rocks. Wentworth and Williams (1932) developed the first version using a φ scale (see Glossary). However, the most commonly used version of this classification is the scheme of Fisher (1961) shown in Table 1. A slightly modified version of this classification was accepted by the Subcommission on the Systematics of Igneous Rocks of the International Union of Geological Sciences (IUGS) (Schmid, 1981).

TABLE 1. GRAIN-SIZE LIMITS AND TERMS FOR PYROCLASTIC ROCKS.

		Pyroclastic Deposit	
Clast Size (mm)	**Pyroclast**	**Mainly unconsolidated: tephra**	**Mainly consolidated: pyroclastic rock**
	Bomb, block	Agglomerate	Agglomerate, pyroclastic breccia
64 mm			
	Lapillus	Layer, bed of lapilli (lapilli tephra)	Lapilli tuff (Lapillistone)
2 mm			
	Coarse ash	Coarse ash	Coarse (ash) tuff
1/16 mm			
	Fine ash	Fine ash	Fine (ash) tuff

There are many classifications of tephra on the basis of the constituent composition (pyroclast compositions). For communication purposes, Cook's classification (1965) is very useful (Fig. 1). Within other petrographic classifications, including that of the IUGS, most tuffs are classified as "vitric" and are not easily separated; this problem can be avoided by using Cook's classification scheme.

Walker (1971) has separated pyroclastic flow from pyroclastic fall deposits on the basis of median grain size and sorting. This approach is very useful in unraveling the genesis of stratigraphic sequences of pyroclastic rocks. Further subdivision of pyroclastic flow, surge, and fall deposits has been made by Wright et al. (1980).

The genetic classification used in this book is similar to that used by Heiken (1972; 1974). The classification is based mainly on two categories: magmatic and phreatomagmatic; i.e., explosive eruptions caused primarily by release of magmatic gases and those caused primarily by interaction with surface or ground water. As in most classifications, there are overlaps; both processes may be responsible for the explosive activity of a single volcanic eruption. Within these two categories there are also many variations, depending upon the physical and chemical properties of the magma.

For each example in this book the grain-size classification (median grain size and sorting) is given. Sorting characteristics are described below:

$$\sigma_\varphi = <1\varphi \text{ — well sorted}$$

$$\sigma_\varphi = 1 \text{ to } 2\varphi \text{ — moderately sorted}$$

$$\sigma_\varphi = >2\varphi \text{ — poorly sorted}$$

$$\text{where: } \sigma_\varphi = \frac{\varphi 84 - \varphi 16}{2}$$

Further classification involves the identification of the magmatic composition. All of the above classifications may be combined easily into one phrase to describe a single sample: e.g., "Well-sorted rhyolitic vitric-lithic lapilli-bearing ash from a pyroclastic fall deposit of phreatomagmatic origin."

REFERENCES: Cook, E., 1965. Stratigraphy of Tertiary volcanic rocks in eastern Nevada. Nev. Bur. Mines, Rept. **11**: 66 pp.

Fisher, R., 1961. Proposed classification of volcaniclastic sediments and rocks. Geol. Soc. Amer. Bull., **72**: 1409-1414.

Heiken, G., 1972. Morphology and petrography of volcanic ashes. Geol. Soc. Amer. Bull., **83**: 1961-1988.

Heiken, G., 1974. An atlas of volcanic ash. Smithsonian Contr. Earth Sci. **12**: 101 pp.

Schmid, R., 1981. Descriptive nomenclature and classification of pyroclastic deposits and fragments: Recommendations of the I.U.G.S. Subcommission on the Systematics of Igneous Rocks. Geol., **9**: 41-43.

Walker, G., 1971. Grain-size characteristics of pyroclastic deposits. Jour. Geol., **79**: 696-714.

Wentworth, C. and Williams, H., 1932. The classification and terminology of the pyroclastic rocks. Nat. Acad. Sci. Nat. Res. Council, **89**: 19-53.

Wright, J., Smith, A., and Self, S., 1980. A working terminology of pyroclastic deposits. Jour. Volcanol. Geotherm. Res.: **8**: 315-336.

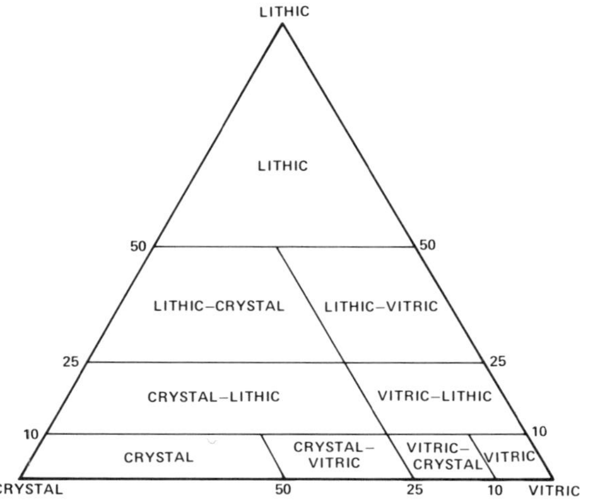

Figure 1. Pyroclastic classification of Cook (1965).

Summary of Physical Methods for Analysis of Tephra and Pyroclastic Rocks

For information about size parameters of unconsolidated tephra and pyroclastic rocks, see *Pyroclastic Rocks* by Fisher and Schmincke. References for detailed physical analysis of tephra are those designed for clastic sediments (Krumbein and Pettijohn, 1938; Pettijohn, Potter, and Siever, 1973; Mueller, 1967).

Grain-size analysis of tephra samples requires considerable judgement on the part of the investigator because of the delicacy of the pyroclasts. If the ash is composed mostly of lithic fragments, for example, a conventional stack of sieves and a ro-tap sieve shaker may be used as one would to analyze river sands.

However, vitric fragments in most ashes are usually very delicate and easily abraded. Vigorous use of a ro-tap can easily turn a coarse vesicular ash into a fine ash by particle abrasion. Wet-sieving and gentle manual shaking of the sieve stack may be necessary to avoid destroying the sample. For fine-grained ashes, a sonic sifter, or other size analysis method (such as a Coulter Counter) may be used.

If size analyses are done, splits of each size fraction should be taken for binocular and thin-section optical inspection as well as for SEM mounts. If samples are not sieved, conventional splitting techniques are adequate to provide samples for SEM, thin sections, and bulk chemical analyses.

Before SEM inspection, the samples should be cleaned to improve image quality. This technique is especially important when samples are partly coated with loosely adhering fine material or a cement such as calcite. By removing this material, one is able to see the original pyroclast surfaces. A "dirty" sample may also make it difficult to obtain the uniform conductive coat that is necessary to counteract sample charging in an electron beam. The most common cleaning technique consists of washing the sample in dilute HCl at low to moderate heat for 3 to 5 minutes, followed by a rinse in distilled H_2O and a drying period that may be accelerated by a second rinse with acetone. This method is not suitable, however, if the dilute acid reacts with cementing agents or alteration products on the pyroclast surfaces. If acid cleansing does not produce good results, washing the sample in acetone or distilled H_2O with ultrasound can loosen unwanted material. Ultrasound cleansing should be done with great care because it may damage the pyroclast surface, especially if it is applied for longer than 30-second intervals.

Mounting samples for SEM examination is simple: (1) coat a sample stub with a thin smear of cement (dilute duco cement is effective) or double-sided adhesive tape, and (2) drop the sample split onto the stub and shake off the excess ash, or place individual grains on the stub with tweezers or a probe. After the sample is mounted, apply a thin carbon coat in a vacuum evaporator. If energy-dispersive x-ray analyses are to be made of grain surfaces, apply a second carbon coat. If the SEM analysis is to produce images only, a gold or gold-palladium coating should be applied over the carbon coat.

A very good polished thin section or grain mount of each sample is absolutely necessary for tephra studies. Either may be used for petrographic descriptions, grain counts (for classification of pyroclasts), and analysis of individual pyroclasts by electron microprobe. If pyroclast textures are finely crystalline or hyalocrystalline, chemical- or ion-etching techniques will enhance the contrast between individual phases. A thin section may then be studied by SEM, reflected-light microscopy, or phase-contrast microscopy.

REFERENCES:

Fisher, R. V. and Schmincke, H.-U., 1984. *Pyroclastic Rocks.* Springer-Verlag, Heidelberg, 472 pp.

Krumbein, W. C. and Pettijohn, F. J., 1938. *Manual of Sedimentary Petrography.* Appleton-Century-Crofts, New York, 549 pp.

Mueller, G., 1967. *Methods in Sedimentary Petrology.* Hafner Pub. Co., New York, 283 pp.

Pettijohn, F., Potter, P., and Siever, R., 1972. *Sand and Sandstone.* Springer-Verlag, Heidelberg, 618 pp.

Lava fountain—1959 eruption of Kilauea Iki, Hawaii.

MAGMATIC ERUPTIONS

Magmatic eruptions produce ashes derived directly from a vesiculating magma that has not interacted with meteoric water. Their characteristics are related to the magma strength and viscosity and the thermodynamic behavior of gas phases within the magma.

Hawaiian Eruptions (Lava Fountaining)

Eruption of low-viscosity basaltic magma often occurs as lava fountains. This behavior is characteristic of eruptions on the island of Hawaii where relatively mild lava fountaining produces tephra. Ash particles from lava fountains generally have fluidal shapes, including droplets and strands.

ALOI-ALAE, HAWAIIAN VOLCANOES NATIONAL PARK

Introduction: The sample, collected 1 km west of Alae Crater, was deposited during lava fountaining from fissures between Aloi and Alae Craters, which are located along the east rift zone of Kilauea Volcano.

The lava fountains of this eruption were similar to many others (e.g., Richter et al., 1970): vigorous, gas-driven fountains, a few meters to hundreds of meters high, of low-viscosity basaltic lava. Pyroclasts ranging from droplets only a few micrometers in diameter to meter-sized bombs were erupted in the fountains. Wind swept away the finer grained fractions described here and deposited them over areas of a few tens of square kilometers. The area of ash fall is considerably larger if multiple fountains erupt along fissures, as is often the case in Hawaii.

Grain Size: Because there is a remarkable variety of pyroclast shapes (especially Pele's hair that may only be 200 μm in diameter but several meters long), grain-size analyses of Hawaiian tephra are nearly impossible. The remarkably delicate nature of these pyroclasts also prevents accurate grain-size measurement. Particle sizes range from fragments of basaltic glass only a few micrometers in diameter to bombs over a meter in diameter that fall around the vents.

Tephra Description: Nearly all the tephra from the Aloi-Alae eruption consist of basaltic glass (sideromelane) with no microlites and only rare olivine phenocrysts. Most pyroclasts have some smooth, curved surfaces from surface tension effects on melt particles from the fountains.

Glass spheres in these tephra deposits are typical of eruptions of basaltic melts with low viscosities. The spheres range in size from a few micrometers to several hundred micrometers. Each sphere has a low bulk density, is highly vesicular, and has numerous (40 to 70% by volume) spherical vesicles.

Many of the angular glass pyroclasts are broken glass spheres or ovoids; therefore, particles from the grain edges of pyroclasts have one smooth, curved surface, whereas interior fragments are remarkably angular. Vesicles are well developed and vesicle walls are thick.

Glass ovoids with pointed tails and "dumbbells" formed by the spinning of melt droplets are transitional to the fibers of Pele's hair. In some cases, these strands are curved back on themselves, forming closed loops.

Pele's hair, in this case, appear to be mostly simple cylindrical strands, a few micrometers to 50 μm in diameter. Some of the coarser hairs have grooved surfaces with furrows located along the strands over parallel vesicles. Vesicles within these strands are highly elongate.

Comments on Pyroclast Formation: All the pyroclasts from this sample of the Aloi-Alae eruption were formed as a spray of basaltic liquid in lava fountains driven by expansion of gases. Viscosities are low, usually less than 10^3 poise. Unbroken particles all have smooth, rounded surfaces formed by surface tension on individual melt droplets. These particles are quenched during cooling by thermal radiation and flow of air over their surfaces before they hit the ground surface. Larger bombs, which land near the vent, are deformed and form spatter ramparts. Finer grained tephra are swept away by wind or ballistically thrown beyond these ramparts.

Some melt particles are "spun" into long, thin, hair-like strands by expanding gases and thermally induced turbulence over the fountain (Pele's hair).

All the basaltic glass particles are delicate and most break upon hitting the ground. Particles with smooth surfaces formed by surface tension are here simply called droplets. A new term *achnelith* (GR., achne = spray) was used by Walker and Croasdale (1972) for this type of pyroclast.

REFERENCES:

Heiken, G., 1974. An atlas of volcanic ash. Smithsonian Contr. Earth Sci., **12**: 101 pp.

Heiken, G. and Lofgren, G., 1971. Terrestrial glass spheres. Geol. Soc. Am. Bull., **82**: 1045-1050.

Richter, D., Eaton, J., Murata, K., Ault, W., and Krivoy, H., 1970. Chronological narrative of the 1959-60 eruption of Kilauea Volcano, Hawaii, U.S. Geol. Surv. Prof. Paper, **537-E**: 73 pp.

Walker, G. P. L. and Croasdale, R., 1972. Characteristics of some basaltic pyroclastics. Bull. Volcanol., **35**: 303-312.

FIGURE 1a. SEM of a 140-μm-diameter spherical pyroclast. The basaltic glass sphere has a smooth surface. Small particles adhering to the surface are partly basaltic glass and partly soluble sublimates from gases driving the eruption.

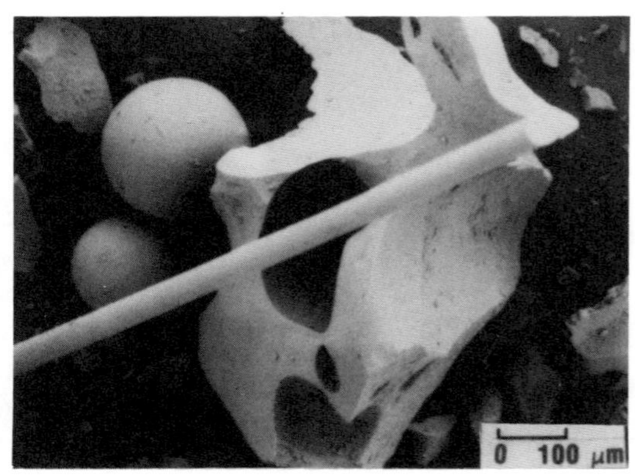

FIGURE 1b. Broken fragments of vesicular droplets, small glass spheres, and a single strand of Pele's hair.

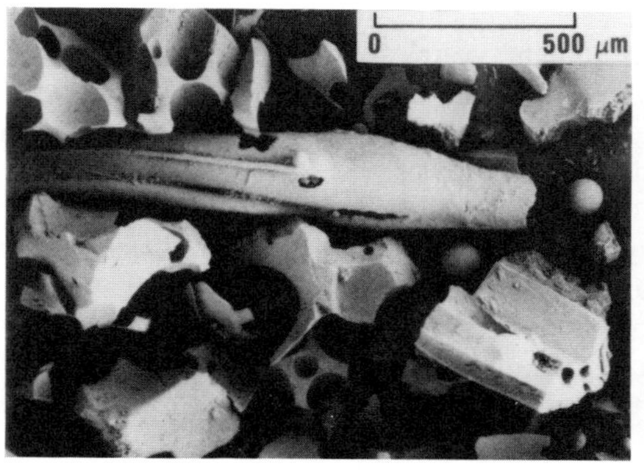

FIGURE 2a. SEM of a typical variety of pyroclasts from this eruption, including spherical droplets, Pele's hair, and broken, angular pyroclasts. The Pele's hair (*center*) has a groove in the surface between two highly elongate vesicles. Angular pyroclasts are mostly interior fragments from broken droplets with large, well-developed vesicles and very thick walls between vesicles.

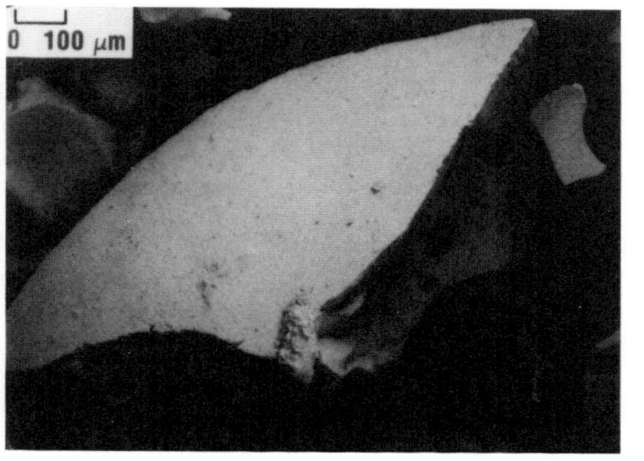

FIGURE 2b. Angular pyroclast; broken droplet fragment with one smooth, curved surface.

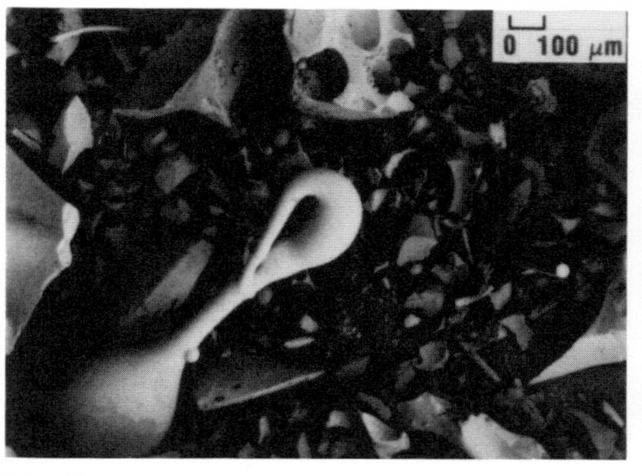

FIGURE 3a. SEM of a representative sample of tephra from Aloi-Alae. Dominating the image is a droplet with an elongate, pointed tail that was bent into a loop and annealed to itself while still molten. A smaller sphere is also welded to this pyroclast; it probably collided with the larger droplet while still partly molten. This particle may represent a transition from spheres and ovoids to strands of Pele's hair. Other particles in the image include vesicular droplet fragments, simple, cylindrical Pele's hair, and spheres.

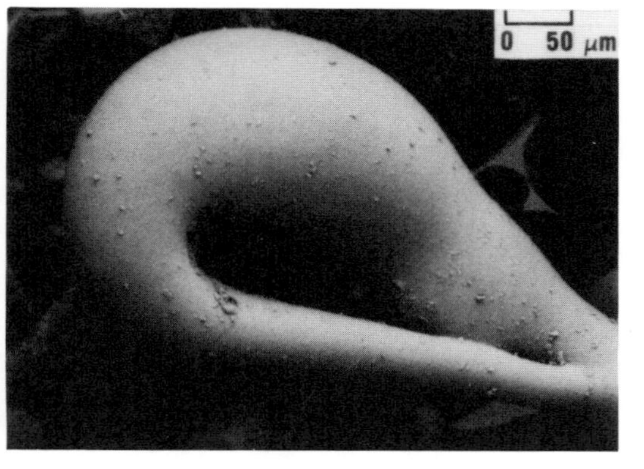

FIGURE 3b. Close-up of the droplet "tail" in *3a*. Bulges in the surface may be over vesicles that continued to expand after the pyroclast's formation.

PELE'S HAIR

Introduction: One of the more intriguing and unusual pyroclastic forms produced during vigorous fountaining of basaltic lava is Pele's hair. Extremely thin (tens to hundreds of micrometers wide) but long (up to several meters) strands are never found complete—they are delicate, breaking up in the wind or when they hit the ground surface. Usually, they are found along a wide zone downwind from the source in deposits consisting of tangled bundles of pale brown fibers.

The only paper dedicated solely to the genesis and characterization of Pele's hair is by Duffield et al. (1977), from which much of the text in this section is condensed.

As discussed in previous sections on basaltic tephra, Pele's hairs are but one pyroclast type formed by the mildly explosive eruption of low-viscosity ($<10^3$ poise) basaltic melt. They are chilled streamers of melt. Pele's hair is, however, also formed in lava cascades and falls (one of our examples is from a 120-m-high lava fall into the Makaopuhi Pit Crater). Strong air currents are induced over hot lava and sweep bits of melt off and away from cascades and falls. Masses of Pele's hair trap small particles and are often coated with sublimates from the fume clouds associated with eruptions.

Tephra Description: Most Pele's hairs are simple cylindrical forms, 1 to 500 μm in diameter. Smaller fibers have no vesicles, but most pyroclasts over 50 μm in diameter are vesicular, with vesicles of high aspect ratio (L:W) parallel to hair axes. Whenever the tips of these hairs are visible, they are tapered. Some forms are twisted into loops and welded to themselves.

Thicker Pele's hairs have multiple, elongate, parallel vesicles; V-shaped, elongate depressions in pyroclast surfaces represent spaces between multiple vesicles. Highly elongate vesicles, circular in cross section, indicate that a melt droplet stretched into a fiber during vesicle growth. Chilling of hair surfaces to form a thin (5- to 10-μm thick) flexible "skin" while the pyroclast interior is still molten may produce Pele's hairs with rumpled surfaces (Heiken, 1972).

Chemistry: Pele's hair is generally associated with low-viscosity, Fe- and Mg-rich melts of basaltic composition (Table 1). The example here, from Duffield et al. (1977), is from the 1969-1971 Mauna Ulu eruption.

TABLE 1. CHEMICAL COMPOSITION OF PELE'S HAIR.*

	(Wt. %)
SiO_2	50.6
Al_2O_3	13.7
Fe_2O_3	0.60
FeO	10.0
MgO	7.6
CaO	11.3
Na_2O	2.2
K_2O	0.51
H_2O^+	0.27
H_2O^-	0.06
TiO_2	2.3
P_2O_5	0.30
MnO	0.18
CO_2	0.02
TOTAL	99.60

*Duffield et al. (1977).

Comments on Pyroclast Formation: Streamers of melt are chilled by thermal radiation from clear, glassy pyroclasts and by air passing over the melt particles; cooling and increased viscosity accompanying the cooling ensures preservation of the thin filaments. Were they not cooled, the streamers of melt would most likely break up into droplets (Isard, 1971).

REFERENCES:

Duffield, W. A., Gibson, Everett K., Jr., and Heiken, G., 1977. Some characteristics of Pele's hair. J. Res. U.S. Geol. Survey, **5**: 93-101.

Heiken, G. H., 1972. Morphology and petrography of volcanic ashes. Geol. Soc. Am. Bull., **83**: 1961-1988.

Isard, J. O., 1971. The formation of spherical glass particles on the lunar surface. Proc. Lunar Sci. Conf., 2nd. Geochim. Cosmochim. Acta, Suppl. 3, **3**: 2003-2008.

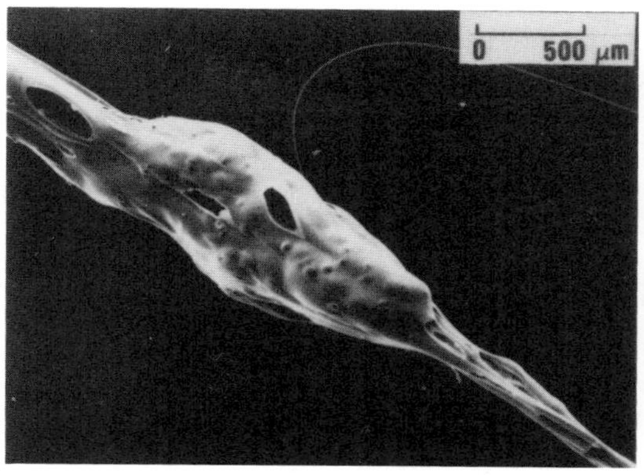

FIGURE 1a. SEM of a Pele's hair from the July 1975 eruption of Mauna Loa, Hawaii. The highly vesicular (50 to 60%), light brown basaltic glass droplet has a smooth surface. Broad, low bulges are over elongate vesicles; some vesicles have ruptured, leaving 100- to 400-μm-long ovoid exterior cavities. Small bulges (30- to 50-μm diameter) are over small olivine and plagioclase phenocrysts, which make up 10% of the glassy pyroclasts.

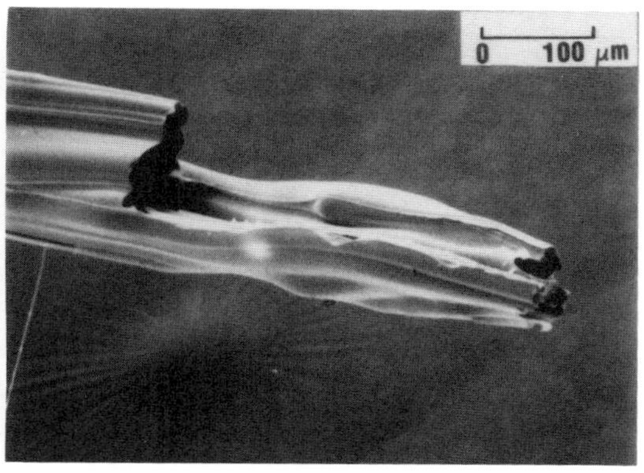

FIGURE 1b. Close up of (*a*); broken tip of the drawn-out, tapered "hair." Vesicles are 3 to 15 μm in diameter; long axes are much greater but not known. Walls between vesicles are thick: usually around 10 μm.

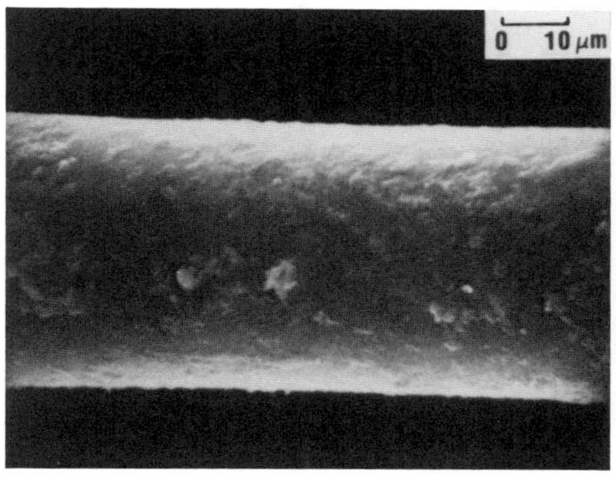

FIGURE 2. SEM of a Pele's hair from Mauna Ulu, Hawaii, October 1, 1970 (collected by W. Duffield). The hair was generated by turbulence over a cascading lava stream. It is a simple, cylindrical Pele's hair coated with patches of sublimates (rough, irregular patches 1 to 2 μm thick). Volatile constituents released upon heating included S, carbonates, sulfates, CO_2, and H_2O (Duffield et al., 1977).

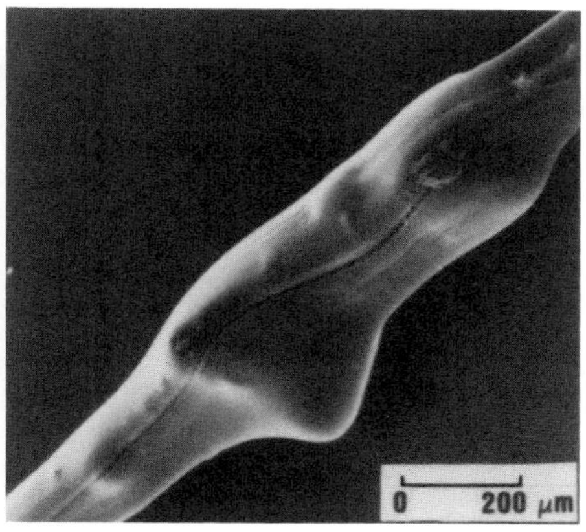

FIGURE 3a. SEM of a Pele's hair from the August 14, 1971 eruption, Kilauea Caldera, Hawaii. Grooves parallel to the pyroclast surface with sharp edges are collapsed vesicles near the pyroclast surface. Note the enormous aspect ratio (L:W) of these vesicles that were stretched during "hair" formation. Large lumps are over phenocrysts (compare with Figs. *4* and *5b*).

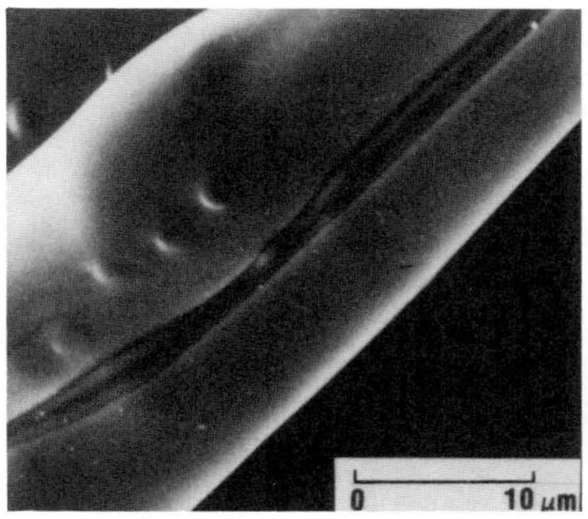

FIGURE 3b. Detail of Pele's hair in Fig. 3a.

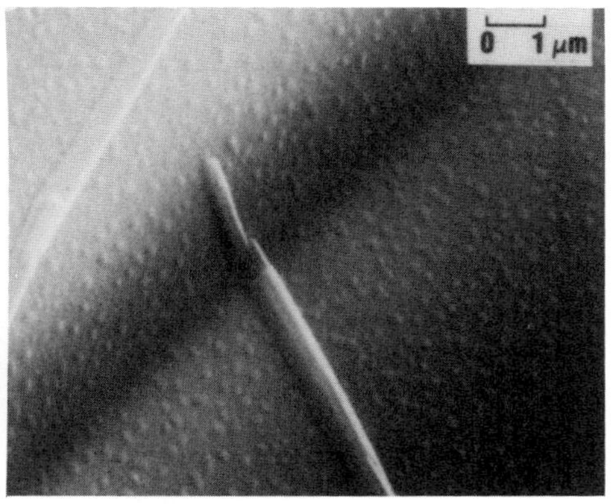

FIGURE 3c. Enlargement of the surface of pyroclast in (*a*), showing 0.5-μm-wide Pele's hair bonded to a larger hair by sublimates. Sublimates are visible as 0.1- to 0.4-μm-diameter mounds.

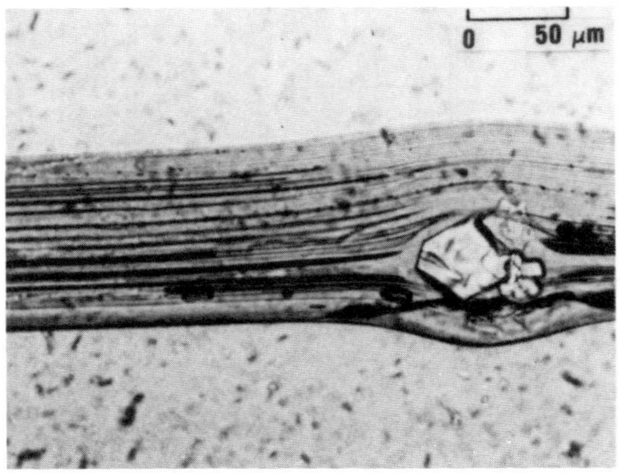

FIGURE 4. Photomicrograph (transmitted light) of a 70-μm-diameter Pele's hair spun from a lava fall by whirlwinds over Makaopuhi Crater, June 11, 1972. The extremely long, very narrow vesicles are similar to those generated in lava fountains and are not similar to most hairs from lava falls. Bulge on the surface is over an olivine phenocryst.

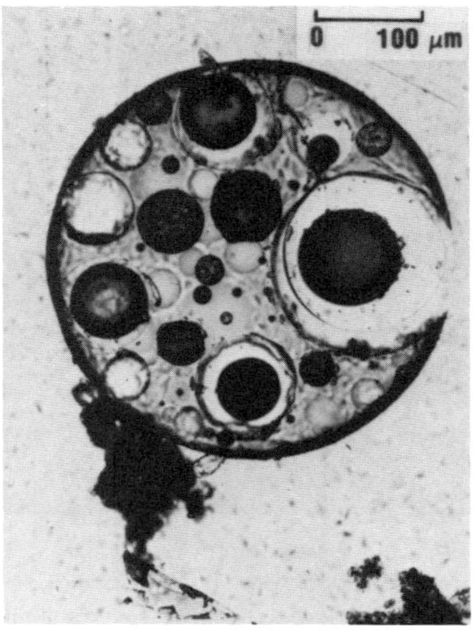

FIGURE 5a. Photomicrograph (transmitted light) of a cross section of Pele's hair from the 1959 eruption of Kilauea Iki, Hawaii. The hair is 460 μm in diameter. Vesicles are circular in cross section only. Note the thick walls of clean basaltic glass (sideromelane) between the vesicle and the thin rim of dark, sheared and jointed glass at surface. Dark patches in the vesicles are grinding compound.

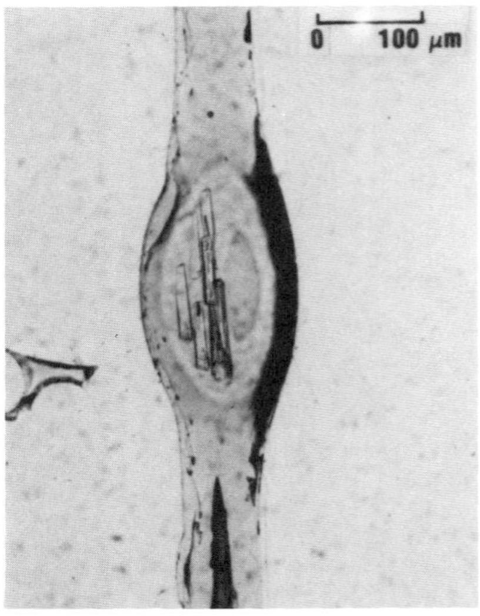

FIGURE 5b. Photomicrograph (transmitted light) of a 50-μm-diameter Pele's hair. Two-thirds of the fiber visible in this photo is vesicle-free; the bulge is over a cluster of small plagioclase phenocrysts.

RETICULITE

Introduction: A very unusual type of pyroclast is reticulite (also known as thread-lace scoria or limu). It is associated with gas-rich basaltic eruptions of lavas that have very low viscosities; there are rare examples of rhyolitic reticulite (S. Self, personal communication). Reticulite is characterized by extreme vesicularity (perhaps as high as 98%; the specific gravity is 0.3). It is formed in lava fountains (pyroclastic) and as thin crusts on shelly pahoehoe flows.

This sample was collected from the "reticulite" bed located at the base of a 2-m-thick sequence of pyroclastic rocks that overlies lava flows on the west rim of Kilauea caldera. Numerous sideromelane droplets (spatter) are also present in the reticulite bed.

Grain Size: Because of the delicate nature of the samples, no attempt was made to sieve them.

Tephra Description: This vitric ash is composed of fragments of sideromelane froth. The fragments are part of a lattice work made up of triangular glass rods. Some thin (<1-μm) glass membranes remain between vesicles, but most of them are broken. These thin, flexible sheets are also present as fragments adhering to the glass rods. However, there is generally a 100-μm to several-millimeter-wide hole between vesicles. These "holes" are rimmed by very sharp, delicate ridges where the thinly stretched membranes had burst or were broken.

Chemical Composition: A chemical analysis of this basaltic reticulite is shown in Table 1.

TABLE 1. CHEMICAL COMPOSITION OF BASALTIC RETICULITE.*

	(Wt. %)
SiO_2	49.29
TiO_2	2.51
Al_2O_3	14.12
Fe_2O_3**	12.79
MnO	0.23
MgO	7.33
CaO	10.44
K_2O	0.35
P_2O_5	0.23
S	0.09
Total	97.38

*XRF analysis by M. Rhodes.
**Fe as Fe_2O_3.

Comments on Pyroclast Formation: Eruption of gas-rich basaltic magma with low viscosity allows maximum development of vesicles. This produces curved, triangular rods at triple junctions (that indicate hexagonal closest packing of vesicles in the melt) and thin membranes between adjacent vesicles. The delicate membranes are mostly broken; they are either burst during the eruption or during weathering or abrasion after deposition.

REFERENCE: Wentworth, C. K., 1938. *Ash Formations of Hawaii*, 3rd Spec. Rpt. of the Hawaiian Volcano Obs.: 183 pp.

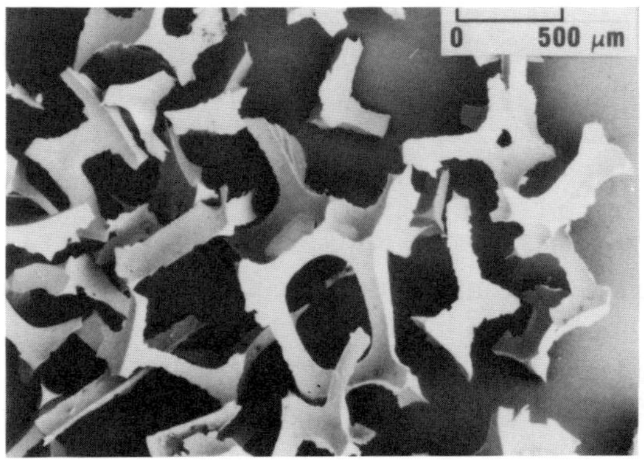

FIGURE 1. Reticulite and reticulite fragments. Triangular rods that make up the lattice are sideromelane and are formed at mutual intersections between three vesicles. Vesicle walls are thin glass membranes (mostly broken after cooling; they now have ragged, sharp edges).

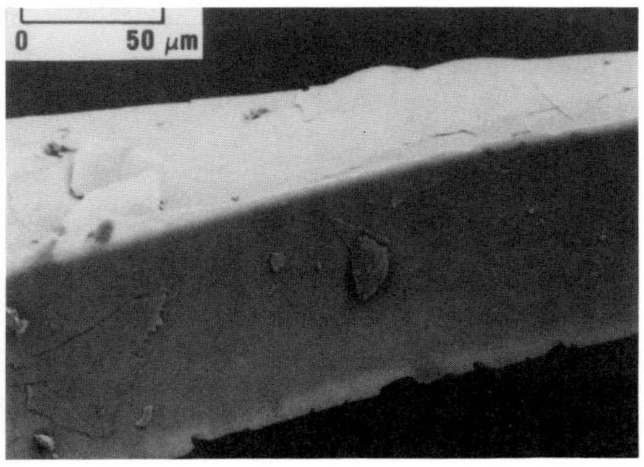

FIGURE 2. Close-up of one of the rods that make up the reticulite lattices. Thin, angular sheets adhering to the rods are pieces of glass membranes that once separated adjacent vesicles.

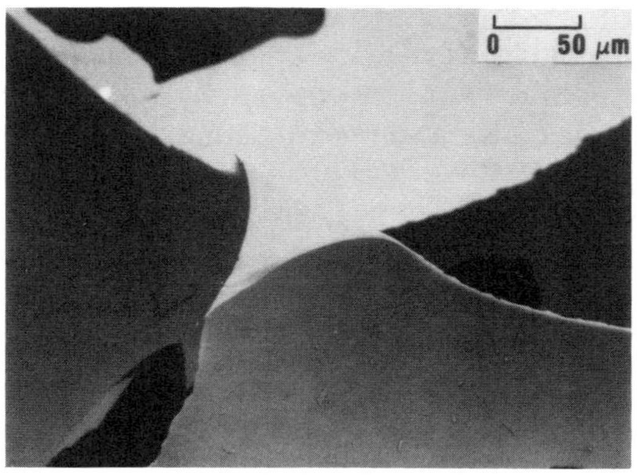

FIGURE 3. Broken rod, showing the typical triangular, curved cross section.

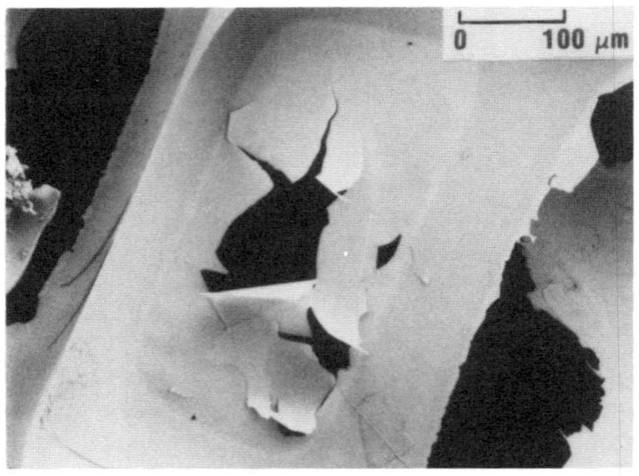

FIGURE 4. Broken glass membrane between adjacent vesicles. Membrane is less than 1 μm thick, quite flexible, but easily torn.

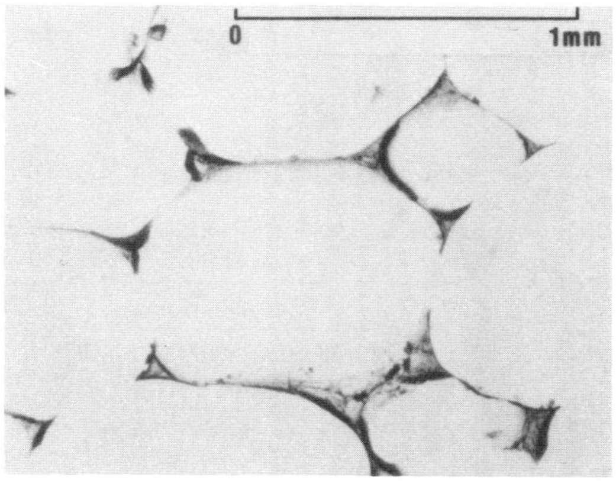

FIGURE 5a. Transmitted-light photomicrographs of reticulite. This thin section demonstrates the remarkable porosity of this tephra. Note the extremely thin vesicle walls.

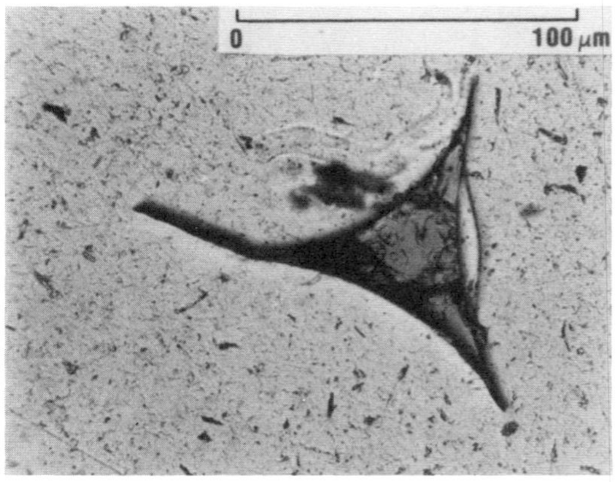

FIGURE 5b. Cross section of a curved, triangular rod. These pass into the extremely thin (<1-μm) basaltic glass membranes that separate adjacent vesicles.

LUNAR PYROCLASTIC ROCKS; "DARK MANTLE DEPOSIT," APOLLO 17 LANDING SITE

Introduction: The "dark mantle deposits" of the Moon have been interpreted as being of pyroclastic origin (e.g., McGetchin and Head, 1973). These deposits are located along the rims of several large mare basins and were sampled during the Apollo 17 mission. In the Valley of Taurus-Littrow (the Apollo 17 site), very dark and orange-tinted regolith samples were evident. Within the rim of Shorty Crater, a 120-m-diameter impact crater, sections of the orange and black clastic deposits were sampled. These are fine-grained clastic rocks consisting of small orange glass droplets and their partly crystalline (quench crystal) equivalent; the partly crystalline droplets are black. These were unlike any regolith soil samples from previous Apollo missions. Although fresh and unaltered, these deposits have an age of about 3.6 billion years and were contemporaneous with interbedded basalt flows at the same site. The samples were collected by Apollo 17 astronauts H. H. Schmitt and E. Cernan.

Grain Size: Mean grain size ranges from 40 to 43 μm (4.64 to 4.53 φ); grains are moderately sorted ($\sigma_\varphi = 1.69$ to 1.58 φ).

Tephra Description: Pyroclasts in these samples consist of spheres, ovoids, and compound droplets. Spheres and ovoids consist of orange glass and partly crystalline orange glass; they have smooth surfaces covered only with sublimates (<1-μm micromounds). Compound forms consist of a central droplet with smaller droplets on the surface that had hit the droplet while still molten. Throughout the stratigraphic sequence sampled by coring, 35 to 74% of the droplets are spheres, 28 to 38% are ovoids, and 4 to 41% have compound shapes. Between 60 and 78% of the droplets are broken

(Heiken and McKay, 1978). Up to 6% of the droplets contain single vesicles; the others are vesicle-free.

The percentage of partly crystalline droplets decreases from 93% near the base to 24% near the top of the 68-cm-long core. Only 0.3 to 2.7% of the droplets contain olivine phenocrysts (composition = Fo_{81}).

Partly crystalline droplets exhibit quench textures. Variations in distinctive olivine quench textures have been used to determine cooling rates and degrees of supercooling of lavas (Donaldson, 1976). Within these droplets there is a progression from the most rapidly to most slowly cooled olivines as follows: (1) branching, dendritic phases, (2) thin, tabular, parallel phases, (3) acicular skeletons and, (4) subequant skeletons.

Chemical Composition: An average chemical analysis of orange glass is shown in Table 1.

TABLE 1. CHEMICAL COMPOSITION OF ORANGE GLASS.*

	Average (Wt. %)	(Variation)
SiO_2	38.63	(0.33)
TiO_2	8.96	(0.47)
Al_2O_3	5.87	(0.36)
Cr_2O_3	0.55	(0.3)
FeO	22.00	(0.32)
MgO	14.79	(1.15)
CaO	7.31	(0.40)
Na_2O	0.37	(0.13)
K_2O	0.08	(0.02)

No. of samples analyzed = 80.
*Microprobe analyses from Reid et al. (1973).

Comments on Pyroclast Formation: Most lunar-sample investigators agree that the ultramafic clastic rocks of the Apollo 17 site are part of a pyroclastic deposit and were formed by lava fountaining. Arguments for and against this hypothesis are summarized by Heiken et al. (1974). Glass droplets from this site are chemically and texturally homogeneous; *none* contain shock-damaged crystals and heterogeneous glass that are typical of glass clasts formed by meteorite impact. Sulfur-rich coatings on pyroclast surfaces appear to have been deposited as sublimates from the volcanic gases that drove the explosive eruptions (Butler and Meyer, 1976).

It is probable that the magma involved in lava fountaining had a very low viscosity; a calculated viscosity of this magma, at a temperature of 1250°C, is 3.5 poise (using the method of Shaw, 1972), which is lower than any determined for Hawaiian eruptions. Such low viscosities would account for the abundance of fine-grained spherical droplets and a dearth of vesicles. Lack of vesicles may be the result of volatiles escaping from a low-viscosity fluid into the lunar vacuum.

Variations in droplet morphology can be explained by variations in the lava fountaining. Fountains with a narrow dispersion angle will contain many spray droplets that fall back into the fountain and subsequently form compound droplets. These droplets remain hot enough for quench crystals to grow. In a broadly dispersed fountain, individual droplets are thrown clear, cool rapidly, and hit the lunar surface as simple glass spheres.

REFERENCES:

Butler Jr., P. and Meyer Jr., C., 1976. Sulfur prevails in coatings on glass droplets: Apollo 15 green and brown glasses and Apollo 17 orange and black (devitrified) glasses. Proc. Lunar Sci. Conf., **7th**: 1561-1581.

Donaldson, C., 1976. An experimental investigation of olivine morphology. Contrib. Mineral. Petrol., **57**: 187-213.

Heiken, G., McKay, D., and Brown, R., 1974. Lunar deposits of possible pyroclastic origin. Geochim. Cosmochim. Acta, **38**: 1703-1718.

Heiken, G. and McKay, D., 1977. A model for eruption behavior of a volcanic vent in eastern Mare Serenitatis. Proc. Lunar Sci. Conf., **8th**: 3243-3253.

Heiken, G. and McKay, D., 1978. Petrology of a sequence of pyroclastic rocks from the Valley of Taurus-Littrow (Apollo 17 landing site). Proc. Lunar Sci. Conf., **9th**: 1933-1943.

McGetchin, T. R. and Head, J. W., 1973. Lunar cinder cones. Science, **180**: 68-70.

Reid, A., Ridley, W., Donaldson, C., and Brown, R., 1973. Glass compositions in the orange and gray soils from Shorty Crater, Apollo 17. Trans. Am. Geophys. Union, **54**: 607-609.

Shaw, H., 1972. Viscosities of magmatic silicate liquids: an empirical method of prediction. Am. J. Sci., **272**: 870-893.

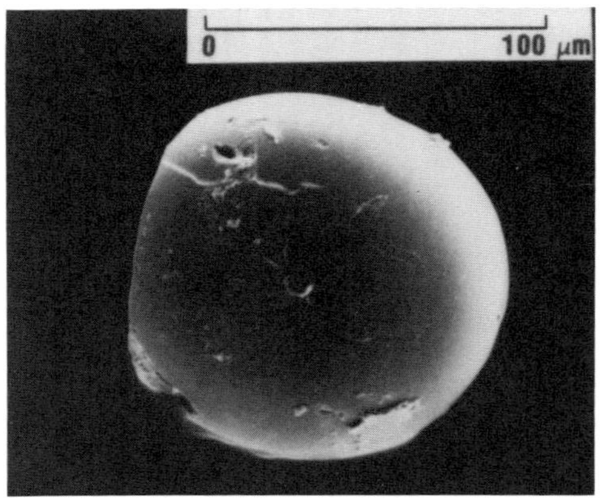

FIGURE 1. SEM of an orange glass sphere. The smooth surface is marred only by spalls and chips caused by low-velocity impacts, perhaps during collision with other particles or when hitting the lunar surface. It is flattened on one side, perhaps upon collision with another particle or the lunar surface.

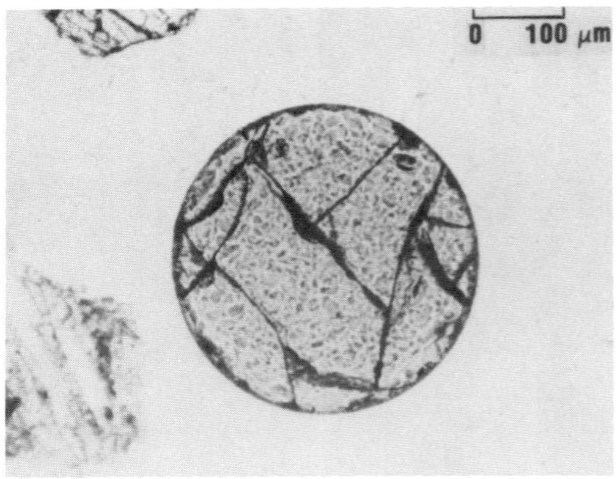

FIGURE 2. Transmitted-light photomicrograph of a 300-μm-diameter orange glass sphere. The glass is homogeneous; fracturing occurred when the thin-section was made.

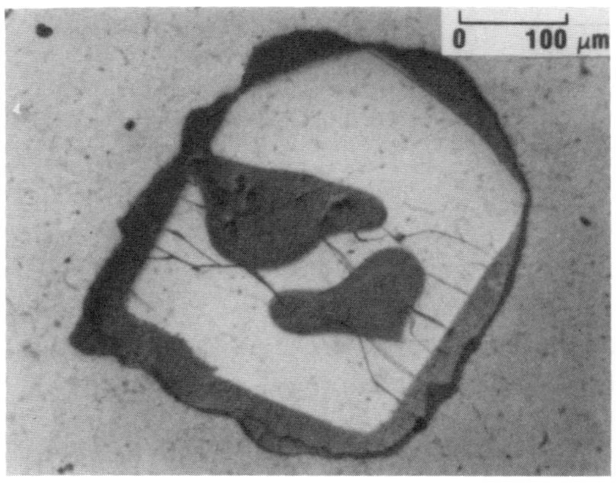

FIGURE 3. Reflected-light photomicrograph of a partly crystalline compound droplet containing an olivine phenocryst (composition = $Fo_{80\text{-}81}$). There are 0.3 to 2.7% olivine phenocrysts in these samples. Note radial growth of dendritic quench-crystal olivine and ilmenite phases from the phenocryst surface.

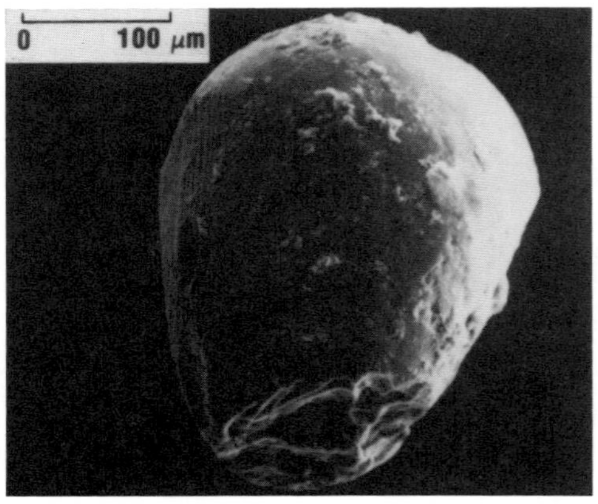

FIGURE 4. SEM of a partly crystalline broken droplet that consists of parallel growths of ilmenite and olivine in orange glass. Parallel lines near the bottom of the droplet are ilmenite phases intersecting the pyroclast surface. Grainy coatings consist of a mixture of sublimates (vapor condensates) and flattened glass particles of the same composition. From Heiken et al. (1974).

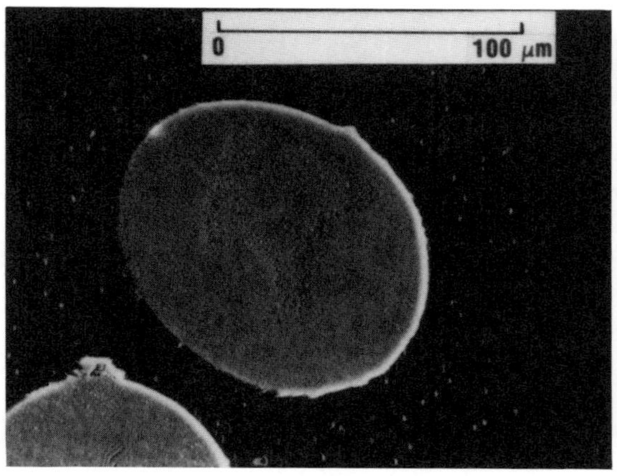

FIGURE 5a. SEM of a polished, etched thin section of a partly crystalline orange glass droplet. Clumps of finely crystalline dendrites (most phases are less than 0.5 μm long) consist of intergrown, feathery olivine and ilmenite phases in an orange glass matrix. The central area of this droplet is somewhat less crowded with dendrites.

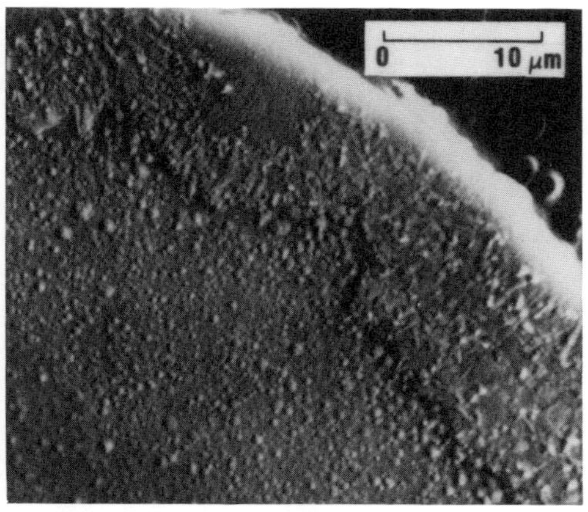

FIGURE 5b. SEM of a polished, etched thin section of a partly crystalline orange glass droplet. Feathery dendrites have grown toward the center of the sphere from the droplet surface to form an irregular, 6- to 15-μm-thick "front." Darker gray phases are olivines and the brighter phases are ilmenite. Mounds in the center are 0.2- to 0.5-μm-diameter dendrites separated by orange glass (low relief).

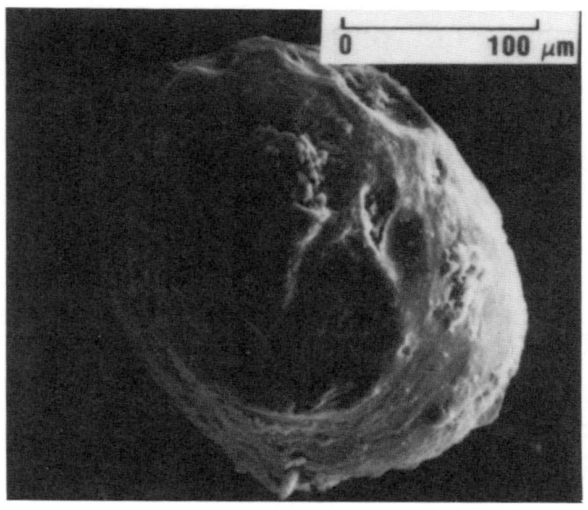

FIGURE 6. SEM of a compound droplet that consists of a partly crystalline, ovoid core with multiple droplets adhering to it. Most of the smaller droplets were flattened when they collided with the larger droplet. From Heiken and McKay (1977).

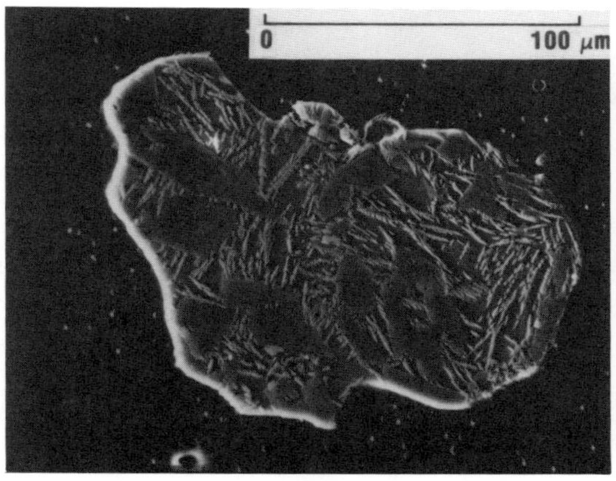

FIGURE 7a. SEM of polished, etched thin sections of compound, partly crystalline droplets. Irregular, broken droplet attached to a sphere (lower). The high-relief, low-contrast crystals are subequant, skeletal olivines (Fo_{76-78}). Ilmenite is the bright, feathery phase and Cr-spinel is the rare, bright, rectangular phase. Areas of low relief are orange glass. From Heiken and McKay (1977).

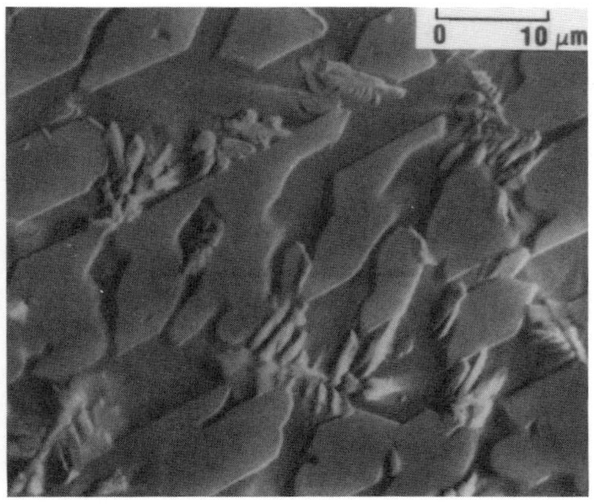

FIGURE 7b. Detail of an ovoid droplet. The ladder-like phases are acicular skeletal olivines (Fo_{70-78}). Areas of low relief are orange glass and the bright, feathery phases are ilmenite. From Heiken and McKay (1977).

Strombolian Eruptions

Stromboli Volcano, called the "lighthouse of the Mediterranean," is characterized by nearly continuous activity. Typical Strombolian activity consists of explosive bursts of solidified and partly solidified ash and scoria. These pyroclasts fall around the crater, building a cinder (scoria) cone, the most common pyroclastic volcanic landform on Earth.

CINDER CONE—LASSEN VOLCANIC NATIONAL PARK, CALIFORNIA

Introduction: Strombolian activity is responsible for construction of cinder cones (also known as scoria cones). The cone-building process is one mainly of ballistic deposition of coarse scoria and ash followed by slumping; some of the finer grained tephra is carried away by the wind (Fries, 1953; McGetchin et al., 1974; and Self et al., 1974). The example used here, Cinder Cone (the geographic name), is located in the northeast corner of Lassen Volcanic National Park, California, on the flank of Prospect Peak. Several periods of activity at Cinder Cone occurred during the last 400 years, and the latest eruption was in 1851 A.D. Eruptions of basalt produced several block lava flows, a cinder cone with two concentric craters (the cone makes up 54% of the volume of tephra produced during these eruptions), and a tephra blanket (scoria fall deposit). Total volume of Cinder Cone tephra is 70×10^6 m^3. Most of the samples studied from this eruption sequence were collected from the tephra deposit surrounding Cinder Cone (Heiken, 1978).

The major tephra units associated with cone-building activity are described in Figures 1 and 2. The lowest unit (Unit 1) consists of poorly bedded, black to dark brown, medium- to coarse-grained vitric ash. Unit 2 is coarsely bedded and composed of highly vesicular, light brown sideromelane bombs and lapilli, which are interbedded with thin layers of dark gray coarse ash. Unit 3 is the most voluminous deposit and makes up 71% of the tephra blanket. It occupies an 18- by 10-km, NNE-trending oval. It consists of graded beds of gray and brown coarse ash. Nearly all beds in Unit 3 are normally graded from vesicular sideromelane lapilli upward into slightly finer grained gray ash.

Grain Size: All samples used for size analysis were collected 1 km SSE of Cinder Cone. Mean grain size ranges from 0.3 φ (0.8 mm = coarse ash) to -1.53 φ (3.1 mm = lapilli) and are moderately to poorly sorted. In plots of mean grain size versus sorting, all of the samples are within the Strombolian "scoria fall" category of Self et al. (1974).

Tephra Description: Although they grade into one another, three categories of pyroclast types occur in this deposit (Table 1).

TABLE 1. PYROCLAST TYPES FROM CINDER CONE.

Cinder Cone Pyroclast Types (%)	Unit 1	Unit 2	Unit 3 (lower)	Unit 3 (upper)
Light brown sideromelane	28.2	47.5	37.0	25.5
Medium brown sideromelane	38.6	28.0	45.0	46.3
Tachylite	32.2	22.0	14.0	25.9
Phenocrysts				
Olivine	0.5	2.5	2.5	2.3
Plagioclase	---	---	0.5	---
Quartz xenocrysts	---	---	1.0	---
Andesite fragments	0.5	---	---	---

Light Brown Sideromelane. These smooth-skinned, light-brown, vesicular sideromelane droplets have vesicularities ranging from 20 to 70%. There are numerous coalesced vesicles, which form irregular, partly open frameworks. Most vesicles are deformed ovoids 20 to 500 μm long. Sideromelane pyroclasts range from highly vesicular, delicate small bombs with low bulk densities to fine-grained ash consisting of broken droplets. There are 2 to 5% olivine phenocrysts and traces of plagioclase laths and quartz xenocrysts within the coarser pyroclasts.

Medium Brown Sideromelane Pyroclasts. These pyroclasts are generally less vesicular (30 to 40%) and exhibit a greater degree of crystallization than do the light brown pyroclasts. There are fewer rounded, glassy surfaces and more rough surfaces caused by fragmentation of partly crystalline vesicular pyroclasts. The groundmass ranges from glass with a few 4- to 20-μm-long feldspar laths to dense networks of finely crystalline phases with <50% glass.

Tachylite Pyroclasts. Completely crystalline, these pyroclasts exhibit a diktytaxitic texture that consists of a network of finely crystalline, glass-coated feldspar and pyroxene crystals separated by irregular vesicles. Pyroclast surfaces are very rough, reflecting the crystalline matrix.

Relative portions of these phenocryst types vary throughout the stratigraphic section. The significance of these variations will be discussed in the following section.

Chemical Composition: Chemical compositions shown in Table 2 represent two types of material: (1) composition of glass between quench crystals in tachylite, and (2) composition of glass in a sideromelane pyroclast.

TABLE 2. CHEMICAL COMPOSITION OF GLASS FROM CINDER CONE.*

	1** (Wt. %)	2*** (Wt. %)
SiO_2	52.95	53.19
TiO_2	1.15	1.00
Al_2O_3	15.95	15.40
FeO	8.37	7.63
MgO	3.95	4.12
CaO	6.72	7.99
Na_2O	4.28	3.91
K_2O	1.13	0.77
P_2O_5	0.16	0.20
Total	94.67	94.20

*Analyses are by Cameca electron microprobe.
**Between quench crystals in a tachylite pyroclast.
***A sideromelane pyroclast.

Comments on Pyroclast Formation: The tephra sequence at Cinder Cone is similar to those described by McGetchin et al. (1974) for Strombolian eruptions at Northeast Crater, Etna, Italy, and at Heimaey, Iceland (Self et al., 1974). MacDonald (1972) describes Strombolian activity as "weak-to-violent ejection of partly fluid blobs." According to McGetchin et al. (1974), clots of partly fluid lava and ash are produced when jets of melt, accelerated by gas release, break up into clusters of pyroclasts.

The type of pyroclast dominating any particular tephra layer appears to have been controlled by clogging of the vent and the lava flux. If the vent were clear and the lava flux were high, an abundant supply of lava at high temperature and lower viscosity could produce highly vesicular, light brown sideromelane pyroclasts typical of Unit 2. If the vent were blocked, either by pooling of lava or slumping of talus into the vent, explosive activity would be suppressed. During periods when the vent was blocked, Strombolian activity could have occured with intermittent, small tephra eruptions, producing tephra similar to those in Units 1 and 3; lava within the crater could have cooled and lost some of its volatile phases. If the crater were blocked by cooling lava, intermittent gas release could disrupt it, breaking the partly molten mass into poorly vesicular tachylite pyroclasts mixed with sideromelane droplets.

REFERENCES:

Fries, Jr., C., 1953. Volumes and weights of pyroclastic material, lava, and water erupted by Paricutin Volcano, Michoácan, Mexico. Trans. Amer. Geophys. Union., **34**: 603-616.

Heiken, G., 1978. Characteristics of tephra from Cinder Cone, Lassen Volcanic National Park, California. Bull. Volcanol. **41**: 119-130.

MacDonald, G. A., 1972. Volcanoes. Prentice-Hall, Englewood Cliffs, N. J.: 510 pp.

McGetchin, T. R., Settle, M., and Chouet, B. A., 1974. Cinder cone growth modeled after Northeast Crater, Mount Etna, Sicily. J. Geophys. Res., **79**: 3257-3272.

Self, S., Sparks, R. S. J., Booth, B., and Walker, G. P. L., 1974. The 1973 Heimaey Strombolian scoria deposit, Iceland. Geol. Mag., **111**: 534-548.

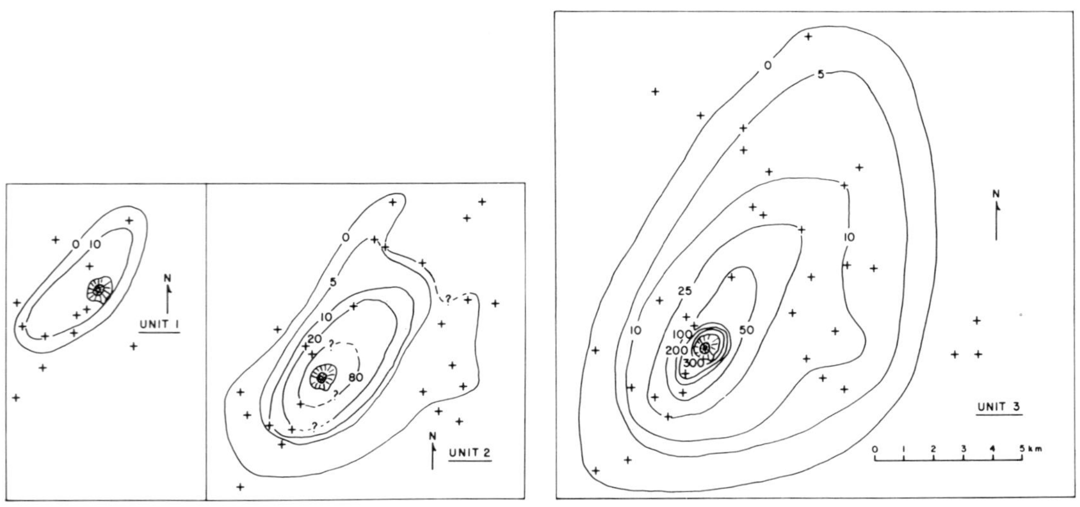

FIGURE 1. Isopach maps of Cinder Cone scoria fall Units 1, 2, and 3. Thicknesses in centimeters.

FIGURE 2a. Tachylite pyroclasts, some of the most common particle types in this unit, are irregular, blocky, and vesicle-poor. Pyroclast surface: coalesced, irregular vesicles are common. Diktytaxitic textures result in rough grain surfaces.

FIGURE 2b. Detail of (*a*). Framework of 10- to 80-μm plagioclase and pyroxene laths surrounded by glass.

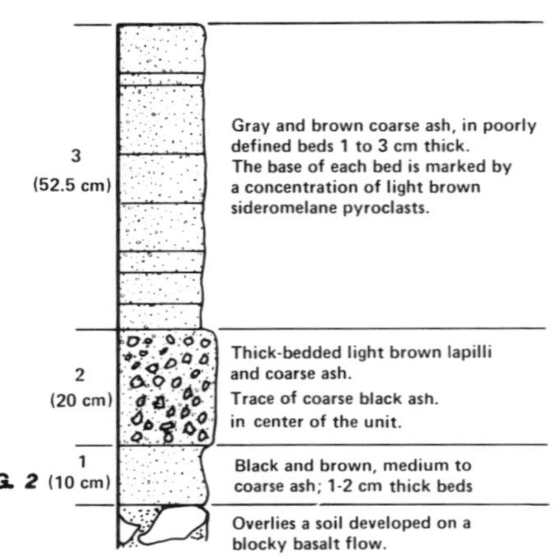

FIG. 2 Tephra sequence at Cinder Cone, 0.5 km N of the cone.

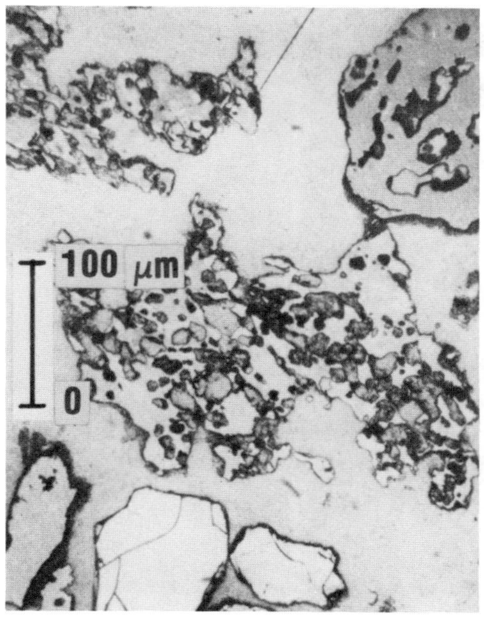

FIGURE 3a. Unit 2 is characterized by mostly light brown, highly vesicular sideromelane pyroclasts. Reflected-light photomicrograph of a representative ash sample from Unit 2. *Center* - 0.32-mm-long, highly vesicular sideromelane pyroclast. Note the irregular, delicate, glassy framework and smooth vesicle and grain surfaces. Other pyroclasts include similar sideromelane droplets, tachylite, and olivine phenocrysts.

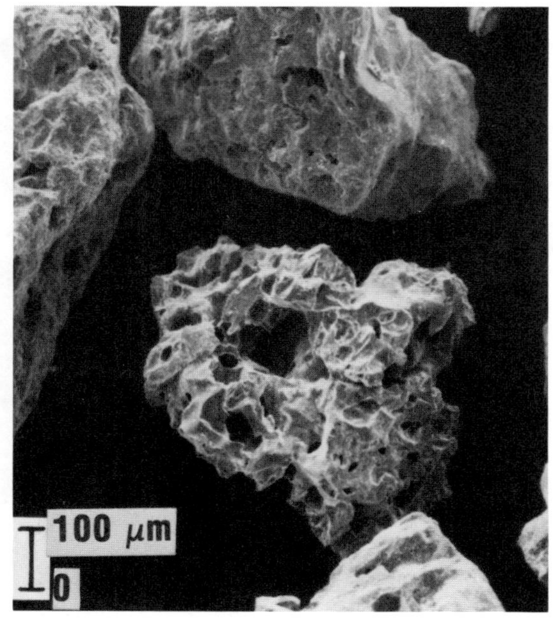

FIGURE 3b. SEM of ash from Unit 2. *Center* - subequant sideromelane pyroclast with smooth, glassy surfaces. Highly vesicular, with 20- to 150-μm-diameter vesicles. Other pyroclasts are mostly tachylite with low vesicularities.

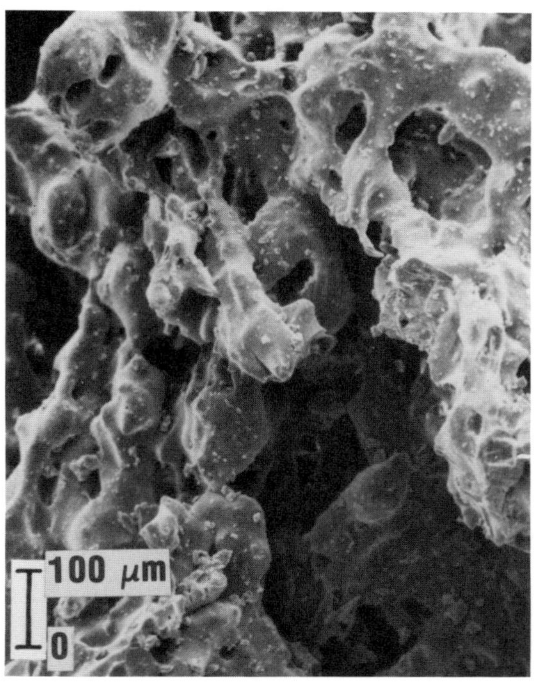

FIGURE 3c. SEM of surface of a light brown sideromelane pyroclast. Traces of small, angular particles are welded to the pyroclast surface.

FIG. 3

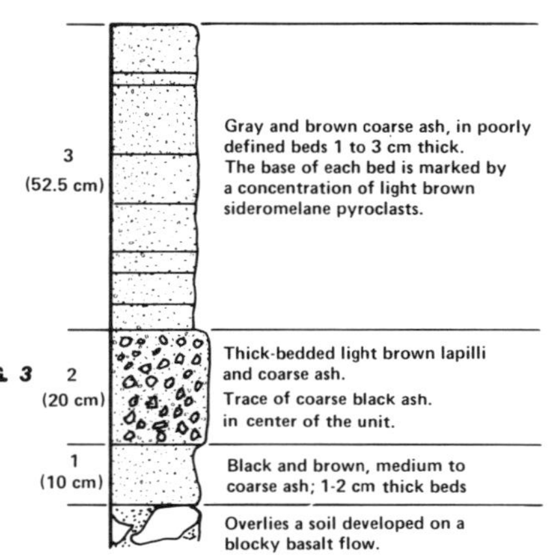

Tephra sequence at Cinder Cone, 0.5 km N of the cone.

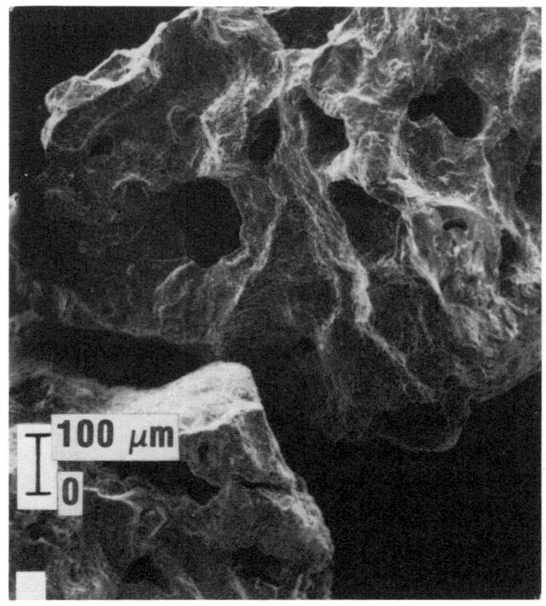

FIGURE 4a. Medium brown sideromelane and tachylite pyroclasts. SEM of a pyroclast with partly crystalline groundmass. It has a low vesicularity but larger vesicles than those in the light brown sideromelane pyroclasts.

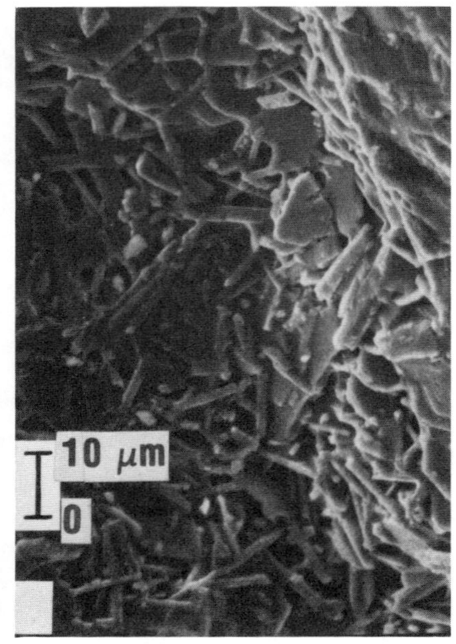

FIGURE 4b. Detail of grain surface in (*a*). The surface is hyalocrystalline and consists of a network of 5- to 15-μm-long plagioclase and pyroxene laths in a glassy matrix.

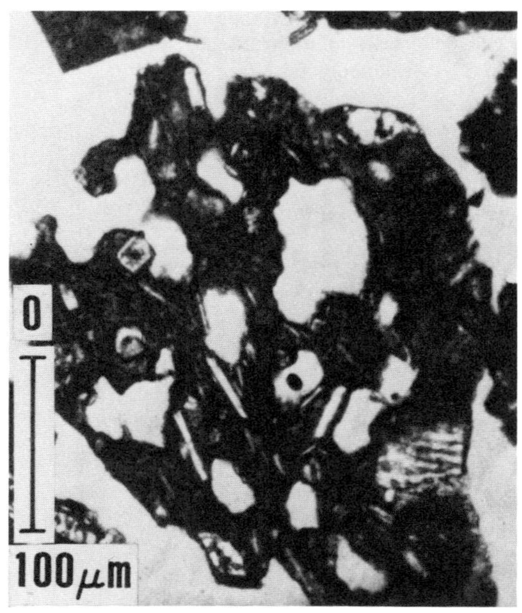

FIGURE 4c. Transmitted-light photomicrograph of a pyroclast similar to the one in (*a*) and (*b*), showing the well-developed, coalesced vesicles.

FIG. 4

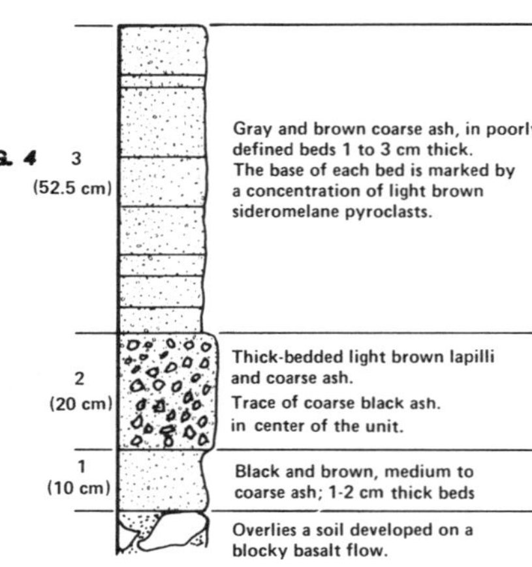

Tephra sequence at Cinder Cone, 0.5 km N of the cone.

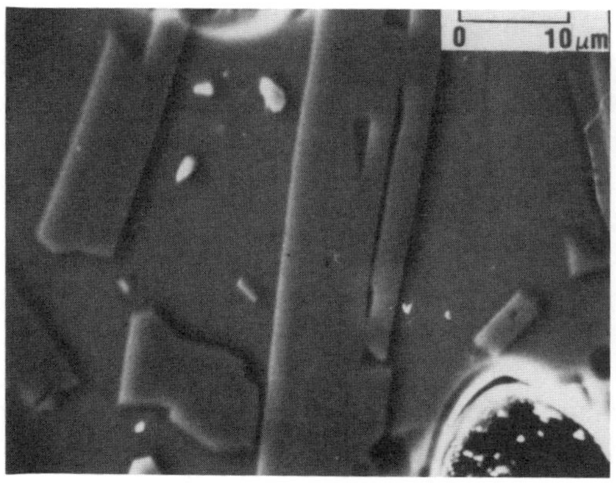

FIGURE 5a. Comparison of sideromelane and tachylite textures. SEM images of ion-etched polished thin sections across pyroclasts. Sideromelane—plagioclase and pyroxene laths, exhibiting quench textures (skeletal forms) are the high-relief phases. The smooth, low-relief phase is basaltic glass.

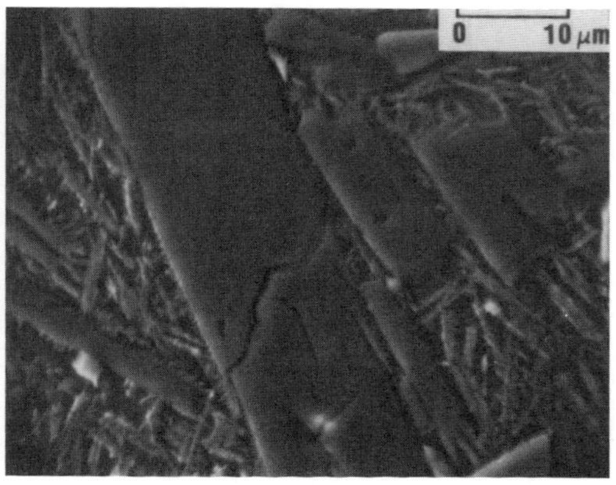

FIGURE 5b. Tachylite—well-developed pyroxene and plagioclase laths surrounded by dendritic growths of 0.2- to 2-µm-wide pyroxene and Fe-Ti oxides. There is a trace of glass (low-relief phase) between dendritic phases.

CERRO NEGRO, NICARAGUA; 1968 ERUPTION

Introduction: Cerro Negro is one of four clustered cinder cones located along a N-S fault line in the central Marabios Range, Nicaragua. Strombolian eruptions are characteristic of this cone, which has been active intermittently since 1850; the most recent eruptions were in 1968, 1969, and 1971. The 1968 eruption ejected abundant blocks and ash, and an ash cloud rose to several thousand meters above the central crater. At the same time, there was lava fountaining on the flank of the cone. Lavas and ash from this eruption are all olivine basalts. This sample was provided by W. I. Rose, Jr.

Grain Size: A weighted average of grain-size estimates was made by Rose et al. (1973). The ash has a mean grain size of 180 µm and is very poorly sorted.

Tephra Description: The vitric ash consists of mostly broken sideromelane droplets and tachylite pyroclasts that contain abundant augite, olivine, and plagioclase phenocrysts. Sideromelane droplets have smooth, botryoidal or lumpy surfaces that are broken by depressions of outermost vesicles. There are cavities (broken vesicles) with both smooth and sharp, irregular rims. Tachylite pyroclasts are less vesicular and have irregular grain surfaces over the outermost groundmass minerals.

Chemical Composition: Chemical analyses of Cerro Negro basalts are given in Table 1.

TABLE 1. CHEMICAL ANALYSES OF OLIVINE BASALT FROM CERRO NEGRO, 1968 ERUPTION.

	Ash* (Wt. %)	Lava** (Wt. %)
SiO_2	49.20	49.50
TiO_2	1.28	0.80
Al_2O_3	19.38	19.50
Fe_2O_3	4.42	4.60
FeO	5.55	4.20
MnO	0.14	n.d.
MgO	5.41	7.25
CaO	11.89	11.50
Na_2O	1.76	2.35
K_2O	0.85	0.30
P_2O_5	0.05	0.19
TOTAL	99.93	100.1

*Analyzed by M. Rhodes.
**Analyzed by J. G. Viramonte (Viramonte and DiScala, 1970).

Comments on Pyroclast Formation: Pyroclast shapes are related to vesicle growth within droplets that are formed as a spray in lava fountains. Viscosities are higher than those of lavas erupted in Hawaiian lava fountains. There are no vesicular spheres or ovoids; instead, there are highly irregular droplets with smooth, glassy surfaces. There are some surface tension effects on outer pyroclast surfaces (sideromelane scoria); some of the outermost depressions with smooth, curved rims were formed when vesicles broke while the pyroclast was still molten. Broken droplets with some angular, sharp edges were formed when chilled solid pyroclasts hit the ground or other pyroclasts. Some may have been broken by rupture of overpressured vesicles within chilled, brittle, vesicular droplets.

Tachylite fragments typically have hyalocrystalline groundmass textures. They may have formed by the disruption of crusts on lava lakes that are present between explosive bursts, or they may have been introduced as accidental debris from the slumping of crater walls.

REFERENCES:

Rose, W. I., Jr., Bonis, S., Stoiber, R. E., Keller, M., and Bickford, T., 1973. Studies of volcanic ash from two recent Central American eruptions. Bull. Volcanol., **37**: 338-364.

Viramonte, J. G. and DiScala, L., 1970. Summary of the 1968 eruption of Cerro Negro, Nicaragua. Bull. Volcanol., **34**: 347-351.

FIGURE 1a. Sideromelane scoria lapilli. The unbroken 800- by 700-μm scoria fragment is heavily coated with finer grained ash particles. This scoria pyroclast may be heterogeneous, consisting partly of basaltic glass (sideromelane) and partly of hyalocrystalline groundmass (tachylite).

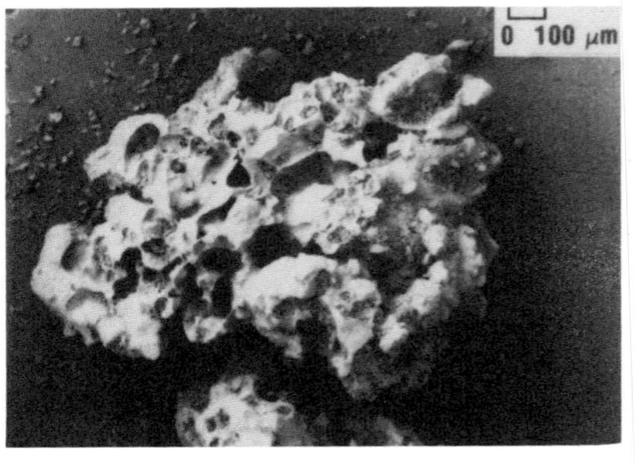

FIGURE 1b. A 1.1- by 0.8-mm sideromelane scoria pyroclast (medium to dark brown). As can be seen in the broken face of this pyroclast, all irregular projections are associated with vesicles or clusters of vesicles. Unbroken surfaces are smooth glass (where not coated with finer grained pyroclast fragments). Vesicles range in length from 10 to 230 μm; most are irregular ovoids deformed at contacts with adjacent vesicles. Highly variable vesicle-wall thicknesses also show the heterogeneity of vesicle size and shape.

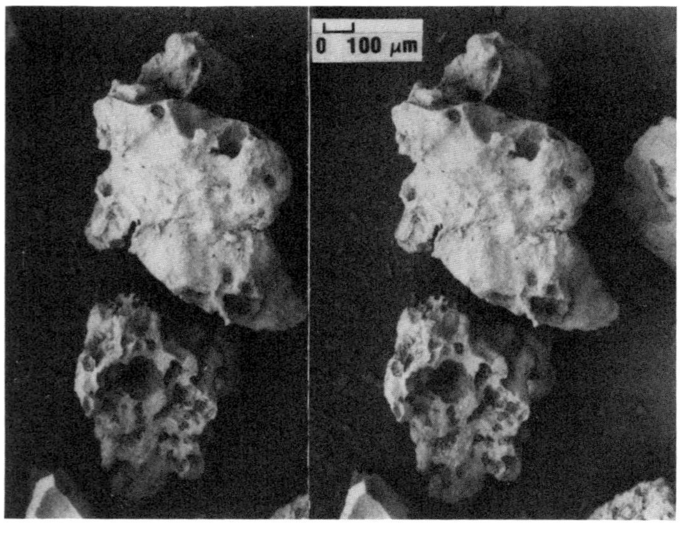

FIGURE 2. Stereophotomicrograph of two broken scoria lapilli. Note the heterogeneity of vesicle size and shape and the lack of deformation by flow (no highly elongate vesicles). The upper lapillus is tachylitic and the lower is composed of mostly sideromelane.

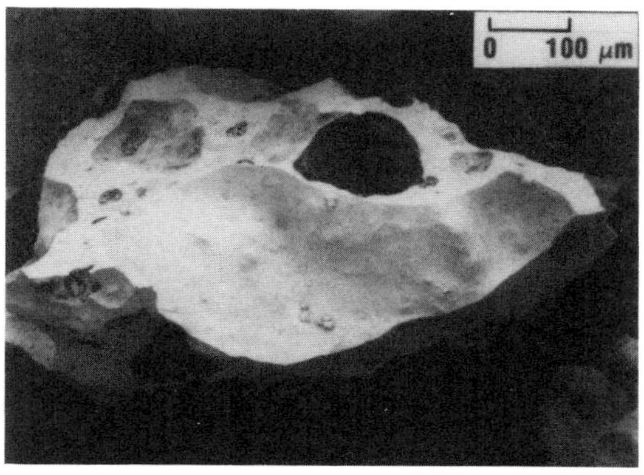

FIGURE 3. Large shard broken from a scoria lapillus; all depressions are vesicle remnants.

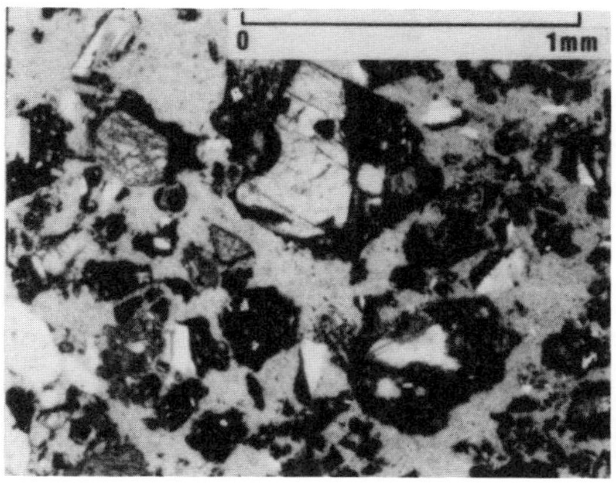

FIGURE 4a. Photomicrograph of tephra from Cerro Negro. Partly plane-polarized transmitted light. Irregular, blocky, phenocryst-rich tachylite pyroclasts make up most of the grains visible here (black, opaque pyroclasts). Some hyalocrystalline sideromelane pyroclasts are present in the lower left corner of the photograph.

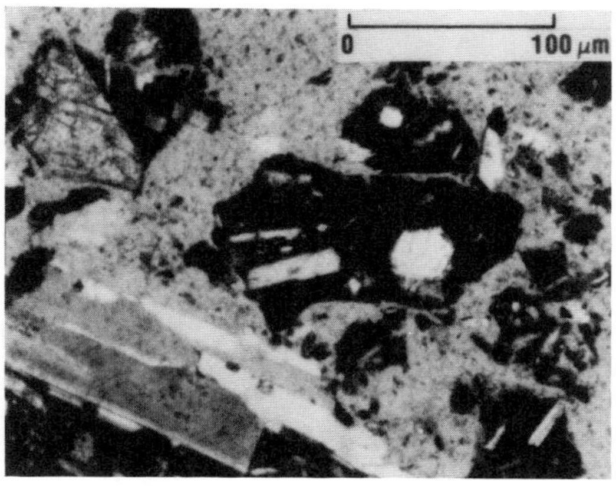

FIGURE 4b. All pyroclasts visible here are irregular, blocky tachylite grains. All are phenocryst-rich, mostly plagioclase and clinopyroxene.

PACAYA, GUATEMALA; 1965-1969 ERUPTION

Introduction: Pacaya is a composite cinder cone located 40 km SSW of Guatemala City. It lies within the crater of an older composite cone (Cerro Chino), which is, in turn, located on the southern rim of Amatitlan caldera (Mooser et al., 1958). Pacaya was formed during recent intermittent Strombolian activity. This particular eruptive phase was most active from 1965 to 1969 and has continued intermittently.

Grain Size: These samples, collected at a distance of 1 km west of the summit vent, have a median grain size of 340 μm (1.57 φ) and are moderately sorted ($\sigma_\varphi = 1.3$ φ).

Tephra Description: In this deposit, 97% of the pyroclasts consist of sideromelane (basaltic glass) droplets or droplet fragments. The pyroclasts are irregularly shaped, yet smooth skinned; the skin is broken by depressions over the outermost vesicles. Vesicles are large (30- to >100-μm) ovoids and spheres, and they have thin walls. Some pyroclasts contain phenocrysts of plagioclase (An_{70}), olivine, and clinopyroxene. Phenocrysts partly coated with basaltic glass are present as individual pyroclasts.

Chemical Composition: The chemical composition of Pacaya lava is shown in Table 1.

TABLE 1. CHEMICAL COMPOSITION OF HIGH-ALUMINA BASALT FROM PACAYA.*

	(Wt. %)
SiO_2	51.80
TiO_2	1.15
Al_2O_3	17.73
Fe_2O_3	2.96
FeO	6.49
MnO	0.22
MgO	4.03
CaO	9.52
Na_2O	4.05
P_2O_5	0.26
S	0.06
TOTAL	99.1

*M. Rhodes, analyst (from Heiken, 1972).

Comments on Pyroclast Formation: Intermittent Strombolian activity produced spray bursts consisting of small molten droplets and bombs. Each droplet vesiculated rapidly before quenching and is irregular because of heterogeneous vesicle growth. Viscosity before quenching was low enough to allow smoothing of pyroclast surfaces, but high enough to prevent formation of vesicular glass spheres. Most of the pyroclasts formed in this way are very delicate and break upon impact with one another or the ground. Some may have broken during rupture of pressurized vesicles, but this is difficult to prove. Slightly curved to curved thin-glass shards are vesicle-wall fragments from broken basaltic lapilli and coarse ash-size droplets.

REFERENCES: Heiken, G., 1972. Morphology and petrography of volcanic ashes. Geol. Soc. Am. Bull., **83**: 1961-1988.

Mooser, F., Meyer-Abich, H., and McBirney, A., 1958. *Catalogue of the active volcanoes and solfatara fields of Central America. Part VI of the Catalogue of Active Volcanoes of the World.* Int'l. Volcanol. Assoc., Napoli: 146 pp.

FIGURE 1. Basaltic tephra from Strombolian eruptions at Pacaya, Guatemala. Nearly all the pyroclasts visible here are light brown (in thin section) sideromelane droplets. Megascopically, these tephra are dark brown. Most are characterized by high vesicularity, smooth surfaces (except where there are lumps caused by underlying vesicles and phenocrysts), and some breakage.

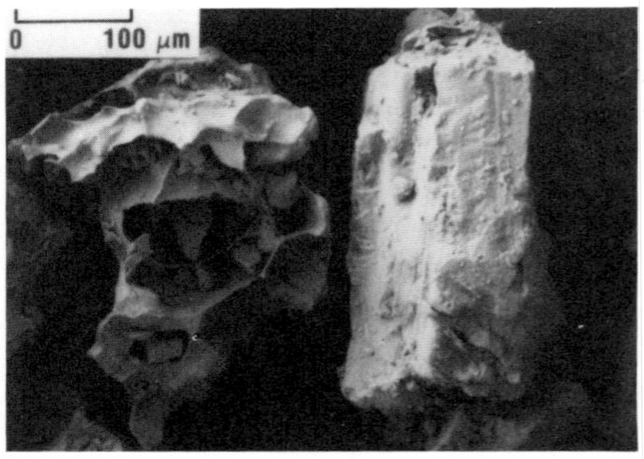

FIGURE 2. Right - Plagioclase phenocryst partly coated with basaltic glass; some twin planes are visible near its center. *Left* - Broken sideromelane droplet with high vesicularity. Vesicles are large (30- to >100-μm) and thin walled (generally ≤ 10 μm).

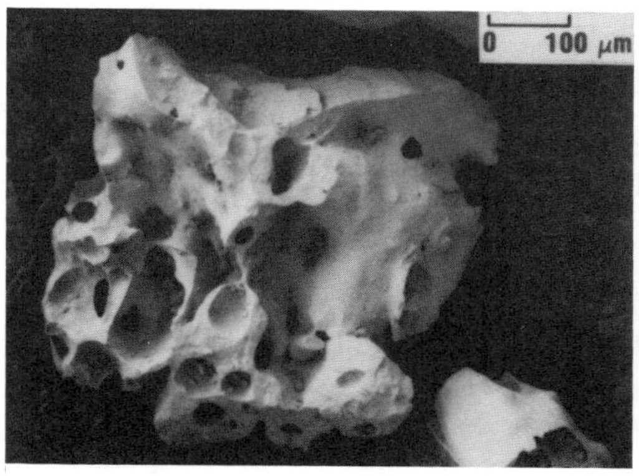

FIGURE 3. Sideromelane lapillus. The outermost vesicles collapsed while the droplet was still molten, leaving smooth-rimmed depressions and smooth glass pillars between vesicles (*center*). Other surfaces are broken, exposing the highly vesicular interior.

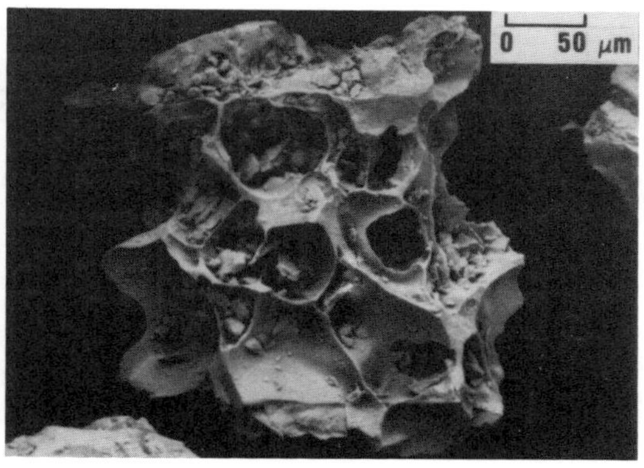

FIGURE 4. Broken sideromelane lapillus. Many vesicles are now polygonal as a result of mutual interference of vesicle walls during vesicle growth. Small clasts adhering to the lapillus surface are angular, broken vesicle-wall fragments (shards).

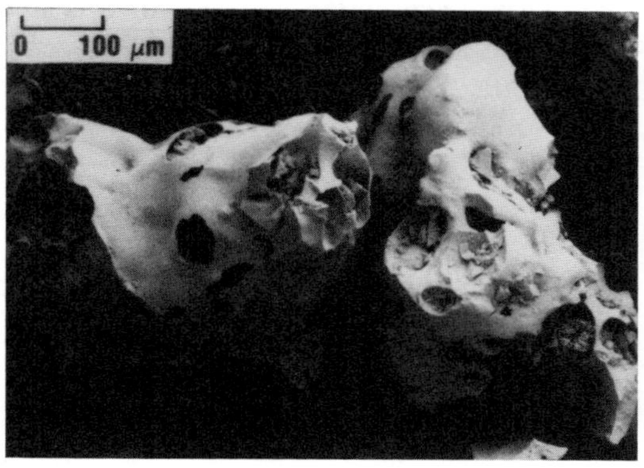

FIGURE 5. Two sideromelane droplets. These pyroclasts have irregular, twisted forms, yet have smooth surfaces. These formed by surface tension within still-molten, vesiculating droplets that were a spray in lava fountains.

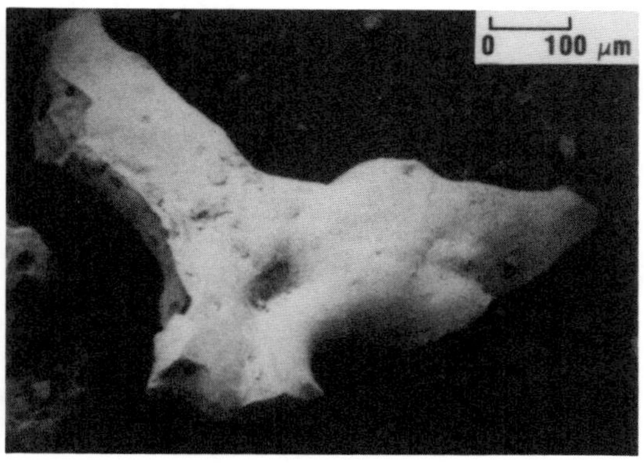

FIGURE 6. Large shard that was broken from a sideromelane droplet.

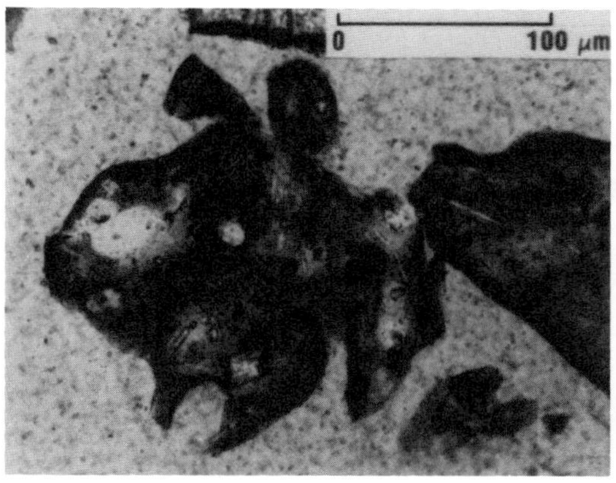

FIGURE 7a. Transmitted-light photomicrographs of Pacaya tephra. Irregular sideromelane droplets. There are small plagioclase phenocrysts (>25 μm) and abundant quench-texture plagioclase and pyroxene phases (generally less than 20 μm long). A rind, only a few micrometers thick, is darker brown than the light brown glass of the interior.

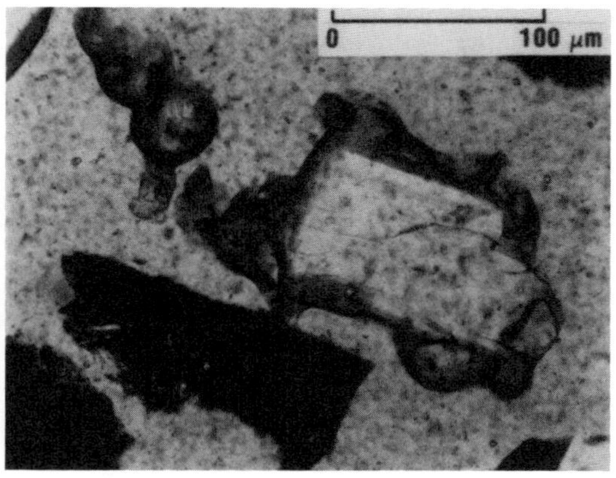

FIGURE 7b. Sideromelane-coated plagioclase phenocryst. Black grains are tachylite pyroclasts.

MT. ETNA, ITALY; 1971 ERUPTION

Introduction: Mt. Etna is a 60- by 40-km, 3300-m-high composite volcano (a shield of lava flows capped with interbedded lava flows and minor pyroclastic deposits) located on the eastern coast of Sicily, Italy. Frequent historical eruptions have produced lava flows and numerous chains of cinder cones at the summit and on the flanks. The April-May 1971 eruption was from SE and SSE radial fissures at about the 3000-m level (Tanguy, 1971; Rittmann, 1971). Continuous explosive activity built new, 30- to 50-m-high cinder cones along the two fissures. Rittmann (1971) noted that there may have been some phreatomagmatic interaction because the snow was up to 4 m deep at the time. Near the end of this eruption, a new explosion crater was formed (May 4 to May 26) several hundred meters NE of the fissures; Booth and Walker (1973) noted that activity at this crater was both Strombolian and phreatomagmatic.

This sample was collected in April 1971 at the 2400-m level (first cable car station) by R. Funiciello, University of Rome.

Grain Size: There is no grain-size data available for this sample. However, McGetchin et al. (1974) note that ash samples from the 1971 vents have median sizes of 100 μm to 1 mm. Ash falls from the explosion crater of 1971, collected 850 m NE of the crater, have median grain sizes of 110 μm (3.2 φ) and 125 μm (3.0 φ) and are moderately sorted ($\sigma_\varphi = 1.25$ φ) (Booth and Walker, 1973).

Tephra Description: The sample is a vitric-lithic ash that consists of mostly angular, poorly vesicular, blocky sideromelane fragments. Also present in this ash are blocky tachylite pyroclasts that exhibit both aphanitic and diktytaxitic textures (the latter texture consists of interwoven, finely crystalline feldspar and pyroxene grains (1- to 2-μm long and 0.2- to 0.3-μm wide) and irregular, broken, vesicular sideromelane droplets. Void spaces between crystalline phases in the tachylite pyroclasts occupy 30 to 50% of the rock by volume. There are also smaller amounts of broken droplets, glassy agglutinates, and clinopyroxene and plagioclase pyroclasts.

Chemical Composition: The chemical composition of a 1971 Etna lava is shown in Table 1.

TABLE 1. CHEMICAL COMPOSITION OF ETNA LAVA.*

	(Wt. %)
SiO_2	47.01
Al_2O_3	17.21
Fe_2O_3	6.83
FeO	4.44
MgO	5.54
CaO	10.77
Na_2O	3.40
K_2O	1.72
TiO_2	1.66
H_2O	0.11
H_2O^-	0.18
TOTAL	98.87

*Lava flow from upper east vent, May 4, 1971 (from Tanguy, 1971); Analyst, A. Lenoble, University of Paris.

Comments on Pyroclast Formation: Published observations concerning the 1971 eruption indicate that activity along the fissures was Strombolian and that chains of cinder cones were formed. Most of the pyroclasts in this sample indeed appear to be of Strombolian origin—irregular, broken sideromelane droplets and tachylite pyroclasts. There are, however, many pyroclasts that appear to be of phreatomagmatic origin—blocky, sideromelane clasts with low vesicularities. This suggests that minor phreatomagmatic explosions were mixed with Strombolian activity and confirms Rittmann's (1971) observation that there may have been lava/water interactions caused by deep snow cover on the mountain.

REFERENCES: Booth, B. and Walker, G. P. L., 1973. Ash deposits from the new explosion crater, Etna, 1971. Phil. Trans. R. Soc. Lond., **A. 274:** 147-151.

McGetchin, T. R., Settle, M., and Chouet, B. A., 1974. Cinder cone growth modeled after Northeast Crater, Mount Etna, Sicily. J. Geophys. Res., **79:** 3257-3272.

Rittmann, A., 1971. The Mt. Etna volcanic eruption of 1971. Smithsonian Inst. Center for Short-Lived Phenomena, Cambridge, Massachusetts, 9 July, 1971: 23 pp.

Tanguy, J. C., 1971. Mt. Etna volcanic eruption, 5 April—12 June 1971. Smithsonian Inst. Center for Short-Lived Phenomena, Cambridge, Mass., 15 December, 1971: 5 pp.

FIGURE 1. Vitric-lithic basaltic ash from the 1971 eruption of Mt. Etna. Most pyroclasts are angular, blocky, and subequant with low vesicularities. Pyroclasts with relatively smooth fracture surfaces are sideromelane; scalloped surfaces are broken vesicles (30- to 250-μm-long vesicles). Particles with rough surfaces are mostly tachylite fragments. The lumpy particle in the lower left corner of the photo is an agglutinate made up of welded sideromelane droplets.

FIGURE 2. There are mostly angular sideromelane pyroclasts in this photograph. Both smooth and conchoidal fracture surfaces are visible. Outermost vesicles are quite broad and flattened (disc shaped). The shard in the lower left corner of the photo appears to have been broken from a highly vesicular foam with large (>500-μm), well-developed vesicles. Distribution and the size of vesicles in these pyroclasts is heterogeneous.

FIGURE 3. Subangular sideromelane pyroclast. This particle has a low vesicularity, but a few flattened vesicles are visible at the surface. In the center of the pyroclast is a small phenocryst that has deformed the wall of a vesicle.

FIGURE 4. A mixture of sideromelane and tachylite pyroclasts. *Left bottom* - Broken sideromelane droplet; smooth, curved surface and few vesicles. *Left top* - Broken, highly vesicular sideromelane fragment with large vesicles. *Right top and bottom* - Subangular to subrounded tachylite pyroclasts, with diktytaxitic texture (*top*) and aphanitic texture and hackly fracture surfaces (*bottom*).

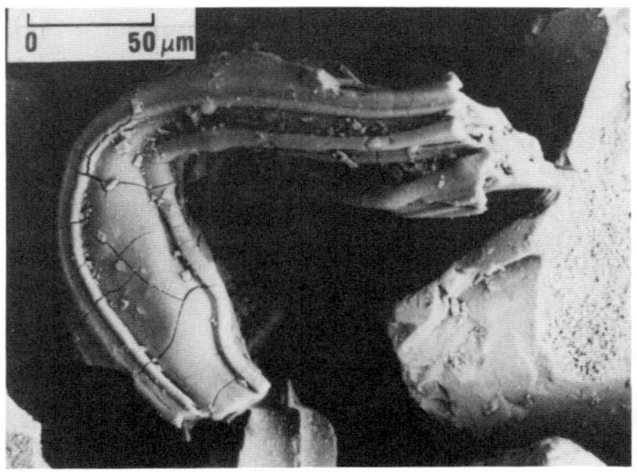

FIGURE 5. Twisted sideromelane droplet. The ribs are over elongate vesicles. Note the vesicles on broken ends of the droplet. This could be classified as Pele's hair.

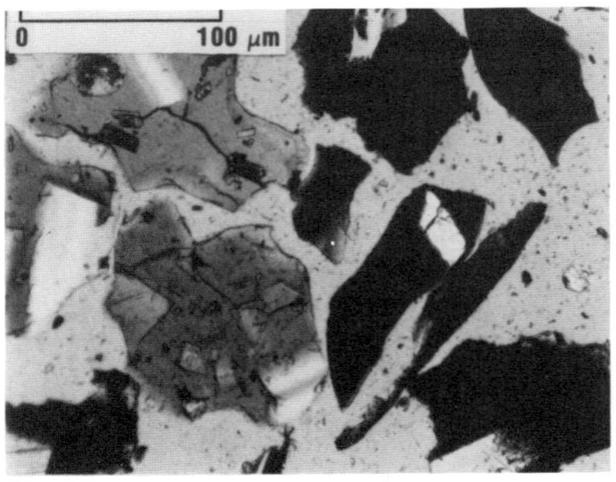

FIGURE 6. Photomicrograph of Etna tephra. On the *right* are mostly broken tachylite droplets and angular, blocky clasts. On the *left* are hyalocrystalline, phenocryst-bearing, blocky, sideromelane phenocrysts.

Vulcanian Eruptions

This activity is typified by the historic, intermittent explosions of ash from Vulcano, Aeolian Islands, Italy. The distinguishing characteristics of this type of ash are not well defined and reflect a mixture of magmatic and phreatomagmatic eruption mechanisms. Typically, Vulcanian ash is composed of varying amounts of juvenile and nonjuvenile fragments. The nonjuvenile glass pyroclasts are generally very angular, reflecting their origin by explosive pulverization of older lava flows or plugs within the vent. Juvenile glass particles may also be vesicular or blocky, depending upon the amount of magmatic and meteoric water involved in the eruption. Vulcanian eruptions can occur at silicic domes where viscous lava is extruded. During explosions, both magmatic and phreatomagmatic, much of the preexisting vent plug as well as new lava may be shattered by shock waves generated by rapidly expanding gases. Vulcanian ash deposits often resemble those of phreatomagmatic eruptions and the reader is refered to that section in this book for further discussion of depositional textures.

VOLCAN FUEGO, GUATEMALA; 1966 ERUPTION

Introduction: Volcan Fuego is a large, composite cone in Guatemala; its activity has been characterized over the last 450 years by frequent, short-lived eruptions. Much of the activity during the last several decades has consisted of extremely energetic Vulcanian eruptions of sulfur-rich, high-alumina basalt (Anderson, 1975; Rose, 1977). The short but intense eruptions of the last 20 years have frequently sent eruption plumes into the stratosphere; these have often been accompanied by small pyroclastic flows that travel down topographic lows on the volcano's flanks.

Activity at Fuego generally fits descriptions of Vulcanian activity: "moderate to violent ejection of solid or very viscous hot fragments of new lava" (Macdonald, 1972). Abundant lithic clasts are also ejected by these eruptions. The extremely energetic eruptions may be related to a high volatile content (Harris, 1979; Rose, 1977).

Grain Size: The sample collected 4 km N of Fuego's active vent, from the 1971 eruption, has a median grain size of 2.5 mm (-0.3 φ) and is moderately sorted ($\sigma_\varphi = 1.3$ φ). Downwind, median grain size decreases from >0.5 mm (within 40 km) to <0.1 mm (beyond 120 km) (Rose et al., 1973).

Tephra Description: The ash mostly consists of angular, fractured, dark brown sideromelane fragments with 30% tachylite and phenocryst fragments. None of these pyroclasts has the smooth grain surfaces typical of tephra from Hawaiian and Strombolian eruptions.

Most vitric pyroclasts are equant to slightly elongate with hackly or conchoidal grain surfaces. A few irregular patches of smooth glass are preserved on some fragments. Tachylite pyroclasts have very irregular, hackly surfaces. All pyroclasts have few vesicles; most have fewer than 30% vesicles, and these are deformed ovoids ranging from 10 to 40 μm.

Chemical Composition: Microprobe analyses of glass inclusions in crystalline phases of tephra from the 1974 eruption of Fuego indicate a sulfur content of 2500 ppm (Harris, 1979; Rose, 1977). The bulk chemical composition of some basaltic ash from Fuego is shown in Table 1.

TABLE 1. CHEMICAL COMPOSITION OF HIGH-ALUMINA BASALTIC ASH FROM THE APRIL 22, 1966 ERUPTION OF FUEGO.*

	(Wt. %)
SiO_2	49.42
Al_2O_3	19.09
Fe_2O_3**	10.61
MgO	6.73
CaO	10.07
Na_2O	3.09
K_2O	0.60
TiO_2	0.87
P_2O_3	0.41
TOTAL	100.83

*A. K. Baird, analyst (from Stoiber and Rose, 1970).
**Total Fe expressed as Fe_2O_3.

Comments on Pyroclast Formation: Little is known about the formation of tephra in Vulcanian eruptions of basaltic magma. Viscosities at the time of eruption are fairly high, as is indicated by the blocky, fragmented nature of the tephra. The large percentage of pyroclasts consisting of basaltic glass shows that the tephra were rapidly quenched. Although the original volatile content of the magma was high, as indicated by analyses of glass inclusions (Harris, 1979), vesicularities of the vitric pyroclasts are quite low. A high, pre-eruption magma viscosity suppressed vesicle growth. It is possible that total pressure within those vesicles that developed was very high, resulting in the short but energetic eruptions. Pyroclasts may have formed during passage of a very pronounced pressure gradient down from the vent surface into a somewhat vesicular melt.

REFERENCES:

Anderson, A., 1975. Some basaltic and andesitic gases. Rev. Geophys. Space Phys., **13**: 37-53.

Harris, D., 1979. Pre-eruption variations of H_2O, S, and Cl in a subduction zone basalt. Trans. Am. Geophys. Union, **60**: 968.

Macdonald, G., 1972. *Volcanoes.* Prentice-Hall, Inc., Englewood Cliffs, N.J.: 510 pp.

Rose, W. Jr., 1977. Scavenging of volcanic aerosol by ash: atmospheric and volcanologic implications. Geology, **5**: 621-624.

Rose, W. Jr., Bonis, S., Stoiber, R., Keller, M., and Bickford, T., 1973. Studies of volcanic ash from two recent Central American eruptions. Bull. Volcanol., **37**: 338-364.

Stoiber, R. and Rose, W. Jr., 1970. The geochemistry of Central American volcanic gas condensates. Geol. Soc. Am. Bull., **81**: 2891-2912.

FIGURE 1a. Sideromelane (basaltic glass) pyroclasts. Subequant, angular pyroclast with low vesicularity. There are a few patches of smooth outer skin (*left side*, *right center*), but most surfaces are conchoidal or hackly.

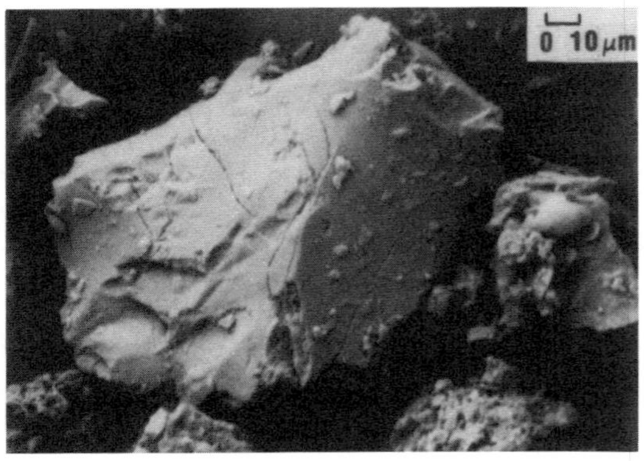

FIGURE 1b. Nonvesicular sideromelane pyroclast with a smooth surface that is broken in places by cracks and conchoidal spalls.

FIGURE 2a. Vitric and crystal pyroclasts. Angular vitric pyroclast with a variety of small, angular, vitric and crystal clasts adhering to the surface. Note the twin planes in plagioclase pyroclast (*upper left*).

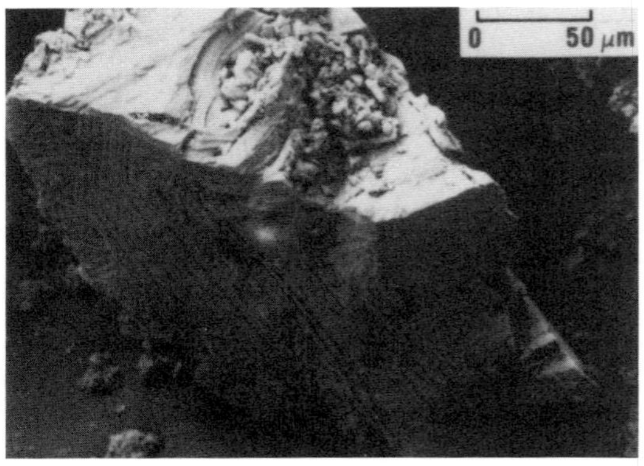

FIGURE 2b. Fractured plagioclase phenocryst. This pyroclast is free of glass (with the exception of aggregates of fine-grained pyroclasts loosely adhering to the surface). Note the complexity of oscillatory zoning that is visible on the face of this phenocryst.

FIGURE 3. Aggregate of the fine-grained fraction of 1966 Fuego tephra. Although mostly very fine grained (<2-μm) angular clasts, there are several rounded glass droplets visible: a rod and small (2-μm) droplet on the right and some delicate, irregular droplets throughout the matrix.

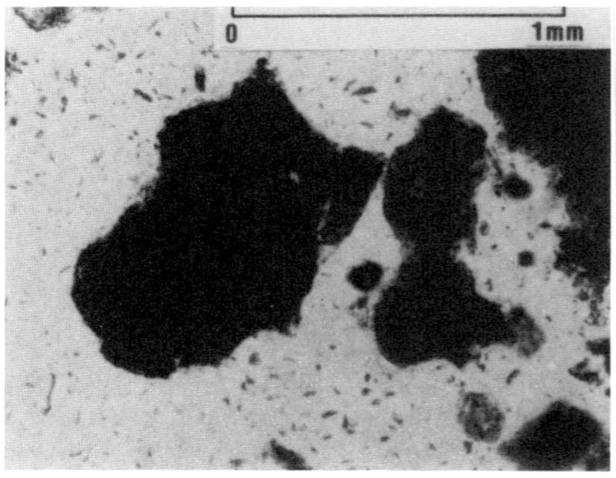

FIGURE 4. Transmitted-light photomicrograph of Fuego tephra. On the *left* is an angular, nonvesicular tachylite pyroclast, and on the *right* is a pyroclast with a tachylite core and sideromelane at the surface.

VULCANO, ITALY; 1888-1890 ERUPTION

Introduction: Vulcano is the southernmost active volcano in the Aeolian archipelago; it is located north of Sicily in the Tyrrhenian Sea (Keller, 1980). Numerous explosive eruptions have occurred at Vulcano in historic times; the most recent occuring from 1888 to 1890 during which Mercalli and Silvestri (1891) made classical observations of Vulcanian activity. Characteristic activity consisted of staccato explosions followed by quiet periods that lasted several minutes to several days. Initial explosions emitted clasts of solidified older lava derived from the vent; later explosions ejected juvenile blocks, bombs, and ashes.

Characteristically, Vulcanian explosions produce a vertically directed eruption cloud that may reach elevations of up to several kilometers. Base surges may emanate from this eruption cloud as directed blasts accompanied by ballistic fall of material onto the slopes of the cone. These eruptions produce coarse layers of ash fall that are interbedded with layers of fine surge ash on the cone flanks. Lahars develop near the base of the cone, usually as the result of heavy rains just after an eruption. The collapse of the eruptive column may occasionally produce pyroclastic flows. Sheridan et al. (1981) and Frazzetta et al. (1983) have described and modeled other eruptive phenomena. Samples described in this report were taken from the 1888 to 1890 eruptive phase (Sheridan and Marshall, 1981).

Grain Size: Size analyses (Table 1) of the most recent eruption products give the average median diameter and sorting for principal bed forms on the flanks of the volcano. These bed forms are: (1) sandwave- and massive-surge beds; (2) fall, coarse and fine layers (sets of fall and surge ash in one depositional unit); and (3) wet-surge deposits that have bed forms transitional between surges and lahars. Most samples consist of two populations, with dominant modes at 0.5 to 1.5 φ and 2.5 to 3.5 φ. Ash falls, however, commonly show an additional mode in the range of -2.0 to -1.0 φ, and wet-surge deposits have a mode finer than 4.5 φ.

TABLE 1. GRAIN-SIZE ANALYSES.

Bed Form*	No.**	Md_{φ} (mm)	σ_{φ}
S	35	1.96 (0.26)	1.08
M	10	1.63 (0.32)	1.03
WS	10	2.92 (0.13)	1.26
F	4	1.26 (0.42)	1.40
C+F	6	1.62 (0.32)	1.12

*Bed forms are: S - sandwave surge, M - massive surge, WS - wet surge, F - fall, and C + F - coarse and fine unit.
**Number of samples analyzed.

Tephra Description: The 1888-1890 eruption products form a gray to dark gray mantle of pyroclastic deposits resting on the lighter colored slopes of the Fossa cone. These deposits consist of fresh glass and crystal contituents that show little secondary alteration. The lithic contributuion to the 1888-1890 deposits is small and is not discussed here.

Glass. Glass particles are poorly vesicular to almost nonvesicular. Although the grains are blocky in appearance, they have sharp, acute edges that have suffered little abrasive action. Solution and precipitation effects on some grains, however, give an uneven appearance to curviplanar surfaces. For the most part, grains have broad, planar faces and dish-shape concavities that are typical of fragmentation of the viscous, silicic magma; smooth, ropy surfaces of the more fluid, basic magma are rare.

Crystals. Phenocrysts of clinopyroxene, calcic plagioclase, and olivine are entirely free of glass coatings. Crystal and cleavage faces are modified by stepped fractures and some shallow, dish-shaped concavities. These surface modifications appear to be mostly cleavage controlled; the grains are angular, and their sharp corners show little effect from transport rounding. A small amount of sugary coating appears on some crystal surfaces and may be related to solution and preciptation.

Chemical Composition: Frazzetta et al. (1983) suggest that two magmas of different compositions mixed before and during each of Vulcano's eruptive periods. One is an evolved aphyric rhyolite that was derived from a sanidine trachyte; the other appears to be a tephrite, as inferred from mafic melt inclusions with reaction rims in the rhyolite. This mafic component is also represented by phenocrysts of clinopyroxene, calcic plagioclase, olivine, and magnetite and by the fluidal-shaped pyroclasts mentioned earlier.

Comments on Pyroclast Formation: The blocky, nonvesicular nature of most of the 1888-1890 pyroclasts and their mixed magma glass and phenocryst compositions has led to a complex eruption model proposed by Frazzetta et al. (1983). In that model, eruptive cycles at Vulcano are initiated by tectonic movement, which promotes mixing of the evolved rhyolite and tephrite magmas. The mixing causes viscosity changes and volatile separation that allow the mixed magma to move to the surface. When magma reaches the water-saturated zone beneath the volcano, hydrofracture occurs and promotes the contact of magma and water, which leads to hydrovolcanic surge-producing eruptions. As these eruptions proceed, water access to the vent decreases and vesiculation of the magma begins because of the reduced confining pressure above the magma column. Degassing of the magma by vesiculation causes eruption of pumice falls. After the magma column has degassed, a viscous, silicic, and often trachytic lava flow plugs the vent.

Repetition of this cycle begins with the initial phreatic eruptions that fracture and expel the vent-blocking obsidian flow—"ultra-Vulcanian activity." The stratigraphic column in Fig. 1 illustrates the Vulcanian eruption cycles for recent activity at Vulcano. Characterisitic "Vulcanian" activity produces a complex assemblage of pyroclast types that range from lithic materials of fragmented old lavas to hydrovolcanic materials, which consist of mixed magmas and pumiceous debris of vesiculated lava. This significant component of nonjuvenile material is characteristic of Vulcanian deposits in general. The 1888-1890 materials represent the middle-stage activity of the eruptive cycle at Vulcano.

REFERENCES:

Frazzetta, G., LaVolpe, L., and Sheridan, M. F., 1983. Evolution of the Fossa vent, Vulcano. In: Explosive Volcanism, M. F. Sheridan and F. Barberi (Eds.), J. Volcanol. Geotherm. Res., **17**: 329-360.

Keller, J., 1980. The island of Vulcano. Rend. Soc. Ital. Min. Petrol., **36(1)**: 369-414.

Mercalli, G. and Silvestri, O., 1891. Le eruzioni dell'isola di Vulcano, incominciate il 3 Agosto 1888 e terminate il 22 Marzo 1890. Relazione Scientifica 1891, Ann. Uff. Cent. Meteor. Geodin., **10(4)**: 213 pp.

Sheridan, M. F. and Marshall, J. R., 1981. SEM surface-textural analysis of volcanic ejecta: an interpretation of base-surge behavior on Vulcano. Int. Assoc. Sed., Proc. 2nd European Reg. Meet., Bologna, Italy: pp. 185-188.

Sheridan, M. F., Moyer, T. C., and Wohletz, K. H., 1981. Preliminary report on the pyroclastic products of Vulcano. Mem. Soc. Astron. Ital., **52**: 523-527.

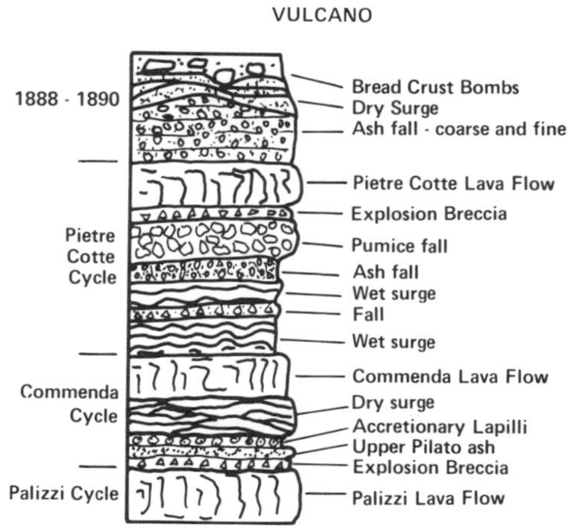

FIGURE 1. Stratigraphy of recent eruption cycles at Vulcano, adapted from Frazzetta et al. (1983) and Sheridan et al. (1981). Each cycle begins with pyroclastic eruptions that trend from hydrovolcanic to magmatic and end with lava-flow emplacement. Samples described are taken from the 1888-1890 deposits.

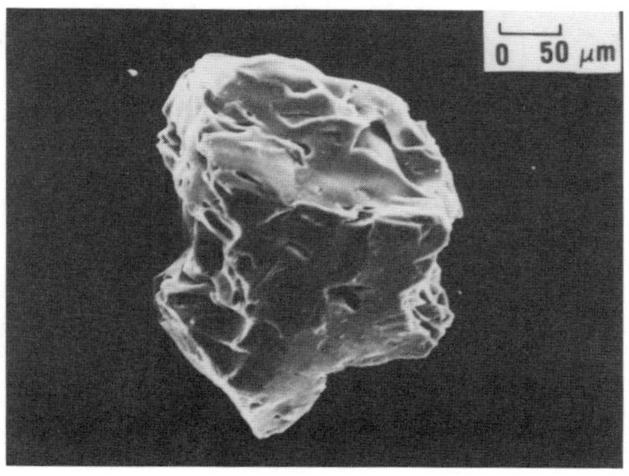

FIGURE 2. This SEM of a nonvesicular, blocky pyroclast shows nearly planar surfaces that have irregularities caused by poorly developed vesicles. The elongate, curving cracks of the hollow areas are likely caused by flow foliations in this obsidian, which developed when small vesicles stretched and bent during flow.

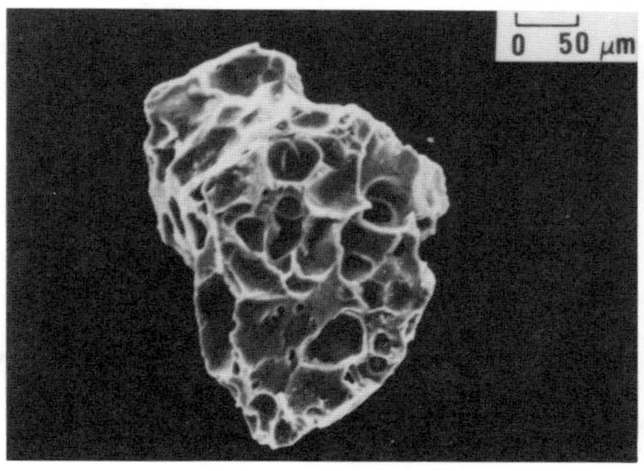

FIGURE 3. SEM of a vesicular pyroclast from magma of more mafic composition. Many of the vesicles are cut by curviplanar fractures as a result of hydrovolcanic fragmentation of the melt.

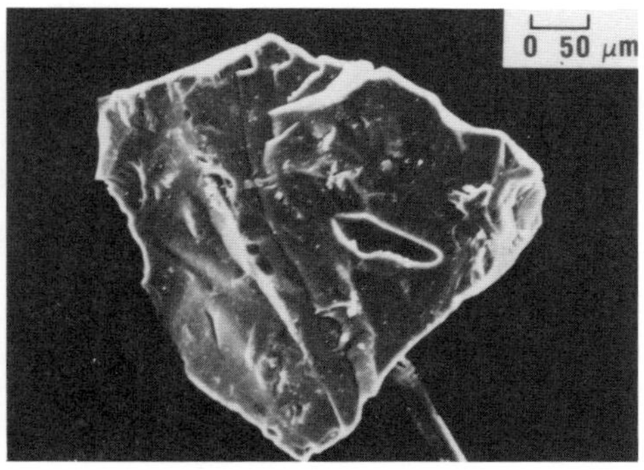

FIGURE 4. SEM of an obsidian pyroclast. The sharp, angular edges show that few chips and scratches developed during transport. Small sets of nearly parallel fractures are caused by conchoidal-like fracture behavior of the viscous glass.

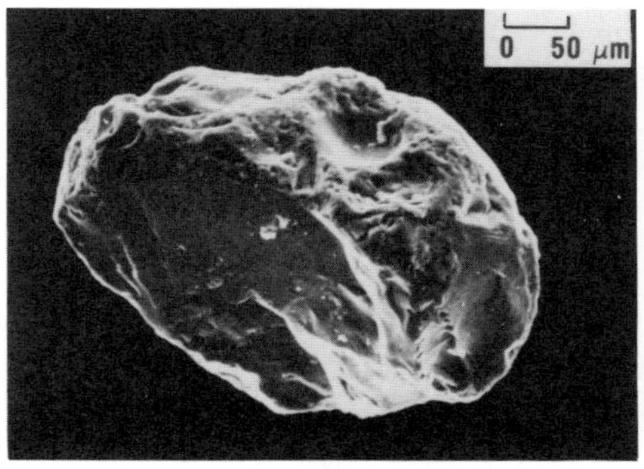

FIGURE 5. This SEM of a pyroclast shows two large, dish-shaped, curviplanar fractures and a frosted surface formed by solution and precipitation.

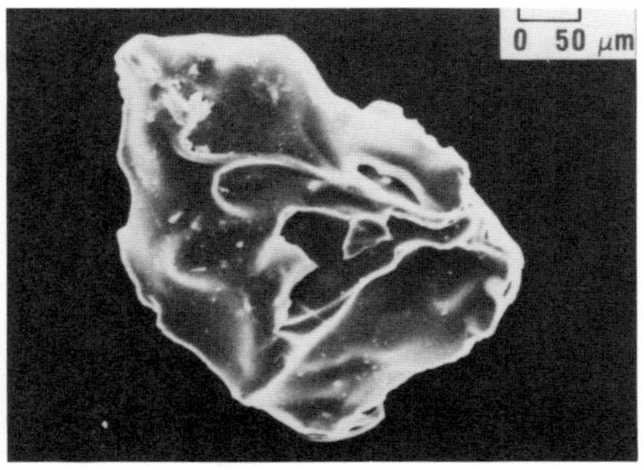

FIGURE 6. SEM of a fluidal-surfaced pyroclast of the mafic magma. Note the smooth, curved, and ropy skin texture, deformed vesicle embayments, and sharp edges of this grain.

FIGURE 7. SEMs of crystals showing stepped and dish-shaped fracture faces dominantly related to cleavage planes. Note the lack of a glassy coating and the sharpness of fracture surfaces.

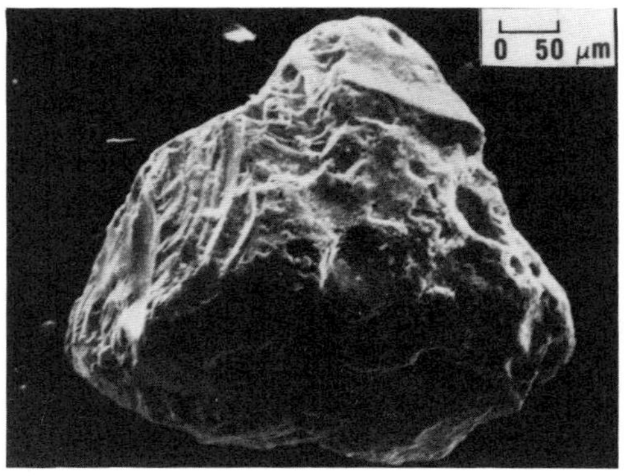

FIGURE 8. SEM of a partly frosted crystal; this grain shows solution pitting and precipitation coating on one side. The uneven surface has a sugary appearance that may be the result of contact with a grain of different composition.

MAYON VOLCANO, PHILIPPINES; 1968 ERUPTION

Introduction: Mayon is a symmetrical composite cone with an elevation of 2462 m. It has had 45 eruptions since 1616—all explosive and many with pyroclastic flows (Simkin et al., 1981). The 1968 eruptions were described by Moore and Melson (1969).

All eruptions were from the summit crater. Incandescent blocks and ash were ejected to a height of 600 m above the crater. Collapse of the eruption column was followed by pyroclastic flows in ravines on all sides of the volcano. Buoyant ash clouds rose to heights of 3 to 10 km (Moore and Melson, 1969). Ash falls, in measurable quantities, covered several hundred square kilometers; at distance, the ash is fine grained. Accretionary lapilli are present in the thicker ash fall deposits (Moore and Melson, 1969).

The sample examined here was supplied by A. Alcarez of the Philippine Commission on Volcanology.

Grain Size: The median grain size of six samples from pyroclastic flows ranges from 0.4 to 1.1 mm (1.3 to -0.1 φ); all are poorly sorted (Moore and Melson, 1969). The sample examined by the authors (distal ash fall) has a median grain size of 90 μm and is poorly sorted.

Tephra Description: The crystal-lithic ash consists of broken augite, hypersthene and plagioclase phenocrysts, and equant to slightly elongate angular glass shards. Scoria pyroclasts make up a few per cent; they have low vesicularity and thin, brown, oxidized rims.

Most pyroclasts are angular and blocky with few or no vesicles and are the products of brittle fracture. A few per cent of the pyroclasts are droplets. They have smooth, curved surfaces formed by surface tension within melt droplets. Some are agglutinated and some have wrinkled surfaces.

Chemical Composition: The chemical composition of Mayon andesitic pyroclasts is shown in Table 1.

TABLE 1. CHEMICAL COMPOSITION OF MAYON ANDESITE.

	Ash*	Core of Breadcrust Bomb**
SiO_2	56.2	54.90
TiO_2	0.68	0.64
Al_2O_3	19.13	19.04
Fe_2O_3	7.81	2.96
FeO	---	4.81
MnO	0.20	0.15
MgO	3.71	4.05
CaO	8.54	8.62
Na_2O	4.15	3.40
K_2O	1.16	1.13
P_2O_5	0.28	0.26
S	0.12	n.d.
TOTAL	102.10	100.00

*M. Rhodes, analyst.
**E. Jarosewich, analyst (from Moore and Nelson, 1969).

Comments on Pyroclast Formation: The formation of this tephra is not well understood. Observations of the eruption by Moore and Melson (1969) indicate that explosions produced vertical ejections of incandescent material, much of which fell back into the slopes forming pyroclastic flows. These deposits were welded within 1 km of the vent and consisted of crystal-rich ash, lapilli, and scoriaceous blocks (with some breadcrust bombs). From the observations, all indications are that this eruption was similar to energetic lava fountaining of intermediate viscosity melts; thus the ash contains droplets similar to those produced by some Strombolian and Vulcanian eruptions. However, droplets form only a few per cent of the ash fall tephra. The vesicularity of pyroclasts is very low, which indicates that disruption of expanding gases within a gas-charged magma probably did not occur (the interpretation by Heiken, 1972, is probably not correct).

Perhaps heterogeneities within the magma are responsible for disruption and fragmentation of the magma. A rarefaction wave (associated with an atmospheric shock wave) could have passed through the rising magma, fracturing more viscous portions to form the fine ash and less viscous portions to form large blobs of molten material. Much of the finer grained material could have been winnowed out or carried higher in the eruption column and dispersed as ash falls. The molten blobs would have fallen back to form most of the agglutinated pyroclastic flow deposits. This hypothesis can account for the extremes in pyroclast types present in these deposits. Another possible cause is similar to that described for Vulcanian eruptions at Ngauruhoe, New Zealand (Nairn and Self, 1978): comminution of brittle material, mixed with molten ejecta occurs during the pressure build-up beneath a partly cooled, solid plug in the vent of a composite core. The heterogeneities are vertical in this case and could account for characteristics of the ash from Mayon.

REFERENCES:

Heiken, G., 1972. Morphology and petrography of volcanic ashes. Geol. Soc. Am. Bull., **83**: 1961-1988.

Moore, J. G. and Melson, W. G., 1969. Nuées ardentes of the 1968 eruption of Mayon Volcano, Philippines. Bull. Volcanol., **33**: 600-620.

Nairn, I. A. and Self, S., 1978. Explosive eruptions and pyroclastic avalanches from Ngauruhoe in February 1975. J. Volcanol. and Geotherm. Res., **3**: 39-60.

FIGURE 1. Equant to slightly elongate angular shards. A few grain surfaces are smooth and curved; they are broken across vesicles. Most are irregular, with numerous conchoidal fracture surfaces. Some of the smaller shards are wedge or needle shaped.

FIGURE 2. The whole spectrum of pyroclast types is visible in this SEM. The largest equant fragments (*center*) are crystal pyroclasts (plagioclase or pyroxene). Most grains are angular equant to subequant vitric pyroclasts with flat or conchoidal surfaces. A few droplets are present. Pyroclasts are nonvesicular or have rare vesicles although, in thin section, some of the small scoria clasts are vesicular.

FIGURE 3. A closer view of angular shards. One (*lower center*) is broken from the intersection of several vesicles. All others are nonvesicular. A broken droplet surface is visible (*lower right*). There are numerous equant, angular particles in the <5-µm range.

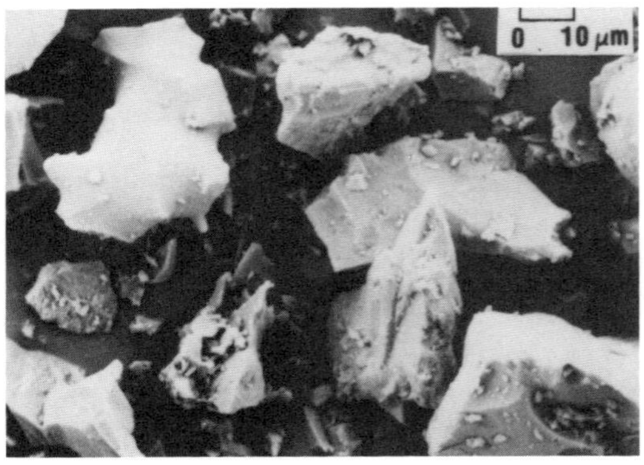

FIGURE 4. All except one pyroclast are subequant, angular shards. There is a broken droplet (*upper left corner*) whose smooth, curved surfaces are caused by flow after comminution. Most clasts are nonvesicular.

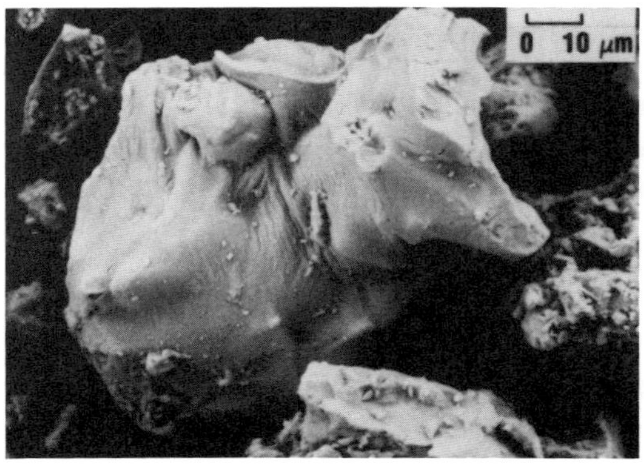

FIGURE 5. Broken droplet with a wrinkled "skin;" this occurs when the pyroclast surface has cooled while the interior was still molten. Another droplet was annealed to the surface (*top-center*) while still molten.

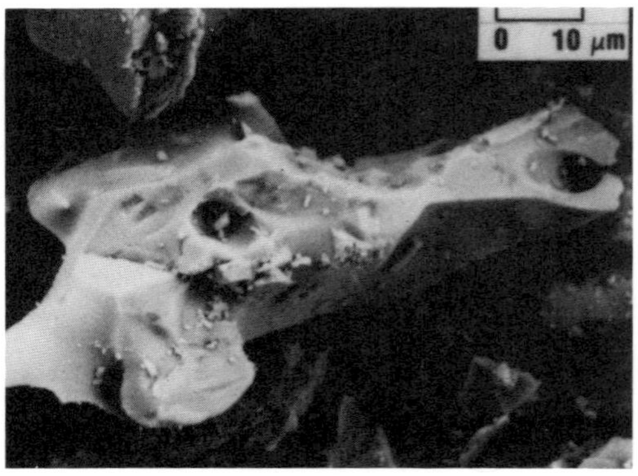

FIGURE 6. Angular, vitric fragment with only two well-developed vesicles visible in this image. It is typical of the vitric pyroclasts with very low vesicularities in this sample.

NGAURUHOE, NEW ZEALAND; 1974 ERUPTION

Introduction: Ngauruhoe Volcano is located in the central North Island of New Zealand. It is part of the Taupo volcanic zone, which extends 250 km SW-NE from Ruapehu to White Island and marks the continuation of the Tonga-Kermadec arc. Recent documented eruptive activity at Ngauruhoe from 1972 through 1975 was typical for that andesitic stratovolcano and consisted mostly of periodic, small Vulcanian explosions. The sample characterized here was donated by S. Self, who witnessed the explosive activity in 1974 (Self, 1975). Although this eruption was relatively small and produced only a small, thin fallout, Nairn (1976) and Nairn and Self (1978) documented highly explosive, cannon-like eruptions in 1975 that produced pyroclastic avalanche deposits.

This 1974 Ngauruhoe ash sample is described by Self (1975) as part of a 0.74×10^6 m^3 fallout (0.5×10^{-3} km^3 dense rock volume) deposited from a plume of fine ash that erupted to 3600 m above the vent and was dispersed by winds to the NW. Most of this deposit was less than several millimeters thick. By comparison, the 1975 eruption produced 2.4×10^6 m^3 of fallout and 1.0 to 1.5×10^6 m^3 of avalanche deposits from ash columns reaching nearly 11,000 m above the vent (Nairn and Self, 1978).

Grain Size: Table 1 lists grain-size characteristics of Ngauruhoe ash collected from the 1974 and 1975 eruptions. Ash particles from the 1974 eruption are generally less than 1 mm in diameter and consist dominantly of moderately sorted, fine ash. The more explosive eruptions of 1975 produced deposits that are much coarser grained; the fallout is well sorted and the avalanche deposits are very poorly sorted. It is likely that much of the fine material from the 1975 eruption was separated from the coarser ash in the high eruptive column before emplacement.

TABLE 1. SIZE CHARACTERISTICS OF NGAURUHOE 1974 AND 1975 DEPOSITS.*

Sample	Md_φ	(mm)	σ_φ
1974 (Fallout)	4 - 5	(0.03-0.06)	1.5
1975 (Fallout)	1.1	(0.46)	0.4
1975 (Avalanche)	−4 - 0	(16.0-1.0)	3.5

*Data taken from Self (1975) and Nairn and Self (1978).

Tephra Description: Mostly nonjuvenile, glassy pyroclasts are represented in the sample illustrated. The nonvesicular, blocky particles may be distinguished from vesicular juvenile ones by their vesicularity and a greater degree of surface alteration, which is manifested as rough surfaces. The rough surface texture is a thin coating of 1- to 5-µm particles and crusty material that gives grains a frosted appearance. Gross surface structure is marked by uneven or hackly fractures. Some surfaces are smooth where phenocrysts have been plucked during fragmentation. Other smooth surfaces are slightly curved and have stepped fractures that appear as faint striations; these features are characteristic of brittle breakage. Many of these nonjuvenile clasts show rounded, smooth edges that appear to be remelted. The v-shaped surfaces marking the edges of many of these grains resulted from subsequent abrasion of grains within the vent after fragmentation.

Juvenile glassy particles show mostly delicate prong-like edges of vesicle walls. The tips of the prongs also show rounding caused by abrasion. Vesicle surfaces are smooth and relatively free of alteration materials and adhering dust grains. Some solution pitting is evident with high magnification of vesicle surfaces, and a frosted appearance of the vesicle prongs and blades is also evident.

Small (10- to 25-µm) grains appear to be vesicular but also blocky; surfaces are planar, cutting across vesicle-wall projections. Fragmentation of these small grains likely involved vesicle bursts followed by brittle fracture. Grains less than 5 µm in diameter are rare.

Chemical Composition: As mentioned above, the 1974 fallout deposit is composed of both essential (juvenile) and accessory (nonjuvenile) components. Data from Self (1975) show that 3 to 4% of the ash is dark, glassy, slightly vesicular juvenile andesite. The nonjuvenile component is mainly from the pulverized 1954 andesite lava plug. This accessory ash is a nonvesicular labradorite-pyroxene andesite. Although bulk chemical analyses are not available for the 1974 material, in Table 2 chemical analyses of selected grain surfaces from that eruption are compared to those of the 1954 andesite lava and the White Island andesite. The variations in oxide abundances for 1974 ash, especially apparent for Al_2O_3, are likely the result of grain-surface alteration.

TABLE 2. CHEMICAL COMPOSITION OF NGAURUHOE ANDESITES.*

	1	2	3	4	5
SiO_2	56.23	58.87	59.87	55.90	62.16
TiO_2	0.30	0.60	0.67	0.76	0.71
Al_2O_3	22.78	17.65	14.96	16.90	14.32
$FeO + Fe_2O_3$	5.07	8.20	10.22	8.40	5.97
MnO	---	0.01	---	0.15	0.15
MgO	3.59	4.23	5.20	5.20	3.97
CaO	7.58	5.29	3.66	8.40	5.89
Na_2O	3.64	3.59	3.40	2.60	2.49
K_2O	0.79	1.53	2.03	1.00	2.37
Total	99.98	99.97	100.01	99.31	98.01

*Analyses 1, 2, and 3 are standardless energy dispersive spectral values of 1974 ash grain surfaces; analysis 4 is of a 1954 andesite lava from Ngauruhoe (Steiner, 1958); and analysis 5 is a hypersthene andesite from White Island (Black, 1970).

Comments on Pyroclast Formation: A high contribution of nonjuvenile ash is typical of Vulcanian eruption products. Self (1975) explained that the 1974 ash was the result when new magma degassed in the volcano's vent conduit. Pulverization of the older lava plug contributed to the ash volume. Nairn and Self (1978) showed that the crater produced in the 1975 eruption required an excavation energy equivalent to about 5 kt of high explosive. At 1000°C and 2 kbar (7.5-km burial depth), water has nearly the same energy density as high explosive. For a phreatomagmatic mechanism of ash formation and ejection, at least 5 kt of water (probably several times more) must have been vaporized to produce the observed crater; this figure is well within annual precipitation limits for the Ngauruhoe cone.

A more recent evaluation of Vulcanian eruption at Ngauruhoe by Self et al. (1979) explains the characteristic volumes and velocities of ash expelled during discrete cannon-like explosions (5×10^5 m^3 dense rock at 400 to 600 m s^{-1}). From geometric considerations, 4 to 6 wt. % water was necessary for explosions; this suggests that both magmatic volatiles and small amounts of vaporized ground water were involved; however, the high pressures (3 to 5 kbar) characteristic of some magma/water interactions were not required for the explosions. Vapor pressures of 100 to 200 bars, which correspond to the tensile strength of an andesite plug several tens of meters thick, satisfy the volume constraints; this indicates that ground water heated in pores and fractures constituted the small phreatic component.

Because most of the Vulcanian sample illustrated here represents fragmented nonjuvenile lava, it is difficult to assess the eruptive contribution of ground water. Both vesicular and nonvesicular clasts are present. Those that appear to be the youngest and freshest are vesicular; however, there is clear evidence that curviplanar fractures, characteristic of ground water mixing with magma, do form surfaces on some vesicular grains.

REFERENCES:

Black, P. M., 1970. Observations of White Island volcano, New Zealand. Bull. Volcanol., **34**: 158-167.

Nairn, I. A. 1976. Atmospheric shock waves and condensation clouds from Ngauruhoe explosive eruptions. Nature, **277**: 440-443.

Nairn, I. A. and Self, S., 1978. Explosive eruptions and pyroclastic avalanches from Ngauruhoe in February 1975. J. Volcanol. Geotherm. Res., **3**: 39-60.

Self, S., 1975. Explosive activity of Ngauruhoe, 27-30 March 1974 (note). N. Z. J. Geol. Geophys., **18(1)**: 189-195.

Self, S., Wilson, L., and Nairn, I. A., 1979. Vulcanian eruption mechanisms. Nature, **277**: 440-443.

Steiner, A., 1958. Petrologic implications of the 1954 Ngauruhoe lava and its xenoliths. N. Z. J. Geol. Geophys., **1**: 325-363.

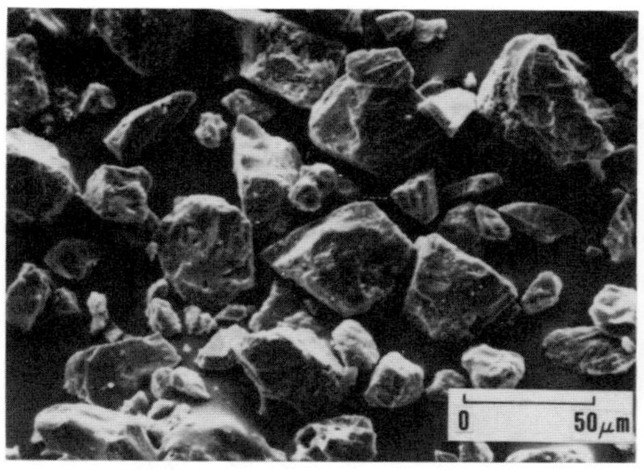

FIGURE 1. SEM of glassy nonjuvenile pyroclasts. These particles are mostly nonvesicular and blocky, with many angular surfaces. These particles may have been formed when an older lava plug was pulverized within the vent during eruption.

FIGURE 2. SEM of nonjuvenile pyroclasts that show frosted surface textures caused by a thin alteration coating of hydrated glass. Some curviplanar fracture surfaces have tiny adhering particles, whereas others are relatively clean.

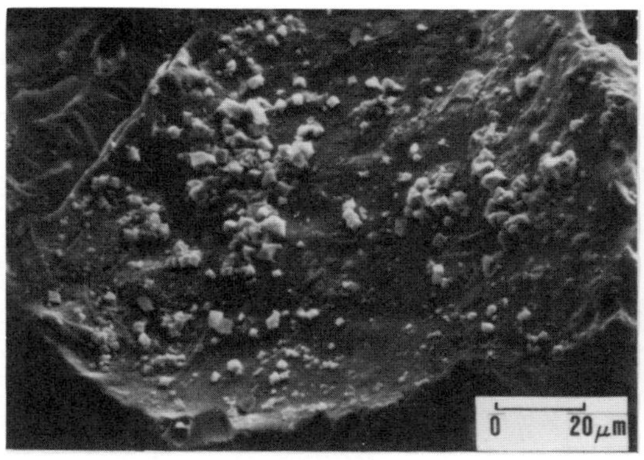

FIGURE 3. SEM closeup of a grain surface from Fig. 2. Small particles on the grain surface are probably crystalline alteration materials of a zeolite structure.

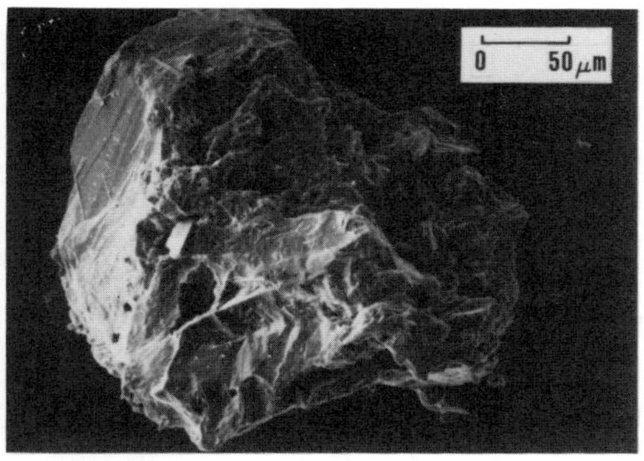

FIGURE 4. SEM of a nonjuvenile particle shows a large crystal mold near a smaller crystal still embedded in the glass.

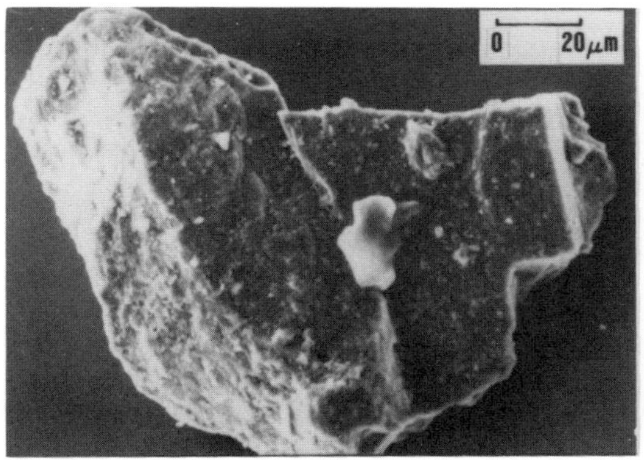

FIGURE 5. SEM showing frosted surface of nonjuvenile, twinned, clinopyroxene crystals. Note the rounded, abraded edges, rough surfaces, and fused piece of glass stuck to the crystal surface.

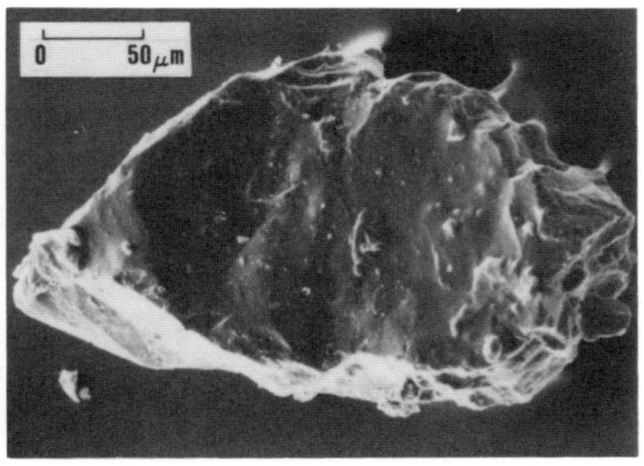

FIGURE 6. SEM of a glassy nonjuvenile particle that shows smooth surfaces indicative of remelting. Angular edge fractures have also been subdued by remelting.

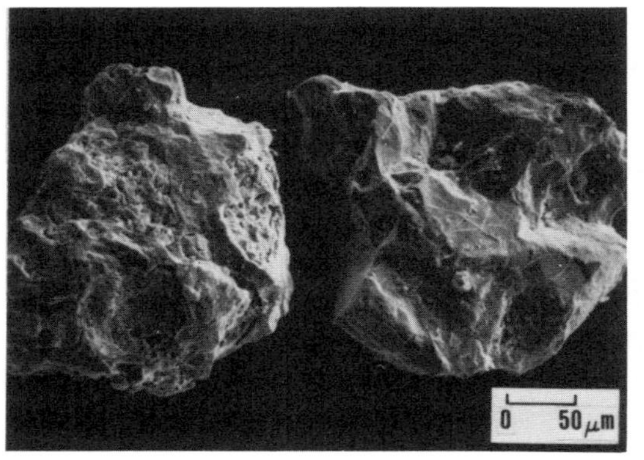

FIGURE 7. SEM of two glassy, nonjuvenile particles. Particle on the left shows extensive alteration and hydration cracks, whereas the other grain shows less alteration. Small secondary crystals form needle-like structures on some surfaces of the less altered grain.

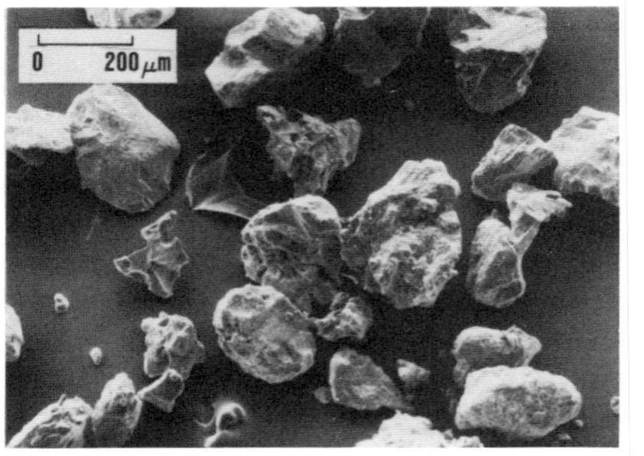

FIGURE 8. SEM field of view showing a few vesicular juvenile particles with other nonjuvenile ones. Juvenile pyroclasts constitute only a few per cent of the total population of this sample.

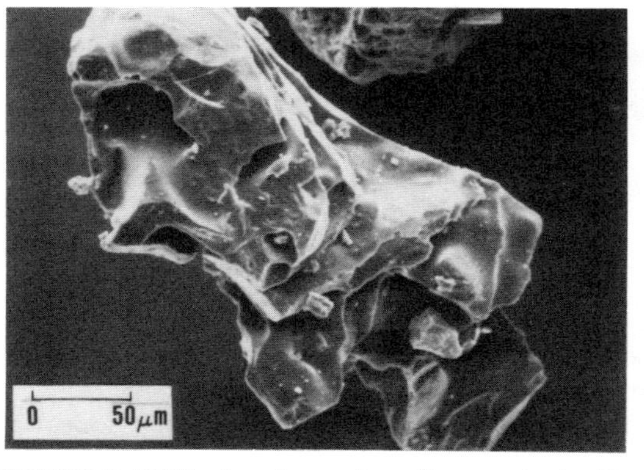

FIGURE 9. SEM of a glassy, juvenile pyroclast with characteristic fresh and embayed surfaces and delicate prong- and blade-like edges.

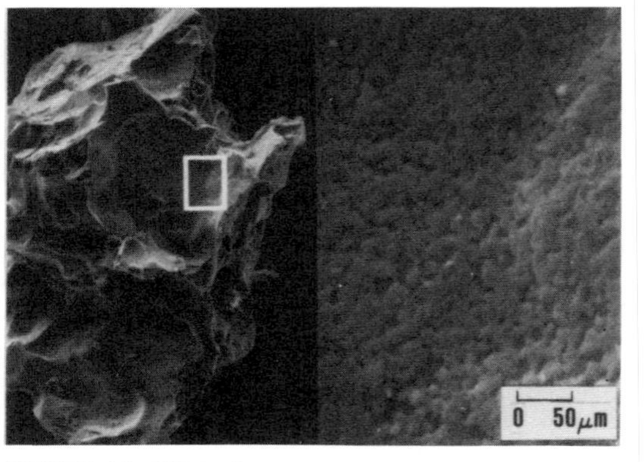

FIGURE 10. SEM of a glassy juvenile particle with a 10X enlargement of the vesicle surface. The enlargement details surface pitting caused by solution and precipitation of amorphous material.

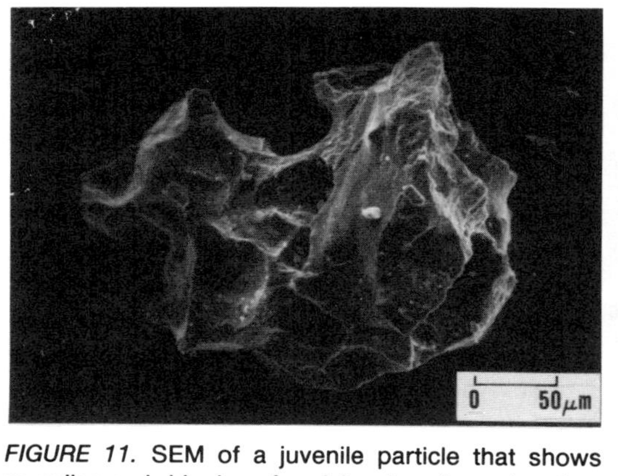

FIGURE 11. SEM of a juvenile particle that shows rounding and chipping of vesicle edges from abrasion.

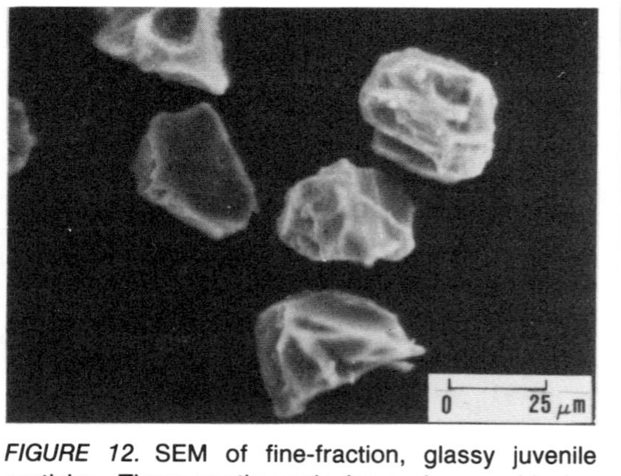

FIGURE 12. SEM of fine-fraction, glassy juvenile particles. These partly vesicular grains are blocky, with surfaces that cut across vesicle embayments; this indicates that brittle fracture occurred after an initial fragmentation caused by vesicle burst.

SANTIAGUITO, GUATEMALA

Introduction: Santiaguito is a dome complex in southwestern Guatemala that consists of dacite domes and flows. Its eruption, which began in 1922, followed the catastrophic eruption of Volcán Santa Maria in 1902 by 20 years and has been the site of constant volcanic activity since that time (Rose, 1972; Rose et al., 1976).

Accompanying the slow growth of this dome complex are sporadic small explosions from several craters located near or over vent areas (Caliente and El Brujo vents). Small ash clouds from these vents rarely exceed 2 km above the vent but occasionally may reach an altitude of 5 km above sea level (Rose et al., 1980). In 1973, a small nuée ardente issued from both the El Brujo and Caliente vents (Rose et al., 1976).

This sample was a fresh ash fall from the Caliente vent and was collected July 28, 1970. Rose et al. (1980) described Santiaguito tephra that was sampled by flying through eruption clouds.

Grain Size: This tephra has a mean grain size of 0.5 mm (1φ) and is very poorly sorted. The grain size is similar to that of nuée ardente deposits from Santiaguito mentioned by Rose et al. (1976). The ash clouds, sampled by aircraft, have a bimodal distribution, with peaks at 2 to 10 μm and 0.2 to 0.8 μm (Rose et al., 1980).

Tephra Descriptions: The ash consists of mostly hyalocrystalline, equant, blocky dacite fragments. Most grain surfaces have hackly to conchoidal surfaces and have low vesicularities or are nonvesicular. Some are glassy enough to be classified as vitric pyroclasts, but most have abundant phenocrysts and microlites in a glassy matrix. There are 25% crystal pyroclasts that are mostly plagioclase with lesser amounts of quartz. Individual phenocrysts are often fractured and in some cases broken along twinning planes. The more glassy pyroclasts are vesicular with sparse, but large, irregular elongate vesicles.

Rose et al. (1976), in their examination of tephra from Santiaguito nuée ardente, found a similar variety of pyroclasts. However, there was a larger fraction of pumiceous pyroclasts in the nuée ardente deposits than in ash fall from routine small explosions of the dome complex. Angular, nonvesicular particles are predominant in eruption cloud samples.

Chemical Composition: Table 1 is a chemical comparison of the slightly more silica-rich Caliente Dome lava with that of the ash fall.

TABLE 1. CHEMICAL COMPOSITION OF SANTIAGUITO ASH AND LAVA.

	Ash Fall (7/28/70)*	Lava (Caliente Dome)**
SiO_2	62.44	64.3
TiO_2	0.68	0.36
Al_2O_3	16.69	16.8
Fe_2O_3	2.98	1.6
FeO	2.08	2.4
MnO	0.20	n.d.
MgO	2.48	1.48
CaO	5.02	4.5
Na_2O	4.56	4.9
K_2O	1.54	1.70
P_2O_5	0.20	0.17
S	0.07	n.d.
Total	98.75	98.63

*M. Rhodes, analyst.
**From Rose (1972).

Comments on Pyroclast Formation: Most of the pyroclasts were formed by vesiculation and disintegration of highly viscous magma near vent areas. The same process occurs occasionally near the ends of blocky dacite flows, when it generates nuées ardentes (Rose et al., 1976). Vesiculation and fragmentation by pressure release may be the main cause of tephra formation. However, as the region is characterized by high rainfall, it is possible that there may be some magma/water interaction, which results in phreatomagmatic eruptions.

REFERENCES: Rose, W. I., Jr., 1972. Santiaguito volcanic dome, Guatemala. Geol. Soc. Am. Bull., **83**: 1413-1434.

Rose, W. I., Jr., Pearson, T., and Bonis, S., 1976. Nuée ardente eruption from the foot of a dacite flow, Santiaguito Volcano, Guatemala. Bull. Volcanol., **40**: 1-16.

Rose, W. I., Jr., Chuan, R., Cadle, R., and Woods, D., 1980. Small particles in volcanic eruption clouds. Am. J. Sci., **280**: 671-696.

FIGURE 1. Blocky, angular, hyalocrystalline dacite pyroclasts typical of tephra from Santiaguito. Vesicularity is usually less than 5%. Conchoidal or hackly fracture surfaces are common. The pyroclast with smooth, planar fracture surface (*lower left*) is a plagioclase phenocryst.

FIGURE 2. Lithic pyroclast with smooth, hackly, and conchoidal fracture surfaces.

FIGURE 3. Hyalocrystalline dacite pyroclast. Most grain surfaces are blocky and angular, but one surface (facing) is smooth and curved; perhaps it bounded a broad, irregular vesicle before comminution. All grain surfaces are coated with sublimates.

FIGURE 4. Lithic pyroclast similar to the one described in Fig. *5*. Grain surfaces include angular, hackly fractures and smooth, curved vesicle walls.

FIGURE 5. Hyalocrystalline lithic pyroclasts with large, irregular vesicles.

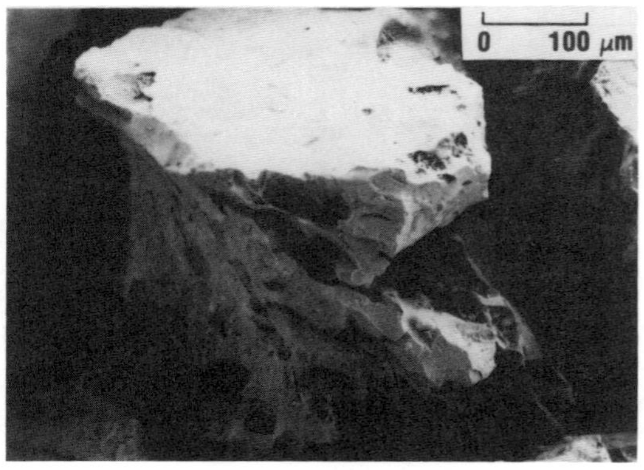

FIGURE 6. Blocky, hyalocrystalline lithic pyroclast with well-developed, lenticular vesicles. The large vesicles have curved yet lumpy walls. Comminution of a highly viscous melt, such as one from the Santiaguito dacite flows with heterogeneous patches of well-developed vesicles, could account for the mixture of blocky, angular pyroclasts and pyroclasts with one or two curved yet lumpy surfaces; the latter are particles broken from vesicle walls.

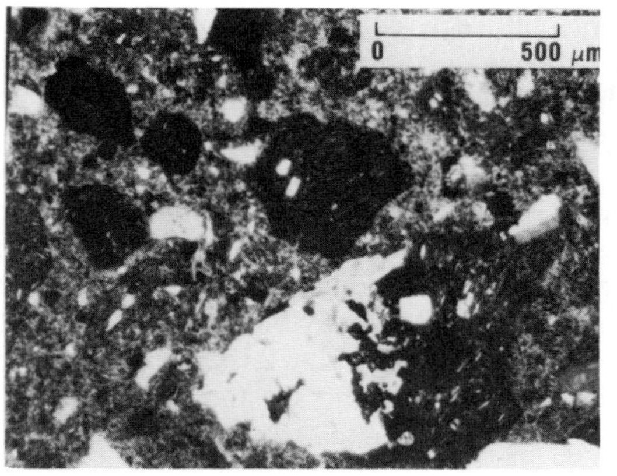

FIGURE 7. Photomicrograph (partly polarized light) of tephra from Santiaguito. There are mostly hyalocrystalline dacite pyroclasts: subequant, blocky forms with abundant, zoned plagioclase phenocrysts. This sample is bimodal, with 0.2- to 1-mm-long dacite pyroclasts in a matrix of <100-μm glass, crystal, and lithic clasts of the same composition.

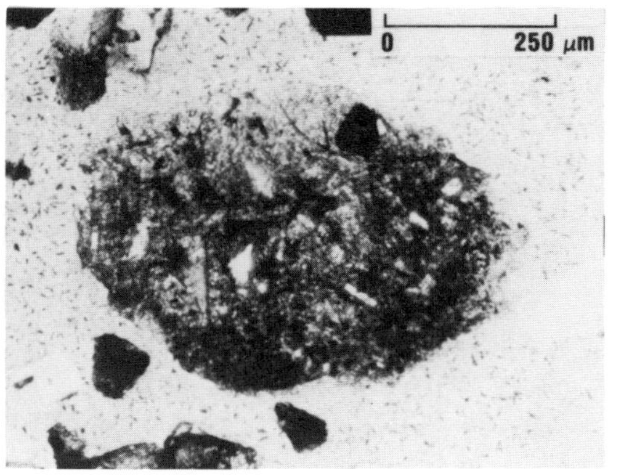

FIGURE 8. Photomicrograph of a hyalocrystalline dacite pyroclast; the groundmass is glassier along the rims (*top*).

Plinian and Peléean Eruptions

Plinian pumice falls and associated pyroclastic flows range from small-volume eruptions associated with domes to some of the truly great volcanic eruptions that produce widespread blankets of tephra and are associated with the collapse of large calderas. They are most often associated with rhyolitic or dacitic magmas that are charged with volatiles. The explosive decompression of the gas-charged magma forms highly vesicular ash that rises to great heights before settling to the Earth's surface. Peléean eruptions are generally associated with the explosive disintegration of a lava dome.

MEDICINE LAKE HIGHLAND, CALIFORNIA

Introduction: The Medicine Lake Highlands, California, is a 25-km-diameter shield volcano located 55 km E of Mount Shasta. Holocene volcanic activity there has been bimodal, producing basaltic eruptions along the flanks of the shield and rhyolitic eruptions near the rim of an 8- by 6-km shallow caldera. Eruptions during the last 1,100 years have generally consisted of highly explosive Plinian activity that mantled the countryside with pumice and was followed by extrusion of thick lava domes and flows. Nearly all of this pumice came from two sources, Little Glass Mountain and Glass Mountain. There is no evidence for development of pyroclastic flows during these eruptions.

Measurable tephra layers from Little Glass Mountain cover about 200 km^2 and have a volume of 0.046 km^3. Within a distance of 3 km from the vent, there are five tephra units that consist mainly of reversely graded, pale-gray pumice blocks and lapilli. These units range in thickness from 20 to 50 cm, and each contains 5 to 20% lithic fragments. At greater distances only one or two beds are visible. The tephra are overlain by the lava flows of Little Glass Mountain, which have a volume of 0.3 km^3

The nearly contemporaneous Glass Mountain eruptions covered a larger area (320 km^2) and have a volume of 0.09 km^3. There are 11 pumice falls with thicknesses of 5 to 102 cm; most beds are reversely graded.

Grain Size: All samples consist of very poorly sorted lapilli (near the vent) to medium to coarse ash (distal). All samples are within the field of tephra fall deposits discussed by Walker (1971).

Tephra Descriptions: Aphyric pumice pyroclasts make up 78 to 100% of each sample. All are remarkably homogeneous in composition and morphology; tephra from Little Glass Mountain and Glass Mountain are nearly identical (Table 1). All are equant to slightly elongate, blocky fragments with average L/W ratios of 1.4 to 1.6. All have sharp, angular grain surfaces either parallel to or normal to long axes of vesicles. Pyroclast surfaces do not exhibit the postfragmentation deformation or glazing that is characteristic of larger blocks or bombs in near-vent pumice cones. Vesicles are small and highly elongate; they range from 5 to 90 μm wide and 20 to 700 μm long. Some are twisted into "vortex" patterns that are interlayered with the highly elongate forms. Nearly all vesicles consist of several coalesced vesicles; vesicle cavities are crossed by threads and ribs of glass that remained after coalescence.

TABLE 1. TEPHRA DESCRIPTIONS OF GLASS MOUNTAIN TEPHRA, MEDICINE LAKE HIGHLANDS.

Composition and Texture	Unit				
	7	8	9	10	11 (Top)
TEXTURE					
Pyroclast shape	equant to slightly elongate	equant to platy	equant to highly elongate	equant to platy	equant to slightly elongate
Length/width ratio	1.4	1.4	1.5	1.2	1.5
Vesicle shapes	ovoid to highly elongate	ovoid to highly elongate	spherical to highly elongate	spherical to elongate ovoid	spherical to elongate
Coalesced vesicle shapes	spherical to highly elongate	ovoid to highly elongate	ovoid to highly elongate	spherical to elongate ovoid	spherical to highly elongate
Lithic fragments	yes	yes	no	no	yes
Vesicularity (%)	40-50	40-50	40-50	40-50	55
Mode	all glass	all glass	n.d.	all glass	all glass
Pumice surface	A (angular)	A	A	A	A
COMPOSITION (%)					
Highly vesicular pumice	77	78	n.d.	80	20
Basalt fragments	---	---	n.d.	---	40
Dacite fragments	33	---	n.d.	---	---
Banded, nonvesicular glass fragments	---	22	n.d.	20	40
Unit thickness (cm)	102	26	33	20	60

Chemical Composition: All samples were analyzed using an electron microprobe (Table 2). There are no mineral phases in Glass Mountain tephra and only traces of plagioclase and pyroxene phenocrysts in Little Glass Mountain tephra. All glasses analyzed were of rhyolitic composition. The compositions are very similar to rhyolite flows erupted following the Plinian activity.

TABLE 2. GLASS ANALYSES OF GLASS MOUNTAIN TEPHRA, MEDICINE LAKE HIGHLANDS.*

Unit	1	2	3	4
SiO_2	73.61	73.17	73.19	73.30
TiO_2	0.28	0.27	0.32	0.27
Al_2O_3	13.73	13.91	14.31	13.80
FeO	1.53	1.55	1.82	1.62
MgO	0.27	0.26	0.36	0.25
CaO	1.22	1.17	1.42	1.13
Na_2O	2.53	3.19	3.02	3.30
K_2O	4.20	4.23	3.99	4.34
F	0.11	0.18	0.12	0.08
Total	97.48	97.93	98.55	98.11
Number of Analyses	4	5	4	5

*Pumice Pit section (2.8 km east). Analyses by Cameca electron microprobe, Model Camebax.

Comments on Pyroclast Formation: The high vesicularity and lack of phenocrysts indicate rapid rise of magma and eruption without much volatile loss. All the pumice pyroclasts contain parallel vesicles with high aspect ratios. It is likely that initially the highly elongate vesicles grew in the magma as spheres and then were stretched into their present shapes by laminar flow of the highly viscous rhyolite magma through dikes. Most exhibit some degree of vesicle coalescence in their ruptured mutual vesicle walls; thin rounded threads and ridges within these vesicles are remnants of burst vesicle walls.

Most vesiculation appears to have occurred before fragmentation of the magma. The majority of pumice pyroclasts here have blocky, angular forms with surfaces parallel to and normal to elongate vesicles. If, indeed, elongation of vesicles were caused by flow within the vent, then fracture surfaces perpendicular to this flow may have formed when an expansion wave moved downward from the surface, much like pyroclast formation models developed by Rittmann (1936), Szekely and Reitan (1971), and Bennett (1974). The angularity of the pyroclasts indicates derivation from a highly viscous, brittle foam. After fragmentation and eruption of pyroclasts, none of the pumices were deformed or "inflated," which would have been the case if expansion of gases had continued after their eruption. After the volatile-rich top of the magma body had erupted, rhyolitic lavas with low (<20%) vesicularities were erupted.

REFERENCES: Anderson, C. A., 1941. Volcanoes of the Medicine Lake Highland, California. Univ. Calif. Berkeley Publ., Geol. Sci., **23**: 215-276.

Bennett, F. C., 1974. On volcanic ash formation. Am. J. Sci., **274**: 648-661.

Chesterman, C. W., 1956. Pumice, pumicite and volcanic cinders in California. Calif. Div. Mines Bull., **174**: 119 pp.

Heiken, Grant, 1978. Plinian-type eruptions in the Medicine Lake Highland, California and the nature of the underlying magma. J. Volcanol. Geotherm. Res., **4**: 375-402.

Rittmann. A., 1936. Vulkane und ihre Tätigkeit. Ferdinand Enke, Stuttgart, 188 pp.

Szekely, J. and Reitan, P. H., 1971. Dike filling by magma intrusion and by explosive entrainment of fragments. J. Geophys. Res., **76**: 2602-2609.

Walker, G. P. L., 1971. Grain-size characteristics of pyroclastic deposits. J. Geol., **79**: 696-714.

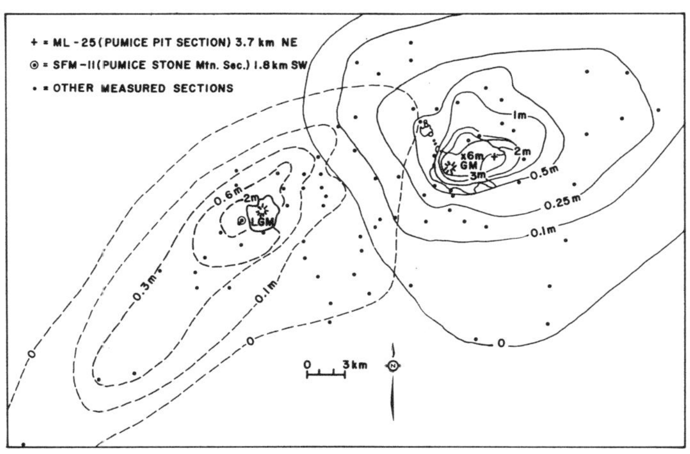

FIGURE 1. Isopach map of Little Glass Mountain (LGM - dashed lines) and Glass Mountain (GM - solid lines) tephra. Stippled areas are flows at the two vent areas. Dots are the locations of measured stratigraphic sections. The cross near Glass Mountain is the location of the thickest section measured.

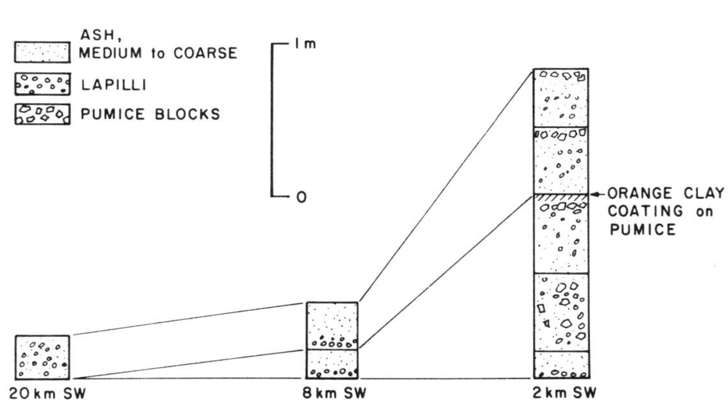

FIGURE 2. Tentative correlation of tephra units along a line SW from Little Glass Mountain. There may have been two episodes of explosive activity at LGM, since near-source units are separated by a thin layer of orange clay (wind blown, infiltrated dust?).

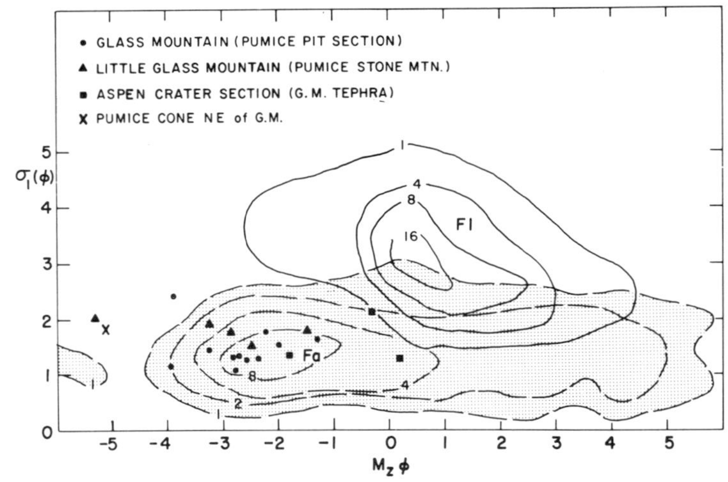

FIGURE 3. Mean grain size Mz_{φ} plotted against sorting σ_{φ} of tephra from Little Glass Mountain and Glass Mountain. Outlines of tephra fall (Fa) and tephra flow (FL) deposits are from Walker (1971). Contours represent percentages. Nearly all the samples are within the tephra fall category.

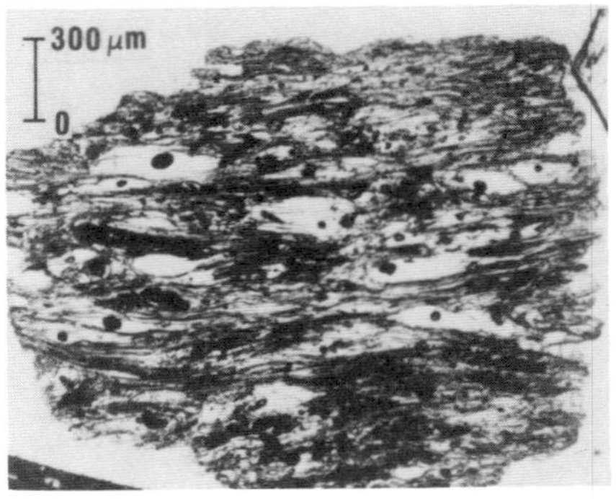

FIGURE 4. Photomicrograph of a typical pumice pyroclast with highly elongate, parallel vesicles. Note the fracture surfaces oriented normal to the long axes of vesicles.

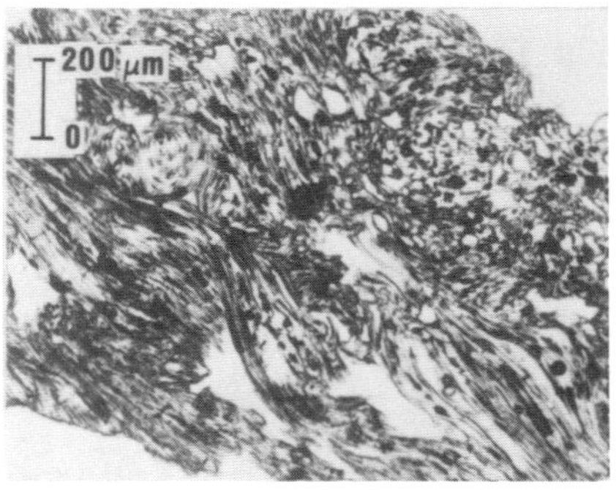

FIGURE 5. Photomicrograph of a pumice pyroclast; it shows deformed elongate vesicles and "vortices."

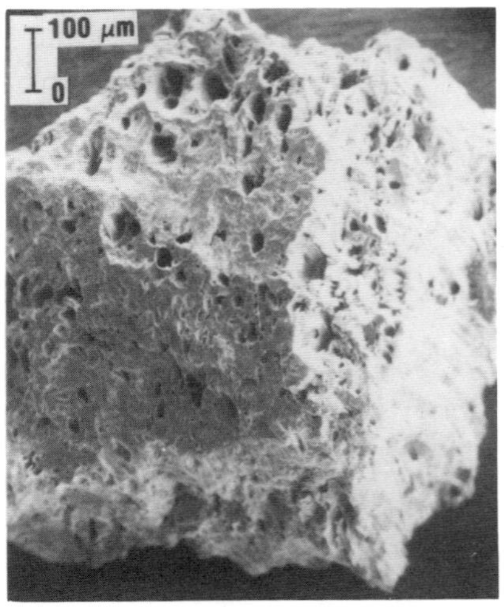

FIGURE 6. End view of a pumice pyroclast that is normal to elongate vesicles.

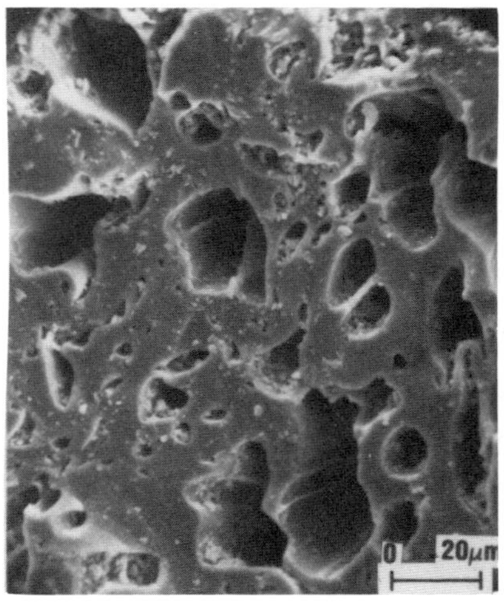

FIGURE 7. Closer view of the pyroclast in Fig. 6, showing cross sections of single and coalesced vesicles. Septae that remain from collapsed walls between coalesced vesicles are visible within the vesicles.

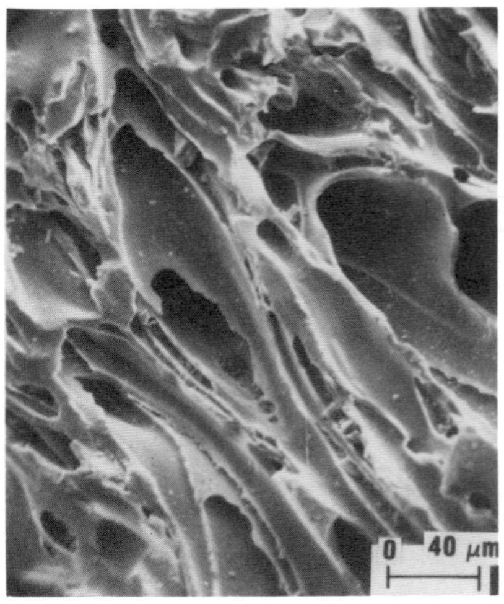

FIGURE 8. Surface of a pumice pyroclast with thin vesicle walls. Tapered ends of highly elongate vesicles are visible.

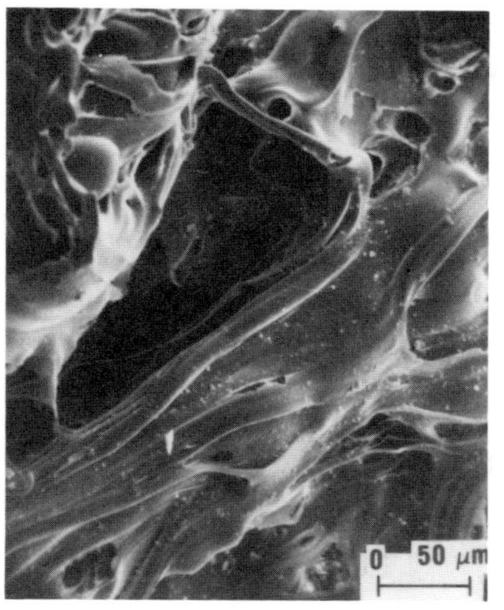

FIGURE 9. A few of the pumice pyroclasts in this deposit exhibit features that are indicative of continued vesicle growth and plastic deformation after comminution and eruption. In this example, the outer surface is deformed and outermost vesicles are intact. Some vesicle walls have collapsed but were drawn into thin glass threads. This texture is similar to that of many coarser pumice bombs found near the vents.

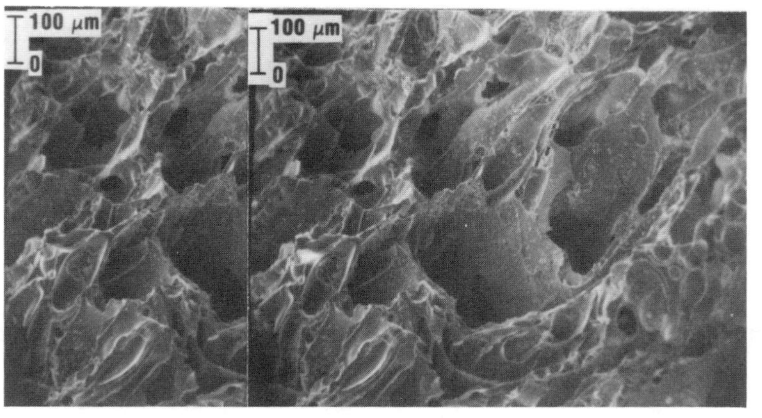

FIGURE 10. Stereo-pair image of a pumice pyroclast surface similar to that in Fig. *8*. Many of the vesicles are not only elongate but flattened. Note the extremely thin, delicate walls between vesicles. All the larger vesicles are compound; within them are septae or wall remnants of coalesced vesicles.

CHAOS CRAGS, LASSEN VOLCANIC NATIONAL PARK, CALIFORNIA

Introduction: The Chaos Crags are a group of silicic lava domes and associated tephra deposits located on the edge of the central plateau of Lassen Volcanic National Park, California. About 0.15 km^3 of rhyodacitic pyroclastic flows were erupted 1,100 years ago (Crandell et al., 1974) from vents beneath the Crags and moved down the valleys of Lost and Manzanita Creeks. Three episodes of pyroclastic flows were defined at stratigraphic sections in Lost Creek Valley, 3 km ENE of the vent.

Nearly all the pyroclastic-flow units consist of massive, nonwelded coarse ash that sometimes contains large pumice blocks and lapilli. The larger blocks have breadcrusted surfaces. Examples of the ash shown here are from a stratigraphic section in the Lost Creek drainage. Units A, B, and C-D are separated by thin silt layers that are possibly from windblown dust deposited between pyroclastic flows.

A section of ash fall tephra that is associated with the pyroclastic flows extends northeast from the Crags. Tephra deposits are overlain by a short, stubby flow and a three-lobed dome (the Chaos Crags). These domes were extruded from or near the vents that produced the tephra and consist of 1.25 km^3 of dacite lava.

Tephra Description: The pyroclastic flows consist of pumice block bearing medium to coarse, vitric-crystal or crystal-vitric ash (Tables 1 and 2). They contain pumice, shards, plagioclase, hornblende, Fe-Ti oxide, and quartz pyroclasts; there are also andesitic and dacitic lithic clasts. There is little variation in pyroclast types from Unit A to Unit D. Plagioclase content decreases from 42% in Unit A to 25% in Unit B and stays at that level throughout. Plagioclase pyroclasts (average composition of An_{30}) generally have a rind of rhyolitic glass.

Pumice pyroclasts are angular and slightly elongate; most pyroclasts have sharp, angular surfaces, but a few have smooth surface "skins." The smooth outer surfaces are only a few µm thick. These pumices have vesicularities of less than 30%. Vesicle shapes range from spheres to highly elongate "tubes," but most are ovoids 10 to 200 µm long. In many pumice pyroclasts, elongate vesicles are radial to phenocryst surfaces. This texture is associated only with larger phenocrysts or glomeroporphyritic clots.

Shards, consisting of vesicle walls broken from pumice pyroclasts, are present in some of the pyroclastic flows. The most common crystal pyroclast is plagioclase; some have bubble-wall textures and traces of vesicular glass adhering to grain surfaces. Lithic pyroclasts make up 2 to 4% of the tephra and are mostly angular to subrounded andesite fragments.

TABLE 1. PYROCLAST TYPES, PYROCLASTIC FLOWS OF LOST CREEK, CHAOS CRAGS.*

							Xenoliths (Lithic Fragments)		
Unit	**Pumice**	**Shards**	**Plag**	**Hb**	**Bio**	**Qtz**	**Basalt**	**Andesite**	**Dacite**
D	53.0	27.5	7.0	tr	1.5	1.0	---	5.5	1.5
C	100.0	---	---	---	---	---	---	---	---
B (upper)	72.7	---	15.0	0.7	2.7	1.7	tr	6.7	0.7
B (lower)	69.1	6.3	10.2	1.6	1.3	2.6	---	3.0	5.6
A**	62.7	3.7	20.6	2.3	3.0	1.7	2.7	1.3	0.3

*Volume per cent determined by grain counts.
**Also contains 0.7% Fe-Ti oxides, 0.3% resorbed plagioclase, and 0.6% pyroxene.

TABLE 2. PETROGRAPHIC CHARACTERISTICS OF TEPHRA, PYROCLASTIC FLOWS OF LOST CREEK, CHAOS CRAGS.

		Mode*											
Unit	**Thickness 3 km E of vent (m)**	**Glass**	**Plag**	**Hb**	**Bio**	**Qtz**	**Fe-Ti Oxide**	**Glomerocrysts**	**Resorbed Plag**	**Vesicularity (%)**	**Vesicle Shape**	**Vesicle Size (µm)**	**Ash Classification****
D	0.82	67.5	19.9	2.4	5.1	3.4	0.9	0.4	0.9	15.5	spherical to ovoid	40-650	pumice block (near vent)
D	0.82	64.6	27.9	4.3	3.2	---	---	---	---	46.0	irregular ovoid	5-2100	pumice block (at 3 km)
D	0.82	58.5	27.1	6.2	2.2	3.0	0.9	---	---	24.0	ovoid to highly elongate	5-100	vitric-xl
C	0.6	63.4	21.1	8.9	3.7	0.4	1.9	---	---	30.0	ovoid	10-600	vitric
B	5.7	52.3	31.4	3.8	2.1	8.4	1.3	---	0.4	24.0	ovoid	10-1000	vitric-xl
B	5.7	62.5	24.6	4.7	2.4	4.0	2.0	---	1.2	16.0	spherical to ovoid	5-30	vitric-xl
A	1.2	42.2	42.7	4.6	7.4	1.6	1.6	---	---	19.0	ovoid	20-600	xl-vitric

*Percentages determined from thin section.
**See introduction for classification used here.

Chemical Composition: All glass fractions are rhyolitic; they were analysed by a Cameca electron microprobe (Table 3).

TABLE 3. CHEMICAL COMPOSITION OF TEPHRA, CHAOS CRAGS.*

	Unit A	**Unit B**	**Unit B lower**	**Unit C upper**	**Unit D**
SiO_2	77.07	77.67	78.72	77.19	75.77
Al_2O_3	12.60	12.53	12.34	12.51	12.17
FeO	0.73	0.74	0.78	0.74	0.71
CaO	0.81	0.83	0.74	0.82	0.82
Na_2O	2.53	1.98	1.58	3.03	2.82
K_2O	2.53	1.78	1.86	3.98	4.02
Total	96.27	95.53	96.02	98.27	96.31
Number of Analyses	7	6	10	8	5

*Glass fraction of tephra for Lost Creek pyroclastic flow.

Comments on Pyroclast Formation: Elongate vesicles, oriented perpendicular to surfaces of large phenocrysts or glomeroporphyritic clots, are common within pyroclasts of this tephra sequence. This texture is anomalous within pumice pyroclasts that are dominated by small, ovoid vesicles. It is present only on phenocrysts larger than 0.7 mm. This texture is associated with plagioclase, hornblende, biotite, quartz, and magnetite phenocrysts. The number of phenocrysts with radial vesicular texture decreases from the bottom to the top of the section (e.g., 45 to 60% of the phenocrysts in Unit C are surrounded by radial vesicles, whereas within Unit D, 0 to 19% of the phenocrysts are associated with this texture).

This texture may have formed in several ways. The most commonly accepted is that vesicle growth was initiated at phenocryst surfaces, and subsequent vesicle growth was normal to those surfaces. Another possible explanation is that of pool boiling. Pool boiling may occur where a hot surface is submerged in a cooler liquid; closely spaced bubbles nucleate at the surface, then move away from it (Tong, 1972). Before vesiculation, phenocrysts and surrounding liquid are in thermal equilibrium, and when vesiculation begins, the liquid is cooled by adiabatic expansion. If the vesiculation process is rapid enough, the temperature gradient between phenocrysts and liquid may be great enough for vapor exsolution near the crystal/melt interface. The process is fast: a 1-mm-diameter phenocryst reaches thermal equilibrium within a vesiculating melt in about 1 second.

Radial vesicle patterns in pumice must form as the melt is being comminuted and accelerated out of the vent; the pattern is not deformed by flow, as would be the case if this process occurred at depth in a melt that was rising to the surface before eruption.

REFERENCES: Crandell, D. R., Mullineaux, D. R., Sigafoos, R. S., and Rubin, M., 1974. Chaos Crags eruptions and rockfall-avalanches, Lassen Volcanic National Park, California. J. Res. U.S. Geol. Surv., **2**: 49-59.

Heiken, G. and Eichelberger, J., 1980. Eruptions at Chaos Crags, Lassen Volcanic National Park, California. J. Volcanol. Geotherm. Res., **7**: 443-481.

Tong, L. S., 1972. Boiling crisis and critical heat flux. U.S. At. Energy Comm. Crit. Rev. Ser., TD-25897: 82 pp.

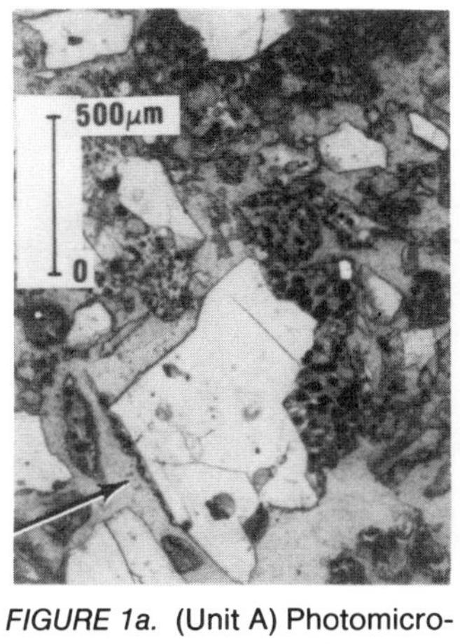

FIGURE 1a. (Unit A) Photomicrograph of crystal-vitric ash, consisting of mostly small, angular pumice pyroclasts. Abundant phenocrysts (mostly plagioclase) are present in the pumice and as separate pyroclasts. Partly crossed polarizer.

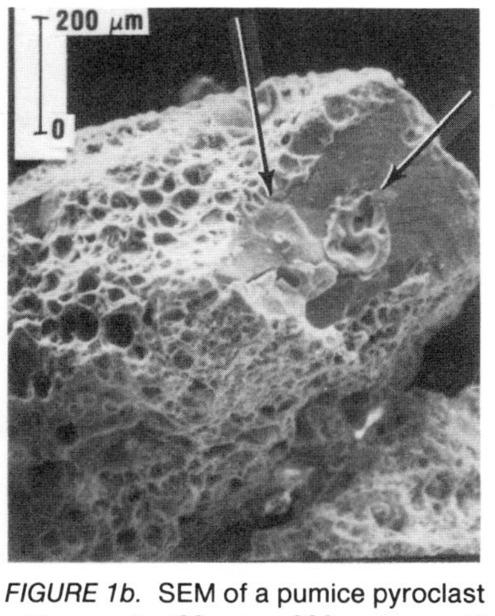

FIGURE 1b. SEM of a pumice pyroclast with small (20- to 600-µm) ovoid pyroclasts that are typical of the tephra in this unit. Within the core is a phenocryst surrounded by vesicular glass, and on one end, a "bubble-wall" texture similar to the texture visible on the left-lower edge of the largest phenocryst visible in *1a* (arrow).

FIG. 1 —
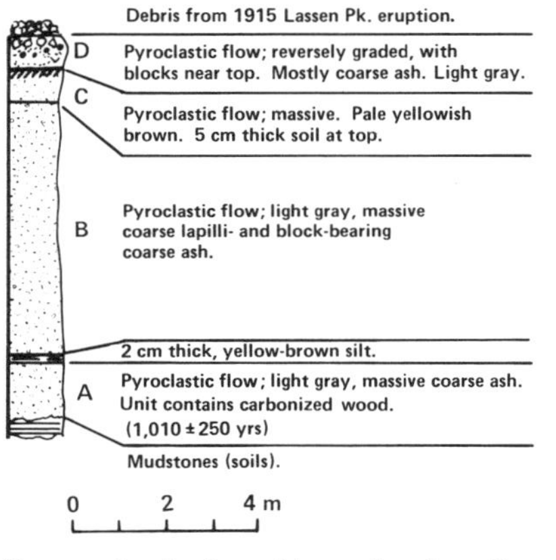

Composite stratigraphic section from the floor of Lost Creek Valley, 3.0 km E and 550 m below the vent. Carbon-14 data from Crandell et al. (1974).

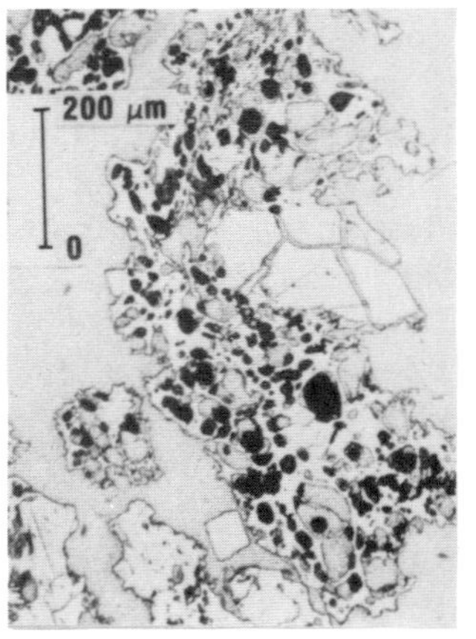

FIGURE 2a. (Unit B) Reflected-light photomicrograph of pumice pyroclast with a relatively high vesicularity. Although it is generally angular, there is some degree of rounding.

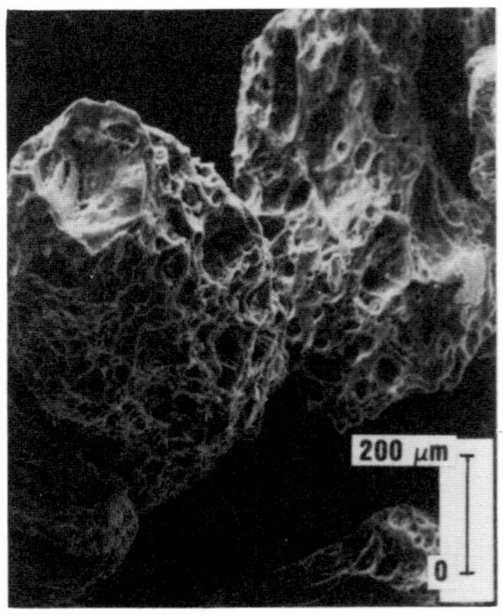

FIGURE 2b. SEM of pumice pyroclasts similar to those in Fig. *2a*. It has mostly ovoid vesicles. Although most pumice surfaces are irregular and angular, some are rounded by viscous deformation of vesicle walls after comminution (arrow).

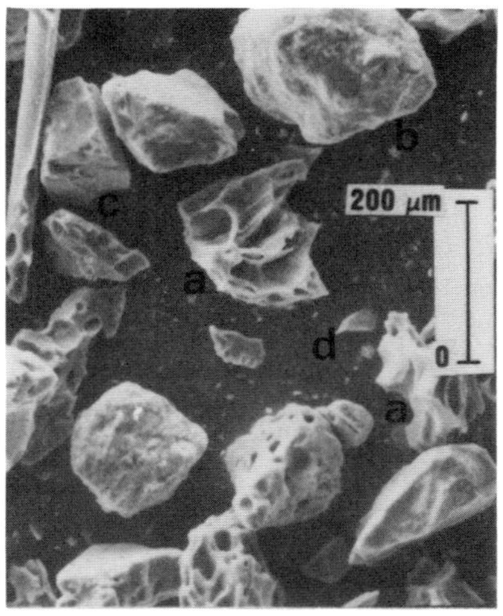

FIGURE 3. (Unit C) SEM of fine-grained tephra that consist of small pumice (*a*), lithic (*b*) and crystal (*c*) pyroclasts, and shards (*d*).

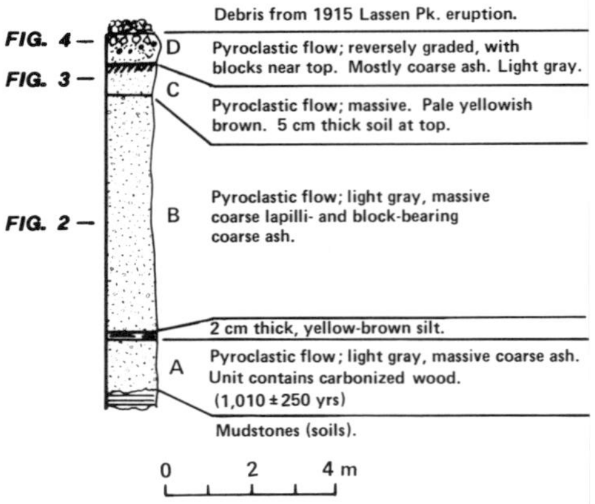

Composite stratigraphic section from the floor of Lost Creek Valley, 3.0 km E and 550 m below the vent. Carbon-14 data from Crandell et al. (1974).

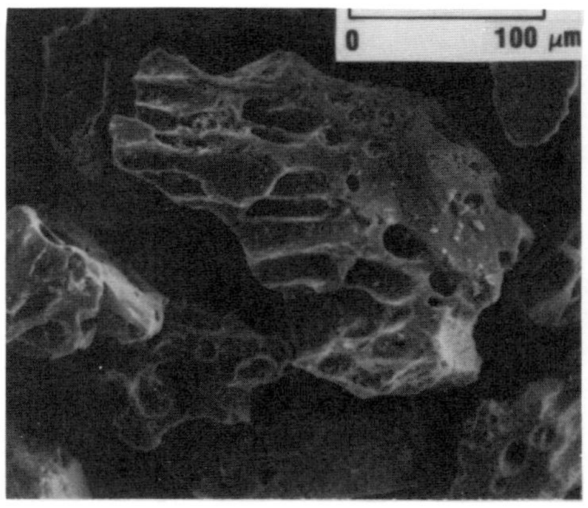

FIGURE 4a. (Unit D) SEM of pumice pyroclasts that show low vesicularity (20 to 40%) with slightly elongate vesicles. Most vesicles are separated by thick (10- to 200-μm) vesicle walls. These pyroclasts have sharp, angular surfaces.

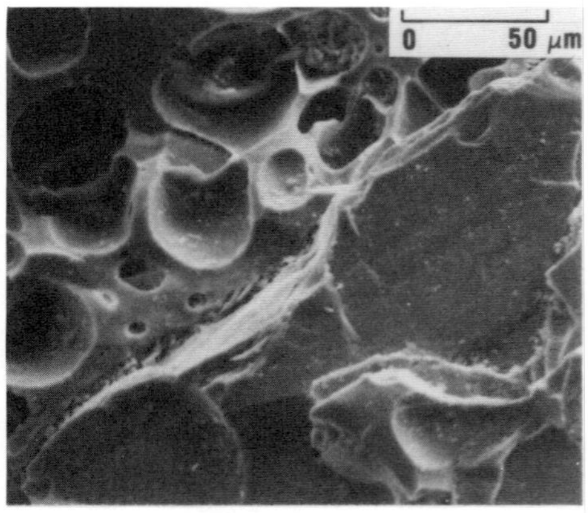

FIGURE 4b. SEM of a phenocryst/glass boundary. In this particular case there appears to be no relation of phenocryst surfaces and vesicle shapes, as discussed in the text and in Fig. *5*. There has been some flattening and deformation of very small vesicles (1 to 2 μm by 10 to 13 μm) that are located at the phenocryst surface. Spherical to ovoid vesicles that are typical of pumices from this eruption sequence are visible 10 μm beyond the phenocryst surface.

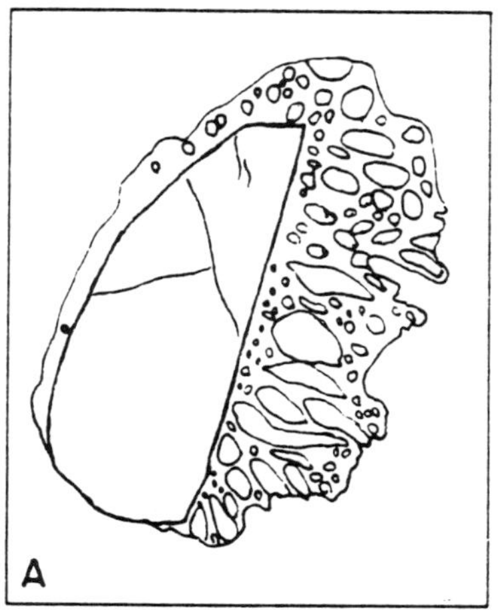

FIGURE 5a. Elongate vesicles oriented radially around phenocrysts. Pumice pyroclast from a pyroclastic flow in Manzanita Creek. Long axes of vesicles are up to 220 μm long and oriented normal to the surface of a plagioclase phenocryst. Base of the sketch is 0.8 mm wide.

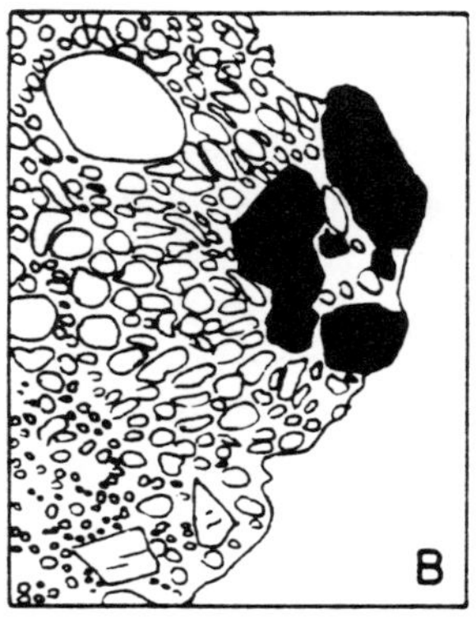

FIGURE 5b. Long axes of vesicles are radially oriented around a cluster of magnetite phenocrysts in this pumice pyroclast. Base of the sketch is 1.2 mm wide.

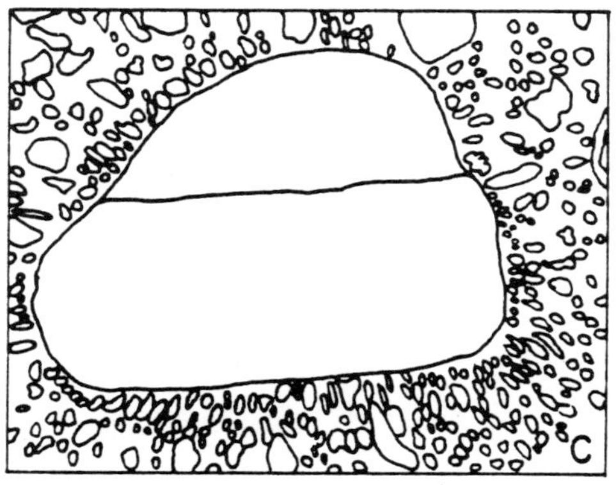

FIGURE 5c. Vesicles are clustered and oriented radially around a plagioclase phenocryst. Base of the sketch is 2.1 mm wide.

TAMBORA (LESSER SUNDA ISLANDS), INDONESIA; APRIL 1815 ERUPTION

Introduction: Tambora is located on the Sangaar Peninsula of the Island of Sumbawa, Lesser Sunda Islands, Indonesia. It is well known for its disastrous eruption of 1815, which resulted in great loss of life and a vast plume of ash and gases in the stratosphere. The overall effect on the Earth's atmosphere in 1816 was large; it caused the "year without a summer" (Rampino and Self, 1982).

The eruption began April 5, 1815, with earthquakes, eruption clouds, and rumbling (van Padang, 1951). Climactic activity began April 10, 1815, with a Plinian eruption that produced large, dense, ash clouds and pumice falls and lasted for 24 to 36 hours. The Island of Madura 500 km away was in total darkness for 3 days (van Padang, 1951). Plinian ash falls were followed by pyroclastic flows and surges. Petroeschevsky (1949) estimated that the ejecta had a volume of 100 km^3, but Self (personal communication) has calculated a bulk volume of 150 to 175 km^3. A 6-km-diameter, 1-km-deep caldera was formed during the eruption.

Grain Size: Samples from the 1815 deposits were provided by S. Self. Grain-size data given in Table 1 show both a decrease in median diameter and sorting from the Plinian phase products to the ignimbrites and surges.

TABLE 1. GRAIN-SIZE DATA FOR TAMBORA 1815 ERUPTION PRODUCTS.

Plinian Phase							
Md_φ	−.90	−.80	−.23				
σ_φ	1.65	2.0	2.14				
Ignimbrite							
Md_φ	−.50	−.80	+2.6	−.44	+1.0	+1.3	−.80
σ_φ	3.05	2.00	1.75	2.7	2.05	2.45	2.55
Surges							
Md_φ	+1.0	−.60					
σ_φ	4.1	3.4					

Tephra Descriptions: Tephra characteristics indicate eruption of a physically heterogeneous, volatile-rich magma. Within the Plinian pumice fall deposits there are at least three kinds of pumice. Two pumice types developed in a very vesicular, viscous (brittle) melt: one has large, ovoid vesicles and thin vesicle walls and the other has parallel, thin, highly elongate vesicles. Both pyroclast types are very angular and have sharp, thin broken vesicle walls that form grain surfaces. Most vesicles are composite.

Coalescence and better development of vesicles around phenocrysts form lens-shaped "pockets" within pumices. In contrast to these highly vesicular pumices, another type of pumice has fewer vesicles and is characterized by thick vesicle walls and rounding of all surfaces by viscous flow after vesiculation.

Tephra from the ignimbrite of 1815 is characterized by pyroclasts with larger, better developed vesicles than those of pumices from the Plinian phase. Overall vesicularity is lower and vesicle walls are thicker in pyroclasts from the ignimbrite. Shards are thick, curved plates or equant forms. All pyroclasts in this deposit are coated with small (2- to 16-μm), equant shards; these adhere to grain surfaces but are not welded to them.

Cross-bedded pyroclastic flow deposits overlying the ignimbrite contain a variety of pyroclasts similar to a mixture of those described for both the Plinian pumice fall and the ignimbrite. Small pyroclasts cling to grain surfaces but are not as pervasive as in the ignimbrite.

Chemical Composition: The chemical composition listed in Table 2 is from S. Self et al. (1984). These two analyses, representative of the compositional range of the 1815 eruption products, are normalized to 100% and are volatile-free.

TABLE 2. CHEMICAL COMPOSITION, TAMBORA.*

Element	1	2
SiO_2	56.00	57.88
TiO_2	0.63	0.58
Al_2O_3	19.78	19.82
Fe_2O_3	5.78	5.36
MgO	1.79	1.34
MnO	0.18	0.21
CaO	4.55	3.94
Na_2O	5.01	5.03
K_2O	5.83	6.01
P_2O_5	0.45	0.33

*Note: Total Fe as Fe_2O_3. From Self et al. (1984).

Comments on Pyroclast Formation: The eruption sequence is a classic example of decreasing volatile content as the eruption proceeded from a Plinian phase to mainly pyroclastic flows and surges. Vesiculation and comminution appear to have developed within a heterogeneous melt; angular, thin-walled pumices were derived from a more viscous melt, and less vesicular pyroclasts with fluidal surfaces came from a less viscous portion. Pyroclasts are clean; the finer grained fraction (if it existed) was winnowed out and carried farther away.

Pyroclastic flow and surge deposits contain a wider range of grain sizes; small particles cling to surfaces of larger pyroclasts of all types and sizes. Pumices and shards were derived from a melt with larger vesicle sizes but lower overall vesicularity. There is less elongation caused by flow of vesicles within these pyroclasts than within those from the Plinian deposits.

There is little evidence of abrasion of pyroclasts within any of the deposits; only minor abrasion of thin, delicate shard edges within the surge deposit can be seen.

REFERENCES:

Rampino, M. and Self, S., 1982. Historic eruptions of Tambora (1815), Krakatau (1883) and Agung (1963); their stratospheric aerosols and climatic impact. Quat. Res., **18**: 127-143.

Self, S., Rampino, M., Newton, M., and Wolff, J., 1984. A volcanological study of the great Tambora eruption of 1815. Submitted to Geology.

van Padang, M. N., 1951. Catalogue of the active volcanoes of Indonesia. Part I of *Catalogue of the Active Volcanoes of the World.* Int'l. Assoc. Volcanol.,Napoli: 271 pp.

Petroeschevsky, W. A., 1949. A contribution to the knowledge of the Gunung Tambora (Sumbawa). Tidschr. Kan. Ned. Aardr. Gen., **66**: 688-703.

FIGURE 1a. Pumice pyroclast from a fall unit with trachyandesitic pumice from the base. Stereo pair showing the slightly elongate, subrounded pumice. Most vesicles are parallel and have large L:W ratios. There has been considerable coalescence of larger vesicles around phenocrysts, forming irregular "pockets." Elongate vesicles are bent around the "pockets." Pumice surfaces are parallel to and normal to elongate vesicles.

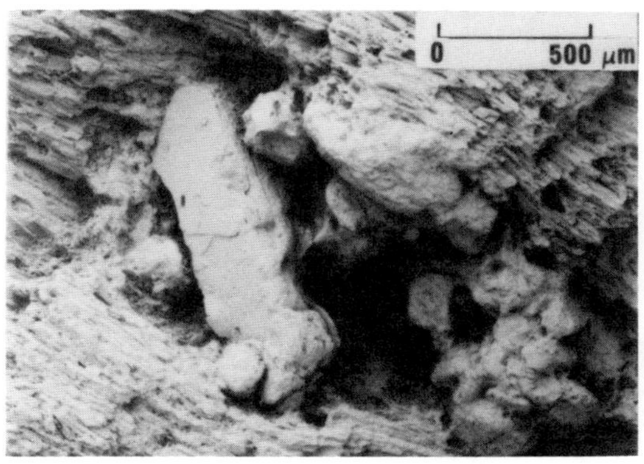

FIGURE 1b. Close-up of Fig. *1a*, showing the "pocket" of coalesced vesicles that developed around a cluster of phenocrysts, including biotite (*center-left*).

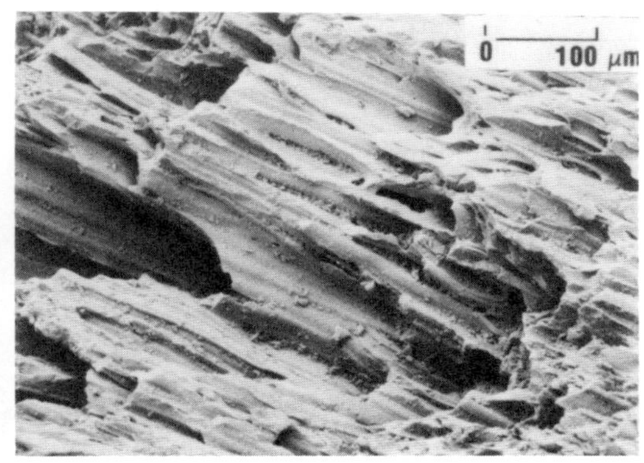

FIGURE 1c. Detail of a pumice surface, showing elongate, parallel vesicles with fairly thick (>3-μm) vesicle walls. Many of these elongate vesicles show considerable coalescence. Note the rounded parallel ridges that are collapsed vesicle walls.

FIGURE 1d. Detail of a pumice surface, showing remarkable smoothness and thickness of vesicle walls. Ridge-like septae are indicative of vesicle coalescence.

FIGURE 2a. Plinian pumice fall sample from near the base. Pumice pyroclast with large, well-developed ovoid vesicles. Most of the ovoid vesicles show collapse of mutual vesicle walls and coalescence to form larger composite vesicles. This is a very angular pyroclast with sharp edges along thin, broken vesicle walls.

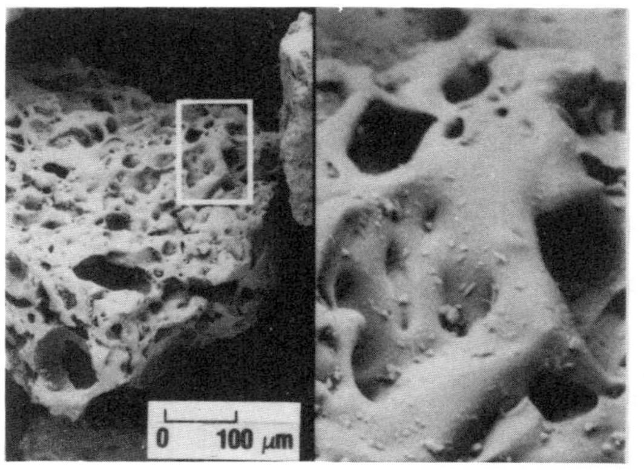

FIGURE 2b. In contrast with the sharp, angular, brittle pumice pyroclast of Fig. *2a*, this pumiceous pyroclast shows evidence of viscous deformation by the smoothing and rounding of all surfaces that must have occurred after vesiculation and separation of individual pyroclasts. Irregular, ovoid vesicles. *Inset* - view of surface, showing smooth, curved surfaces. Note that even margins of vesicles open to the grain surface have been smoothed, perhaps by movement during flow.

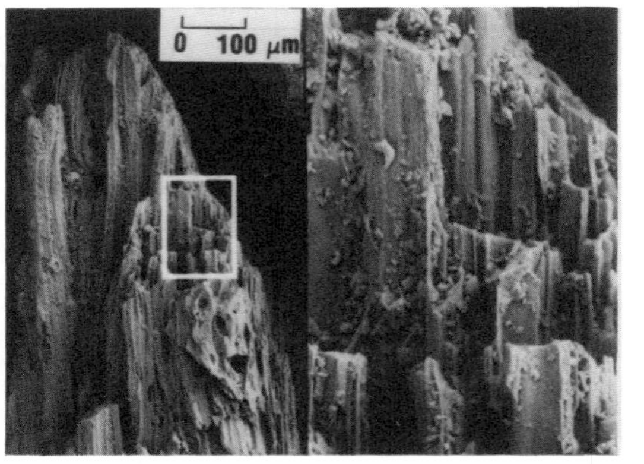

FIGURE 2c. Angular pumice pyroclast with closely spaced, parallel, elongate vesicles and thin, angular vesicle walls. *Inset* - close-up of surface; the composite vesicles with rounded and angular parallel ridges within vesicles exceed the volume of single vesicles.

FIGURE 2d. Stereomicrograph of a pumice fall sample. Note the diversity of pumice types, which illustrates magma heterogeneity. Some rounded or subrounded lithic fragments are present.

FIGURE 3a. The ignimbrite of 1815. Fine-grained size fraction of the ignimbrite. In contrast with the Plinian phase pumices, these have large, better developed vesicles but lower overall vesicularity. Vesicle walls are thick (>10 µm). Shards are thick, curved plates or equant forms that were broken from a pumiceous melt with ovoid or spherical vesicles and low overall vesicularity. Many small, angular particles cling to pyroclast surfaces.

*FIGURE 3b.*Slightly closer view of a sample similar to that in Fig. *3a*. Note the 2- to 10-µm particles that cling to shard surfaces.

FIGURE 3c. Rounded pumice pyroclasts from coarser size fraction. Vesicularity is low, vesicles are mostly ovoids, and grain surfaces are coated with fine-grained particles.

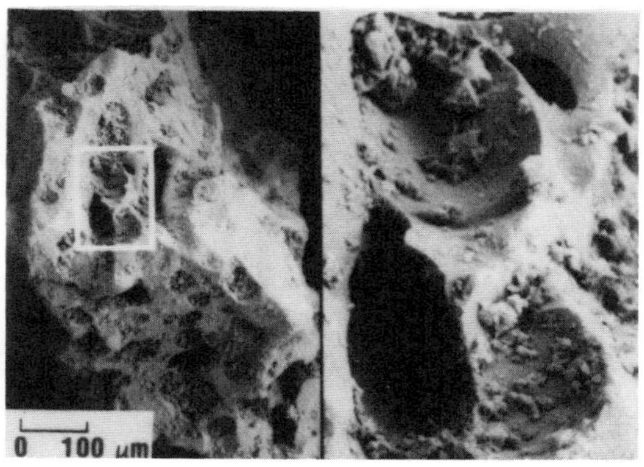

FIGURE 3d. Subangular pyroclast showing relatively low overall vesicularity of the pumices in this deposit. *Inset* - view of ovoid vesicles at and near the grain surface. Small (2- to 16-µm), equant particles adhere to grain surfaces.

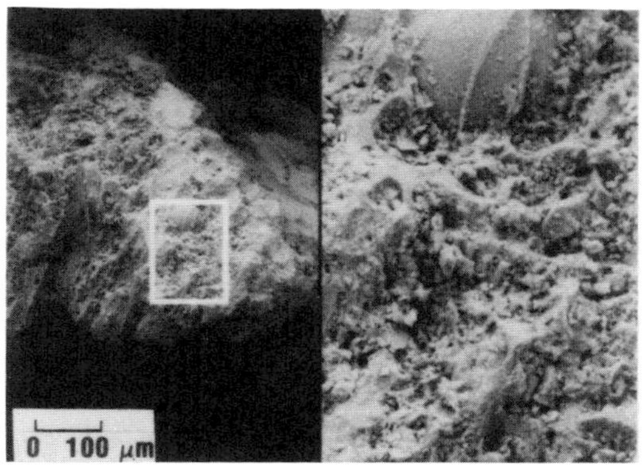

FIGURE 3e. Pumice pyroclast partly coated with fine-grained particles.

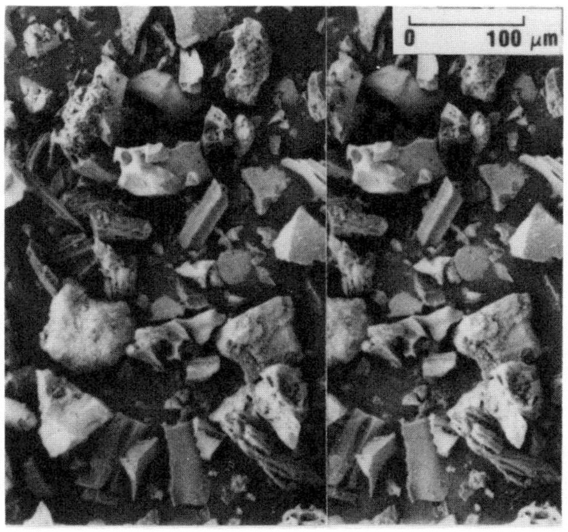

FIGURE 4a. Pyroclasts from a cross-bedded pyroclastic flow deposit *above* the 1815 ignimbrite. Stereomicrograph of the fine-grained size fraction. As in the Plinian fall deposit, this deposit contains a variety of pyroclast types including: 1. Shards derived from pumice containing highly elongate, composite vesicles and thin vesicle walls. 2. Equant or subequant angular blocky pyroclasts with low vesicularity, large ovoid vesicles, and thick vesicle walls. 3. Smooth, rounded droplets. 4. Aggregates of fine-grained (<10-µm), equant shards.

FIGURE 4b. Angular, subequant pumice pyroclast. *Inset* - surface of pumice. Irregular, composite vesicles bounded by both thin and thick vesicle walls. Some small, equant particles cling to pyroclast surfaces.

FIGURE 4c. Coarser grain-size fraction of the surge deposit. The variety of pyroclast types is the same as that described in Fig. *4a*. *Lower center* - a rounded lithic fragment.

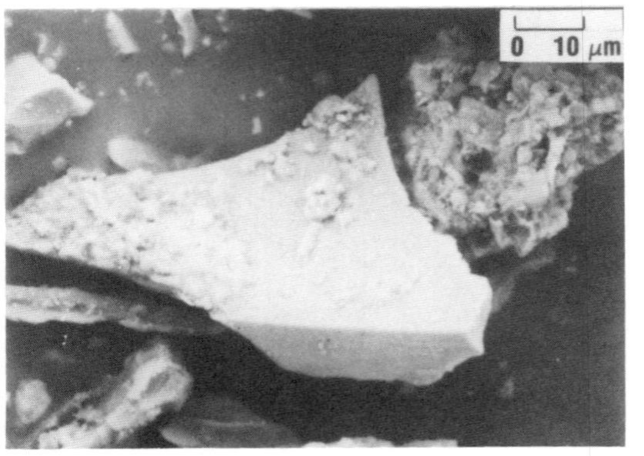

FIGURE 4d. Close-up of fine grain-size fraction. *Center* - angular, curved shard with fine-grained particles adhering to the grain surface. A few of the more delicate grain edges have been abraded. *Upper right* - aggregate of fine-grained shards.

HEKLA, ICELAND; H_4 TEPHRA

Introduction: The volcano Hekla, located along SW-NE-trending fissures in south-central Iceland, is that country's most famous volcano. Five major eruptions over the last 7000 years and 15 eruptions since 1104 A.D. (the first recorded activity) have spread tephra over much of Iceland and occasionally over northern Europe. Nearly all of Hekla's eruptions have begun with explosive eruptions of rhyolitic or rhyodacitic pumice. As these eruptions proceed, the magma becomes more mafic and less explosive and eruptions end after extrusion of basaltic or basaltic-andesite lava flows. Thorarinsson (1981) estimates that Hekla has erupted between 6 and 7 km^3 of silicic tephra over the last 7000 years. The example of Hekla tephra shown here is from the H_4 tephra layer, which is about 4000 ^{14}C years old. It is one of the most widespread Hekla tephra layers, covering 78,000 km^2 (on land), along a zone trending NNE of Hekla (Larsen and Thorarinsson, 1977). The samples for this example came from the Thjórsádalur Valley and from near Hraunfoss Dam, west of Hekla.

Tephra Descriptions: The fine-grained base of H_4 consists of finely comminuted rhyolitic vitric ash. The ash consists of small (25- to 50-μm-long) pumice pyroclasts and shards (1 to 15 μm long). Pumice pyroclasts contain: (1) narrow, elongate and (2) flattened ovoid vesicles. Most shards are slightly curved, but many are flat plates.

Pumices from the base of the Plinian pumice fall (collected in the Thjórsádalur Valley) are subrounded to subangular and equant to subequant. Vesiculation was heterogeneous; "pockets" of coalesced vesicles, 60 to 350 μm long, are surrounded by zones of slightly elongate vesicles that are 1 to 25 μm wide.

Pumice pyroclasts from the top of the pumice-fall are similar in shape to those from the base. They too are characterized by a heterogeneous vesicle distribution. In contrast with the base, however, vesiculation is more advanced; there are more pockets of coalesced vesicles. The pockets are better developed and consist of irregular, 100- to 500-μm-diameter lenticular tubes lined by layers of collapsed vesicle walls.

Chemical Composition: This is one of the most silicic tephras erupted from Hekla; near the base, SiO_2 = 74%, but it gradually decreases toward the top to 57%. Hekla is a classical case of a zoned magma chamber; each eruption begins as a gas-rich rhyolite and ends as a basaltic andesite. These variations are reflected in the nature of the tephra erupted.

Comments on Pyroclast Formation: A thin layer at the base of the H_4 unit appears to be an excellent example of phreatoplinian tephra. It is a highly vesicular rhyolitic magma that mixed with surface water (or ice and snow) and has been broken into extremely fine-grained pyroclasts.

Pumice pyroclasts in the pumice bed show some variation from the base to the top: not in the size or appearance of the coarse-grained pumice pyroclasts but in the size and shape of vesicles. The less rhyolitic pumices in the upper part of the fall unit have better developed vesicles and show more vesicle coalescence. This variation may be related to the lower silica content (and lower viscosity) of the last pumices to be erupted.

REFERENCES: Larsen, G. and Thorarinsson, S., 1977. H_4 and other acid Hekla tephra layers. Jökull, **27**: 28-46.

Thorarinsson, S., 1981. The application of tephrochronology in Iceland. In *Tephra Studies* (S. Self and S. Sparks, editors), Reidel Publ. Co., Dordrecht: 109-134.

FIGURE 1a. Fine-grained base of H_4 (collected near Hraunfoss Dam). Angular, colorless glass shards. These range from 1-μm-wide slivers to 15-μm irregular forms.

FIGURE 1b. Shards and small pumice pyroclasts. Elongate, thin vesicles and ovoid, flattened vesicles within the pumices provide some idea of the heterogeneous vesicularity that existed in the magma. As would be expected, these fragments are quite similar (in the nature of their vesicles) to pumices from the base of the pumice fall (Fig. *2a, b*).

FIGURE 2a. Pumice from the base of the H_4 pumice fall (from the Thjórsádalur Valley). Subequant, subrounded pumice pyroclast.

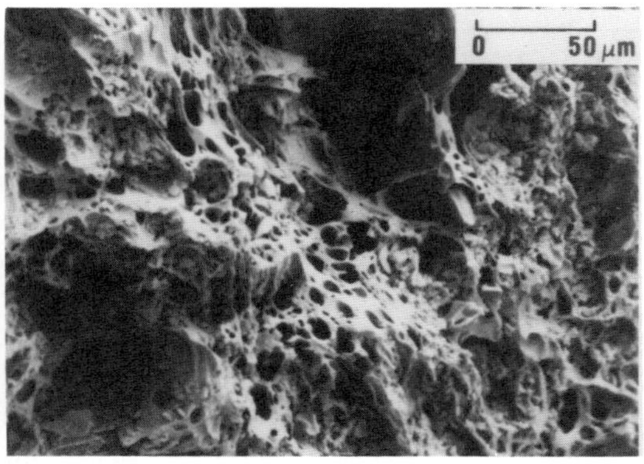

FIGURE 2b. Close-up of the pumice in Fig. *2a*. The heterogeneous nature of the vesicles is obvious: there are large pockets composed of coalesced vesicles and small, elongate vesicles surrounding the pockets; these are separated by relatively thick vesicle walls. Remnant vesicle walls in the "pockets" are thin and tapered.

FIGURE 3a. Pumice pyroclast from the top of the H_4 Plinian pumice-fall. Surface of a subangular pumice pyroclast. Pockets of coalesced vesicles form long, tapered tubes. These are surrounded by well-developed, ovoid, flattened vesicles. This pyroclast is partly covered with fine-grained shards.

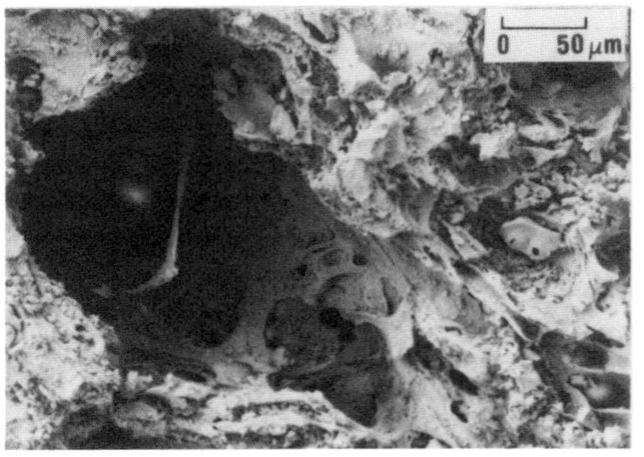

FIGURE 3b. Close-up of one of the tapered, tube-like vesicle pockets. It was formed by inflation and coalescence of many vesicles. Remnant vesicle walls are visible as flattened membranes and twisted fibers.

Peralkaline and Carbonatite Eruptions

Unusual volcanic rocks include the *peralkaline* rocks in which the abundance of Na_2O and K_2O is greater than that of Al_2O_3 and the *carbonatites*, which are composed of mostly carbonate minerals. Typically, the relatively low abundance of Al_2O_3 in peralkaline lavas causes magma viscosities that are much lower than their calcalkaline counterparts with the same silica content; hence pyroclasts from peralkaline eruptions have fluid textures. The abundance of CO_2 in carbonatites causes these magmas to be volatile-rich and eruptions to often be explosive.

PANTELLERIA, ITALY

Introduction: The island of Pantelleria is located in the Channel of Sicily, and its total area of 85 km^2 is underlain by mostly persodic volcanic rocks. Pantelleria is well known for its peralkaline silicic rocks (pantellerites), which were described by Washington (1914) and Villari (1974). Welded tuffs (Borsi et al., 1963) make up nearly half of the volcanic rocks of which the youngest, the Green Tuff, is best studied (Villari, 1969; Wright, 1980; Wolff and Wright, 1981a,b). Schmincke (1974) attributed the Green Tuff to a pyroclastic flow origin; Barberi et al. (1969) and Villari (1974) suggest, based on radiometric dates, that it was erupted within the last 0.12 m.y. (50,000 y. b. p.; Cornette et al., 1983). The widespread distribution of this well-sorted pumice, its mantle-bedding on steep slopes, and alternations of welded and nonwelded layers, are evidence that Sparks and Wright (1979), Wright (1980), and Wolff and Wright (1981a) have given for an air fall (tephra fall) emplacement origin. In fact, both modes of eruption and deposition could account for different parts of the Green Tuff.

The Monastero caldera, which is near the central part of the island, formed over the vent area of the Green Tuff. This vent area, a possible central vent, is older (Orsi, 1982 personal communication) than the Cuiddia di Mida pumice cone vent cited by Mahood and Hildreth (1980, 1983). The samples described here were obtained from G. Orsi and M. Sheridan: they were collected near the north rim of the Monastero caldera on the north side of Bagno dell'Acqua. The samples were from a pumice fall and welded ash flow just below and in contact with the Green Tuff (M. Sheridan, 1983 personal communication); they are pantellerites and may be part of the Green Tuff. In addition to the following descriptions, Schmincke (1974) describes ash from the Green Tuff in reference to overall characteristics of tephra from peralkaline, welded ash flow tuffs.

Tephra Description: The tephra consists of glass, crystal, and lithic fragments; the glass and crystals described here range in size from 0.1 to 0.5 mm in maximum dimension.

Glass. Because of the lower viscosity of peralkaline silicic magmas (as much as 3 orders of magnitude less than that of calcalkaline rhyolite), glass particles show both fluid-formed textures and those of brittle breakage. Typically, the glass is vesicular as a result of its magmatic fragmentation and shows partly ruptured bubbles (see Schmincke, 1974, p. 609). Some vesicles are elongated and half of the studied samples are hydrated; glass surfaces are covered by clay minerals as well as abundant feldspar laths. The posteruptive fluidity of the magma is evident in deformed vesicle surfaces and globular shards (Johnson, 1968; Hay et al., 1979) that resemble Pele's tears formed in Hawaiian lava fountains. Vesicle-wall shards and chips are common in finer size fractions and several generations of vesiculation are evident. Stretched and deformed vesicles developed during postemplacement (rheomorphic) movement of still-fluid pyroclasts are a characteristic of pantelleritic tuffs. Vesicular tricuspate (Y-shaped) shards that resemble calcalkaline Plinian ash are also evident. However, extreme aspect ratios of vesicles and fluidal surfaces along vesicle walls betray their fluid nature.

Crystals. Euhedral to subhedral crystals of anorthoclase, hedenbergitic clinopyroxene, aenigmatite, fayalite, apatite, and magnetite/ilmenite are abundant (>10 wt. %). Clinopyroxenes are distinct because of the glass encrustations that form bubble-wall textures. Feldspars, however, have clean surfaces with no adhering glass.

Chemical Composition: Energy-dispersive spectral (EDS) analyses of samples studied here are compared to previous analyses of fresh samples of the Green Tuff (Table 1). Hydration has affected the studied samples to some degree so that elemental abundances are somewhat altered. However, these analyses also suggest that the tephra at the base of the Green Tuff may be derived from the silica-enriched top of the magma chamber that erupted the Green Tuff.

TABLE 1. CHEMICAL COMPOSITION OF GLASS FROM THE GREEN TUFF.*

	1	2	3	4	5	6
SiO_2	70.0	69.2	74.74	68.30	74.67	72.24
TiO_2	0.5	0.44	0.61	1.04	0.47	0.73
Al_2O_3	7.4	7.8	8.62	14.68	8.75	10.99
$FeO + Fe_2O_3$	8.1	8.4	7.30	4.68	7.00	6.20
MnO	---	0.35	0.04	0.29	0.15	0.17
MgO	---	0.2	0.36	0.66	0.43	0.49
CaO	0.3	0.34	0.46	1.37	0.55	0.88
Na_2O	6.5	6.4	1.64	2.04	1.76	1.83
K_2O	3.6	4.5	6.23	6.94	6.21	6.45

*Analysis of sample 1 is taken from Wolfe and Wright (1981) and that of sample 2 is from Wright (1980), based upon data from Korringa and Noble. Normalized standardless EDS analyses of samples 3, 4, and 5 are those of partly altered pyroclasts of this report, and 6 is an average analysis of all samples studied in this report.

Comments on Tephra Formation: Schmincke (1974) discussed the physical and chemical aspects of peralkaline magmas and the formation of tephra from them. General inferences made from these physio-chemical aspects are that the magmas are hotter, lower in H_20, and of lower viscosity than calcalkaline silicic magmas.

The formation of spherical shards with hollow centers and subspherical globular particles may be the result of several processes: (1) Surface tension effects minimize surface free energy by spherical geometry for fluid fragments—a mechanism occurring during dispersal before deposition. (2) Posteruptive expansion of gas inside volatile-rich shards is especially possible for fluid melt fragments and is manifested by spherical vesicles that occur in collapsed pumice fragments; postdepositional vesicles transect earlier formed vesicles. (3) Finally, postdepositional movement of ash deposits may distort the original spherical shape of vesicles and blade-like shapes of shards (Wolff and Wright, 1981a,b).

All the above processes that contribute to the distinctive shape of peralkaline ash particles are strongly functions of viscosity. Abundant chlorine and fluorine, as well as the small amounts of alumina and the hot eruption temperatures, are responsible for the fluidity of these pyroclasts. That the initial fragmentation of these particles in the vent was entirely magmatic is not yet clear. The existence of both brittle, fractured shards and fluid-deformed particles, the inferred low-volatile content of the peralkaline magma, and the abundance of hydrated glass sugggest that the eruption may have been phreatoplinian or dominantly phreatomagmatic. This hypothesis also explains the uncommonly wide dispersal of the peralkaline deposits by both ash flow and ash fall.

REFERENCES:

Barberi, F., Borsi, S., Ferrara, G., and Innocenti, F., 1969. Strontium isotopic composition of some recent basic volcanites of the southern Tyrrhenian Sea and Sicily Channel. Contr. Mineral. Petrol., **23**: 157-172.

Borsi, B., Marnelli, G., Mazzoncini, F., Mittempergher, M., and Tedesco, C., 1963. Reconnaissance of some ignimbrites of Pantelleria and Eolian Islands. Bull. Volcanol., **25**: 359-367.

Cornette, Y., Crisci, G. M., Gillot, P. Y., and Orsi, G., 1983. Recent volcanic history of Pantelleria: a new interpretation. J. Volcanol. Geotherm. Res., **17**: 361-373.

Hay, R. L., Hildreth, W., and Lambe, R. N., 1979. Globule ignimbrite of Mount Suswa, Kenya. Geol. Am. Spec. Paper **180**: 43-75.

Johnson, R. W., 1968. Volcanic globule rock from Mt. Suswa, Kenya. Roy. Soc. London Trans., **265**: 383-412.

Mahood, G., and Hildreth, W., 1980. Pantelleria, a new interpretation. Trans. Am. Geophys. Union, **61**: 1141.

Mahood, G., and Hildreth, W., 1983. Nested calderas and trapdoor uplift at Pantelleria, Strait of Sicily. Geology **11(2)**: 722-726.

Schmincke, H.-U., 1974. Volcanological aspects of peralkaline silicic welded ash-flow tuffs. Bull. Volcanol., **38**: 594-636.

Sparks, R. S. J. and Wright, J. V., 1979. Welded air-fall tuffs. Geol. Soc. Am. Spec. Paper, **180**: 155-166.

Villari, L., 1969. On particular ignimbrites of the island of Pantelleria (Channel of Sicily). Bull. Volcanol., **38**: 689-724.

Villari, L., 1974. The Island of Pantelleria. Bull. Volcanol., **38**: 680-724.

Washington, H. S., 1914. The volcanoes and rocks of Pantelleria. J. Geol., **22**: 653-670 and 683-713.

Wolff, J. A. and Wright, J. V., 1981a. Formation of the Green Tuff, Pantelleria. Bull. Volcanol., **44**: 681-690.

Wolff, J. A. and Wright, J. V., 1981b. Rheomorphism of welded tuffs. J. Volcanol. Geotherm. Res., **10**: 13-34.

Wright, J. V., 1980. Stratigraphy and geology of the welded air-fall tuffs of Pantelleria, Italy. Geol. Rundsch., **69**: 263-291.

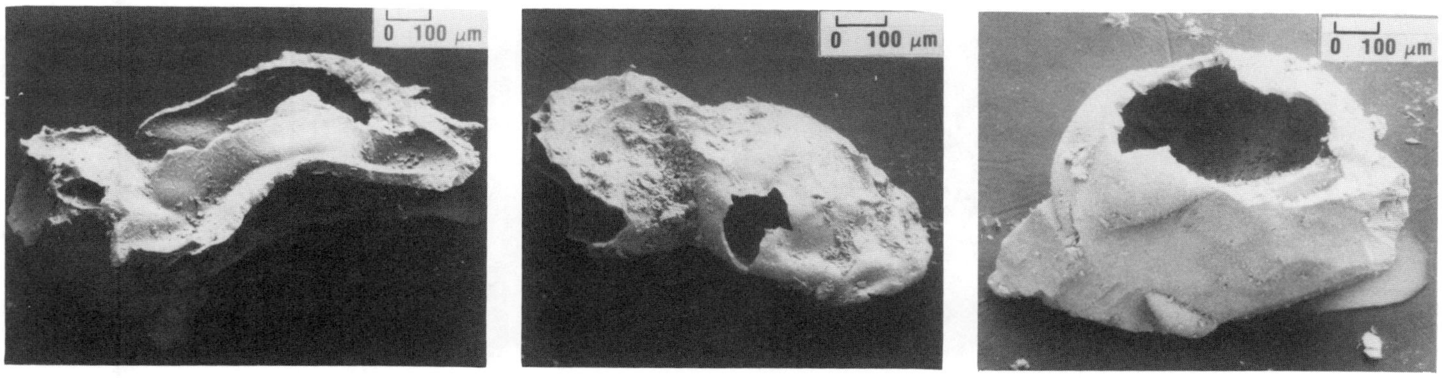

FIGURE 1. SEMs of three vesicular pyroclasts showing partly ruptured bubbles. Note the thin bubble walls; these vesicles may have grown after initial eruptive fragmentation. Spherical or subspherical shards from peralkaline eruptions are often hollow and contain evident secondary vapor-phase minerals (*center*) (Schmincke, 1974).

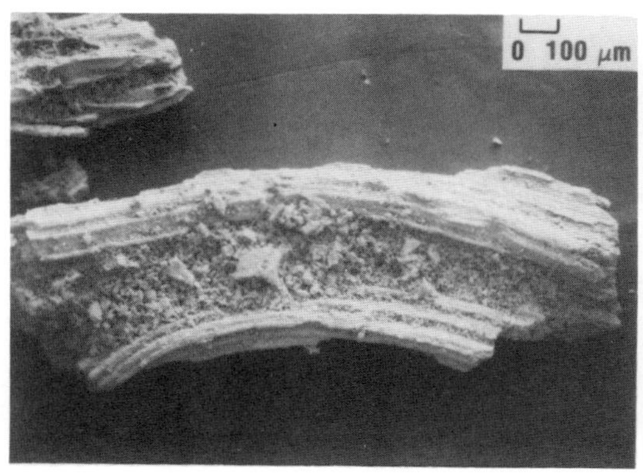

FIGURE 2. SEM of elongate pumice shard with tubular vesicles that have either been stretched during extrusion or flattened during postdepositional compaction. The vesicles are filled with adhering material, most of which is composed of the alteration products, clays and zeolites, that were formed during hydration of the glass. Some of the material is glassy volcanic dust scavenged by shards from the eruptive cloud during dispersal.

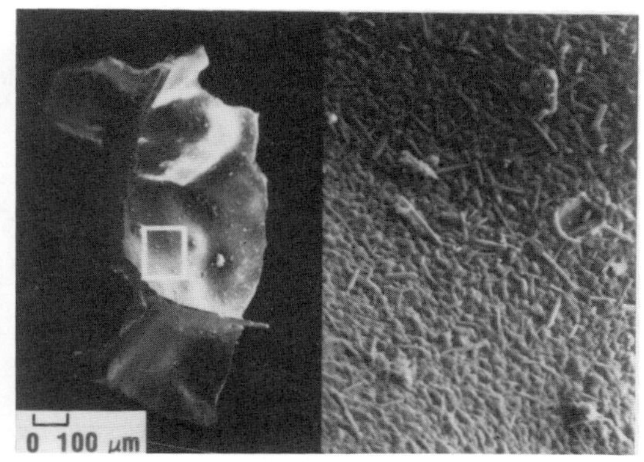

FIGURE 3. Split-image SEM, a 10X enlargement of the surface of a vesicular shard. Note the intricate crust of interlocking feldspar laths, many of which result from vapor-phase crystallization on the glass.

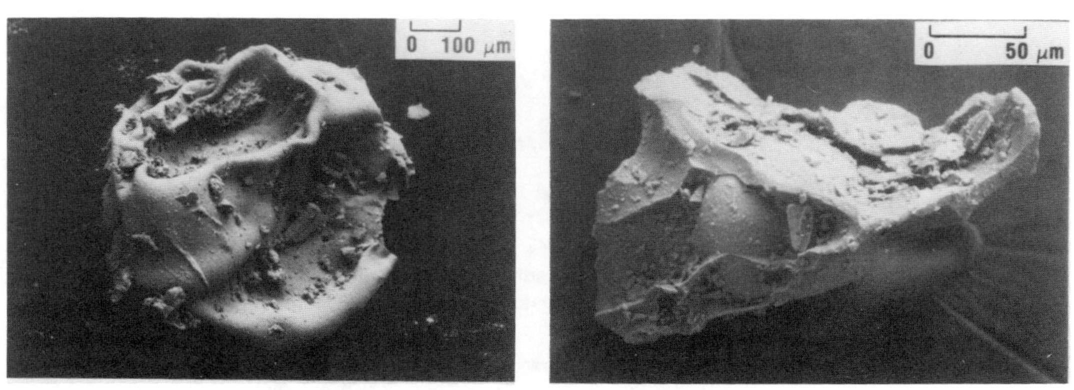

FIGURE 4. SEMs of globular shards. The distorted vesicle surfaces of the shard (*left*) and its ropy surface texture indicate the fluidity of the glass and its rheomorphic behavior after eruption. *Right*—a shard consisting of a subspherical globule surrounded by prong- and blade-like shards. Prongs of the vesicular shard appear to have broken in a brittle fashion, which is unlike the more ductile behavior of the globules.

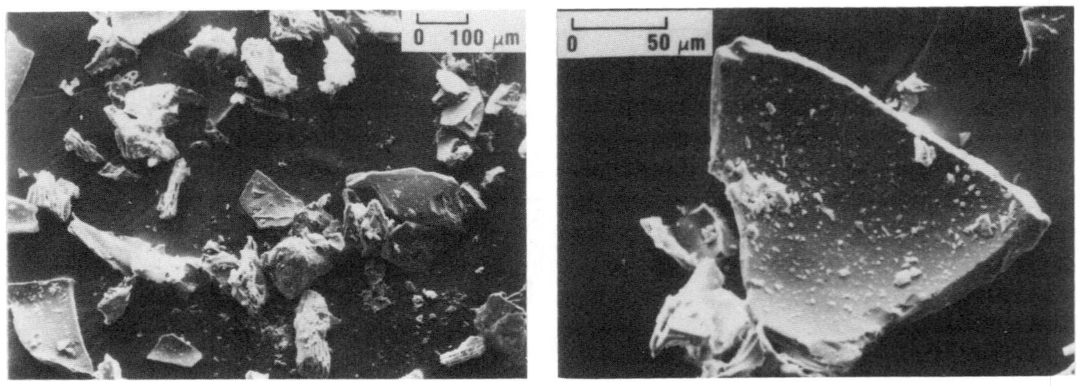

FIGURE 5. SEMs of fractured bubble walls that form curved splinters and chips and constitute much of the fine ash fraction. Globules are absent in the fine debris, and the material appears to have brittly fractured. Fine, adhering dust particles cover the surface of plates and splinters. Hydration cracks are evident on some shards.

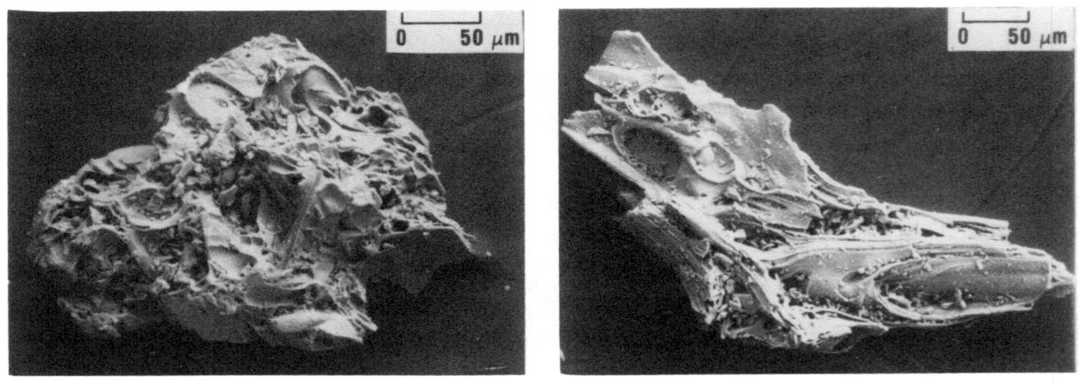

FIGURE 6. SEMs showing multiple vesiculation and deformed vesicles. Secondary vesicle formation is evidenced by bubble walls transecting earlier formed bubbles (*left*) and tubular, stretched vesicles that may be bent as a result of rheomorphic behavior of ash deposits after emplacement.

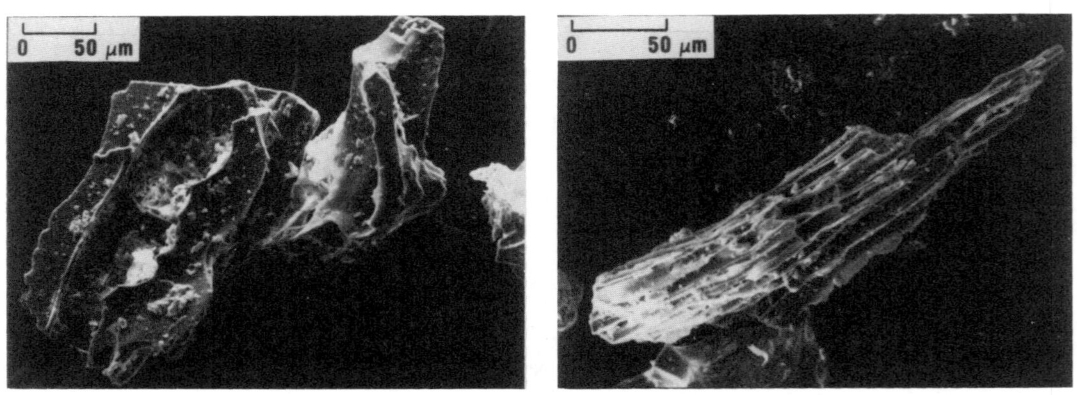

FIGURE 7. SEMs of vesicular shards that resemble typical calcalkaline silicic ash from Plinian eruptions. Abundant, fine, adhering dust is evident on the surface of the vesicle walls. The pumice pyroclast (*right*) shows the typical elongate vesicles.

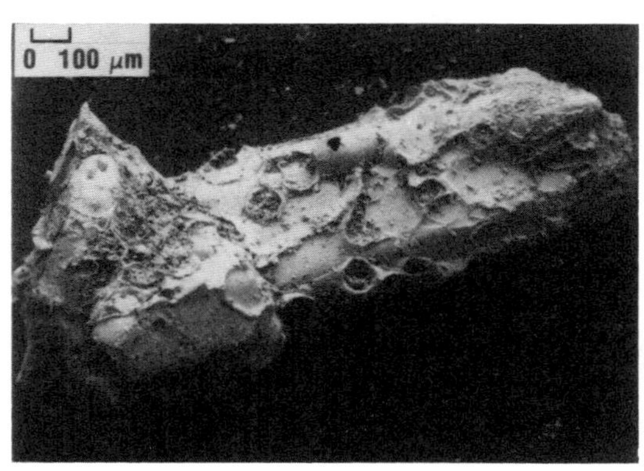

FIGURE 8. SEM of a clinopyroxene phenocryst surrounded by a glassy rind with bubble-wall texture. Note the burst vesicle that formed in the glass coat; it is evidence that the glass was still fluid after the fragment separated from the magma.

FIGURE 9. Sem of a feldspar phenocryst with a clean surface that shows no glass encrustation. Cleavage traces are apparent on this backscattered electron image.

OLDOINYO LENGAI, TANZANIA; 1966 ERUPTION

Introduction: Oldoinyo Lengai is part of the Neogene volcanic province of northern Tanzania and is located 19 km south of Lake Natron. The dominantly pyroclastic cone, with an elevation of 1,920 m, has also produced lavas in which the general trend is phonolite-nephelinite-melanephelinite. Volcanic ash and lavas from the 1966 explosive eruption consist of Ca-Na carbonatites. The eruptions varied from minor emissions of ash to major Plinian and Vulcanian-type behavior. An ash sample from the 1966 eruption was provided by J. B. Dawson.

Grain Size: None available.

Tephra Description: Samples received from the Institute of Geological Sciences, London (Dawson et al., 1968) consist of gray volcanic ash that contains individual mineral pyroclasts and accretionary aggregates of mineral grains cemented by trona. In addition to sodium carbonate (trona) grains, there are phenocrysts of nepheline, pyroxene, wollastonite, apatite, melanite, and pyrrhotite. The ash was originally black but turned white after deposition because of the growth of trona crystals on grain surfaces. The ash consists of mineral fragments (phenocrysts) and aggregates of irregular and tabular trona crystals.

Chemical Composition: The unusual chemical composition of this volcanic ash is shown in Table 1.

TABLE 1. CHEMICAL COMPOSITION OF ASH FROM OLDOINYO LENGAI.*

	(Wt. %)
SiO_2	25.23
TiO_2	0.46
Al_2O_3	5.72
Total FeO	8.71
MnO	0.38
MgO	1.62
CaO	14.10
Na_2O	18.65
K_2O	4.89
P_2O_5	1.49
BaO	1.00
SrO	0.77
SO_3	1.23
S	0.31
Cl	1.13
F	0.62
H_2O^+	1.15
H_2O^-	4.18
CO_2	8.75
Total	100.39

*Analyzed at the Institute of Geological Sciences, London. Analyst was P. Bowden (from Dawson et al., 1968).

Comments on Pyroclast Formation: It is obvious from the Dawson et al. (1968) descriptions of the Oldoinyo Lengai eruption that carbonatite magmas are volatile-rich and may be very energetic; ash analyses support the observation that these magmas are rich in volatile phases. Irregular aggregates of sodium carbonate and phenocrysts were erupted. Carbonatite deposits of the Laetolil Beds of Tanzania near Ngorongoro Caldera are believed to have been ash falls from a volatile-rich eruption of very fluid magma (Hay, 1978). Hay observed numerous carbonatite spheres in these deposits, which implies formation in fountains of low-viscosity magma.

Carbonatite ashes are rapidly altered and might be difficult to recognize. Most of the trona crystals in the Oldoinyo Lengai ash are secondary, having grown on grain surfaces after they were saturated with rain water. Hay (1979) proposes that carbonatite ashes are first cemented by soluble salts such as trona. Diagenetic alteration in the vadose zone forms rocks consisting mainly of calcite, montmorillonite, and phillipsite. Sparry calcite was precipitated as a replacement cement and a pore filling. Carbonatite globules (droplets) are replaced by calcite pseudomorphs. Much of the diagenetic process must be rapid, as footprints, delicate fossils, and raindrop imprints are preserved in these tuff beds. Finer grained carbonatite ash beds are indistinguishable from calcretes of non-carbonatite origin (Hay and Reeder, 1978).

REFERENCES:

Dawson, J., Bowden, P., and Clark, G., 1968. Activity of the carbonatite volcano Oldoinyo Lengai, 1966. Geol. Rundschau, **57**: 865-879.

Hay, R., 1978. Melilitite-carbonatite tuffs in the Laetolil beds of Tanzania. Contrib. Mineral. Petrol., **67**: 357-367.

Hay, R. and Reeder, R., 1978. Calcretes of Olduvai Gorge and the Ndolanya Beds of Northern Tanzania. Sedimentology, **25**: 649-673.

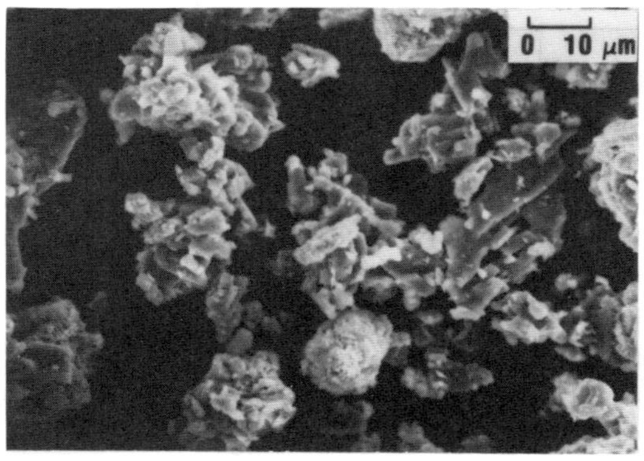

FIGURE 1. Fine-grained fraction of carbonatite ash. There are mostly clusters of sodium and calcium carbonate; these secondary overgrowths have obscured the phenocrysts.

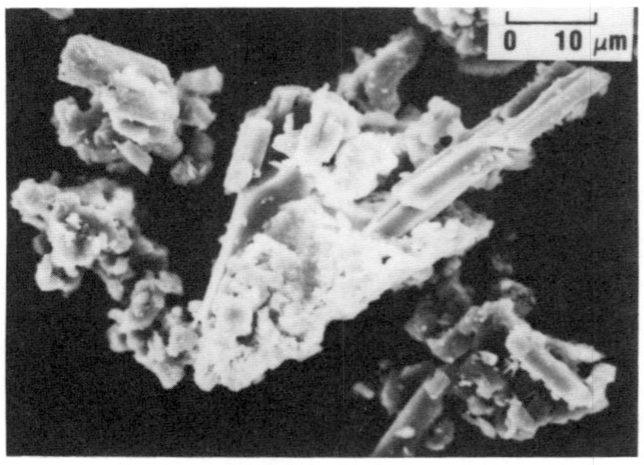

FIGURE 2. Thin, tabular sodium carbonate crystals bonded by an irregular aggregate of the same composition.

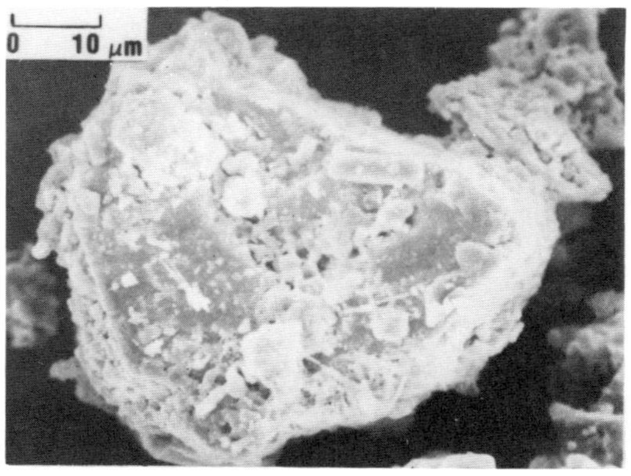

FIGURE 3a. Phenocrysts from the Oldoinyo Lengai tephra. SEM of a nepheline phenocryst coated with both irregular aggregates and well-developed tabular crystals of sodium carbonate.

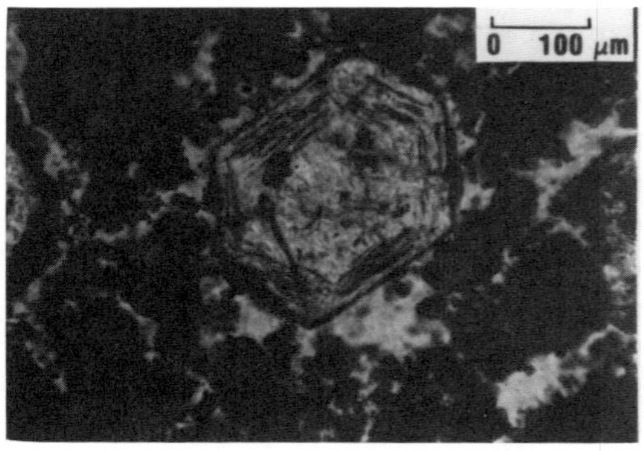

FIGURE 3b. Photomicrograph (transmitted light) of a euhedral nepheline crystal. It is zoned, with apatite crystals growing parallel to the grain rim. Dark clusters around the phenocryst are carbonate aggregates similar to those of Fig. *1*.

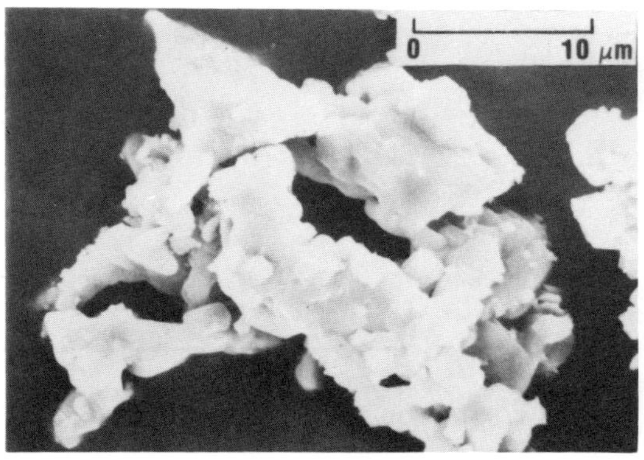

FIGURE 4a. Sodium carbonate (trona) and calcium carbonate phases that grew after ash deposition. Irregular clusters typical of the ash today.

FIGURE 4b. Well-developed trona crystals.

Eruption of Capelinhos, Azores, 1958.

PHREATOMAGMATIC ERUPTIONS

Explosive, ash-producing eruptions may occur when rising magma encounters ground or surface water. Volcanic ash deposits produced during these eruptions are unique and their characteristics may be used to interpret eruptive mechanisms. Generally, two types of vent landforms result from phreatomagmatic eruptions of monogenetic volcanoes. Steep-sloped, cone-shaped vents with wide craters are called *tuff cones* and consist mostly of highly palagonitized and altered tephra that is thickly bedded. *Tuff rings* are vents with a low-profile apron of tephra, which surrounds a wide crater that is often below the general topographic level of the area (termed a *maar* crater); they consist of mostly nonpalagonitized, thinly bedded tephra that are often referred to as a *base surge* or *pyroclastic surge* deposit. Tuff ring and tuff cone deposits may also be associated with polygenetic volcanoes, such as composite cones, and are commonly interbedded with lavas or other pyroclastic debris. Recently, an additional set of terms has been adopted for phreatomagmatic deposits; these terms refer to whether the eruption involved abundant water or not. When phreatomagmatic eruptions occur in lakes or other places where water is abundant, vast amounts of condensing or wet steam are usually erupted with the tephra. Accordingly, these deposits are characterized by bedding and textures referred to as *wet*, such as the *wet-surge* deposits found in tuff cones. Conversely, if little water is involved during eruption, dry or superheated steam is erupted with tephra; hence, *dry-surge* deposits that are usually found in tuff rings are characterized by textures produced by dryer conditions than those of wet surges.

Tuff Cones and Tuff Rings (Mafic Magmas)

Because mafic magmas are generally of low viscosity, their contact with external water results in formation of tephra that are characterized by fluidal textures as well as by the typical blocky textures resulting from brittle behavior. Grain-size distributions of phreatomagmatic ash reflect the energy of the eruption; the strong phreatomagmatic eruptive component produces fine ash (<64 μm in diameter).

SURTSEY, VESTMANN ISLANDS, ICELAND

Introduction: Surtsey is a volcano that erupted in the ocean south of the Icelandic coast from 1963 through 1965. Eventually, the two tuff cones Surtur I and II formed a new island. In his book *Surtsey, The New Island in the North Atlantic*, Thorarinsson (1966) gave day-by-day accounts of the growth of the tuff cone. The main cone, Surtur II, is about 500 m in diameter and has a crater about 60 m below the rim and about 90 m above sea level.

The initial phreatomagmatic eruption of Surtsey excellently characterized this type of activity. The earlier phases of the eruption were typified by rapid cycles of sea water pouring into the vent and alternating with steam explosions as the water came into contact with the rising basaltic magma. This phreatomagmatic activity is now termed *Surtseyan* at other volcanoes (Walker and Croasdale, 1972). Deep explosions produced vertical eruption columns, whereas shallow explosions, caused by interaction of sea water and magma near the surface, produced jets of material that expanded into hemispherically shaped clouds (Thorarinsson, 1966; Thorarinsson et al., 1964). Base surges were associated mainly with the latter events. When the growing cone completely surrounded the vent and shut out sea water, Hawaiian and Strombolian activity began, forming a spatter cone and a small apron of lava flows.

The cone of Surtur II is composed of a number of beds, many of ash fall origin and others that are massive and planar pyroclastic surge deposits. The dips on the beds are variable—nearly horizontal at the base and becoming steeper near the top. Jakobsson (1978) has studied the consolidation and palagonitization of the Surtsey tuffs. He found that shortly after eruption the tuffs were largely unconsolidated and 82 to 88% (by volume) unaltered sideromelane. Consolidation and palagonitization increased over a period of several years.

Recent drilling at Surtsey has revealed a stratigraphy similar to that found at Heimaey (Tomasson, 1967). Pillow lavas occur at depth; they are overlain by hyaloclastites and then by tuffs and breccias. M. F. Sheridan provided these samples of surge ashes from an earlier study (Sheridan, 1971), and S. Thorarinsson provided a fallout sample collected from the deck of a Coast Guard vessel 3 km downwind from Surtsey on December 1, 1963.

Grain Size: Surtsey ash is moderately to poorly sorted in both fall and surge samples. The fall sample (Thorarinsson) has a median size of 0.09 mm (3.5φ), whereas that collected on Surtsey (sample 9) is coarser (1.16 mm; −0.2φ). All surge samples have an abundant fine-ash fraction of 14 wt.% or greater. Figure 2 shows ternary size distributions of the ash from data given in Table 1.

TABLE 1. GRAIN-SIZE DESCRIPTION OF SURTSEY TEPHRA.

Sample	Bed Form*	Md_φ (mm)	σ_φ	Fine Ash (Wt. %)
1	M	1.62 (0.33)	2.48	17.42
2	P	0.57 (0.67)	3.06	14.00
3	M	1.16 (0.45)	2.22	12.46
4	P	0.25 (0.84)	3.15	13.74
5	M	1.48 (0.36)	3.03	26.09
6	M	1.60 (0.33)	2.55	20.33
7	M	1.15 (0.45)	3.01	18.28
8	M	1.26 (0.42)	2.76	16.30
9	F	−0.22 (1.16)	2.17	6.62

*Bed forms are S = sandwave surge, M = massive surge, P = planar surge, and F = fall.

Tephra Description: Vitric pyroclasts can be divided into three types of relatively equal abundance in the samples studied. (1) Vesicular ash dominates in ash fall deposits. Vesicles are spherical and irregularly shaped, and they often determine the overall particle shape. Some vesicular particles, however, show planar breakage surfaces crossing vesicle walls. (2) The second ash type dominates the phreatomagmatic surge deposits. These grains are very angular and poorly vesicular, and they have shapes dominated by curviplanar breakage surfaces. Small grains have some adhering dust, especially in vesicle hollows. Edge modification is developed and manifested as chips, dish-shaped fractures, and hackly or rough areas, especially along grain edges where two planar surfaces intersect. These features are best developed on larger grains that had significant transport abrasion. (3) Finally, a third ash type shows smooth fusiform surfaces between vesicles. These are mostly drop-like pyroclasts.

All samples show varying degrees of alteration to palagonite, which changes the overall appearance of grains. A hydrated skin forms first in vesicles and then over the entire grain surface as palagonitization increases. Note that these samples were collected before extensive palagonitization. Hydrated surfaces tend to crack after extensive alteration. Cracked walls reveal a 1- to 5-μm-thick skin that curls away from the grain along edges in much the same way that desiccation cracks appear in mud.

The sample collected on shipboard has fragments with smooth, planar fracture surfaces that are broken by curved depressions where fractures intersect vesicle surfaces. Vesicularity is low in this sample, suggesting that vesicles must have been widely spaced and poorly developed when fragmentation occurred.

Chemical Composition: A representative analysis of Surtsey basalt from Thorarinsson (1967) is given in Table 2. Table 3 lists a petrographic description of samples 1 through 9; samples from massive deposits show greater amounts of palagonitization than do those from planar beds.

TABLE 2. CHEMICAL COMPOSITION OF SURTSEY TEPHRA.*

	(Wt. %)
SiO_2	46.5
TiO_2	2.3
Al_2O_3	16.8
Fe_2O_3	1.7
FeO	10.8
MnO	0.2
MgO	7.6
CaO	9.4
Na_2O	3.3
P_2O_5	0.3
Total	98.90

*From Thorarinsson, 1967.

TABLE 3. PETROGRAPHIC DESCRIPTION OF SURTSEY TEPHRA.

Sample	500 μm	250 μm	88 μm
1	75% Sideromelane 24% Palagonite	75% Sideromelane 24% Palagonite	97% Sideromelane 3% Crystals* tr. Lithics
2	90% Sideromelane 10% Palagonite tr. Crystal	60% Sideromelane 40% Palagonite	78% Sideromelane 20% Palagonite 2% Crystal and lithic
3	59% Sideromelane 40% Palagonite 1% Crystal	99% Sideromelane 1% Crystal	99% Sideromelane 1% Crystal
4	89% Sideromelane 10% Palagonite 1% Crystal	99% Sideromelane 1% Crystal	99% Sideromelane 1% Crystal
5	79% Sideromelane 20% Palagonite tr. Crystal	59% Sideromelane 40% Palagonite 1% Crystal	58% Sideromelane 40% Palagonite 2% Crystal
6	70% Sideromelane 30% Palagonite	99% Sideromelane 20% Palagonite 1% Crystal	89% Sideromelane 10% Palagonite
7	60% Sideromelane 40% Palagonite	98% Sideromelane 1% Palagonite 1% Crystal	95% Sideromelane 5% Palagonite
8	15% Sideromelane 85% Palagonite tr. Crystal	30% Sideromelane 70% Palagonite tr. Crystal	40% Sideromelane 60% Palagonite
9	24% Sideromelane 75% Palagonite 1% Crystal	38% Sideromelane 60% Palagonite 1% Crystal 1% Lithics	90% Sideromelane 10% Crystal

*Crystals: plagioclase > olivine > clinopyroxene.

Comments on Pyroclast Formation: Surtsey went through stages of phreatomagmatic eruption early in its development; they were followed by Hawaiian and Strombolian ejections and lava flows. Blocky, vitric pyroclasts, showing varying degrees of palagonitization, resulted from sea water interacting with magma as it erupted.

The degree of vesicle development depends in part upon the depth of the magma/water interaction and volatile content of the basalt. Most of the vesicle growth in basaltic magmas apparently occurs within 1 km of the surface (Heiken, 1971). Magma/water interactions deep within the vent can produce nearly vesicle-free tephra, whereas those formed at or near the ground surface may produce scoria. The amount of water pouring into the vent helps determine the energy of vapor explosion and the particle morphology. Where mixing of water is minimal, the magma is fragmented by thermal and acoustic stress waves propagating into the vent. Fusiform particles have drop-like shapes and fluid-formed surfaces between breakage surfaces; these resulted from abundant water physically mixing with magma before quenching and thereby turbulently fragmenting the fluid melt. Some eruptions involved so much water that slurries of ash and water were weakly ejected and flowed from the vent like lahars (Thorarinsson, 1966). Vesicular pyroclasts were ejected during Strombolian phases when access of water to the vent was cut off. This activity dominated the ending phases of explosions when the vent was above sea level and volatiles in expanding vesicles tore apart the lava as it moved out of the vent.

REFERENCES:

Heiken, G., 1971. Tuff rings: examples from the Fort Rock-Christmas Lake Valley, south-central Oregon. J. Geophys. Res., **76**: 5615-5626.

Jakobsson, S. P., 1978. Environmental factors controlling the palagonitization of Surtsey tephra, Iceland. Geol. Soc. Denmark Bull., **27**: 91-105.

Sheridan, M. F., 1971. Textural analysis of Surtsey tephra—a preliminary report. Surtsey Research Progress Report, **6**: 1-4.

Thorarinsson, S., 1967. *Surtsey—The New Island in the North Atlantic.* New York, Viking Press, 47 pp.

Thorarinsson, S., Einarsson, T., Sigvaldasson, G. E., and Ellison, G., 1964. The submarine eruption off the Vestmann Islands, 1963-64. Bull. Volcanol., **27**: 1-11.

Tomasson, J., 1967. On the origin of sedimentary water beneath Vestmann Islands. Jökull, **17**: 300-311.

Walker, G. P. L. and Croasdale, R., 1972. Characteristics of some basaltic pyroclastics. Bull. Volcanol., **35**: 303-317.

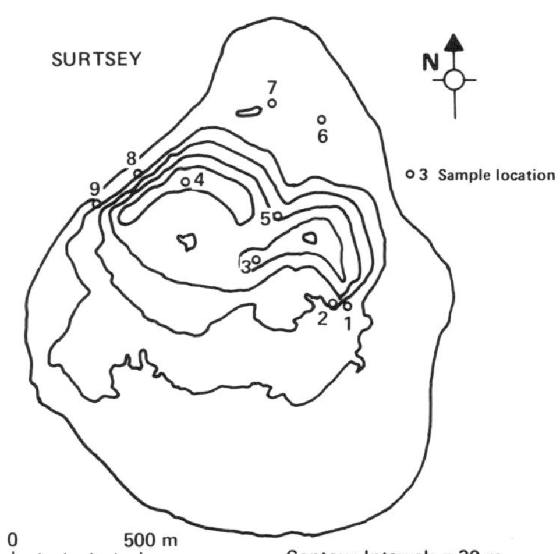

FIGURE 1. Topographic map of Surtsey showing sample locations around the tuff cones.

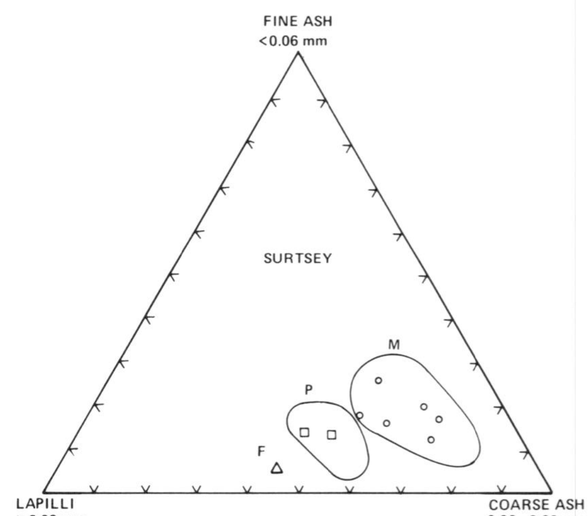

FIGURE 2. Ternary diagram of tephra-size analyses showing fields of planar (P) and massive (M) pyroclastic surge beds.

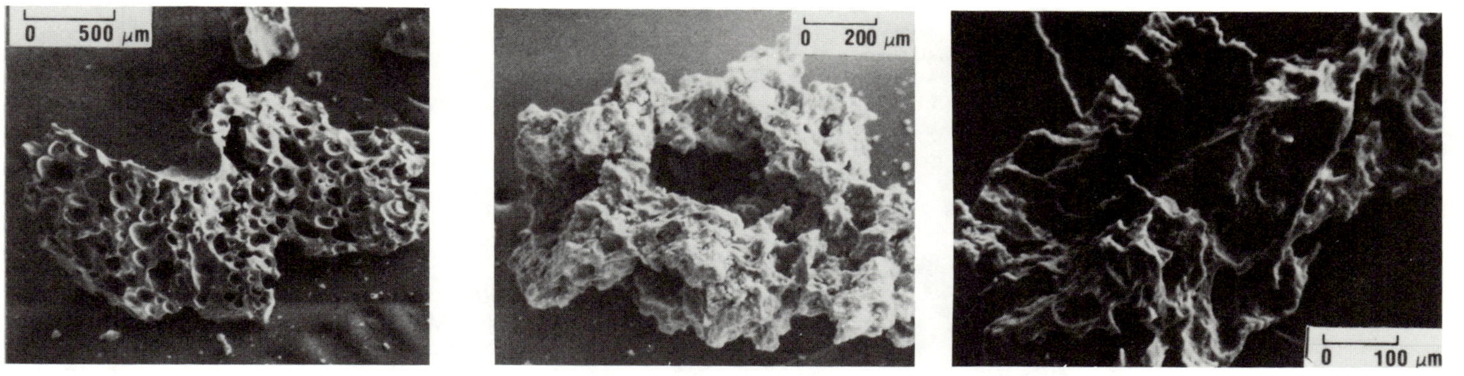

FIGURE 3. SEMs of vesicular particles produced by Strombolian eruptions during which water did not have access to the vent. Note the abundant vesicle surfaces, their irregular shapes, and the jagged outline of particles.

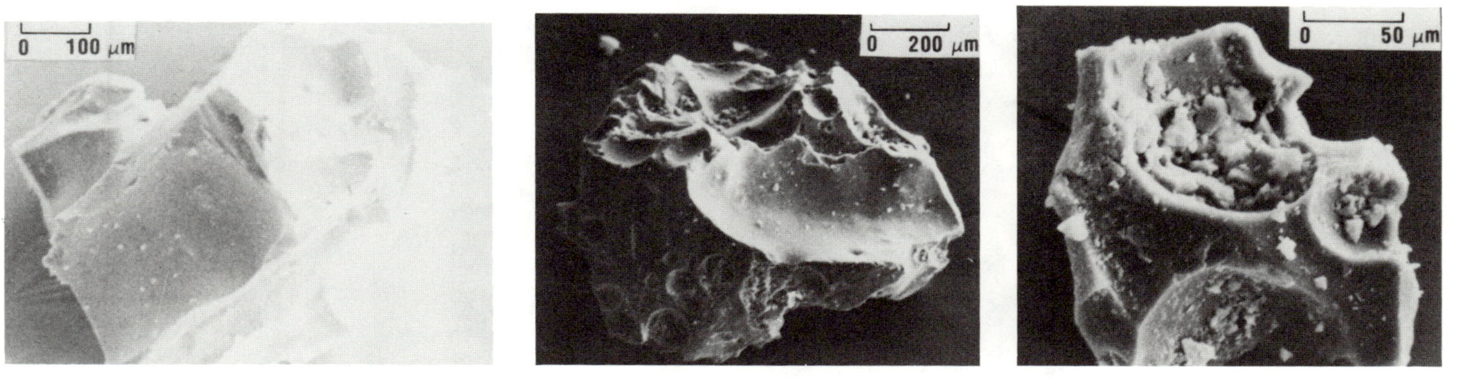

FIGURE 4. SEMs of blocky pyroclasts that have curviplanar fracture surfaces cutting vesicles. Many of the corners where planar surfaces meet have been chipped by transport abrasion. The fine-grained particle *right* shows planar fracture surfaces as well as adhering dust.

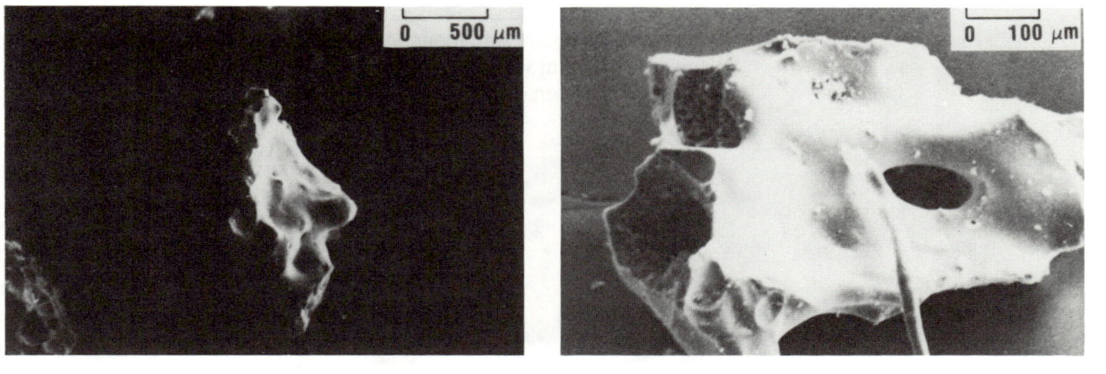

FIGURE 5. SEMs of a drop-like particle *left* and one with fused surfaces between vesicles, which obscure original fracture surfaces. These particles were fragmented by stress waves before they solidified. The smooth, irregular surfaces may also indicate actual physical mixing of particles with water during fragmentation.

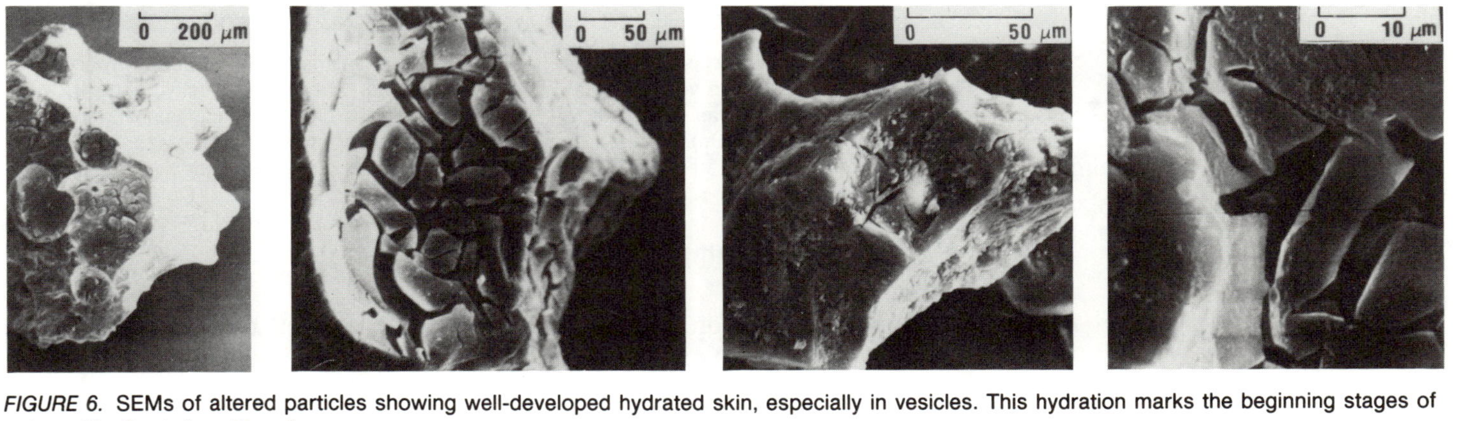

FIGURE 6. SEMs of altered particles showing well-developed hydrated skin, especially in vesicles. This hydration marks the beginning stages of palagonitization in basaltic ash.

FIGURE 7. General SEM view of the Surtsey fallout tephra. Most of the pyroclasts are angular sideromelane (basaltic glass) fragments. Curved surfaces are former vesicle-wall fragments.

FIGURE 8. SEM stereomicrograph of Surtsey fallout tephra. Note both smooth and conchoidal fracture surfaces. Fine-grained (<63-μm) vitric pyroclasts clinging to larger grain surfaces are common in phreatomagmatic tephra.

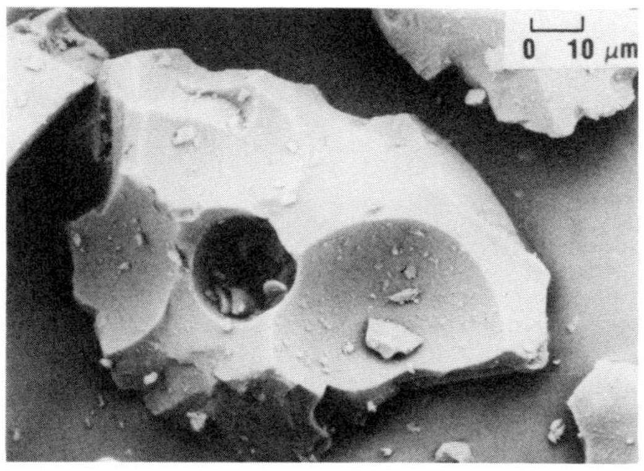

FIGURE 9. SEM closeup of pyroclasts similar to those of Figs. *7* and *8*. There were vesicles in the magma from which this fragment was derived; however, vesicle walls were thick (>20 μm).

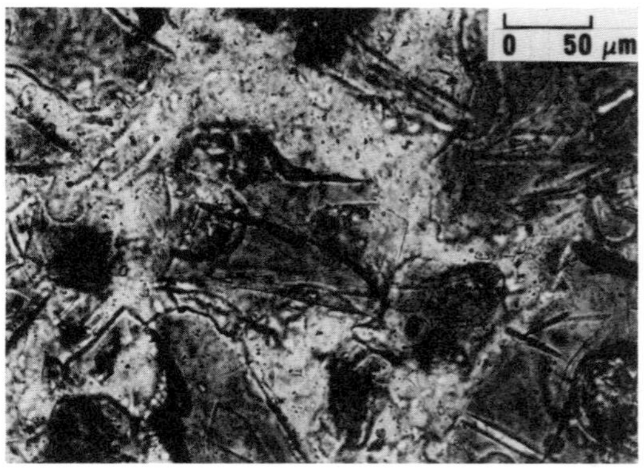

FIGURE 10. Transmitted-light photomicrograph of Surtsey tephra. These are blocky, angular, vitric pyroclasts (light brown basaltic glass) with traces of feldspar laths. Most have remnants of vesicles (curved surfaces).

CAPELINHOS, AZORES; 1957-1958 ERUPTION

Introduction: A submarine eruption began on September 27, 1957, off the west coast of the island of Faial, Azores. All early phases of the activity consisted of phreatomagmatic explosions that constructed a tuff ring. Ash deposition was by base surge and ash fall. Phreatomagmatic activity continued until April 1958, when access of sea water to the crater was cut off. Lava fountaining and lava flows occurred until October 25, 1958 (Servico Geologicos de Portugal, 1959; 1962). The sample described here is from the phreatomagmatic activity and was provided by the Portuguese Geological Survey.

Grain Size: The sample has a median grain size of 132 μm and is poorly sorted. A large sample would most likely include bombs and blocks.

Tephra Description: Much of the tephra consists of small, equant or subequant, basaltic glass fragments. All are angular and bounded by flat, smooth, or conchoidal surfaces or by smooth, curved surfaces where large ovoid vesicles were broken during fragmentation. There are some wedge-shaped and pointed vitric pyroclasts. Most of the sample consists of sideromelane (basaltic glass) pyroclasts with small amounts of tachylite and feldspar, olivine, and clinopyroxene phenocrysts. Vesicularity is very low; only rarely are individual pyroclasts vesicular. Pyroclast surfaces are partly coated with small (0.5- to 4-μm-long), angular vitric shards.

Chemical Composition: Table 1 shows chemical analyses of basaltic pyroclasts erupted from Capelinhos.

TABLE 1. CHEMICAL COMPOSITION OF BASALTIC PYROCLASTS FROM CAPELINHOS, AZORES.*

	Ash (Wt. %)	Bomb (Wt. %)
SiO_2	47.59	46.66
TiO_2	2.65	2.06
Al_2O_3	17.05	19.02
Fe_2O_3	2.97	3.29
FeO	7.44	7.47
MnO	0.20	0.06
MgO	5.47	5.86
CaO	9.09	10.26
Na_2O	4.57	3.28
K_2O	1.85	1.52
P_2O_5	0.50	0.68
Total	99.33	100.16

*Guimares and Viera, analysts, Museu e Laboratorio Mineralogico e Geoligico da Faculdade de Ciencias de Lisboa.

Comments on Pyroclast Formation: Fine-grained ash, consisting of blocky, angular shards with low vesicularity, is formed during interaction of magma and water. Rising basaltic melt with a few well-developed vesicles is mixed with sea water, chilled, fragmented, and erupted as part of a water-steam-tephra mixture. Pyroclast surfaces are mostly fracture planes; uniformity of grain size and particle shape indicates that the process is thorough, with uniform comminution of the glass produced during the magma/water interaction.

REFERENCES:

Servico Geologicos de Portugal, 1959. Le volcanisme de l'isle de Faial et l'eruption de volcan de Capelinhos. Mem. 4 (Nova Serie), 100 pp.

________________________, 1962, Le volcanism de l'isle de Faial et l'eruption de volcan de Capelinhos. Mem. 9 (Nova Serie), 54 pp.

(Note: both monographs listed above provide excellent documentation of the year-long eruption of Capelinhos).

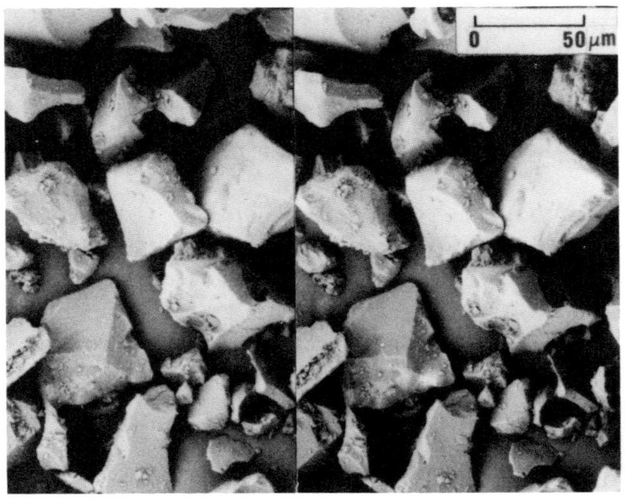

FIGURE 1. Stereomicrograph of ash typical of the phreatomagmatic activity of Capelinhos. All blocky pyroclasts with smooth or conchoidal surfaces are composed of basaltic glass. Smooth, curved surfaces are vesicle-wall remnants.

FIGURE 2. Detailed SEM view of vitric pyroclasts that illustrates surface textures, mainly smooth and conchoidal surfaces. Surfaces are coated, in part, by 0.5- to 5-µm-long, angular, vitric fragments.

FIGURE 3. SEM of vitric (basaltic glass) pyroclasts. Many of these have at least one surface that is part of a vesicle wall. All surfaces are smooth, curved, and partly filled with small, angular pyroclasts similar to those described in Fig. *2*.

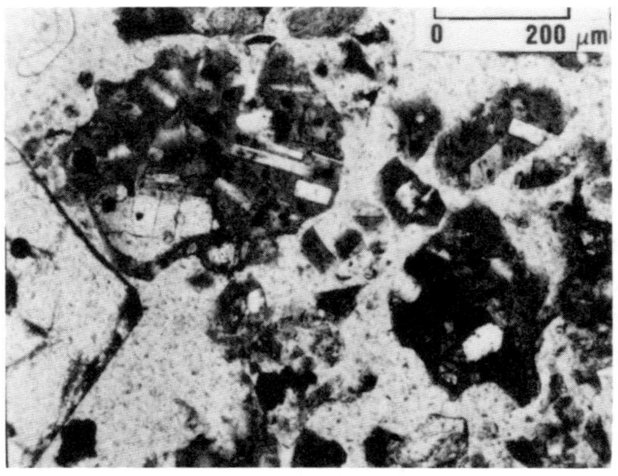

FIGURE 4. Phenocryst-rich, vitric pyroclasts. Photomicrograph in transmitted light. Most pyroclasts are subequant, angular, vesicle-poor sideromelane (basaltic glass). Phenocrysts are feldspar, olivine, and clinopyroxene.

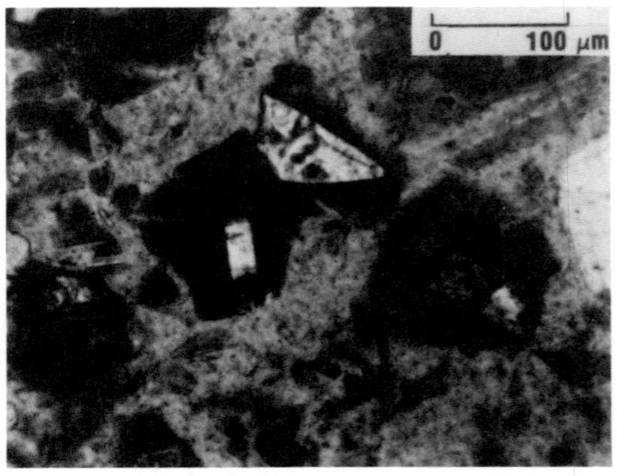

FIGURE 5. Closer view of the ash shown in Fig. *4*. Vitric pyroclasts in this size range are free of vesicles.

KOKO CRATER, HAWAII

Introduction: Koko Crater is a tuff cone that was formed during Quaternary time (Stearns, 1961). It belongs to the Honolulu group, a series of cones formed during at least 30 eruptions of silica-undersaturated basalt. These eruptions are part of the late-stage, posterosional activity that is typical of the Hawaiian Island chain. Other well-known vents in this group are tuff rings and cones such as Diamond Head, Salt Lake Craters, Punchbowl, Manana Island, and Pali. Almost all these tuffs are, to some degree, palagonitized. They have been discussed by Hay and Iijima (1968), Wentworth (1926), Wentworth and Winchell (1947), and Winchell (1947). Fisher

(1977) studied stratigraphic relationships among surge deposits from Koko Crater and other craters in the vicinity.

Koko Crater is 1000 m in diameter and 330 m in elevation from the crater base to the topographic rim. The stratigraphy is complicated by onlapping deposits from Koko Head and Kahauloa Craters, which are located about 1 km to the SE. Beds near the base of the cone dip gently away from the crater at 5 to 10°, whereas beds on the cone dip 20 to 30°. Koko Crater tuffs can be divided into three units: (1) a basal breccia and lapilli fall, (2) a tuff ring sequence of thinly bedded surge deposits, and (3) the cone-building unit composed of highly altered, poorly stratified massive beds.

Grain Size: Table 1 shows the results for the 18 samples analysed. These samples were taken from the dominant bed-form types, which include planar and massive surge beds and ash fall lapilli beds that resulted from short Strombolian bursts during the predominantly phreatomagmatic activity. Massive beds show the finest median grain sizes (0.50 mm = 1φ), and as much as 11 wt.% of the sample is fine ash (<0.06 mm). These beds also show the poorest sorting ($\sigma_\varphi \simeq 2.00\varphi$). Ash fall lapilli beds are better sorted ($\sigma_\varphi \simeq 1.75\varphi$) than surge deposits and have median diameters of 1.00 mm (0.0φ).

TABLE 1. GRAIN-SIZE ANALYSES OF KOKO CRATER TEPHRA.

Sample	Bed Form*	Md_φ (mm)	σ_φ	Fine ash (Wt. %)
1	P	−0.54 (1.45)	1.28	1.89
2	P	−0.11 (1.08)	1.47	2.53
3	M	1.06 (0.48)	2.04	8.43
4	M	1.07 (0.47)	2.09	8.73
5	M	0.76 (0.59)	1.73	4.28
6	M	0.77 (0.58)	1.68	5.14
7	AF	0.31 (0.81)	1.44	2.88
8	AF	−0.21 (1.16)	1.68	2.86
9	M	1.12 (0.46)	2.34	11.03
10	M	1.22 (0.43)	2.27	11.78
11	AF	0.29 (0.82)	1.85	4.93
12	AF	0.14 (0.91)	1.93	5.03
13	AF	−0.27 (1.21)	2.02	2.77
14	AF	−0.35 (1.27)	1.77	2.08
15	M	1.02 (0.49)	1.61	2.65
16	M	0.97 (0.51)	1.61	2.79
17	M	0.96 (0.52)	1.82	6.37
18	M	1.11 (0.46)	1.80	6.41

*Bed forms are AF = ash fall, M = massive surge, and P = planar surge.

Tephra Description: Overall, the tephra sequence at Koko Crater illustrates a typical tuff cone, beginning with alternating phreatomagmatic and Strombolian explosions that produce an explosion breccia and lapilli fall (Unit 1) of medium to dark brown color. The tephra fall beds contain relatively fresh and unpalagonitized fragments with jagged outlines formed by bubble walls, which is typical of Strombolian eruption phases. These lapilli fall beds are 20 to 60 cm thick and show poorly developed normal grading.

Unit 2 resulted from phreatomagmatic eruptions that deposited thinly bedded, massive and planar surge deposits with a few interbedded sandwave beds (dry surge). These beds are a medium gray color where unaltered and unconsolidated or orange-brown where palagonitization has lithified the tuff. Coral fragments are abundant in the tuff of Unit 2 because coral reef and beach material was pierced by the eruption and fragmented within the vent. Typical pyroclasts of this unit are blocky with rounded edges. Some particles have a highly altered, rounded surface that is covered with a cracked, hydrated skin about 1 μm thick.

Unit 3 is phreatomagmatic (wet surge); fragments are totally altered except for lapilli fall beds that contain tephra with vesicle surfaces. Planar bed tephra show less surface alteration than do massive bed tephra; both have few vesicle surfaces and a broken, blocky morphology.

Koko Crater tuffs are composed of sideromelane, tachylite, and palagonitic glass shards; lithic fragments consisting of coral and minor amounts of quartz beach sand; and plagioclase, olivine, and minor pyroxene crystals. Basaltic glass is partly to totally palagonitized and makes up about 90% of the samples; 2 to 4% is crystals, and the rest is lithic materials. Samples taken from near the base of Unit 2 consist of up to 50 wt.% lithic materials from country rock that was comminuted during early vent-clearing eruptions (Table 2).

TABLE 2. PETROGRAPHIC DESCRIPTION OF KOKO CRATER TEPHRA.

Sample	354 μm	177 μm	63 μm
1	95% Sideromelane, highly palagonitized 5% Carbonate grains	90% Sideromelane, partly palagonitized 10% Carbonate grains	80% Sideromelane, partly palagonitized 20% Carbonate grains
3	---	50% Sideromelane, partly palagonitized 50% Quartz and coral beach sand	---
7	85% Sideromelane, highly palagonitized	90% Sideromelane, partly palagonitized	80% Sideromelane highly palagonitized
9	95% Sideromelane 4% Crystals 1% Carbonate	85% Sideromelane 5% Crystals 5% Carbonate	80% Sideromelane 10% Crystals 5% Carbonate
11	100% Sideromelane, totally palagonitized	98% Palagonitized sideromelane 2% Coral fragments	25% Sideromelane 5% Carbonate
15	98% Palagonite 1% Carbonate	---	---
17	---	70% Sideromelane 30% Palagonite and Carbonate	---

Comments on Pyroclast Formation: Koko Crater tuffs illustrate a sequence in which the contribution of hydromagmatic fragmentation increases throughout the eruption. Fresh ash deposited from Strombolian explosions in Unit 1 is magmatic; there is, however, a phreatomagmatic component. Upward in the section, pyroclasts are almost entirely phreatomagmatic and show evidence of brief phases of ash falls produced by Strombolian activity during times when water did not have access to the vent.

REFERENCES:

Fisher, R. V., 1977. Erosion by volcanic base-surge density currents: u-shaped channels. Geol. Soc. Am. Bull., **88**: 1287-1297.

Hay, L. and Iijima, A., 1968. Nature and origin of palagonite tuffs of the Honolulu Group on Oahu, Hawaii. Geol. Soc. Am. Mem., **116**: 331-376.

Stearns, H. T., 1961. Eustatic shorelines on Pacific islands. Zeitschr. Geomorph., **3**: 3-16.

Wentworth, C. K., 1926. Pyroclastic geology of Oahu. Bernice P. Bishop Museum Bull., **30**: 121.

Wenthworth, C. K. and Winchell, H., 1947. Koolau basalt series, Oahu, Hawaii. Geol. Soc. Am. Bull., **58**: 49-78.

Winchell, H., 1947. Honolulu series, Oahu, Hawaii. Geol. Soc. Am. Bull., **58**: 1-48.

FIGURE 1. Photograph of Koko Crater, which rises about 300 m above the surrounding plain and is 1000 m in diameter.

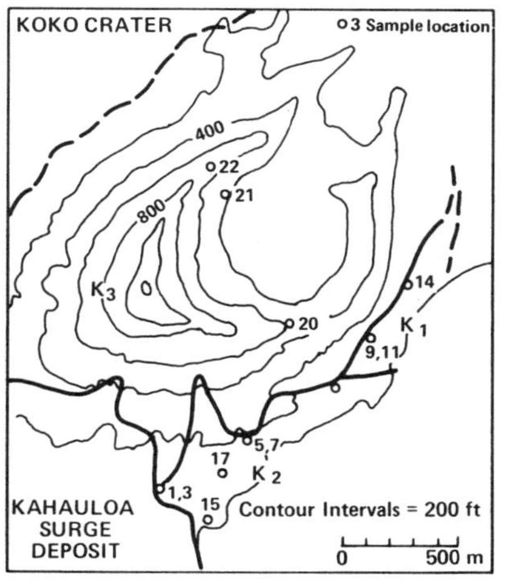

FIGURE 2. Map of Koko Crater that shows distribution of Units 1 (K_1) and 2 (K_2) (early tuff ring), Unit 3 (K_3) (later tuff cone), and sample locations.

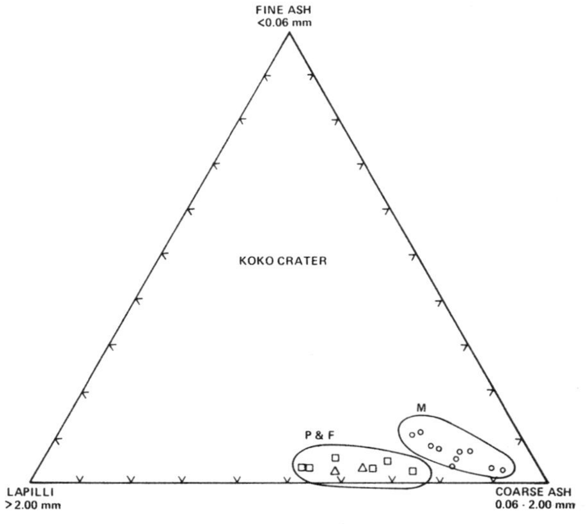

FIGURE 3. Triangular plot of tephra size analyses that shows fields related to ash fall and planar surge beds (F + P) and massive surge beds (M).

FIGURE 4. Explosion breccia and lapilli fall exposed near sample location 10. The base of this unit is not exposed. It is dominantly composed of 5- to 15-cm blocks of basalt with subordinate ash. Note geologic hammer for scale.

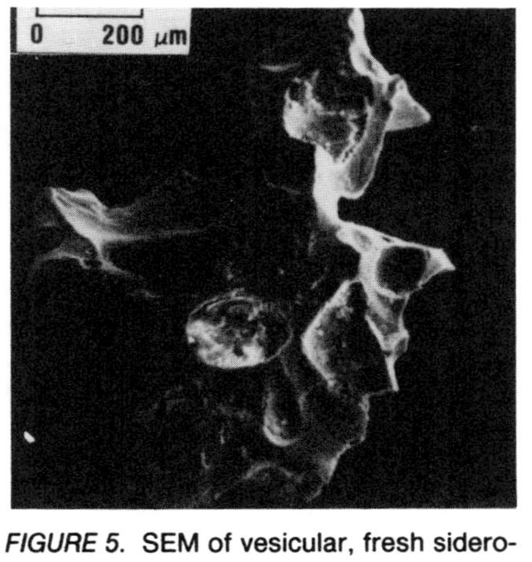

FIGURE 5. SEM of vesicular, fresh sideromelane pyroclast from fall deposit near the top of Unit 1. Note the smooth, curved surfaces and the jagged edges of bubble walls. Alteration products from minor palagonitization fill some vesicles.

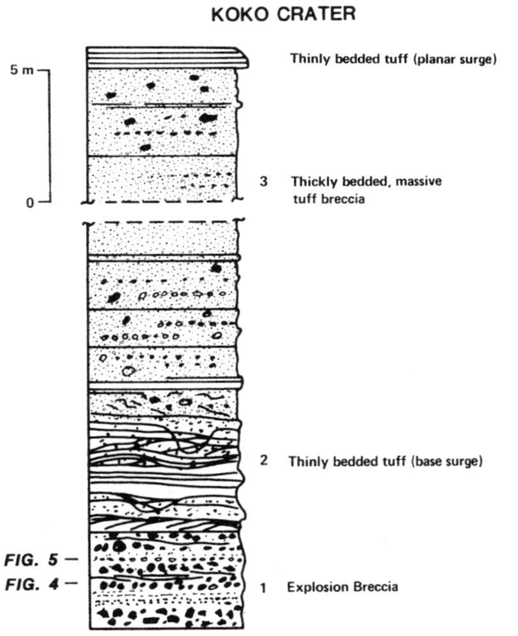

Koko Crater rim stratigraphy measured on the eastern slopes. Parts of Unit 2 and the base of Unit 3 are included in the section described by Fisher (1977).

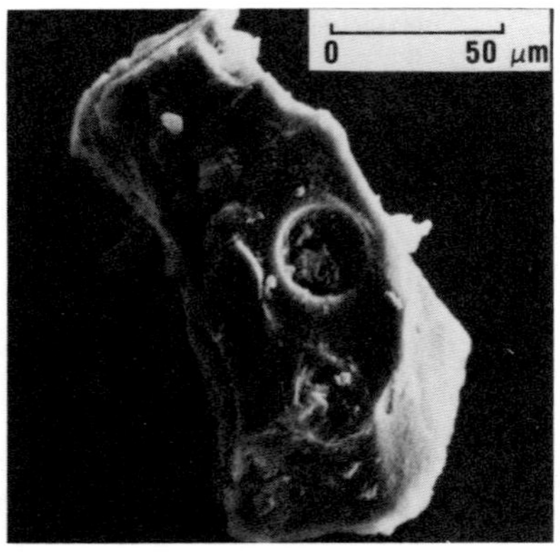

FIGURE 6. SEM of a blocky pyroclast from the opening phases of phreatomagmatic activity. Few vesicle surfaces characterize this grain, which shows curviplanar surfaces that have been chipped and abraded during transport.

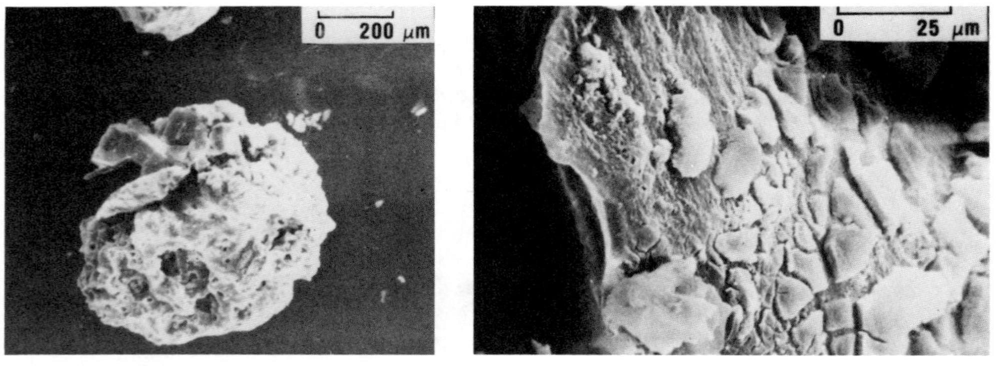

FIGURE 7. SEMs of a phreatomagmatic pyroclast rounded by abrasion during transport in surge (*left*) and covered by hydrated, palagonitic skin, which is shown at higher magnification (*right*).

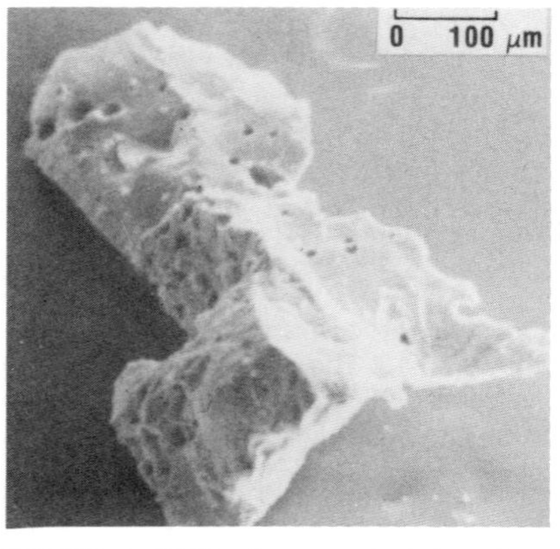

FIGURE 8. SEM of a sample taken from lapilli fall bed; it was deposited from a Strombolian magmatic burst during a period of little water/magma interaction. Note vesicle surfaces that show only slight palagonitization.

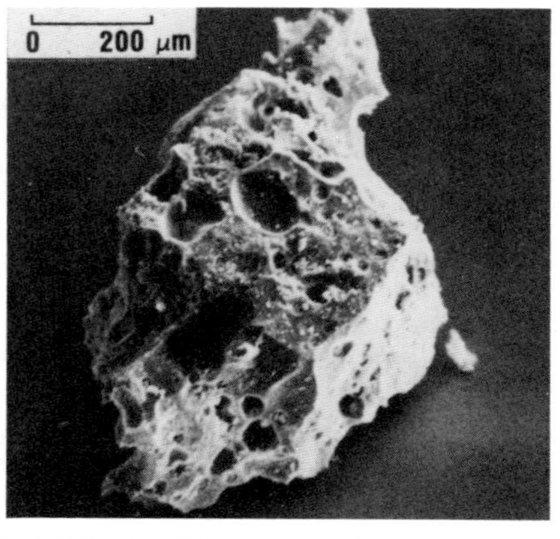

FIGURE 9. SEM of a blocky, partly palagonitized pyroclast from a planar surge bed.

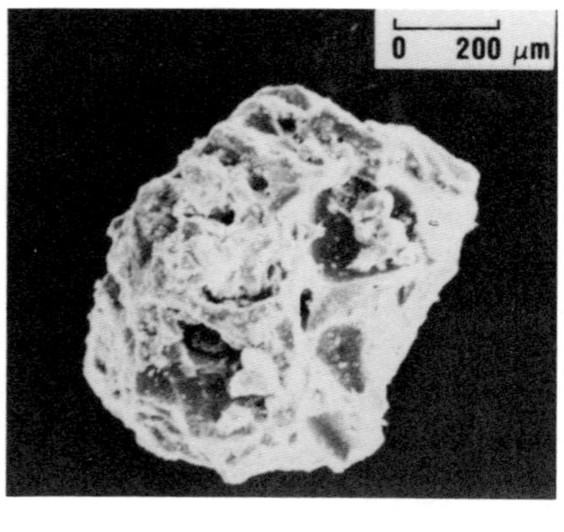

FIGURE 10. SEM of a rounded, highly palagonitized pyroclast from a massive surge bed. Contrast the surface alteration with that shown in Fig. *9*. This contrast results from the massive beds having been emplaced as much wetter deposits than those of planar beds.

KOKO CRATER

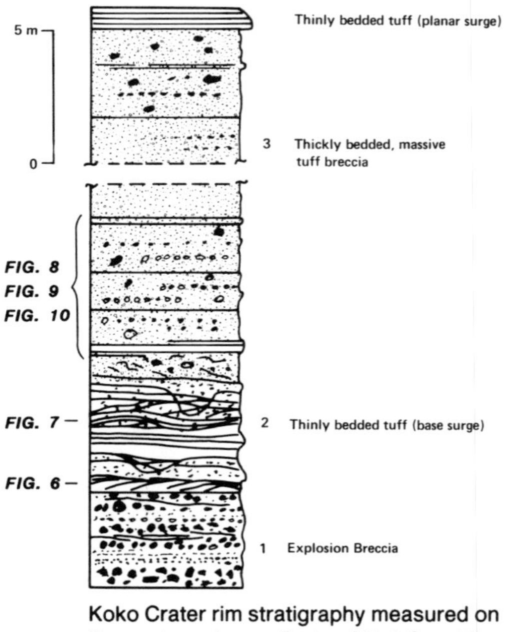

Koko Crater rim stratigraphy measured on the eastern slopes. Parts of Unit 2 and the base of Unit 3 are included in the section described by Fisher (1977).

ZUNI SALT LAKE, NEW MEXICO

Introduction: Zuni Salt Lake maar is located in west-central New Mexico. It is the northernmost of a number of vents, many of which are tuff rings, aligned along a SW-trending fault line that is over 15 km long. Early studies of this maar include those of Herrick (1900) and Darton (1905, 1916). Later studies include a comparison with lunar craters (Shoemaker, 1962), a dissertation on the origin of the salt lake (Bradbury, 1967), and a geologic map (Cummings, 1968). Fisher and Waters (1970) and Wohletz and Sheridan (1983) give descriptions of its phreatomagmatic origin and surge deposits.

The crater is 1.6 km in diameter and about 100 m deep. It is underlain by Precambrian crystalline rocks; the Permian Abo, Yeso, Glorieta, and San Andreas Formations; and the Cretaceous Dakota Sandstone, Mancos Shale, and Mesa Verde Group. Basaltic eruptions pierced these strata, and the resulting tephra deposits contain a significant proportion of lithic sandstone fragments.

Water that interacted with basaltic magma to produce phreatomagmatic eruptions appears to have been derived from spring water structurally trapped along the fault or from an underlying sedimentary aquifer. Eruptions produced a tuff ring with rim beds as much as 30 m thick. The tuff is mostly thinly bedded surge with a subordinate amount of ash fall deposits. The ring is mantled by an ash and lapilli fall that is mappable to 8 km from the crater rim. Maar floor collapse during and after eruption produced an arcuate, concentric fault system that now exposes the tuff sequence on the inner walls of the ring.

Grain Size: Table 1 presents size analyses of 17 samples of Zuni Salt Lake tephra. The tuff is a lapilli- and block-bearing ash to a lapilli- and block-bearing fine ash that is moderately to poorly sorted and light gray to gray. Fall ashes are coarser (Md $\simeq$ 1.2 mm) and better sorted ($\sigma_\varphi \simeq 1.45\varphi$) than are surge ashes (Md $\simeq$ 0.2-0.8 mm, $\sigma_\varphi \simeq 1.90\varphi$). However, the surge deposits contain significantly more fine ash than do the fall deposits.

TABLE 1. GRAIN-SIZE CHARACTERISTICS OF ZUNI SALT LAKE TEPHRA.

Sample	Bed Form*	Md_φ (mm)	σ_φ	Fine Ash (Wt. %)
1	P	−0.16 (1.12)	2.39	2.25
2	P	−1.50 (2.83)	2.24	1.65
3	S	1.84 (0.28)	1.65	7.97
4	S	1.80 (0.29)	1.70	12.42
5	S	2.20 (0.22)	1.98	14.14
6	S	2.25 (0.21)	1.99	14.40
7	S	2.38 (0.19)	1.65	9.37
8	S	2.16 (0.22)	2.04	10.94
9	M	0.26 (0.84)	1.83	4.40
10	M	0.31 (0.81)	1.87	4.22
11	F	−0.32 (1.25)	1.45	0.50
12	M	0.83 (0.56)	1.90	5.00
13	M	0.99 (0.50)	1.88	10.73
14	P	0.27 (0.83)	1.43	1.69
15	P	0.35 (0.78)	1.48	1.88
16	F	−0.32 (1.25)	1.38	0.44
17	F	−0.40 (1.32)	1.46	0.70

*Bed forms are P = planar surge, S = sandwave surge, M = massive surge, and F = fall.

Tephra Description: The stratigraphic section here illustrates the upper third of the total section at Zuni Salt Lake. This interval represents the gradual decrease of magma/water interaction as the eruption progressed. During this interval, phreatomagmatic activity, which produced surges, gave way to dominantly Strombolian activity, which mainly deposited ash and lapilli falls. Samples of the sandwave-, massive-, and planar-surge sequence show a gradual upward increase in vesicularity, although particle shapes are dominated by angular, blocky fractures. These samples may be contrasted with fall-deposit samples whose vesicular particles are shaped by bubble walls. Also, there is greater surface alteration and abrasion on the surge samples, which all show some development of altered surfaces: the least of which is typical of particles from sandwave beds, and the greatest is seen on particles from massive beds, especially those of the fine fraction. The surge ashes have angular edges that are smoothed and rounded by laminar and turbulent transport where grain collisions are abundant. The fall clasts, however, retain delicate shapes.

Table 2 shows results of grain counts for size fractions >63 μm. Palagonite appears in greater abundance in surge samples than in fall samples, and there is also a marked increase within the finer grained fractions. The crystal content of these samples ranges from 5 to 10% (distribution not shown in Table 2) and is concentrated in the finer size fractions. Lithic material consists mostly of quartz grains from comminuted sandstone.

TABLE 2. PETROGRAPHIC DESCRIPTION OF ZUNI SALT LAKE TEPHRA.

Sample	354 μm	177 μm	64 μm
1	85% Sideromelane	60% Sideromelane	30% Sideromelane
	10% Lithic	30% Lithic	60% Lithic
3	1% Sideromelane	10% Sideromelane	15% Sideromelane
	85% Palagonite	75% Palagonite	70% Palagonite
	9% Lithic	50% Lithic	5% Altered glass
5	90% Sideromelane	80% Sideromelane	70% Sideromelane
	5% Palagonite	5% Palagonite	10% Palagonite
	3% Lithic	5% Lithic	10% Lithic
7	45% Sideromelane	40% Sideromelane	---
	40% Palagonite	40% Palagonite	
	10% Lithic	10% Lithic	
10	85% Sideromelane	50% Sideromelane	20% Sideromelane
	5% Palagonite	40% Palagonite	60% Palagonite
	5% Lithic	5% Lithic	10% Lithic
12	5% Sideromelane	2% Sideromelane	1% Sideromelane
	5% Palagonite	90% Lithic	90% Lithic
	80% Lithic		
15	90% Sideromelane	60% Sideromelane	40% Sideromelane
	5% Lithic	39% Lithic	55% Lithic
16 & 17	80% Sideromelane	70% Sideromelane	40% Sideromelane
	10% Palagonite	20% Lithic	50% Lithic
	5% Lithic		

Comments on Pyroclast Formation: The eruption sequence illustrated here in part reflects the overall eruptive activity at Zuni Salt Lake. Early explosive eruptions were dominantly phreatomagmatic, producing blocky and angular tephra particles that were then deposited in base surges. Later explosive eruptions became increasingly Strombolian as water access to the vent was impeded. These eruptions produced vesicular ash, which was deposited as fall layers. The last stages of eruption show little explosive water interaction and are marked by the building of cinder cones in the maar. Intrusion of a ring dike, exposed on the inner walls of the maar, may have been responsible for the cessation of water flow into the vent late in the eruptive history.

REFERENCES:

Bradbury, J. P., 1967. Origin, paleolimnology, and limnology of Zuni Salt Lake maar, west central New Mexico. Ph.D. dissertation, University of New Mexico, 246 pp.

Cummings, D., 1968. Geologic map of the Zuni Salt Lake volcanic crater, Catron County, New Mexico. U.S. Geol. Sur. Map. I-544.

Darton, N. H., 1905. The Zuni Salt Lake. J. Geol., **13**: 185-193.

Darton, N. H., 1916. Explosion Craters. Sci. Mon., **3**: 417-430.

Fisher, R. V. and Waters, A. C., 1970. Base surge bed forms in maar volcanoes. Am. J. Sci., **268**: 157-180.

Herrick, D. L., 1900. Report of a geologic reconnaissance in western Socorro and Valencia Counties, New Mexico. Am. Geol., **25**: 331-346.

Shoemaker, E. M. 1962. Primary structures of maar rims and their bearing on the origin of Kilbourne Hole and Zuni Salt Lake, New Mexico (abs.). Geol. Soc. Am., **68**: 1846.

Wohletz, K. H. and Sheridan, M. F., 1983. Hydrovolcanic explosions II. Evolution of basaltic tuff rings and tuff cones. Am. J. Sci., **283**: 385-413.

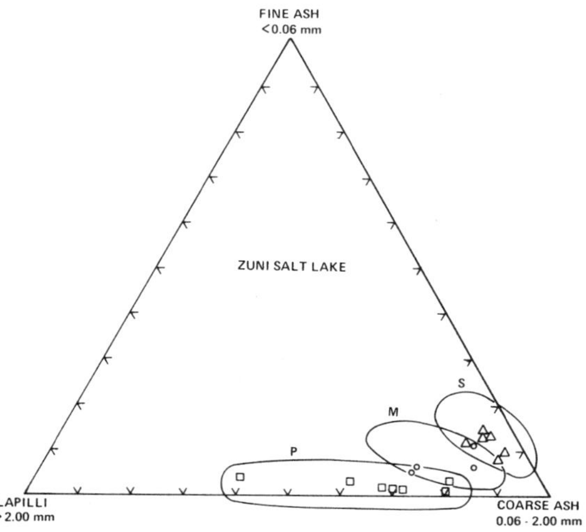

FIGURE 1. Triangular plot of tephra size analyses that shows fields of pyroclastic surge bed forms: S = sandwave; M = massive; P = planar; and F = fall beds.

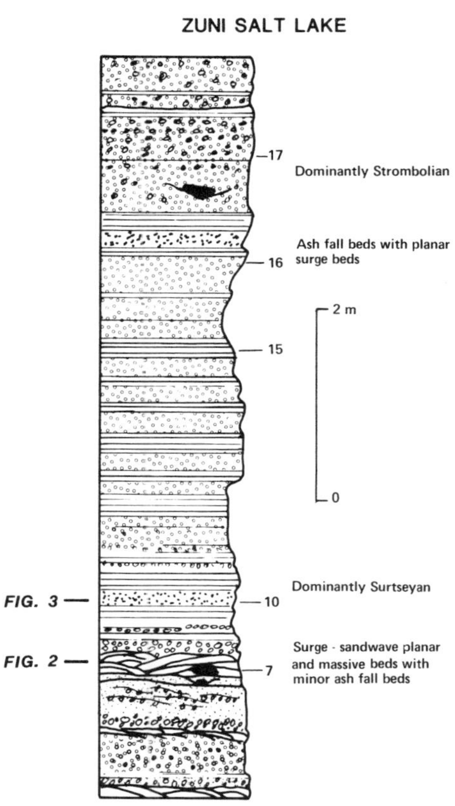

Rim stratigraphic section of Zuni Salt Lake maar. The section shows the upper one-third of the total pyroclastic section. The lower part of this section is dominantly surge beds with intercalated lapilli fall beds. The upper part is dominantly ash and lapilli fall beds with intervals of planar surge deposits.

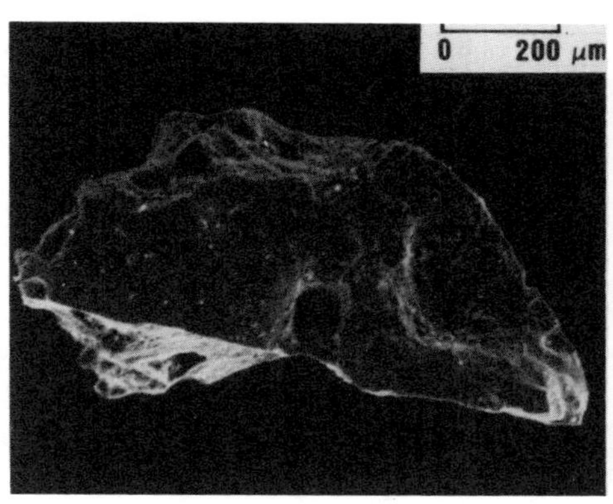

FIGURE 2a. SEMs of pyroclasts from sample 7, from a sandwave surge bed. An angular ash grain with fresh, planar surfaces, chipped edges, and very few vesicles.

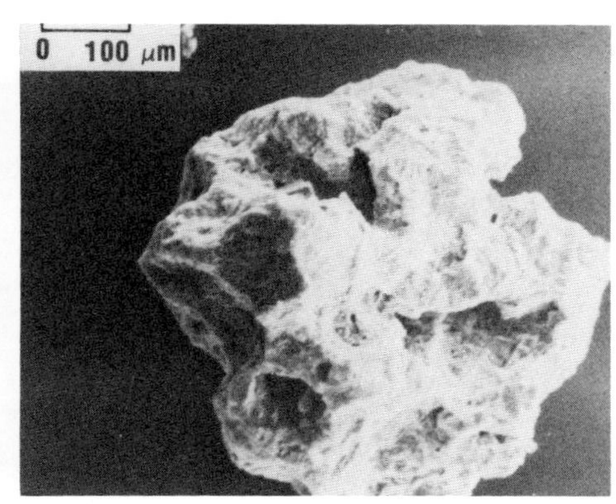

FIGURE 2b. Grain with abrasion pits and chips, which have rounded the original flat surfaces.

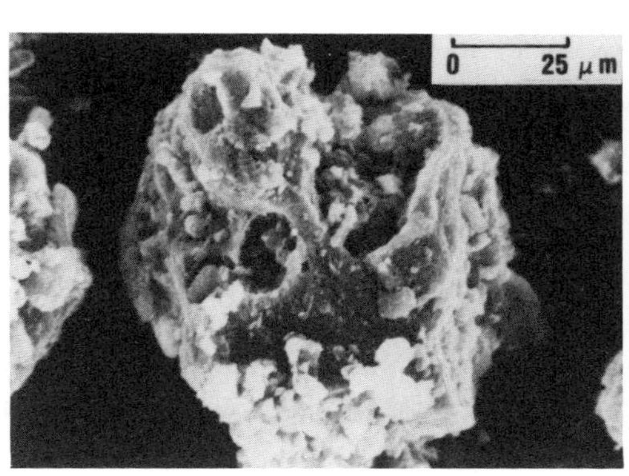

FIGURE 3a. SEMs of pyroclasts from sample 10, from a massive surge bed. Rounded grain covered with adhering alteration material both on flat surfaces and in vesicles.

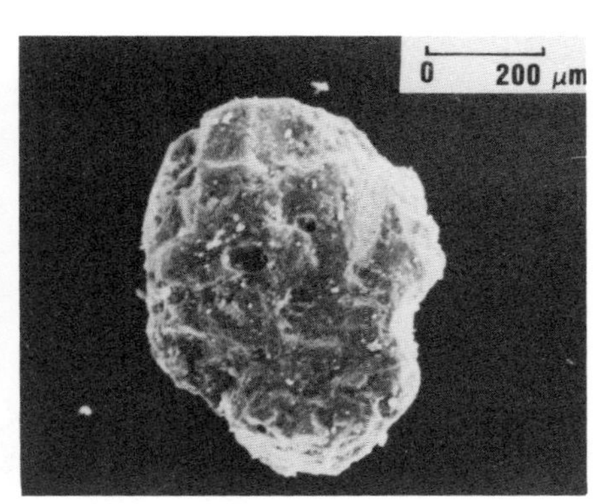

FIGURE 3b. Rounded grain with hydrated surfaces and few vesicles.

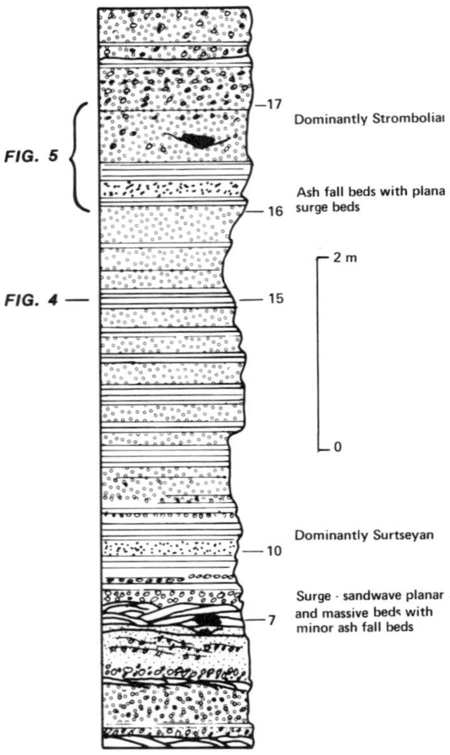

Rim stratigraphic section of Zuni Salt Lake maar. The section shows the upper one-third of the total pyroclastic section. The lower part of this section is dominantly surge beds with intercalated lapilli fall beds. The upper part is dominantly ash and lapilli fall beds with intervals of planar surge deposits.

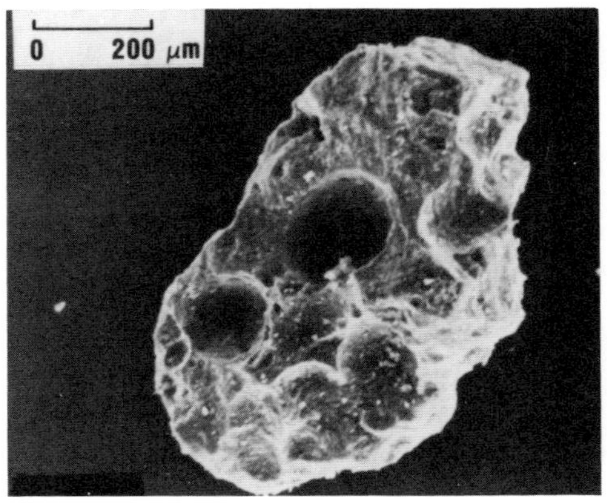

FIGURE 4a. SEMs of sample 15, from a planar surge bed. Blocky, vesicular grain with smoothed edges and pitted surfaces.

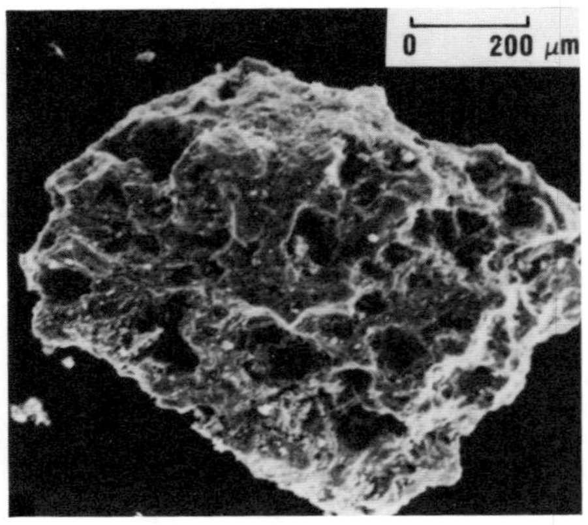

FIGURE 4b. Subangular grain with blocky, vesicular shape and pitted surfaces. Note adhering fine dust.

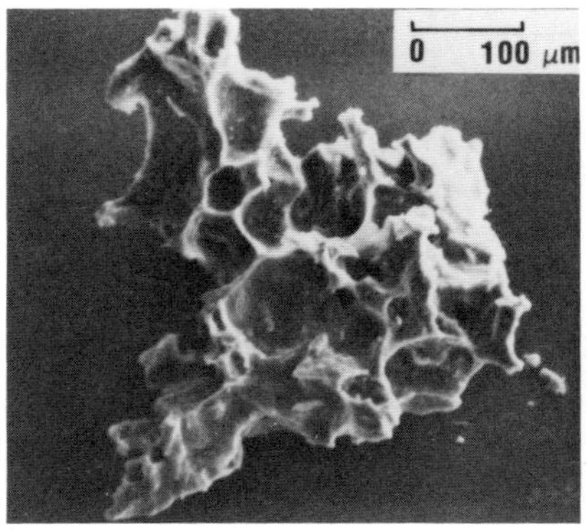

FIGURE 5a. SEMs of pyroclasts from samples 16 and 17, from fall beds. Spongy, highly vesicular grain with a delicate shape. Note the fresh-appearing surfaces and abundance of vesicles.

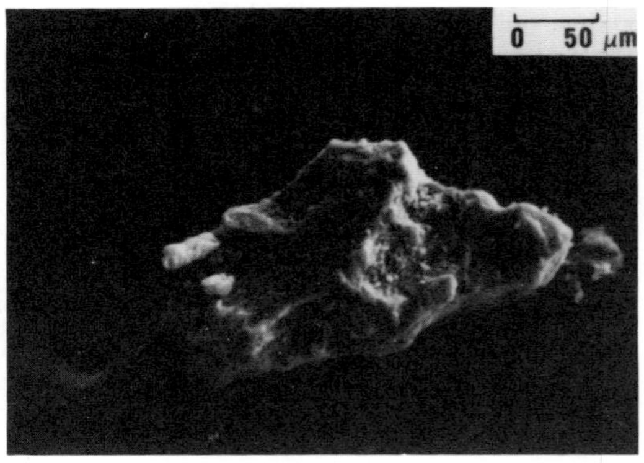

FIGURE 5b. Vesicular grain from the fine fraction. This particle is likely a broken piece of a larger, vesicular grain. Some alteration is apparent in a vesicle hollow on this fine-fraction sample.

KILBOURNE HOLE, NEW MEXICO

Introduction: Kilbourne Hole is one of three maars identified within the Potrillo volcanic field of south-central New Mexico (Hoffer, 1976); the other two are Hunts Hole and Potrillo Maar (Lee, 1907; Darton, 1916; Reiche, 1940; Shoemaker, 1957; DeHon, 1965; Reeves and DeHon, 1965). The maar erupted during Pleistocene time, along the NS-trending Fitzgerald fault system—hence its oblong dimensions of 3.2 by 2.0 km. The maar is about 130 m deep and has a surrounding tuff ring deposit, 30 to 40 m thick, that overlies sediments of the Santa Fe Group and a 5- to 7-m-thick olivine basalt flow. Hoffer (1976) divided the tuff ring into three units: a basal ash fall unit, a middle pyroclastic surge unit, and an upper massive tuff unit that DeHon (1965) considered a lahar. The basal unit is an explosion breccia, approximately 10 to 15 m thick, that crops out on the north and east rims of the maar crater. It consists of angular basalt blocks supported in dark gray, poorly bedded ash and lapilli fall deposits. The middle unit consists of thinly bedded pyroclastic surge beds that are nearly 30 m thick on the east rim. These dry surge beds are a light tan, lithic tuff near the base and grade upward to light gray color as the juvenile constituent increases in abundance. The surge deposits are dominantly sandwave beds near the crater rim; they show repeated gradations from beds of coarse ash and lapilli fall—to sandwave beds—to accretionary lapilli-bearing massive tuff beds. The upper unit, nearly 10 m thick, is composed of dominantly massive beds 3 to 4 m thick and is moderately palagonitized.

Grain Size: The tuff consists of lapilli-bearing ash to lapilli-ash because of the aggregation of fine ash in accretionary lapilli (Table 1). The tuff is moderately sorted and contains very little fine ash.

TABLE 1. GRAIN-SIZE CHARACTERISTICS OF KILBOURNE HOLE TEPHRA.

Sample	Bed Form*	Md_{φ} (mm)	σ_{φ}	Fine Ash (Wt. %)
1	S	2.20 (0.22)	1.10	4.85
2	S	2.16 (0.22)	1.11	6.52
3	F	−0.11 (1.08)	1.98	0.87
4	F	−0.16 (1.12)	1.99	0.87
5	F	−0.74 (1.67)	1.60	0.48
6	F	−0.89 (1.85)	1.58	0.42
7	S	2.17 (0.22)	1.03	3.90
8	S	2.10 (0.23)	1.14	3.67
9	M	1.30 (0.41)	1.64	2.64
10	M	1.26 (0.42)	1.69	2.39
11	M	1.67 (0.31)	1.72	3.80
12	M	1.72 (0.30)	1.70	3.57
13	S	2.19 (0.22)	0.96	5.40
14	S	2.19 (0.22)	0.94	6.76
15	S	1.98 (0.25)	1.12	4.12
16	S	2.00 (0.25)	1.12	3.93

*Bed forms are S = sandwave surge, M = massive surge, and F = fall.

Tephra Description: Kilbourne tuffs are lithic-rich (Table 2). The juvenile component within the lower part of the surge unit (beginning of strong phreatomagmatic eruptions) is only 10%. Lithic material decreases near the top of the surge unit where sideromelane is 90% of the erupted materials. Pyroclasts from sandwave beds near the crater rim are poorly vesicular, subangular to subrounded particles that show well-developed frosting. This frosted coating is the result of extensive abrasion pits and scratches that have rounded the angular, planar surfaces. Tiny grains and irregularities on surfaces, which form by silica precipitation, also contribute to the frosted appearance. The silica is apparently derived from quartz-sand lithic material that surrounds sideromelane grains. The latest phreatomagmatic eruptions produced pyroclasts with somewhat smoother, fluidal surfaces despite the presence of abundant lithic material in the beds. Pyroclasts from massive surge beds are strongly abraded, however, and are distinguishable from those of sandwave beds by surface accumulations of small dust particles and hydrated

TABLE 2. PETROGRAPHIC DESCRIPTION OF KILBOURNE HOLE TEPHRA.

Sample	>354 μm	>177 μm	>64 μm
1	10% Sideromelane 89% Lithic 1% Sideromelane	5% Sideromelane 94% Lithic 1% Sideromelane	5% Sideromelane 95% Lithic
3	10% Palagonite 89% Lithic 1% Crystals	5% Sideromelane 94% Lithic 1% Crystals	2% Sideromelane 98% Lithic
5	35% Sideromelane 62% Lithic 3% Crystals	20% Sideromelane 77% Lithic 3% Crystals	10% Sideromelane 89% Lithic 1% Crystals
7	5% Sideromelane 95% Lithic	2% Sideromelane 98% Lithic	1% Sideromelane 99% Lithic
9	89% Sideromelane 10% Crystals 1% Lithic	40% Sideromelane 20% Palagonite 35% Lithic 5% Crystals	5% Sideromelane 15% Palagonite 75% Lithic 5% Crystals
11	35% Sideromelane 62% Lithic 3% Crystals	8% Sideromelane 2% Palagonite 89% Lithic 1% Crystals	5% Sideromelane 95% Lithic
13	9% Sideromelane 1% Palagonite 89% Lithic 1% Crystals	10% Sideromelane 89% Lithic 1% Crystals	5% Sideromelane 95% Lithic
15	5% Sideromelane 5% Palagonite 90% Lithic	2% Sideromelane 98% Lithic	1% Sideromelane 99% Lithic

glass particles. Surge pyroclasts contrast sharply with those of fall beds. Surge ashes are strongly vesicular, angular-to-subangular particles with relatively unabraded surfaces.

Table 2 shows the persistence of a strong lithic contribution in most samples. The lithic particles are derived from quartzo-feldspathic sands and silts of the underlying Santa Fe Group, which are poorly consolidated. Palagonite, where present, is not strongly developed on glass surfaces, and sideromelane dominates juvenile material.

Comments on Pyroclast Formation: The phreatomagmatic eruptions at Kilbourne Hole fragmented a poorly crystalline basalt magma, which, when ejected with expanding steam, excavated a large crater in strata composed of sand and silt. The sand and silt is generously mixed with basalt pyroclasts to form the lithic ash deposit of the tuff ring. It is likely that sediments of the Santa Fe Group were water saturated, which contributed to the steam explosions. The lack of fine-fraction pyroclasts supports a conclusion that magma fragmentation was never very strong and that much of the activity borders on pure phreatic explosions caused by vaporization of water in sedimentary strata.

REFERENCES:

Darton, N. H., 1916. Explosion craters. Sci. Mon., **3**: 417-430.

De Hon, R. A., 1965. Maare of La Mesa. New Mexico Geol. Soc. Guidebook, 16th Field Conf.: 204-209.

Hoffer, J. M., 1976. The Potrillo basalt field, southcentral New Mexico. New Mexico Geol. Soc. Spec. Pub., **5**: 89-92.

Lee, W. T., 1907. Afton Craters. Geol. Soc. Am. Bull., **18**: 214-215.

Reeves, Jr., C. C. and De Hon, R. A., 1965. Geology of the Potrillo Maar, New Mexico and northern Chihuahua, Mexico. Am. J. Sci., **63**: 401-409.

Reiche, P., 1940. The origin of Kilbourne Hole, New Mexico. Am. J. Sci., **238**: 212-225.

Shoemaker, E. M., 1957. Primary structures of maar rims and their bearing on the origin of Kilbourne Hole and Zuni Salt Lake, New Mexico (abs.). Geol. Soc. Am. Bull., **68**: 1846.

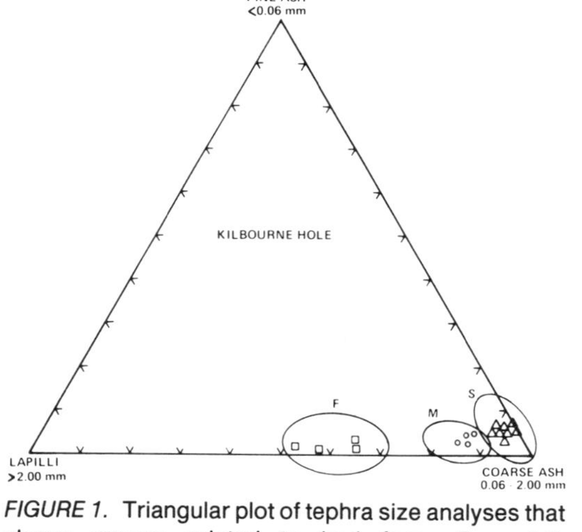

FIGURE 1. Triangular plot of tephra size analyses that shows groups related to bed form: F = fall, M = massive surge, and S = sandwave surge.

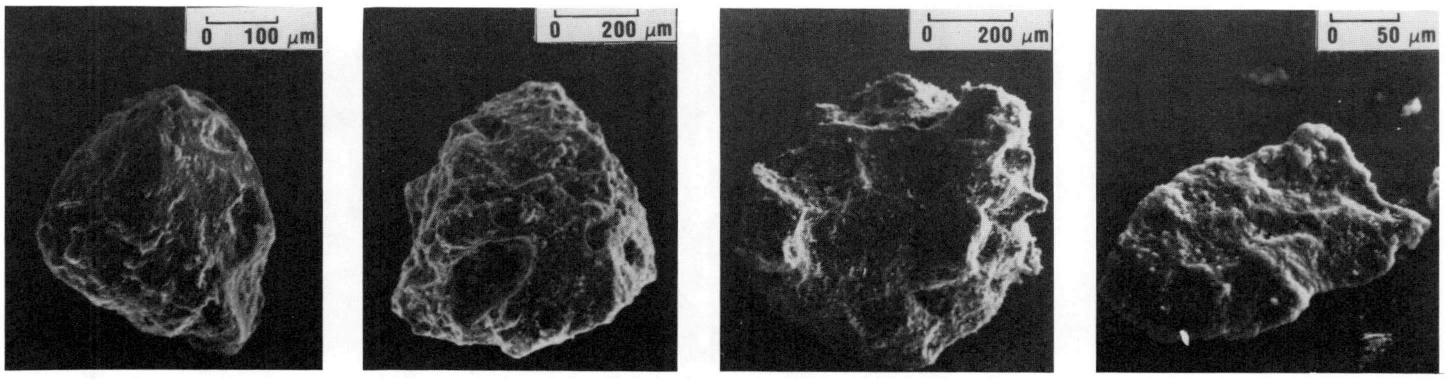

FIGURE 2. SEMs of pyroclasts from sample 1, from a sandwave surge deposit. Note the planar surfaces that have been rounded and the low vesicularity. Chipping and pitting has given the grain and vesicle edges a smooth appearance.

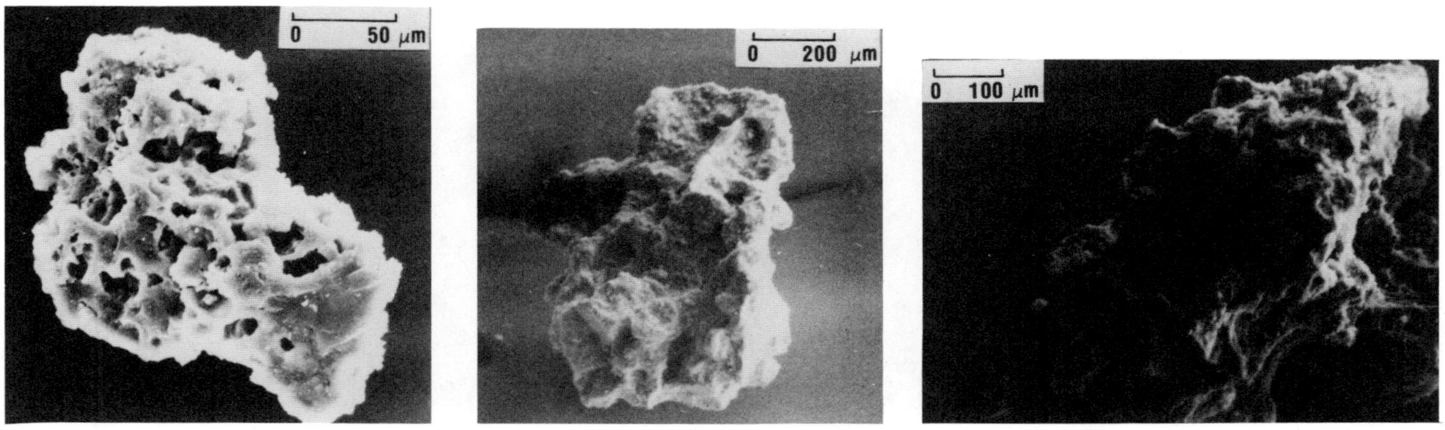

FIGURE 3. SEMs of pyroclasts from samples 3 and 5, from ash-fall deposits. These grains are vesicular and show angular shapes with sharp edges. Silica precipitation on surfaces of the two grains *center* and *right* has produced a somewhat frosted appearance.

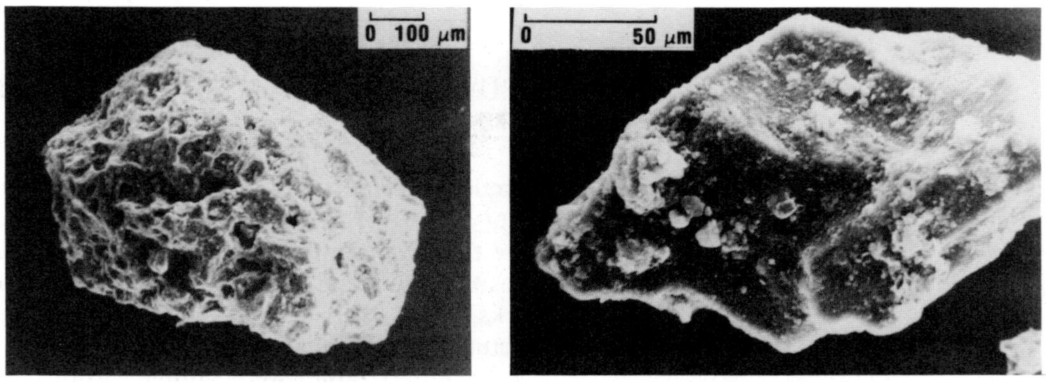

FIGURE 4. SEMs of pyroclasts from sample 7, from a sandwave surge deposit, in which curviplanar surfaces dominate the shapes. Note the edge abrasion on the bottom of the grain on the *right*.

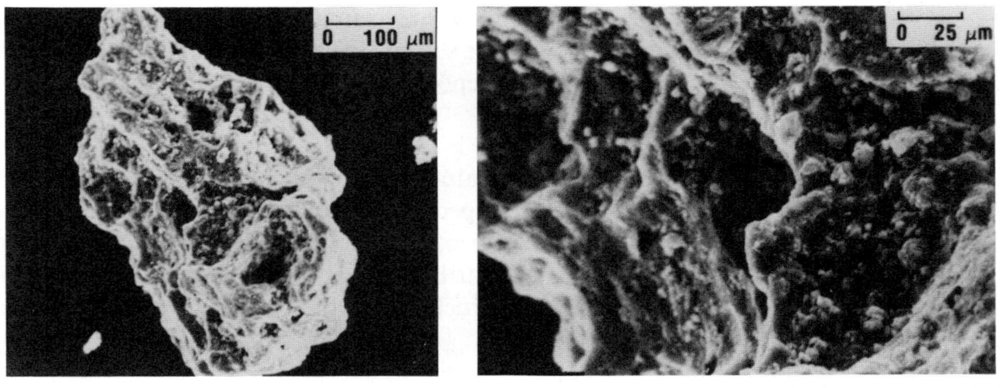

FIGURE 5. SEMs of a pyroclast from sample 9, from a massive surge deposit. *Right* - an enlargement of the central portion of the grain shown on the *left*. Abundant adhering dust particles of fine ash and alteration products exist in vesicle hollows.

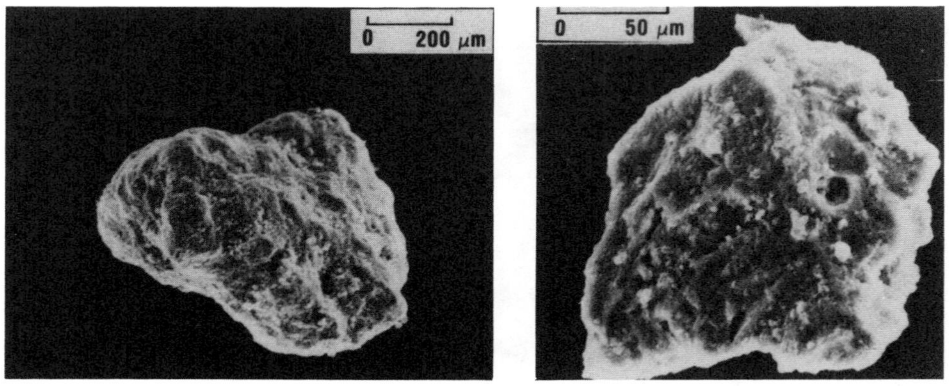

FIGURE 6. SEMs of pyroclasts from sample 11, from a massive surge deposit. The *right* view is of a fine-fraction grain covered with abundant adhering dust particles. The *left* view shows a larger grain with a strongly hydrated surface.

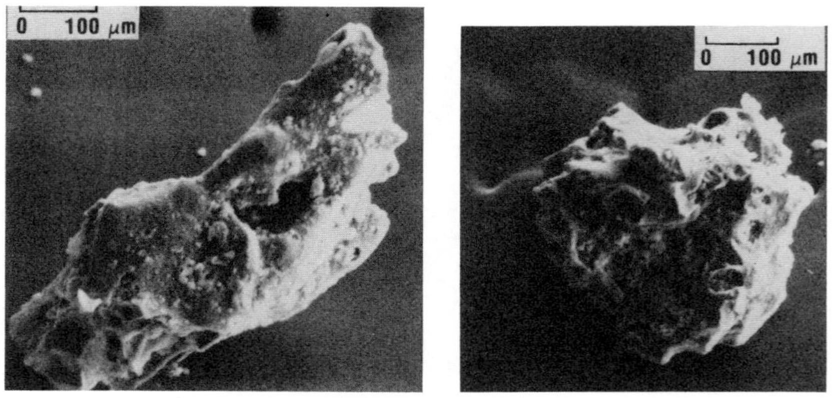

FIGURE 7. SEMs of pyroclasts from sample 15, from a sandwave bed deposited near the end of eruptive activity. Grains show fusiform surfaces between vesicle hollows. These surfaces were likely formed by fluid deformation of particle surfaces after magma fragmentation. *Left* - small vesicles are revealed in the interior of the grain.

VESUVIUS, ITALY; ACCRETIONARY LAPILLI FROM THE ERUPTION OF 79 A.D.

Introduction: Accretionary lapilli have been described in the literature since the turn of the century (LaCroix, 1904; Scrope, 1925; Stearns, 1925). More recent descriptions relate their occurrence to phreatomagmatic eruption of ash falls and base surge deposits (Moore and Peck, 1962; Fisher and Waters, 1970; Heiken, 1971; Self and Sparks, 1978). Sheridan and Wohletz (1983) described in detail the formation and structure of accretionary lapilli from Vesuvius.

Accretionary lapilli were sampled from the Pompeii (A.D. 79) and Avellino (3,500 y.b.p.) Plinian deposits of Vesuvius. The lapilli occur within the fine-ash deposits that rest above thick (>2-m) Plinian pumice fall beds (Fig. 1). Sheridan et al. (1981) proposed that these fine-ash beds are pyroclastic surge deposits erupted from catastrophic hydromagmatic phases of activity shortly after the pumice fall. The samples were studied by optical microscopy and SEM as well as EDS analyses of epoxy-impregnated thin sections.

Grain Size: Grain-size analyses of constituent ash particles were made from SEM images by inspecting approximately 1000 grains from each sample. The median grain size of ash particles ranges from 77 to 24 μm; the common mode is near 45 μm. The distributions are well sorted ($\sigma_\varphi = 1.00\varphi$). These results are characteristic of pyroclasts produced by hydroexplosions (Self and Sparks, 1978; Wohletz, 1983).

Lapilli Description: The lapilli are centimeter-size spheroids consisting of <1-mm fragments of glass, crystals, and lithic clasts. Large clasts (greater than 50-μm diameter) are present in one or two concentric bands or are randomly distributed throughout the lapilli. A matrix of fine-ash particles (less than 1 μm in diameter) has an irregular distribution, but in some lapilli the matrix occurs in distinct layers around the core.

The lapilli have three distinct textural zones. (1) A massive (structureless) core of ash (3- to 6-mm diameter) is surrounded by (2) an intermediate zone with bands of bubbles and larger clasts and (3) a fine-grained rim (0.3 to 1.8 mm thick). Patches of bubbles (50 to 300 μm in diameter) occur in the cores of some lapilli as well as in the bands and patches between cores and rims.

Most clasts in the lapilli have blocky shapes with delicate projections. A small fraction of the clasts have vesicles—2% in Pompeii lapilli and 5% in Avellino samples.

Chemical Composition: Di Girolamo (1968) made chemical analyses of samples from intervals throughout the entire section at Pompeii. These analyses (Table 1) show a significant difference between the composition of samples in the pumice fall and those from the fine-ash horizon. This distinction reflects a general increase in the carbonate lithic constituent of the ash during the eruption. The increase is abrupt from the pumice fall to the fine-ash surge beds. Table 1 also gives pumice analyses of the Avellino and Pompeii deposits (from Barberi et al., 1981).

Table 2 lists representative EDS analyses of samples from selected areas along traverses from the core to the rim of individual lapilli. These analyses also show the general increase in carbonate content.

TABLE 1. CHEMICAL ANALYSES OF POMPEII AND AVELLINO SAMPLES.*

	1	2	3	4	5
SiO_2	53.60	49.44	44.40	52.85	42.88
TiO_2	0.27	0.70	0.56	0.49	0.56
Al_2O_3	21.57	17.33	14.80	18.91	18.84
$Fe_2O_3 + FeO$	2.76	5.66	5.29	4.23	4.51
MnO	0.12	0.13	0.09	0.11	0.12
MgO	0.31	4.51	3.51	2.55	2.12
$CaO + CaCO_3$	2.75	7.65	13.37	6.44	5.68
K_2O	9.71	6.34	5.19	6.56	8.29
Na_2O	5.09	3.28	2.80	5.22	4.42

*Sample descriptions and sources of analyses are as follows:
1 = White pumice from bottom of Pompeii section (see Fig. 1) (di Girolamo, 1968).
2 = Fine-ash sample from directly above pumice fall at Pompeii (di Girolamo, 1968).
3 = Pisolitic (accretionary lapilli) ash sample from 30 cm above pumice fall at Pompeii section (di Girolamo, 1968).
4 = Avellino pumice (Barberi et al., 1981).
5 = Pompeii pumice (Barberi et al., 1981).

TABLE 2. CHEMICAL ANALYSES OF THE CORES AND RIMS OF ACCRETIONARY LAPILLI.*

	Sample 6		Sample 7		Sample 8	
	Core	Rim	Core	Rim	Core	Rim
SiO_2	51.19	50.85	47.57	46.84	44.32	39.49
TiO_2	0.58	0.74	0.59	0.53	0.73	0.72
Al_2O_3	21.70	21.41	19.22	18.31	17.63	17.19
$Fe_2O_3 + FeO$	3.80	4.42	3.38	2.67	3.82	3.51
MgO	2.06	2.59	2.57	2.31	4.60	5.09
$CaO + CaCO_3$	7.84	9.82	17.92	20.15	18.60	24.76
K_2O	5.22	4.14	4.09	4.63	5.49	4.38
Na_2O	6.73	5.18	3.95	3.87	3.96	3.71
Total	99.12	99.15	99.29	99.31	99.15	98.85

*Standardized EDS analyses of samples taken from Pompeii fine-ash deposits at increasing levels above the pumice fall deposit (see Fig. 1).

Comments on Lapilli Formation: The interpretation of lapilli formation is based upon the general model of Plinian eruptions of Vesuvius described by Barberi et al. (1981) and Sheridan et al. (1981). As eruptions proceed, the limestone aquifer walls of the magma chamber disintegrate and allow water to enter the chamber—thus creating hydroexplosions. This phase of activity is reflected by a change from pumice fall deposits to phreatomagmatic surge deposition of fine ash.

The EDS analyses of the lapilli plot on the same trend of compositions (carbonate versus SiO_2) shown in di Girolamo's (1968) ash analyses; this notable increase in carbonate from core to rim for each lapillus indicates that these lapilli accreted in progressively lithic-rich zones of the eruption plume. The chemical, size, and shape analyses of the lapilli suggest the following interpretation of their origin. Cores of the lapilli formed as aggregates in the hot central zone of the eruption plume. As the lapilli moved outward into turbulent margins of the plume, they began to tumble and accrete layers of fine ash that were enriched in lithics from the magma-chamber walls. The turbulent zone of the plume likely contained more water vapor, which promoted the accretion and formed water bubbles in the layered zone of the lapilli. Little accretion occurred after the lapilli moved out of the turbulent plume, and only a thin veneer of carbonate-rich dust was deposited as a rim before the lapilli fell down into advancing surge clouds.

REFERENCES:

Barberi, F., Bizouard, H., Clocchiatti, R., Metrich, N., Santacroce, R., and Sbrana, A., 1981. The Somma-Vesuvius magma chamber: a petrologic and volcanologic approach. Bull. Volcanol., **44**: 295-316.

di Girolamo, P., 1968. Petrografia del somma-Vesuvio: le serie piroclastiche. Rend. Accad. Sci. Napoli, **35**: 1-68.

Heiken, G. H., 1971. Tuff rings: examples from the Fort Rock-Christmas Lake Valley basin, south central Oregon. J. Geophys. Res., **76**: 5615-5626.

Fisher, R. V. and Waters, A. C., 1970. Base surge bedforms in maar volcanoes. Am. J. Sci., **268**: 157-180.

LaCroix, A., 1904. *La Montagne Pelèe et ses eruptions.* Masson et Cie., Paris.

Moore, J. G. and Peck, D. L., 1962. Accretionary lapilli in volcanic rocks of the western continental United States. J. Geol., **70**: 182-193.

Scrope, A. P. 1925. *Considerations on volcanoes.* W. Phillips and G. Yard Pub., London.

Self, S. and Sparks, R. S. J., 1978. Characteristics of widespread pyroclastic deposits formed by the interaction of silicic magma and water. Bull. Volcanol., **41**: 196-212.

Sheridan, M. F., Barberi, F., Rosi, M., and Santacroce, R., 1981. A model for Plinian eruptions of Vesuvius. Nature, **289**: 282-285.

Sheridan, M. F. and Wohletz, K. H., 1983. Origin of accretionary lapilli from the Pompeii and Avellino deposits of Vesuvius. *In*: R. Gooley (Ed.), *Microbeam analysis—1983*. San Francisco Press, pp. 35-38.

Stearns, H. T., 1925. The explosive phase of Kilauea volcano, Hawaii, in 1924. Bull. Volcanol., **5**: 1-16.

Wohletz, K. H., 1983. Mechanism of hydrovolcanic pyroclast formation: size, scanning electron microscopy, and experimental studies. *In*: M. F. Sheridan and F. Barberi (Eds.), *Explosive volcanism*, J. Volcanol. Geotherm. Res., **17**: 31-64.

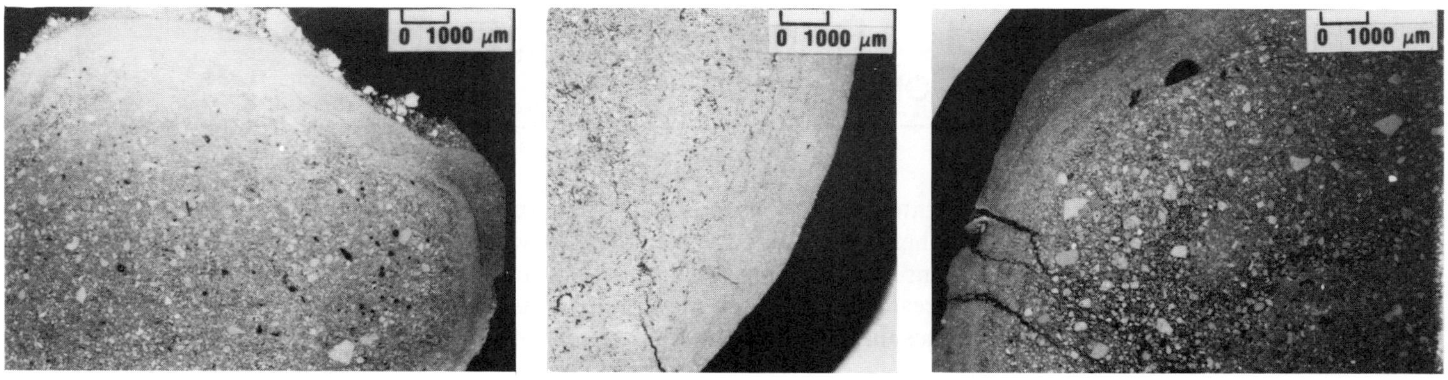

FIGURE 1. SEMs of representative accretionary lapilli that show the coarser grained centers and fine-grained rims.

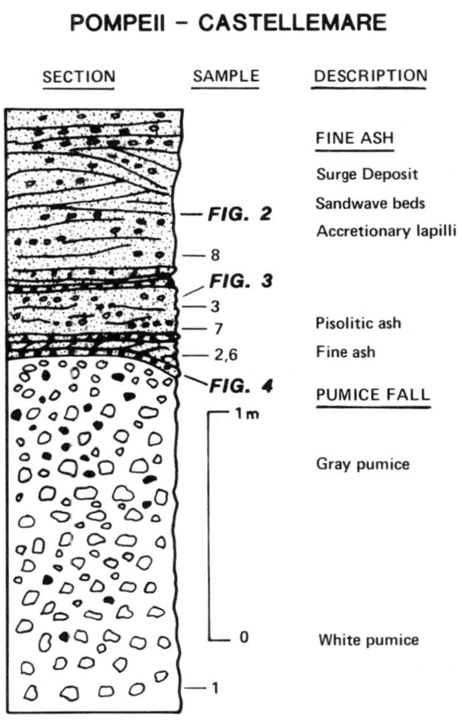

Columnar stratigraphic section of the A.D. 79 Vesuvius deposit at the Pompeii excavation near Castellemare. The lower half of the section is Plinian pumice fall products overlain by fine-ash deposits of pyroclastic surges in which accretionary lapilli occur. Sample numbers 1, 2, and 3 indicate levels from which di Girolamo (1968) took samples. Numbers 6, 7, and 8 indicate approximate levels of samples studied in this report.

FIGURE 2. SEMs of the core (*left*) and rim (*right*) of lapilli from sample 8.

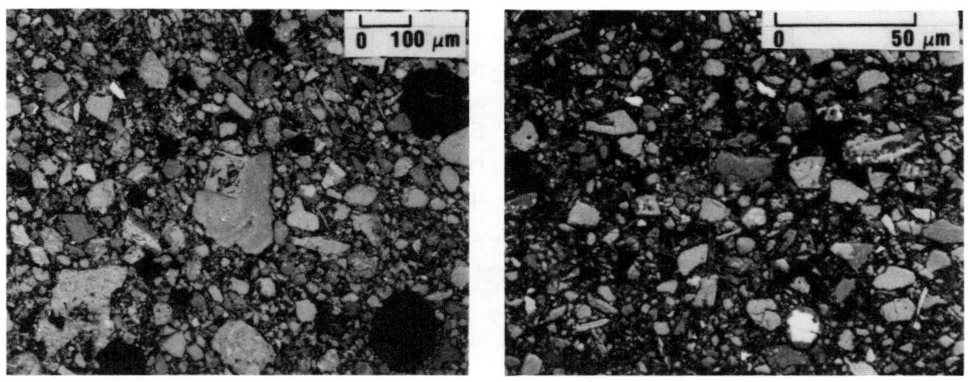

FIGURE 3. SEMs of the core (*left*) and rim (*right*) of lapilli from sample 7.

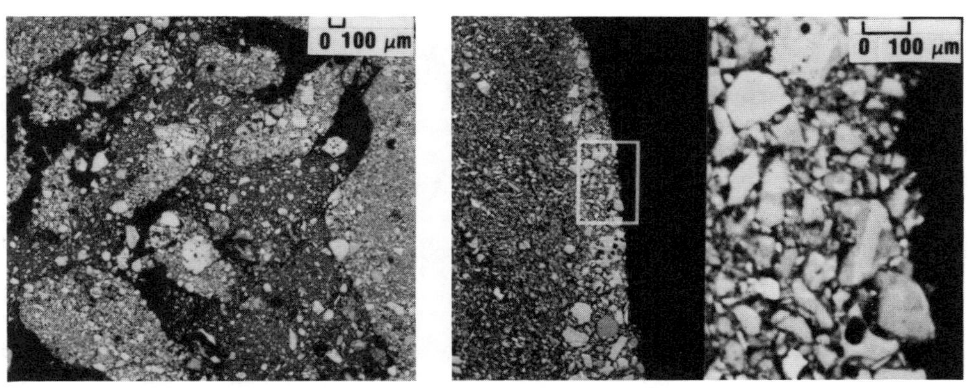

FIGURE 4. SEMs of the core (*left*) and rim (*right*) of lapilli from sample 6.

Tuff Rings (Magmas of Intermediate Composition)

Intermediate-composition (andesitic and dacitic) magmas generally have a higher viscosity than do basalts. The yield strength of such magmas is also higher. Consequently, brittle fracture dominates production of pyroclasts during phreatomagmatic eruption of these magmas. Tephra are typically angular, blocky ashes that are often coarse grained.

CLEAR LAKE VOLCANIC FIELD, CALIFORNIA

Introduction: The Clear Lake volcanic field covers a 400-km^2 area about 150 km north of San Francisco, California. This field has experienced eruptions from the late Pliocene to Holocene time (Hearn et al., 1972). The geology, geophysics, geochemistry, and geothermal potential has been well studied by Anderson (1936), Brice (1953), Hodges (1966), Chapman (1975), U.S. Geological Survey (1981), and Goff et al. (1977). Samples examined in this section were taken from pyroclastic deposits of small tuff rings that developed along the margins of the present-day lake during the last 0.03 m.y. (shown as Holocene pyroclastic deposits in Fig. 1). These pyroclastic deposits range in composition from basalt to dacite within individual tuff rings. This observation supports the hypothesis that mafic and silicic magmas were mixed during evolution of this field (Eichelberger, 1975).

Grain Size: Twenty-three samples taken from pyroclastic fall and surge beds were sieved, and results are shown in Table 1. The samples are moderately to well sorted, coarse ash-to-lapilli size with only a small contribution of fine ash. The small amount of fine ash is unusual for tuff ring deposits and may indicate that eruptions and tuff ring formation occurred beneath the level of Clear Lake. In this situation, fine ash, if produced, may have been segregated from coarse ash within the lake.

TABLE 1. GRAIN-SIZE DESCRIPTION OF CLEAR LAKE TEPHRA.

Sample	Bed Form*	Md_φ (mm)	σ_φ	Fine Ash (Wt. %)
1	M	0.98 (0.51)	1.37	1.51
2	S	−0.66 (1.58)	0.77	0.3
3	F	−2.67 (6.36)	0.54	0.10
4	S	−1.21 (2.31)	1.19	1.36
5	P	−1.00 (2.00)	1.67	2.17
6	S	0.25 (0.84)	1.47	1.46
7	P	−0.24 (1.18)	1.75	1.12
8	F	−1.68 (3.20)	1.88	0.54
9	P	−0.48 (1.39)	1.26	1.34
10	M	−0.81 (1.75)	1.41	2.42
11	F	−0.09 (1.06)	1.42	0.95
12	F	−1.96 (3.89)	1.28	0.34
13	P	0.56 (0.67)	1.30	1.99
14	F	−0.98 (1.97)	0.83	0.45
15	P	−1.78 (3.43)	1.53	0.99
16	S	−0.61 (1.53)	1.15	1.80
17	P	0.19 (0.88)	1.12	2.79
18	F	−1.59 (3.01)	1.22	2.65
19	S	0.72 (0.61)	1.29	4.96
20	F	−0.88 (1.84)	1.68	0.53
21	M	−0.57 (1.48)	1.14	0.45
22	S	−0.08 (1.06)	1.39	3.46
23	F	−0.07 (1.05)	1.36	3.90

*Bed forms are S = sandwave surge, M = massive surge, P = planar surge, and F = fall.

Tephra Description and Composition: Of the glass constituents, five color types were recognized—black, brown, orange, white, and gray (Table 2). Average chemical analyses of these types are shown in Table 3. The analyses partially reflect hydrothermal alteration (e.g., higher SiO_2 abundances). Black and brown glasses are basaltic andesite; the orange is basaltic; and the gray and white glasses are andesitic and dacitic, respectively.

Mixed magma eruption is shown in samples 1, 2, and 3. These samples were collected in ascending stratigraphic order from a tuff ring at Soda Bay. Early eruptions produced dominantly basaltic andesite (black and brown pyroclasts) with a strong lithic contribution (Table 2). These poorly vesicular pyroclasts show typical blocky shapes of phreatomagmatic tephra with fused surfaces. Later eruptions produced vesicular tephra that show planar breakage surfaces that are, in general, chemically altered. These tephra likely represent stronger magma/water interaction and post emplacement hydrothermal alteration of basalt. Final eruptions produced a dacitic tephra composed of white and gray pyroclasts.

Other variations in pyroclast shape are strongly correlated to chemical composition. Gray glass samples commonly show blocky texture with uneven surfaces and no vesicles, whereas orange glass is strongly vesicular with smooth surfaces. This variation is well illustrated by fine-fraction particles in which tiny blocky clasts of gray glass are covered with adhering dust whereas those of an adjacent sample of orange glass are curved splinters and chips of broken vesicle walls.

The Clear Lake fine ashes, where present, show many interesting features. Grain surfaces are commonly covered with adhering dust particles that are from 5 μm to less than 1 μm in diameter. Many of these tiny particles appear to be fused to surfaces, and some have the angular euhedral shapes of crystallites. Overgrowths of clay and zeolite alteration products on glass surfaces, especially in hollows, are very evident on a <1-μm scale. Yet other bulbous particles are stuck together as moss-like aggregates and are attached to grain surfaces; these particles have diameters of <0.05 μm.

TABLE 2. PETROGRAPHIC DESCRIPTION OF CLEAR LAKE TEPHRA.*

			Glass				
Sample	Crystal	Lithic	Gray	White	Black	Brown	Orange
1	---	14.9	---	---	37.3	22.3	---
2	---	---	---	14.9	1.9	---	83.1
3	17.4	9.9	12.6	60.0	---	---	---
4	3.9	---	21.2	---	---	---	74.8
5	4.5	---	27.5	---	---	---	67.9
6	9.9	22.3	15.0	36.6	---	---	16.0
7	22.7	---	48.0	---	---	---	29.2
8	23.3	6.5	45.4	---	---	---	24.6
9	---	---	9.1	79.2	---	---	11.6
10	10.9	11.9	77.1	---	---	---	---
11	---	9.0	20.0	70.0	---	---	---
12	4.5	6.4	---	32.5	56.4	---	---
13	11.0	14.2	59.0	15.5	---	---	---
14	---	---	54.5	---	32.4	---	12.9
15	3.9	---	31.4	---	---	---	64.6
16	6.4	---	21.4	72.1	---	---	---
17	---	4.5	5.6	89.8	---	---	---
18	---	---	80.1	---	5.6	---	14.1
19	8.9	7.5	19.9	---	63.6	---	---
20	---	---	39.6	16.9	---	---	43.4
21	---	---	9.4	33.9	---	---	56.7
22	7.8	--	18.8	64.6	---	8.6	---
23	---	37.1	31.4	31.4	---	---	---

*Sample grain count percentages.

TABLE 3. CHEMICAL COMPOSITION OF GLASSY TEPHRA FROM CLEAR LAKE.*

Element	Gray (65)	White (35)	Black (12)	Brown (7)	Orange (26)
SiO_2	63.30	63.94	60.99	61.38	59.55
TiO_2	1.10	1.18	0.97	1.02	1.06
Al_2O_3	19.03	18.94	20.50	22.24	27.21
FeO	6.69	7.80	4.20	3.93	4.91
MnO	0.80	0.23	0.61	0.15	0.34
MgO	1.92	1.88	3.53	2.57	2.08
CaO	3.05	3.62	3.71	3.90	0.84
Na_2O	0.97	0.69	1.28	1.11	0.61
K_2O	1.94	1.15	1.89	2.16	0.86
Total	98.80	99.42	97.68	98.16	97.46

*Normalized EDS analyses of grain surfaces. Results are the averages for the number of samples shown in parentheses.

Comments on Pyroclast Formation: Eruptions under Clear Lake during recent times have been mostly phreatomagmatic, forming tuff rings. The tuff deposits have mixed pyroclast assemblages consisting of basaltic andesite through dacite, which may be indicative of mixed magmas. Early eruptions produced nonvesicular ash of basaltic andesite composition and tended towards gas-rich eruptions of vesicular basaltic ash. Later products are from a more viscous, crystal-rich magma of dacitic affinity.

These materials generally were explosively erupted because they mixed with lake water. The variation in compositions also resulted in phreatomagmatic pyroclasts of different shapes, from blocky particles to vesicular splinters and broken bubble walls (basaltic andesite) to spongy (hydrated basalt) and angular grains (dacite). The stratigraphic sequence of erupted products suggests that a basaltic to basaltic-andesitic magma body intruded a crystal-rich dacite. The subsequent mixing may have stimulated vesicle growth in the basaltic magma.

REFERENCES:

Anderson, C. A., 1936. Volcanic history of the Clear Lake area, California. Geol. Soc. Am. Bull., **47**: 629-664.

Brice, J. C., 1953. Geology of Lower Lake quadrangle, California. California Div. Mines Bull., **166**: 72 pp.

Chapman, R. H., 1975. Geophysical study of the Clear Lake region, California. California Div. Mines and Geol. Spec. Rept., **116**: 23 pp.

Eichelberger, J. C., 1975. Origin of andesite and dacite: evidence of mixing at Glass Mountain in California and at other circum-Pacific volcanoes. Geol. Soc. Am. Bull., **86**: 1381-1391.

Goff, F. E., Donnelly, J. M., Thompson, J. M., and Hearn, B. C., Jr., 1977. Geothermal prospecting in The Geysers-Clear Lake area, northern California. Geology, **5**: 509-515.

Hearn, B. C., Jr., Donnelly, J. M., and Goff, F. E., 1972. Geology and geochronology of the Clear Lake volcanics, California. Proc. 2nd U.N. Symp. on Develop. Use Geotherm. Res., San Francisco: 423-428.

Hodges, C. A., 1966. Geomorphic history of Clear Lake, California. Ph.D. Dissertation, Stanford University: 210 pp.

U.S. Geological Survey, 1981. Research in The Geysers-Clear Lake geothermal area, northern California. R. J. McLaughlin and J. M. Donnelly (Eds.), U.S. Geol. Surv. Prof. Paper **1141**: 191 pp.

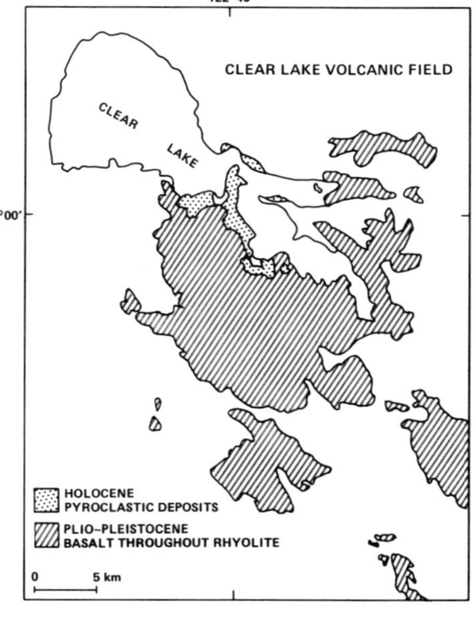

FIGURE 1. Generalized geologic map of Clear Lake showing distribution of the Clear Lake volcanic rocks, including Holocene pyroclastic deposits (adapted from Hearn et al., 1972). The pyroclastic rocks sampled for this section crop out along the shores of the lake as maar deposits.

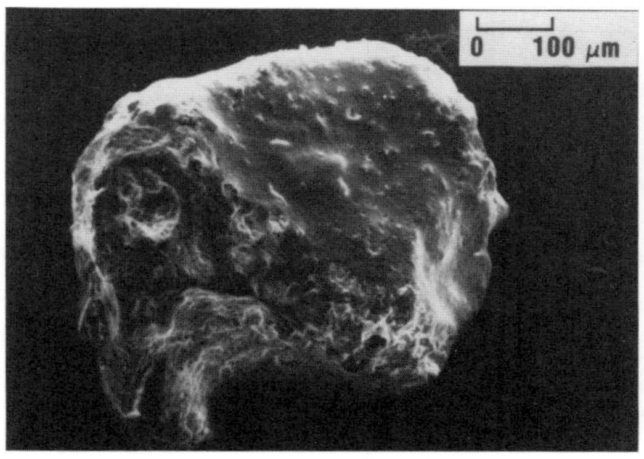

FIGURE 2a. SEMs of sample 1. Pyroclast of basaltic-andesitic black glass. Note the overall blocky shape with smooth, fused edges.

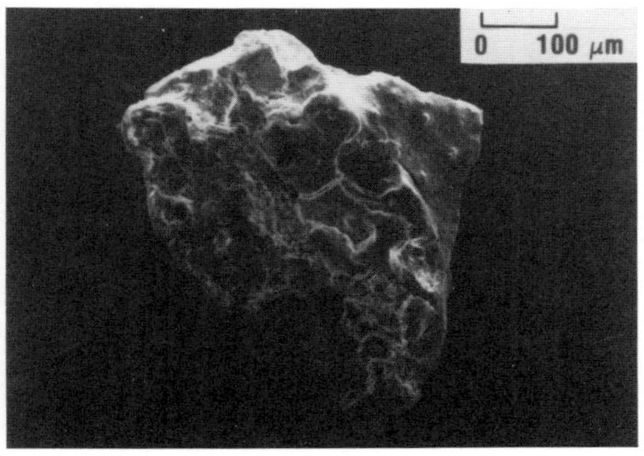

FIGURE 2b. Pyroclast of brown glass. Incipient alteration is shown by cracks on the surface. Smoothed surfaces result from glass quenching after fracture during water interaction.

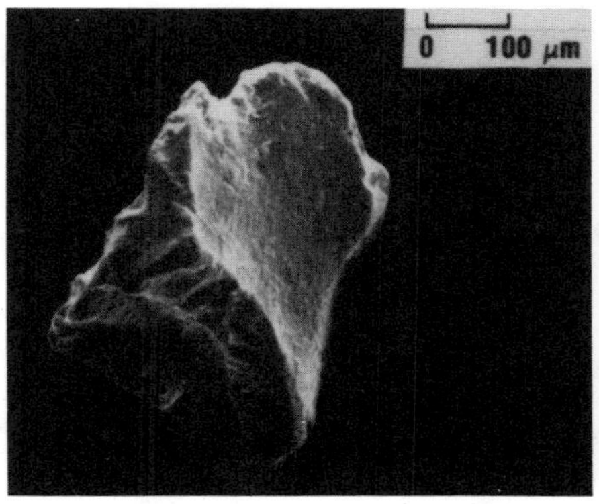

FIGURE 3a. SEMs of sample 2. Angular pyroclast of orange basaltic glass with a smooth vesicle surface that has been cut by a fracture.

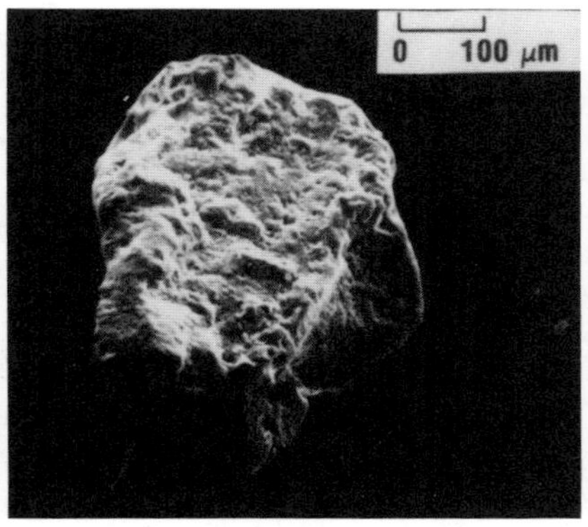

FIGURE 3b. Contrasting pyroclast of white dacitic glass with uneven surface formed by microcrystals.

FIGURE 3c. Fine fraction is composed of angular chips of nonvesicular glass that are splinters from larger vesicle walls.

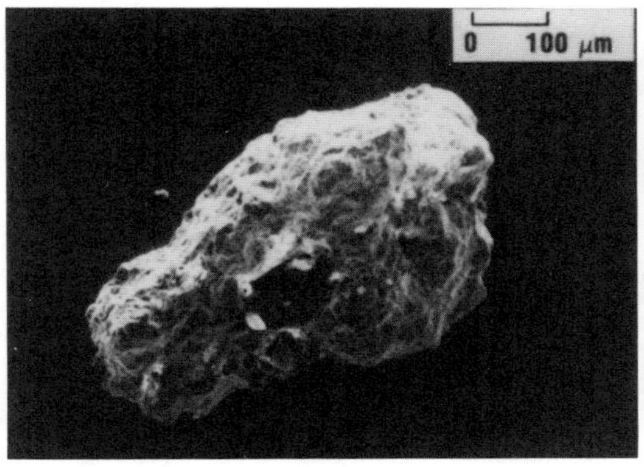

FIGURE 4a. SEMs of sample 3. Gray, dacitic glass with uneven, partly altered surface.

FIGURE 4b. White dacitic glass showing small vesicles and phenocrysts and altered intersertal glass; (*left* view is a secondary electron image; *right* is a backscattered view).

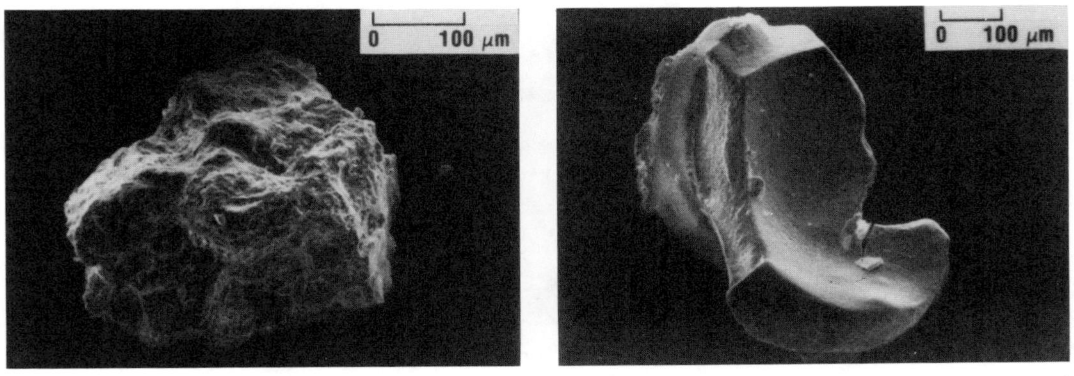

FIGURE 5. Contrasting SEMs of gray (*left*) and orange (*right*) glass showing shape variation between dacitic and basaltic glass. Both samples are from sample 2A and have broken surfaces from phreatomagmatic eruption; however, they show distinct differences in shape (blocky versus vesicular) and crystallinity.

FIGURE 6a. Sample 4 gray glass shows angular, blocky particles of dacitic glass with fine adhering dust.

FIGURE 6b. Sample 5, of dominantly orange glass, shows angular plate-like particles of orange glass that are somewhat finer grained than those of sample 4. Note smooth, curved surfaces with sharp points.

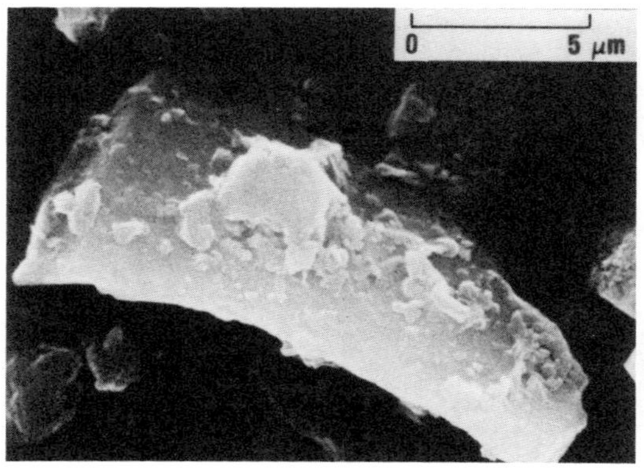

FIGURE 7a. High-resolution SEMs of the fine fraction of sample 16. Plate-like clast with adhering dust particles fused to the surface.

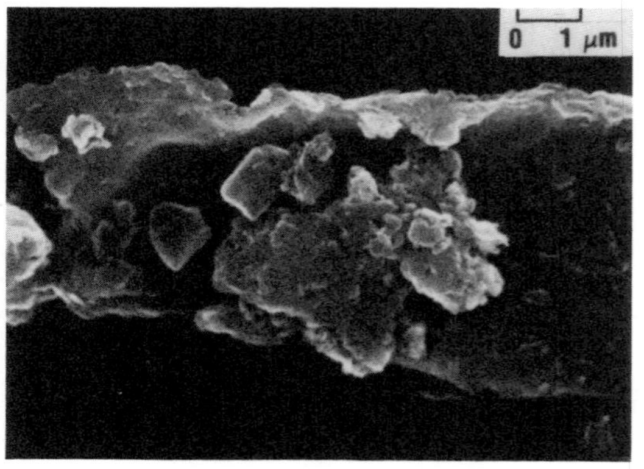

FIGURE 7b. Elongated vesicle hollow filled with clay-like debris and zeolite crystals (euhedral shapes). Vesicle edges are thin and irregular.

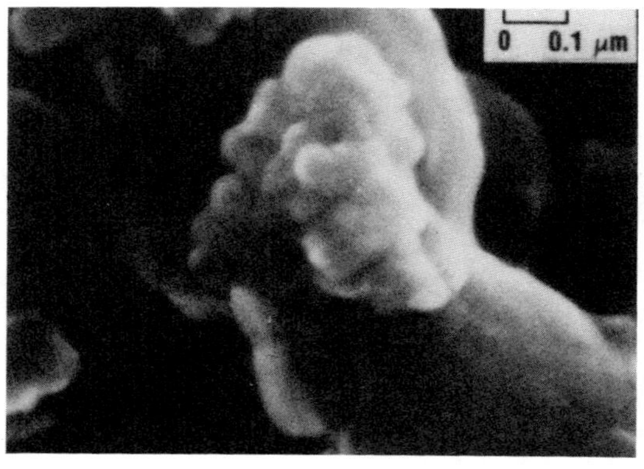

FIGURE 7c. Close-up image of 0.1-μm particle aggregates. The small cracks are likely formed by gold crazing of the conductive coat.

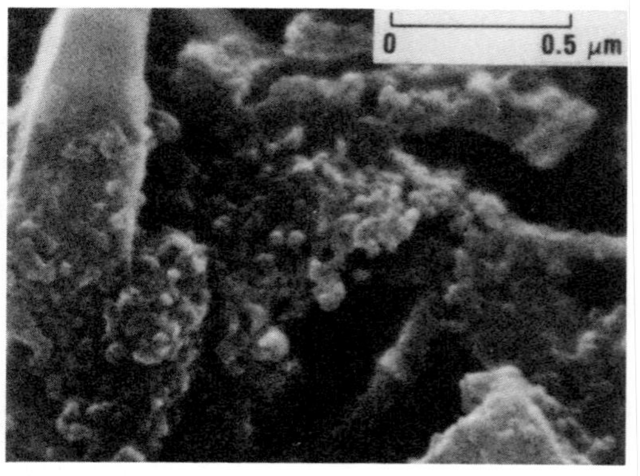

FIGURE 7d. Particle aggregate of 0.05-μm-diameter, sphere-like glass particles attached to flat surfaces of crystals.

FIGURE 7e. Microphenocrysts of plagioclase attached to the surface of a glass particle.

Phreatoplinian Eruptions of Rhyolitic Magma

Phreatoplinian is a term first used by Self and Sparks (1978) for eruptions of extremely fine-grained rhyolitic ash deposited over large areas. This ash is formed by both vesiculation and interaction of magma with meteoric water. It is characterized by the presence of abundant vesicles and planar breakage features. Plinian eruptions at Taupo Volcano occurred within Lake Taupo, New Zealand; field evidence there lends further support to interpretations of the phreatoplinian eruption mechanism.

PANUM CRATER, CALIFORNIA

Introduction: Panum Crater is a Holocene rhyolite dome surrounded by a tuff ring 600 to 700 m in diameter. It is the northernmost volcano of the Mono Crater chain and is located in east-central California (Putnam, 1938; Kistler, 1966). Eruptions, dated at 1190 ± 80 y.b.p. (Wood, 1977), initially were explosive and were caused by extrusion of vesiculating rhyolite through the water-saturated alluvial deposits surrounding Mono Lake. Phreatomagmatic and magmatic explosions were interspersed during the eruptions that formed the tuff ring. Late-stage activity consisted of passive extrusion of a rhyolite dome in the tuff ring crater. Samples are from a pumice quarry on the southeast side of the tuff ring; several samples were provided by Wood (1977), whose tephra #2 was collected from meadow topsoils at several localities in the southern Sierra Nevada. Characteristics of phreatomagmatic activity and the sub-Plinian to Plinian dispersal of pumice and fine ash over more than 500 km^2 in southeastern California indicate that a phreatoplinian-type eruption occurred at Panum Crater.

Grain Size: Five samples taken from the measured stratigraphic sequence were sieved; the analysis results are given in Table 1. Opening activity produced the greatest abundance of fine material (samples 1 and 2). The final stages of pyroclastic activity (samples 3, 4, and 5) resulted in a gradual increase in median grain size, which is interpreted as a decrease in the explosive energy of the eruptions.

TABLE 1. GRAIN-SIZE CHARACTERISTICS OF PANUM CRATER TEPHRA.

Sample	Bed Form*	Md_φ (mm)	σ_φ	Fine Ash (Wt. %)
1	M	1.80 (0.29)	2.09	8.75
2	S	2.83 (0.14)	0.98	10.40
3	F	0.66 (0.63)	2.53	5.59
4	S	0.29 (0.82)	1.18	0.85
5	P	−0.18 (1.13)	2.06	1.69

*Bed forms are M = massive surge, S = sandwave surge, P = planar surge, and F = fall.

Tephra Description: The basal pyroclastic unit is an explosion breccia interstratified with massive surge beds. Grains are vesicular and bounded by nearly flat surfaces that cut bubble walls. Overall, these grains are blocky and have coatings of fine-grained plate-like material, especially in vesicular hollows. Some alteration products are apparent on grain surfaces and sharp vesicle edges have been smoothed by transport abrasion. The fine fraction is nonvesicular and consists of small, plate-like shards.

Above the basal unit is a 2-m-thick sandwave sequence. Grains sampled from this sequence are coarsely vesicular. They have blocky, equant shapes and many are tabular with elongated vesicles. Curved chips and dish-shaped fractures are abundant on edges of the grains, and surfaces are fresh and unaltered. The fine fraction is nonvesicular and consists dominantly of sharp-edged plates.

A sub-Plinian fall deposit above the lower surge sequence is reversely graded. Vesicular pyroclasts from this unit have blocky, equant surfaces with sharp vesicle edges, and they show no rounding from abrasion. The fine fraction consists of blocks and plate-like particles; as in Wood's (1977) tephra #2, vesicles are rare and the particles are bounded by the curved surfaces of vesicle walls.

Late-stage pyroclastic activity at Panum Crater produced blocky ash particles with highly elongated vesicles. Samples from coarse-grained pyroclastic surge deposits show strong grain-edge modification caused by transport. Both fresh and partly altered grain surfaces are evident.

Chemical Composition: The EDS analyses of sample grain surfaces are shown in Table 2. These analyses are similar to those reported by Loney (1968) and Lajoie (1968) and show variations in elemental abundances that are most likely related to variable chemical alteration on grain surfaces. Opening eruptions produced more lithic clasts and fewer crystals than the later eruptions did because of crater excavation and magma differentiation.

TABLE 2. CHEMICAL ANALYSES OF PANUM CRATER PYROCLAST SURFACES IN WT. %.*

	1	2	3	4	5
SiO_2	67.72	72.45	69.90	68.45	70.83
TiO_2	2.14	1.64	2.02	2.22	1.79
Al_2O_3	13.49	12.62	12.92	12.86	13.02
FeO	3.62	1.97	2.64	3.13	1.29
MnO	1.86	1.26	1.62	1.98	1.40
MgO	1.59	1.30	1.54	1.51	1.48
CaO	2.40	1.85	2.23	2.34	2.01
Na_2O	1.37	1.39	1.67	1.50	1.52
K_2O	5.80	5.55	5.94	6.01	5.77
Glass	85.9	74.0	83.0	87.0	83.0
Crystal	4.7	24.0	12.5	8.0	12.0
Lithic	9.4	2.0	4.5	4.8	4.0

*Normalized standardless EDS analyses.

Pyroclast Formation: Pyroclasts of Panum Crater are mostly of hydromagmatic origin, as is shown by their characteristic blocky morphology. High stands of Mono Lake (above the elevation of Panum Crater) occurred during Pleistocene times. Water from the lake and from aquifers in alluvial and glacial gravels underlying Panum Crater probably caused the hydrovolcanic eruptions. Explosive expansion of magmatic gases has also contributed to pyroclast formation, especially in early eruptions. The curviplanar fracture surfaces suggest that the magma fragmented in a brittle fashion. The foliation, caused by elongated vesicles, controlled much of the brittle fracture, and as a result, many of the grains are elongate slabs broken parallel to the foliation. The unaltered grain surfaces suggest that most of the hydrovolcanic eruptions producing the tuff ring were dry (there was little vapor condensation on pyroclasts). Vapor explosions that occurred at the surface of the magma column generated stress waves of sufficient energy to fracture the magma. Stress waves propagated down the vent and initially fragmented the lava. As the viscous particles of fragmented lava approached the surface, vaporized water ejected the particles in base surges.

REFERENCES:

Kistler, R. W., 1966. Geologic map of the Mono Craters quadrangle, Mono and Tuolomne Counties, California. U.S. Geol. Surv. Map GQ-462.

Lajoie, K. R., 1968. Late Quaternary stratigraphy and geologic history of Mono Basin, eastern California. Ph.D. Dissertation., Univ. California, Berkeley: 271 pp.

Loney, R. H., 1968. Structure and composition of the southern coulee, Mono Craters, California. Geol. Soc. Am. Mem., **116**: 415-440.

Putnam, W. L., 1938. The Mono Craters, California. Geog. Rev., **28**: 68-82.

Self, S. and Sparks, R. S. J., 1978. Characteristics of widespread pyroclastic deposits formed by the interaction of silicic magma and water. Bull. Volcanol., **41**: 196-212.

Wood, S. H., 1977. Distribution, correlation, and radiocarbon dating of late Holocene tephra, Mono and Inyo Craters, eastern California. Geol. Soc. Am. Bull., **88**: 89-95.

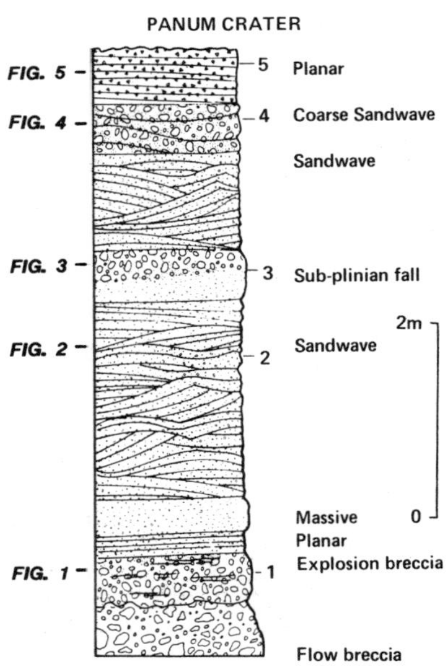

Stratigraphic section of the Panum Crater tuff ring measured in a pumice quarry southeast of the crater. The illustrated pyroclastic sequence overlies a lava breccia and continues nearly 20 m above the level at which sample 5 is shown. The upper section (not included) is covered but appears to be a continuation of the planar bedded deposits from which sample 5 was obtained.

FIGURE 1. SEM of a typical blocky, vesicular pyroclast of sample 1 (taken from a surge layer in the explosion breccia). The surface is slightly hydrated and covered with fine, adhering material, especially in vesicle hollows.

FIGURE 2. SEM of a sample 2 pyroclast. This vesicle-poor sample has broad, flat fracture surfaces and sharp edges. A few curved and dish-shaped spalled surfaces apparently were caused by transport abrasion. These spalled areas resemble conchoidal fractures.

FIGURE 3. SEM of sample 3 fine-fraction pyroclasts. Curviplanar surfaces form from broken vesicle bubble walls as well as from brittle fracture surfaces. Some tube-like vesicles are evident.

FIGURE 4. SEM of a sample 4 pyroclast showing numerous intersecting fracture surfaces that define the blocky morphology. Note the sharp, angular grain edge, unmodified by transport abrasion, that is typical of ash fall tephra.

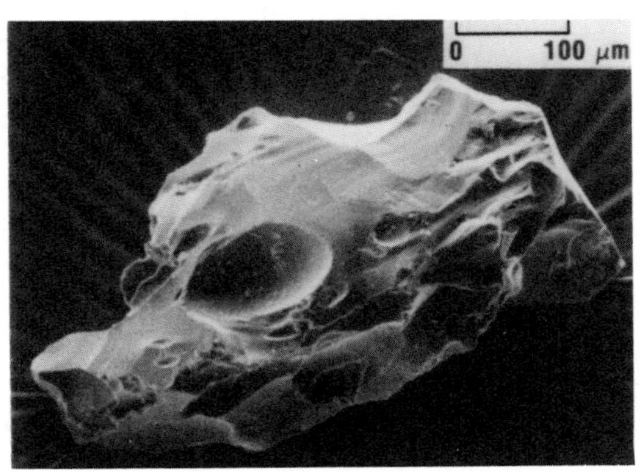

FIGURE 5. SEM of a sample 5 pyroclast. This poorly vesicular, blocky grain has edges that show numerous small chips and scratches resulting from abrasion.

FIGURE 6a. SEM of Wood's (1977) tephra #2 that was collected at several localities in the east-central Sierra Nevada range. The coarse fraction shows a few elongate vesicles that are absent in the fine fraction.

FIGURE 6b. Adhering dust is present on the surfaces of both examples; they strongly resemble grains from sample 3.

WAIRAKEI FORMATION, NEW ZEALAND

Introduction: The Wairakei Formation of North Island, New Zealand, is an excellent example of a phreatoplinian deposit. This deposit was erupted from the Taupo Volcanic center about 20,000 y.b.p. (Vucetich and Pullar, 1969; Self and Sparks, 1978; Self, 1983).

Of the six members of the Wairakei Formation that are exposed near Lake Taupo, four are widespread, fine-grained tephra fall deposits. Accretionary lapilli are common in two of these units. They are characterized by finely stratified, thin beds. The two other units are thick, fine-grained, nonwelded ignimbrites.

The samples here were collected near Lake Taupo by Stephen Self and from cores in peat bogs on the Chatham Islands, 800 km from the source, by New Zealand DSIR Soil Bureau scientists.

Grain Size: Median grain size for samples from the fine-grained tephra of the Wairakei Formation ranges from 1.4 mm to 31 μm (most are 31 to 62 μm). The tephra are mostly poorly sorted (Self and Sparks, 1978). Mean grain size does not change much downwind, which illustrates the uniform fine-grained nature of tephra generated during the eruption and suggests a deposition mechanism dominated by aggregate formation and rain-flushing.

Tephra Description: Most pyroclasts in these samples are fragments of vesicle walls and were produced by comminution of a vesicular rhyolitic melt during magma/water interaction. The vesicles are mostly ovoids; some have flattened to lense-like forms. The shattering of this melt produced curved plates, flat plates, and Y- or double-Y-shaped shards. Small pumice pyroclasts are also present, each with vesicles similar to those described above. Vesicle walls are fairly thick and are mostly greater than 10 μm.

Near-source samples are bimodal with 100- to 200-μm pumice pyroclasts and 12 to 14% phenocrysts (andesine, quartz, hypersthene, hornblende, and Fe-Ti oxides) in a matrix of fine-grained (<30-μm) shards.

Chemical Composition: Microprobe analyses of glass shards from near Lake Taupo and on the Chatham Islands are given in Table 1.

TABLE 1. CHEMICAL COMPOSITION OF GLASS SHARDS FROM THE WAIRAKEI FORMATION.

	Lake Taupo (Wt. %)	Chatham Island (Wt. %)
SiO_2	72.40	72.25
Al_2O_3	11.49	11.40
FeO*	1.65	1.79
MgO	0.09	0.08
CaO	0.85	0.87
Na_2O	3.31	3.54
K_2O	3.00	2.84
TiO_2	0.04	0.07
Total	92.83	92.84
No. of Analyses	7	8

*FeO as total Fe.

Pyroclast Formation: Explosive interaction between a vesiculating rhyolitic magma and surface water has produced a finely comminuted mixture of shards (vesicle-wall fragments) and very small pumice pyroclasts. Self and Sparks (1978) and Self (1983) have provided an initial analysis of these eruption mechanisms.

REFERENCES:

Self, S., 1983. Large-scale phreatomagmatic volcanism: a case study from New Zealand. J. Volcanol. Geotherm. Res., **17**: 433-469.

Self, S. and Sparks, R. S. J., 1978. Characteristics of widespread pyroclastic deposits formed by the interaction of silicic magma and water. Bull. Volcanol., **41**: 196-212.

Vucetich, C. G. and Pullar, W. A., 1969. Stratigraphy and chronology of late Pleistocene volcanic ash beds in central North Island, New Zealand. N. Z. J. Geol. Geophys., **12**: 784-837.

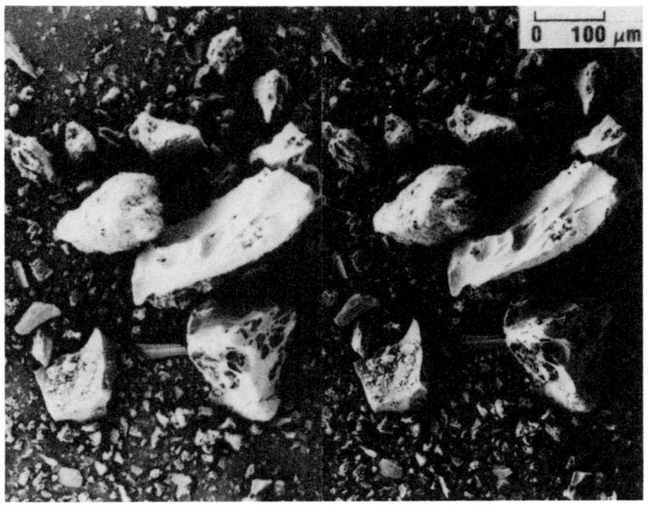

FIGURE 1a. Sample collected near Lake Taupo; location was probably about 15 km from the source. Stereomicrograph of an unsieved sample. The bimodal nature of this fine-grained ash is obvious; the vitric pyroclasts are in the 100- to 200-μm range and <30-μm range. A few rounded lithic fragments are visible. The larger pumice fragments are characterized by well-developed, coalesced vesicles; most are elongate ovoids. Vesicle walls are generally >10 μm thick. The fine-grained matrix consists of mostly finely comminuted fragments of vesicle walls; there are angular plates, wedges, and Y-shaped pyroclasts.

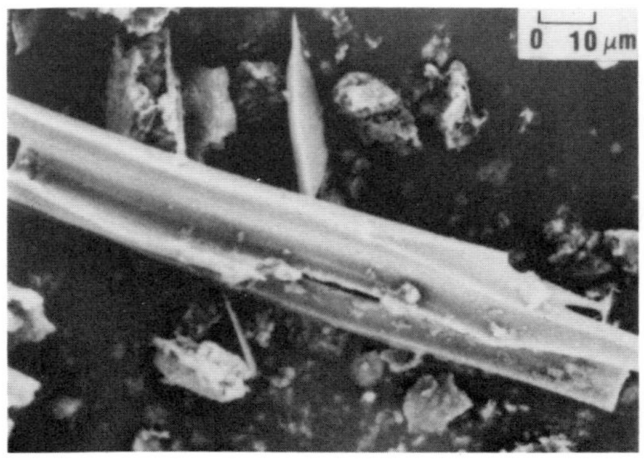

FIGURE 1b. The angular pumice fragment has elongate vesicles. The surrounding particles are angular, wedge-shaped, thin glass shards and clay(?) aggregates.

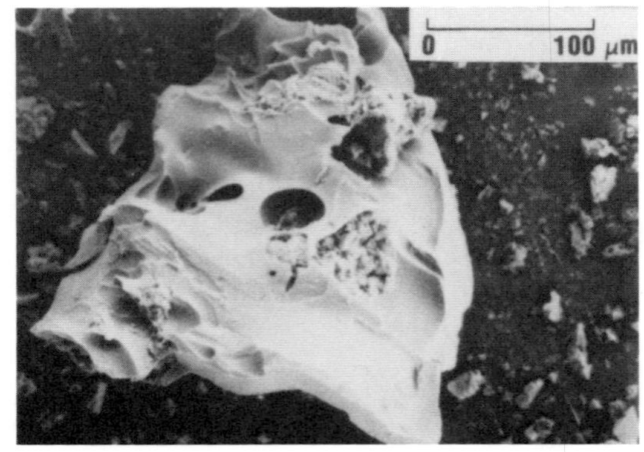

FIGURE 1c. Angular pumice pyroclast. Vesicles are irregular, flattened ovoids exhibiting some degree of coalescence. Most vesicle walls are thick (>10 μm). The pyroclast is surrounded by fine-ash size shards.

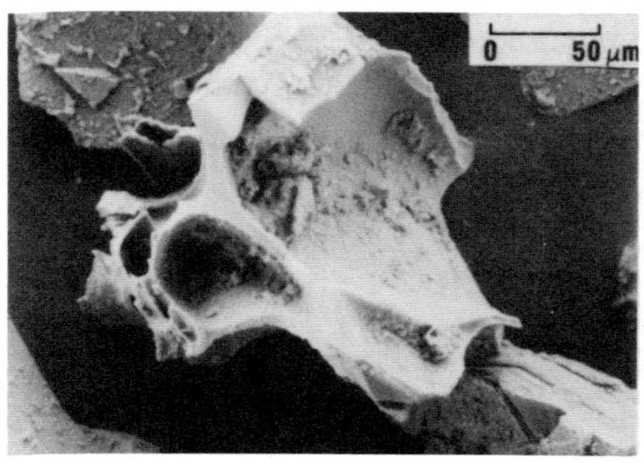

FIGURE 2a. Sample from a peat bog on the Chatham Islands 800 km southeast of the source. Note the better sorting of this sample (although it is similar in grain size) compared with the near-source sample illustrated in Fig. *1*. Angular pumice pyroclast with well-developed large vesicles. Fractures across vesicle walls are both planar and conchoidal. Particles adhering to grain surfaces are mostly smaller shards.

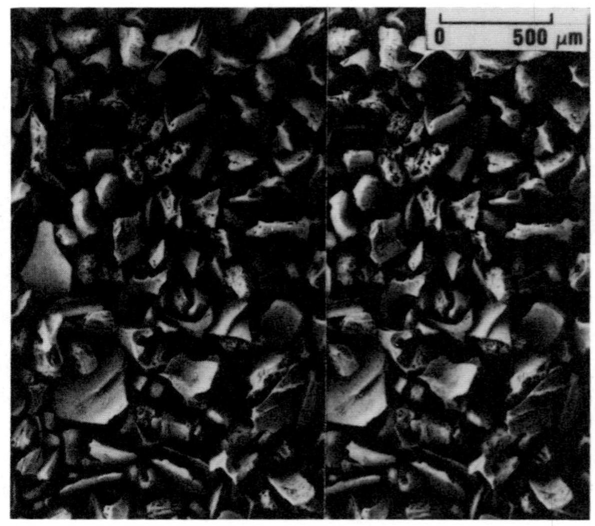

FIGURE 2b. Stereomicrograph of very angular, curved, plate-like, and Y-shaped shards broken from a vesiculated melt with large, well-developed vesicles. Some pyroclasts still retain a few flattened vesicles. Note the well-sorted nature of this sample. Most thin pyroclast edges are not abraded.

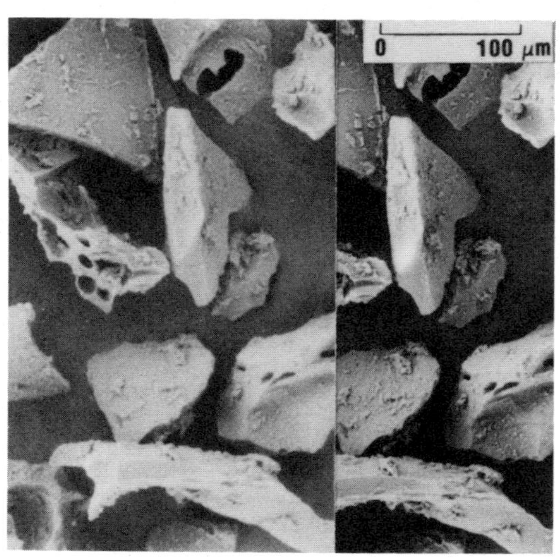

FIGURE 2c. Stereomicrographs of shards and small pumice pyroclasts that are typical of the distal Oruanui ash. These are mostly plate-like shards of 10 to 20 μm.

TAUPO VOLCANO, NEW ZEALAND

Introduction: The Taupo eruption sequence is the most recent explosive activity of the Lake Taupo volcanic area in the central North Island of New Zealand. The eruption, which produced a sequence of five tephra deposits, is dated between 130 and 186 A.D. by ^{14}C and historic atmospheric accounts (Wilson et al., 1980). The total volume of tephra erupted is in excess of 67 km^3. Various proportions of Lake Taupo waters mixed with erupting rhyolitic magma to produce both Plinian pumice falls and ignimbrite (at low lake levels) and phreatoplinian ash falls, as summarized in Table 1.

The late Quaternary ash deposits of this region were described by Healy et al. (1964). Self and Sparks (1978) recognized and discussed phreatoplinian eruptions from the Taupo Volcano. A recent series of papers by Walker and others is devoted to description of the deposits, including the fines-depleted nature of the Taupo Ignimbrite (Walker et al., 1980; Wilson and Walker, 1981, 1982), the exceptional dispersal of the Taupo Plinian deposit (Walker, 1980), the effect of phreatoplinian eruption in dispersing the Hatepe and Rotongaio Ashes (Walker, 1981a), and the Hatepe Plinian deposit (Walker, 1981b). This set of samples was kindly provided by Colin Wilson, University of Auckland.

TABLE 1. PHASES OF THE TAUPO ERUPTION AND THEIR VOLUMES.

Name	Sample No.*	Nature	Erupted Volume (km^3)
Taupo Ignimbrite	5	Nonwelded ignimbrite	≃30
Taupo Plinian	4	Plinian ash fall	≃24
Rotongaio Ash	3	Phreatoplinian ash fall	n.d.
Hatepe Ash	2	Phreatoplinian ash fall	n.d.
Hatepe Plinian	1	Plinian ash fall	6

*Samples were donated by C. J. N. Wilson, University of Auckland. Samples 1 and 4 were collected at the same location, approximately 55 km ENE of the vent; samples 2, 3, and 5 are from the same location 11 km SE of the vent.

Grain Size: Whole-deposit grain-size distributions (Fig. 1) display a fine-grained nature that is characteristic of the pre-Taupo ignimbrite deposits (Md ≃ 60 to 120 μm; Walker, 1981a). The Taupo Ignimbrite varies in median diameter throughout the deposit from 100 to 150 μm (Walker et al., 1981). Like the Taupo Ignimbrite, the phreatoplinian deposits are moderately to poorly sorted ($\sigma_\varphi = 1.6$ to 2.7φ). The Plinian deposits are very poorly sorted ($\sigma_\varphi = 4.0\varphi$).

Tephra Descriptions: Ash particle shapes vary among samples from vesicular and pumiceous to nonvesicular and blocky; minor to abundant very fine ash (<10-μm diameter) adheres to grain surfaces.

The Hatepe Plinian deposit is characterized by highly vesicular, pumiceous pyroclasts with irregular, angular surfaces. Vesicles range from spherical to elongate and, in some places, are stretched and deformed. Vesicles are commonly 50 to 100 μm long and are separated by vesicle walls less than 10 μm thick.

In contrast to tephra of the Hatepe Plinian deposit, the Hatepe Ash is only poorly vesicular; most pyroclasts are blocky. Adhering dust and slight to moderate surface alteration are common on coarser ash particles. The fine ash consists of plate-like shapes that define parts of elongated vesicle walls. Breakage surfaces on the particles are curviplanar and form cuspate shapes. Little of the fine-grained fraction is finer than 10 μm; it is apparently limited by the smallest vesicle sizes.

The Rotongaio Ash is nonvesicular and is dominated by blocky shapes, hydrated surfaces on coarse ash particles, adhering fine dust, and abundant fine material of less than 1 μm. This abundance of aggregated material gives a mossy, irregular appearance to ash particles that are coarser than 10 μm.

Taupo Plinian pyroclasts are very vesicular. Spherical vesicles are 5 to 50 μm in diameter, and elongate, tube-shaped vesicles are 10 to 40 μm in long dimension. Finer grains are blocky and tabular but bounded on at least one face by curving vesicle walls. No adhering dust is present on these pyroclasts.

The Taupo Ignimbrite shows both partly vesicular and blocky pyroclast shapes. Adhering fine-grained ash is moderately abundant and covers either planar surfaces of larger blocky particles or vesicle hollows of poorly vesicular pyroclasts. Fine ash particles are dominated by curviplanar fracture surfaces. Some surface alteration is present on larger particles, and adhering fine-grained ash of less than 1 μm appears as tiny globules. Abrasion features include V-shaped pits and irregular chipped edges on sharp vesicle-wall projections.

Comments on Pyroclast Formation: Volatile supersaturation of a rhyolite magma and interaction of magma with the waters of Lake Taupo have contributed to the variety of ash textures described here. Walker (1981) proposed that the stratigraphic sequence of the Taupo eruption was related to the ratio of magma and water during each phase of activity. Self and Sparks (1978) described the phreatoplinian eruption and fragmentation mechanisms of the deposits of the Oruanui (Wairakei) Formation, which erupted from the Taupo Volcano. They suggested that vesiculation of the magma occurred earlier than the water interaction and thereby facilitated efficient fragmentation of the melt. Their conclusion is supported by SEM descriptions presented here that show fractures cutting across vesicles, apparently induced by magma/water interaction. Most of the particles in the phreatoplinian deposits of the Hatepe and Rotongaio Ashes show shapes indicative of brittle fracture, which may be attributed to the high viscosity of the rhyolitic magma. These same shapes, as well as adhering fine dust and some surface alteration of larger clasts (all indicative of magma/water interaction) are also apparent in the sample of Taupo Ignimbrite; this suggests that waters of Lake Taupo had access to the vent during the last eruption phase. This interpretation differs from that of Walker (1981), who concluded that the Taupo Ignimbrite was formed entirely by magmatic activity. However, limited magma/water interaction during this ignimbrite eruption may explain several features of this deposit: fines depletion caused by scavenging of small particles by accretion and adhesion to larger particles, the unusually wide distribution and formation of a "veneer deposit" (Walker et al., 1981) with cross-laminations typical of surging flow (Fisher et al., 1980), and perhaps, the "valley pond" deposits that could represent the portion of the ignimbrite from which steam had separated.

REFERENCES:

Fisher, R., Smith, A., Wright, J., and Roobol, M.J., 1980. Ignimbrite veneer deposits or pyroclastic surge deposits? Nature, **286**: 912.

Healy, J., Vucetich, C., and Pullar, W., 1964. Stratigraphy and chronology of late Quaternary volcanic ash in Taupo, Rotorua and Gisborne Districts. N.Z. Geol. Surv. Bull., **73**: 1-88.

Self, S. and Sparks, R., 1978. Characteristics of widespread pyroclastic deposits formed by the interaction of silicic magma and water. Bull. Volcanol., **41**: 1-17.

Walker, G., 1980. The Taupo Pumice: product of the most powerful known (ultraplinian) eruption? J. Volcanol. Geotherm. Res., **8**: 62-94.

Walker, G., Wilson, C., and Froggatt, P., 1980. Fines-depleted ignimbrite in New Zealand - the product of a turbulent pyroclastic flow. Geol., **8**: 245-249.

Walker, G., 1981a. Characteristics of two phreatoplinian ashes and their water-flushed origin. J. Volcanol. Geotherm. Res., **9**: 395-407.

Walker, G., 1981b. The Maimihia and Hatepe Plinian deposits from the rhyolitic Taupo volcanic centre. N. Z. J. Geol. Geophys., **24**: 305-324.

Walker, G., Wilson, C., and Froggatt, P., 1981. An ignimbrite veneer deposit: the trail-marker of a pyroclastic flow. J. Volcanol. Geotherm. Res., **9**: 409-421.

Wilson, C., Ambraseys, N., Bradley, J., and Walker, G., 1980. A new date for the Taupo eruption, New Zealand. Nature, **288**: 252-253.

Wilson, C. and Walker, G., 1981. Violence in pyroclastic flow eruptions. *In*: S. Self and R. Sparks (Eds.), *Tephra Studies*, Reidel, Dordrecht: 441-448.

Wilson, C. and Walker, G., 1982. Ignimbrite depositional facies: The anatomy of a pyroclastic flow. J. Geol. Soc. London, **139**: 581-592.

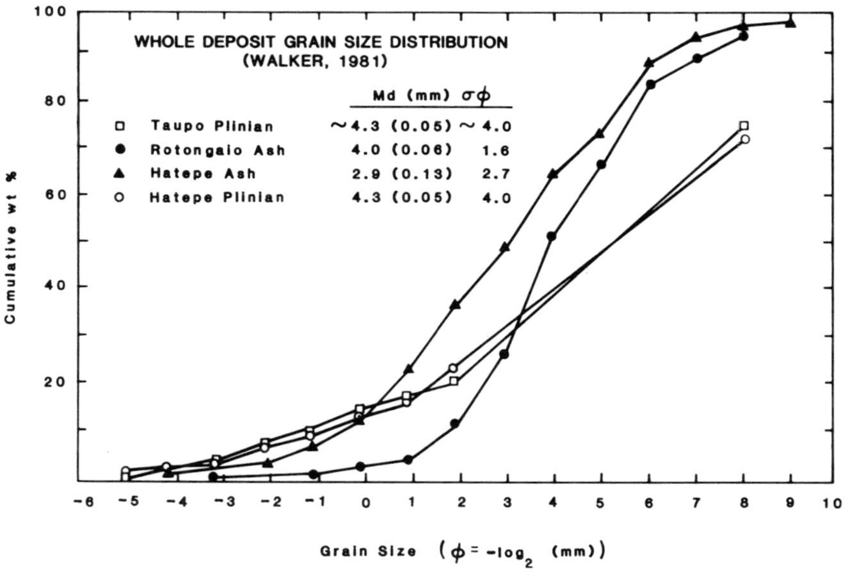

FIGURE 1. Plot of cumulative weight per cent versus grain size for pre-Taupo Ignimbrite deposits. These Plinian deposits are unusually fine grained, which attests to the extreme dispersive power of the eruptions.

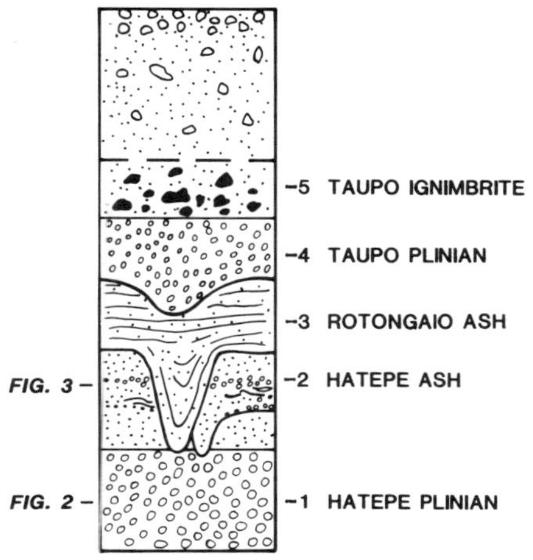

Schematic stratigraphic section of the Taupo eruption deposits; adapted from Walker (1981a).

FIGURE 2. SEM of pumiceous pyroclasts of the Hatepe Plinian deposit. Note the abundance of vesicles and irregular grain surfaces.

FIGURE 3a. SEMs of the Hatepe Ash: poorly vesicular, consisting of chips and splinters of fragmented material.

FIGURE 3b. Blocky grain showing moderate surface alteration.

FIGURE 3c. Finer grained pyroclasts consisting of angular, platy grains. Note the lack of particles less than about 5 μm in diameter.

FIGURE 4a. SEMs of the Rotongaio Ash: blocky, nonvesicular, altered ash with adhering dust.

FIGURE 4b. Fine fraction consisting of mossy and irregularly shaped particles of less than 10 μm—these shapes result from aggregation of <1-μm particles.

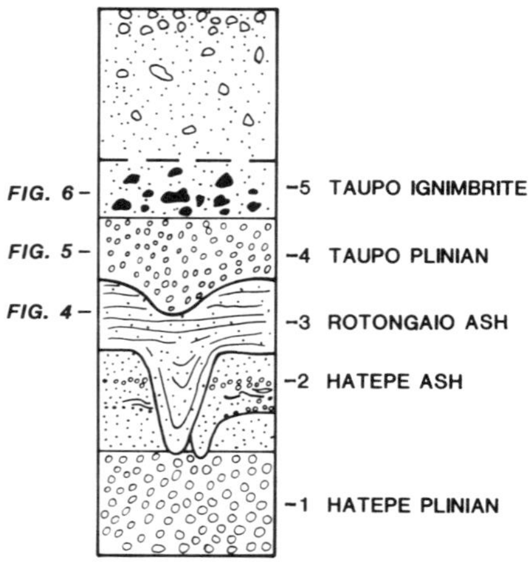

Schematic stratigraphic section of the Taupo eruption deposits; adapted from Walker (1981a).

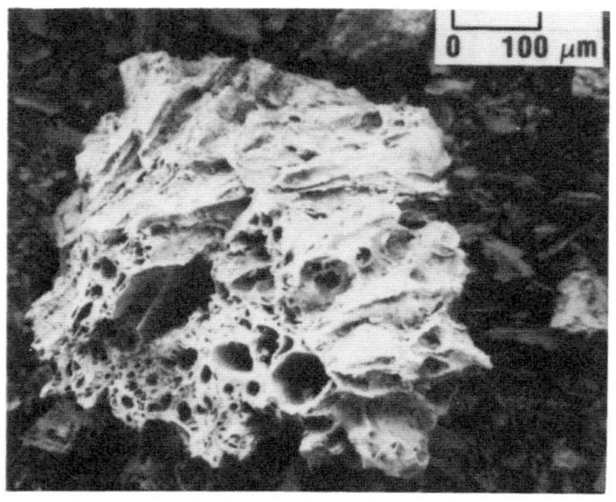

FIGURE 5a. SEMs of tephra from the Taupo Plinian deposit: highly vesicular pumice with elongate vesicles, delicate irregular surfaces, and little adhering dust.

FIGURE 5b. Fine fraction consisting of angular, plate-like, and cuspate fragments of larger particles, each showing at least one vesicle surface. Note the scarcity of <1-μm material.

FIGURE 6a. SEMs of the Taupo Ignimbrite: blocky particle with few visible vesicles, slight surface alteration, and some adhering dust.

FIGURE 6b. Angular, platy fine fraction with curviplanar fracture surfaces.

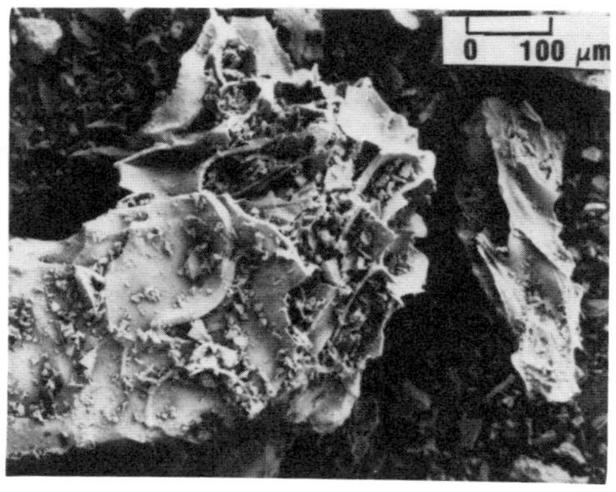

FIGURE 6c. Partly vesicular grain with adhering dust and vesicle-wall edges smoothed by transport abrasion.

Deep-Sea Hyaloclastites

Hyaloclastite is one of the most commonly sampled volcanic rock types from the ocean floor. This glassy, fragmental material is associated with pillow lavas and bedded deposits. The mode of hyaloclastite formation has not yet been fully explained. If hyaloclastite shards were erupted in relatively shallow portions of the ocean floor where there are low hydrostatic pressures, expansion of magmatic gases would be important in their genesis. If they are erupted at depths where vesicle growth is suppressed by hydrostatic pressure, other processes such as thermal spallation would be required to explain the formation of these deposits.

SEAMOUNT "F," EAST PACIFIC RISE

Introduction: Bedded hyaloclastites are associated with deep-sea basaltic lava flows and seamounts. Typically angular, nonvesicular, blocky shards probably formed during energetic mixing of magma and sea water (Carlisle, 1963; Honnorez, 1963; Honnorez and Kirst, 1975). The examples here are from a dredge haul of Research Vessel PLUTO, Leg 7; they were collected in the crater of Seamount "F" near the East Pacific Rise at 21°22′N, 108°37′W. Seamount "F" rises from ≃2900 m depth to a summit at 2100 m depth and is capped by a 2-km-diameter caldera. This caldera was investigated by a dive of the Submersible Vessel ALVIN during 1982 (Batiza et al., 1984); these investigators suggest that thin hyaloclastite pavements postdate the crater and overlie pelagic calcareous ooze, fine-grained crater-wall talus, and pillow basalts. The samples were generously provided by Rodey Batiza, Washington University.

Grain Size: There is a 5-μm to 2.5-mm range of shards within these samples. Within graded units described by Batiza et al. (1984), the median diameter is about 1 mm (0.0φ) for coarser, blocky shards and about 100 μm (3.3φ) for finer parts of the bed.

Tephra Descriptions: As described by Batiza et al. (1984), most of the hyaloclastites in these samples have vesicle-free, blocky, equant forms. These range from equant cubes to complex, angular polyhedrons. The second most common hyaloclastites are angular to slightly curved plates. Also present are rare irregular plates with conchoidal to hackly surfaces. Most of the shards in these samples exhibit some of the surface cracking and foliation that is characteristic of hydration and palagonitization. Some grains have pockmarked surfaces caused by solution.

Batiza et al. (1984) show a subcategory of shards that were broken from a chilled, vesiculated magma and exhibit some effects of surface tension (similar to those in Pele's hair). These are interbedded with the equant, blocky shards. The matrix contains small flakes of basaltic glass, which are mostly in a platy form.

As noted by Batiza et al. (1984), there are rare phenocrysts and hyalocrystalline shards in these samples; deep-sea hyaloclastites are composed of mostly homogeneous basaltic glass.

Chemical Composition: These samples of hyaloclastites are midocean ridge basalts (Table 1). Alteration by interaction with sea water has increased the silica concentrations at grain surfaces.

TABLE 1. CHEMICAL COMPOSITION OF SEAMOUNT "F" HYALOCLASTITE.*

	(Wt. %)
SiO_2	50.14
TiO_2	1.12
Al_2O_3	16.01
FeO**	9.16
MgO	8.45
CaO	12.59
Na_2O	2.42
K_2O	0.06
P_2O_5	0.10
Total	100.05

*R. Batiza, Washington University, analyst.
**Total Fe as FeO.

Comments on Pyroclast Formation: It is likely that most vesicle growth was suppressed and that rapid mixing of magma and water produced blocky and platy glass clasts when these deposits were erupted at a water depth of 2100 m. Batiza et al. (1984) proposed explosive mixing of basaltic magma and seawater for depths of less than 3100 m; boiling and large-volume changes are possible, although most vesicle growth is suppressed. Evidently, some vesicle growth was possible, perhaps within the short period of steam generation, because thin shards derived from vesicular melt have been described in deep-sea hyaloclastites, where they are mixed with nonvesicular, blocky shards.

Batiza et al. (1984) also suggest that hyaloclastites are formed by rapid quenching of magma and *not* by spallation of pillow rinds. Pillow rinds commonly have abundant microlites (quench crystals) and are not as glassy as hyaloclastites. Some of the thin glass slivers may be the result of thermal shock, as proposed by Carlisle (1963).

REFERENCES:

Batiza, R., Fornari, D., Vanko, D., and Lonsdale, P., 1984. Craters, calderas and hyaloclastites on young Pacific seamounts. J. Geophys. Res., **89**: 8371-8390.

Carlisle, D., 1963. Pillow breccias and their aquagene tuffs on Quadra Island, British Columbia. J. Geol., **71**: 48-71.

Honnorez, J., 1963. Sur l'origine des hyaloclastites. Bull. Volcanol., **25**: 253-258.

Honnorez, J. and Kirst, P., 1975. Submarine basaltic volcanism: morphomometric parameters for distinguishing hyaloclastites from hyalotuffs. Bull. Volcanol., **39**: 1-25.

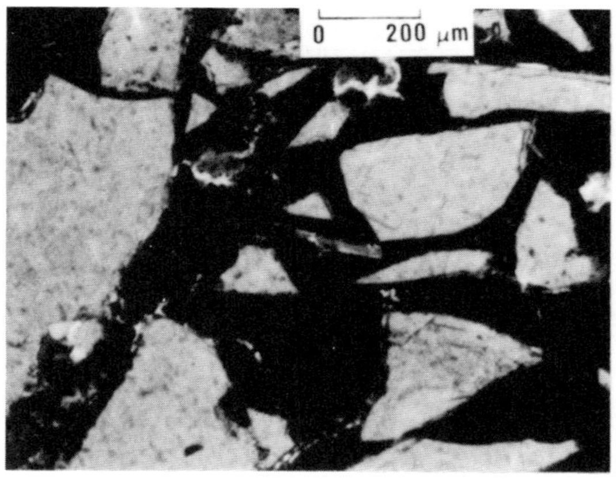

FIGURE 1. Transmitted-light photomicrograph of blocky, equant sideromelane shards.

FIGURE 2. Reflected-light photomicrograph of shards similar to those in Fig. *1*.

FIGURE 3. SEM of hyaloclastite sample 4-13. There are mostly hydrated, equant blocky glass shards of basaltic composition; some show surface pitting. A few irregular, platy shards are visible (*center-right*).

FIGURE 4. Close-up of Fig. 3. Blocky shards are broken by surface cracks and foliations that were caused by hydration and incipient palagonitization of the glass. Irregular clumps are remnants of calcareous cement that bonded the hyaloclastite before disaggregation and sample analysis.

FIGURE 5. SEM of flat, platy shards from sample 4-5 (best examples are near the center). All shards show some surface cracking caused by alteration.

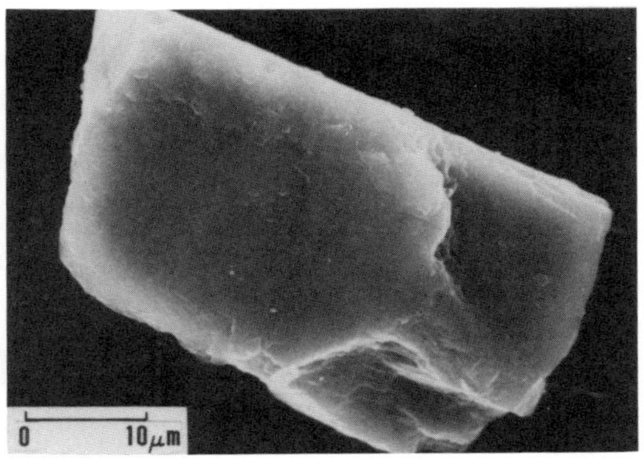

FIGURE 6. SEM of blocky glass shard from sample 4-5.

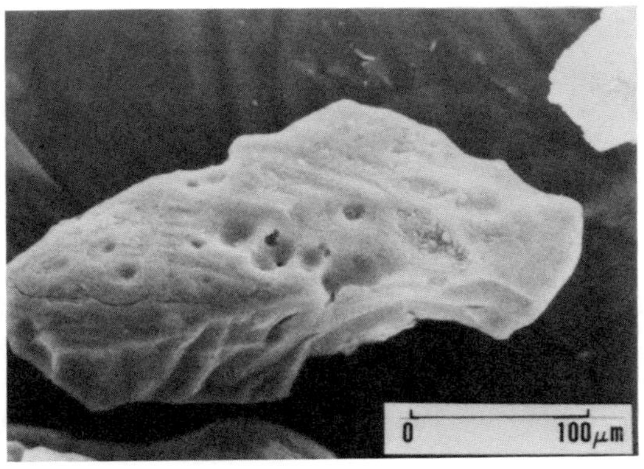

FIGURE 7. SEM of irregular glass shard from sample 4-5. This shard type is rare, with irregular, conchoidal surfaces rather than smooth, blocky surfaces. There is pitting caused by alteration, and irregular patches of carbonate cement are visible on grain surfaces.

Littoral Cones

One of the more curious observations about magma/water contact is that it is not always explosive. For example, when lava from Hawaiian volcanoes enters the sea, there is steam, but most of the activity is passive. A quenched surface allows lava to pour quietly downslope underwater to form pillow structures. However, passive behavior is not always the case, and explosive mixing of lava and water can form a tephra cone in the littoral zone.

PUU HOU AND SAND HILLS, HAWAII

Introduction: Ash similar to that ejected during phreatomagmatic eruptions can be produced when lava flows into a body of water (Moore and Ault, 1963; Fisher, 1968). If explosive mixing of lava and water occurs, small tephra cones (littoral cones) are formed at the water's edge, flanking the lava flow. Littoral cones are characteristically crescent-shaped ridges that, from the ground, appear to be cones without summit craters (Fisher, 1968). The examples described here are from the west coast of the Island of Hawaii, where littoral cones were constructed by basaltic lavas. Puu Hou was formed during the 1868 South Rift eruption of Mauna Loa, and Sand Hills resulted from an eruption of Kilauea in 1840.

Fisher (1968) suggests that the explosive, cone-building activity at Puu Hou occurred when steam was trapped beneath lava crusts. The explosions were in localized areas of the lava streams, and explosion centers migrated seaward as the flow built a lava delta into the ocean (Fisher, 1968). Flow surfaces between the cones were kept free of explosion debris as the flows moved seaward.

Some littoral cones are complete 50- to 100-m-high cones with summit craters and apparently were built above lava tubes (Fisher, personal communication). Peterson (1976) observed the development of a small littoral cone where waves had access to a lava tube through its breached top.

The samples studied here are from the collection of R. V. Fisher.

Grain Size: Mean grain size of the samples ranges from 0.7 to 1.3 mm (0.5 to -0.4φ); all are poorly sorted.

Tephra Description: Within the Puu Hou complex, hyalotuff layers and agglutinated zones (bomb-rich layers) are interbedded. Samples described here are from the hyalotuff layers. Fisher (1968) has distinguished three vitric pyroclast types within these deposits:

(1) angular scalloped fragments (intersecting large vesicles),
(2) brown glass droplets with curved, fluidal edges and a few small vesicles, and
(3) irregular to equant fragments with few or no vesicles.

All the sideromelane pyroclasts listed above have trace to abundant quench crystals (plagioclase, clinopyroxene, and olivine). The 1 to 9% olivine appears as phenocrysts in sideromelane and tachylite pyroclasts and as separate crystal pyroclasts. Tachylite pyroclasts make up 2 to 70% of the deposits. Most are subangular and nonvesicular.

Chemical Composition: All samples are of basaltic composition. Analyses are listed in Table 1.

TABLE 1. CHEMICAL ANALYSES OF PUU HOU AND SAND HILLS TEPHRA.*

	Puu Hou (Wt. %)	Sand Hills (Wt. %)
SiO_2	51.53	50.46
TiO_2	2.02	2.29
Al_2O_3	13.15	12.22
FeO_2O_3**	11.34	11.58
MnO	0.21	0.21
MgO	8.13	11.75
CaO	9.87	9.98
Na_2O	3.17	2.84
K_2O	0.42	0.41
P_2O_5	0.24	0.24
S	0.08	0.11
Total	100.16	102.09

*M. Rhodes and J. Evans, Johnson Space Center, analysts.
**Total Fe as Fe_2O_3.

Comments on Pyroclast Formation: Fisher (1968) determined that the littoral cones he described were formed by steam explosions that resulted from lava in contact with water. Explosion centers were confined to areas where lava was rapidly and continuously delivered. Pahoehoe and aa were the main lava types that reached the sea. Aa does mix with sea water, but its surface texture does not create the confinement necessary to produce explosions. At Puu Hou, pahoehoe and aa were comminuted where steam was trapped beneath rapidly delivered, broken lava crusts.

Comminution of the flow by water/lava interaction and subsequent steam explosions produced a variety of pyroclast types. Angular, chilled sideromelane fragments with large, dispersed vesicles came from the interior of the flow where vesicles were well developed during the lava flow's journey to the sea. Irregular, vesicular and nonvesicular droplets with smooth, fluidal surfaces were bits of melt ejected by the explosion but not chilled during water/lava interaction; they were chilled in flight by thermal radiation and flow of air over grain surfaces. Fragmentation of flow crusts may have produced vesicular and nonvesicular, angular fragments of tachylite.

REFERENCES:

Fisher, R. V., 1968. Puu Hou littoral cones, Hawaii. Geol. Rund. **57**: 837-864.

Moore, J. G. and Ault, W. V., 1965. Historic littoral cones in Hawaii. Pac. Sci., **19**: 3-11.

Peterson, D.W., 1976. Processes of island growth, Kilauea Volcano, Hawaii, 1969-1973. Proc. Symp. Andean and Antarctic Volcanology Problems, Santiago, Chile: 172-189.

FIGURE 1a. Pyroclasts from hyalotuff layers in the Puu Hou littoral cone, Hawaii. Blocky, equant pyroclasts with 10 to 20% spherical or ovoid vesicles are sideromelane with only a trace of quench crystals; these appear to have formed during the lava/water interaction. Broken sideromelane droplets with curved, fluidal edges are drops of melt from the flow interior that were thrown clear by a steam explosion. Ragged, irregular fragments with rough surfaces are bits of tachylite; the flow crust fragmented during the explosions.

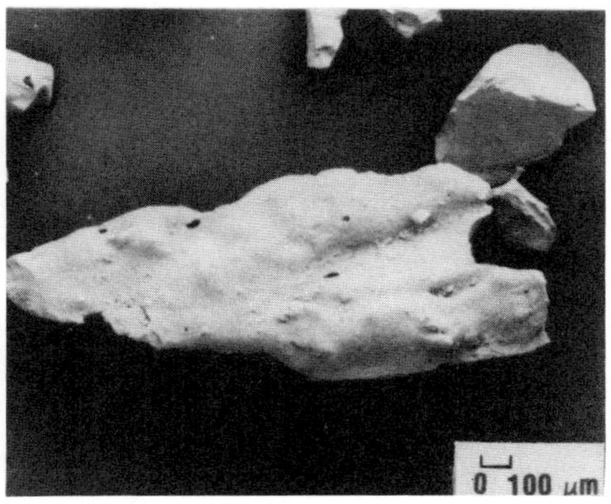

FIGURE 1b. Fragment of a sideromelane droplet characterized by curved, smooth surface (at least on one side). Pyroclasts such as these are vesicular.

FIGURE 1c. Close-up view of the finer grained fraction. It consists of mostly angular, blocky sideromelane pyroclasts with low vesicularity.

FIGURE 2a, b. A sample from Sand Hills littoral cones, Hawaii. General SEM views of hyaloclastites show the variety of pyroclasts, which is the same as for Puu Hou although these samples are predominantly blocky, angular sideromelane fragments with very low vesicularities. Some have curved surfaces where large, well-developed vesicles were intersected during comminution. Some curved sideromelane pyroclasts with smooth surfaces (droplets) are visible.

FIGURE 2c. Center—vesicular sideromelane droplet with elongate, coalesced vesicles. The pyroclast surface is smooth and curved where not broken. Most other pyroclasts are blocky sideromelane pyroclasts. Some of the thin grain edges are abraded, but generally, the grain surfaces show no signs of abrasion.

FIGURE 2d. Left—angular sideromelane fragment with no vesicles. *Center*—ragged, equant tachylite grain. *Right*—fragment of a sideromelane droplet.

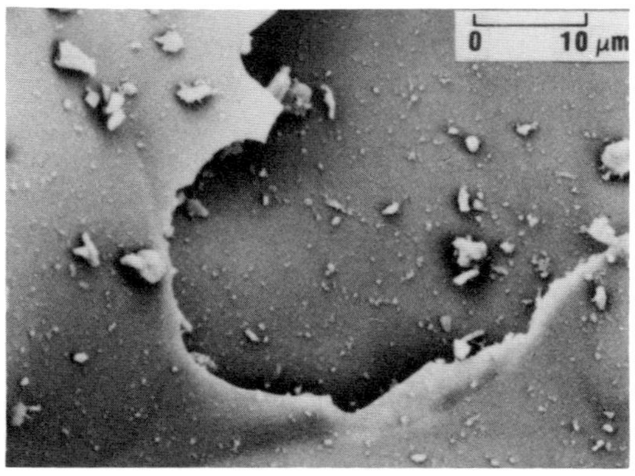

FIGURE 2e. Close-up of the smooth surface of the droplet fragment. A near-surface vesicle was a flattened ovoid before breaking. Note the <1-μm-thick vesicle wall (broken).

Experimental Production of Phreatomagmatic Ash

Experimental investigations provide a unique method for studying the formation of volcanic ash. For example, blast furnace slags that are quenched by water have textures similar to those of hyaloclastites and other phreatomagmatic ashes. This section describes particulates produced in some explosive melt/water mixing experiments.

THERMITE/WATER EXPLOSIONS

Introduction: Explosions caused by rapid vaporization of water during contact with molten material have been noted in industrial accidents for a number of years (Lipsett, 1966; Witte et al., 1970). This class of explosion is often termed *fuel-coolant interaction* (FCI) and has recently been the subject of considerable experimental and theoretical study in connection with nuclear reactor safety (Sandia Laboratories, 1975). The result of vapor explosions is fragmentation of the melt into micrometer and millimeter-size particles that are accelerated outward within a rapidly expanding steam cloud. The FCI mechanism is also applicable to phreatomagmatic (hydrovolcanic) eruptions (Peckover et al., 1973; Colgate and Sigurgeirsson, 1973; Sheridan and Wohletz, 1981).

This section provides a comparison of experimental debris with naturally occurring phreatomagmatic ash. Wohletz and McQueen (1984) described in detail experiments that use a thermite melt as a basalt analog in volcanic models. Additional information on FCI experiments using thermite and corium is given by Nelson et al. (1980). The explosive phenomena of volcano simulations are strikingly similar to a variety of eruptive activities, some of which are not normally considered phreatomagmatic. These activities include Hawaiian lava fountaining, Strombolian eruptions, Surtseyan and Vulcanian explosions, and littoral cone and submarine pillow lava formation.

Grain Size: Experimental phreatomagmatic ash closely resembles ash produced by Surtseyan eruptions. Figure 3 provides a comparison between Surtsey and Capelinhos ashes and those of experimental thermite and corium melts. Mean sizes are 0.13 to 0.14 mm, and particles are moderately to poorly sorted. Median grain size generally decreases with increasing explosive efficiency for thermite/water explosions. Nonexplosive interactions produce median sizes coarser than 2 to 4 mm; finer debris are recovered from explosive interactions. Median particle sizes reach the micrometer range in explosions of maximum energy.

Particle Description: Several dominant particle shapes are evident in experimentally produced ash. The most common are characterized as moss-like, convolute shapes with large surface areas. Less abundant particles have spheroidal or drop-like shapes and strongly resemble industrial fly-ash debris. Another common type of experimental ash shows angular, blocky, and equant shapes that are generally >50 μm in long dimension. Plate-like shapes are angular, flat chips and broken pieces of bubble walls. All the above particles form aggregates, and many irregularly shaped aggregates consist of droplets, blocky particles, and platelets bonded by twisted, deformed melt particles.

Chemical Composition: Thermite (Fe_3O_4 + Al), when reacted, forms iron and aluminum oxides. In volcano simulations, quartzo-feldspathic sand added to the thermite produced compositions more closely resembling a basaltic melt (Table 1). Although these compositions are still very different from natural silicate magmas, their similar density, liquidus temperature, and viscosity make them reasonable models.

TABLE 1. REPRESENTATIVE CHEMICAL ANALYSES OF THERMITE MELT DEBRIS.*

	1	2	3	4	5
SiO_2	10.07	10.73	14.34	36.55	18.25
TiO_2	---	---	2.25	1.44	1.90
Al_2O_3	31.24	25.09	11.36	34.41	42.46
FeO	57.94	64.18	57.01	17.32	23.28
MnO	---	---	1.58	0.96	1.33
MgO	---	---	6.44	3.70	5.84
CaO	0.45	---	2.05	1.65	1.90
Na_2O	---	---	3.07	1.98	3.19
K_2O	0.31	---	1.91	1.99	1.85

*Normalized standardless (EDS) analyses.
Sample descriptions:
1 - Iron-aluminum sphere, large,
2 - Iron-aluminum sphere, small,
3 - Blocky iron particle,
4 - Coating on iron particle,
5 - Iron-aluminum spindle.

Comments on Particle Formation: The blocky, equant pyroclasts from explosive water/magma eruptions have led previous investigators (Peacock, 1926; Carlisle, 1963; Fisher and Waters, 1969; Heiken, 1971; Honnorez and Kirst, 1975) to conclude that the magma is fragmented by thermal contraction and shattering of the glass. The strong similarity between experimentally produced debris and phreatomagmatic ash has provided insight into the mechanisms of ash formation. To create vapor explosions upon contact of magma with water, intimate mixing of the two is required. This mixing occurs when the melt fragments before explosion. Fragmentation increases the contact area of melt and water so that conductive heat exchange occurs at explosive rates. Experiment results indicate that the initial fragmentation of the melt occurs by several mechanisms, including (1) fluid instabilities between the water and melt, which promote turbulent mixing of the two; (2) oscillatory vapor film formation, expansion, and collapse, which produces stress waves at the melt surface and thereby fragments the melt; and (3) large compressive and tensile stress waves produced by vapor explosions—these stress waves propagate into the melt and fragment it. Each of the above mechanisms occurs on a millisecond time scale, increases the water/magma contact area, and leads to catastrophic heat exchange that is followed by vapor explosion (Wohletz, 1983).

REFERENCES:

Buchanan, D. J., 1974. A model for fuel-coolant interactions. J. Phys. D: Appl. Phys., **7**: 1441-1457.

Buxton, L. D. and Benedict, W. B., 1979. Steam explosion efficiency studies. Sandia National Laboratories, SAND79-1399, NUREG/CR-0947: 62 pp.

Colgate, S. A. and Sigurgeirsson, T., 1973. Dynamic mixing of water and lava. Nature, **244**: 552-555.

Carlisle, D., 1963. Pillow breccias and their aquagene tuffs, Quadra Island, British Columbia. J. Geol., **71**: 48-71.

Fisher, R. V. and Waters, A. C., 1969. Bed forms in base surge deposits: Lunar implications. Science, **165**: 1349-1352.

Heiken, G., 1971. Tuff rings: examples from the Fort Rock-Christmas Lake Valley, south-central Oregon. J. Geophys. Res., **76**: 5615-1626.

Honnorez, J. and Kirst, P., 1975. Submarine basaltic volcanism: Morphometric parameters for discriminating hyaloclastites from hyalotuffs. Bull. Volcanol., **32**: 441-465.

Lipsett, S. G., 1966. Explosions from molten materials and water. Fire Tech., May: 118-126.

Nelson, L. S., Buxton, L. D., and Planner, H. N., 1980. Steam explosion triggering phenomena, part 2: Corium-A and Corium-E and oxides of iron and cobalt studied with a floodable arc-melting apparatus. Sandia Laboratories, SAND77-0998, NUREG/CR-0122: 138 pp.

Peacock, M. A., 1926. The basic tuffs. *In*: G. W. Tyrell and M. A. Peacock (Eds.), *The petrology of Iceland*, Royal Soc. Edinburgh Trans., **45**: 51-76.

Peckover, R. S., Buchanan, D. J., and Ashby, D. E. T. F., 1973. Fuel-coolant interactions in submarine volcanism. Nature, **245**: 307-308.

Sandia Laboratories, 1975. Core-meltdown experimental review. SAND 74-0382: 472 pp.

Sheridan, M. F. and Wohletz, K. H., 1981. Hydrovolcanic explosions, the systematics of water-tephra equilibration. Science, **212**: 1387-1389.

Witte, L. C., Cox, J. E., and Bouvier, J. E., 1970. The vapor explosion. J. Metals, **22**: 39-44.

Wohletz, K. H., 1983. Mechanisms of hydrovolcanic pyroclast formation: size, scanning electron microscopy, and experimental studies. *In*: F. Barberi and M. Sheridan (Eds.), *Explosive Volcanism*. J. Volcanol. Geotherm. Res., **17**: 31-64.

Wohletz, K. H. and McQueen, R. G., 1984. Experimental studies of hydromagmatic volcanism. *In: Explosive volcanism: inception, evolution, and hazards*. Studies in Geophysics, National Academy Press, Washington, D.C.: 158-169.

FIGURE 1. Simulation of Surtseyan phreatomagmatic eruption. The vertical plume is 24 m high and composed of steam, sand, and thermite particles. Note the base-surge cloud spreading out horizontally.

FIGURE 2. Interaction of thermite melt with small amounts of water has produced explosions that resemble Strombolian activity. The centimeter-size fragments of melt are being ejected vertically in a billowing steam cloud.

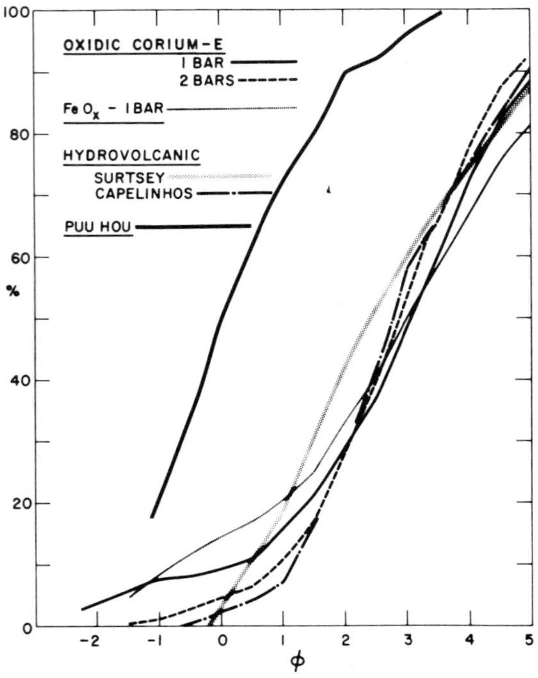

FIGURE 3. Comparison of grain-size analyses of experimentally produced ash with phreatomagmatic ash from Surtsey, Capelinhos, and Puu Hou. Plot shows cumulative weight per cent versus phi size ($\varphi = -\log_2$ *mm).*

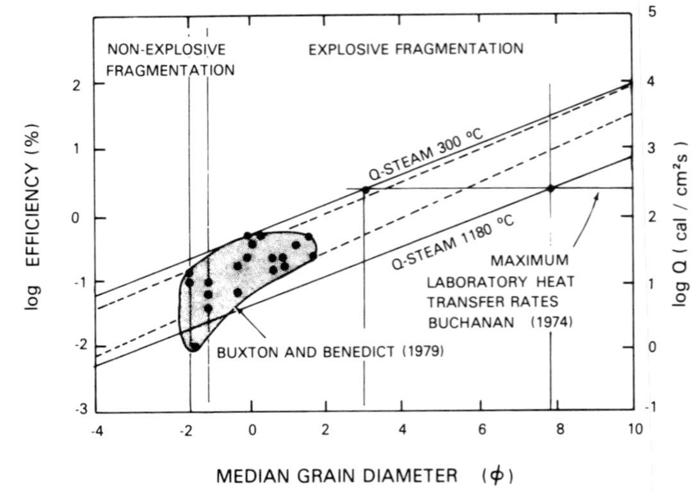

FIGURE 4. Plot of explosive efficiency (mechanical energy/thermal energy) versus median grain size for experimental ash. Points shown are from experiments by Buxton and Benedict (1979). Dashed lines extrapolate the plotted trend toward higher efficiencies and finer grain sizes. Lines plotted for Q, conductive heat transfer rates, are calculated from conduction of heat from a molten basalt sphere at 1200°C to a surrounding jacket of steam at both 300 and 1180°C. The maximum laboratory-measured heat transfer rates for fuel/coolant interactions (Buchanan, 1974) correspond to fragment sizes between 3φ (0.125 mm) and 4φ (0.063 mm).

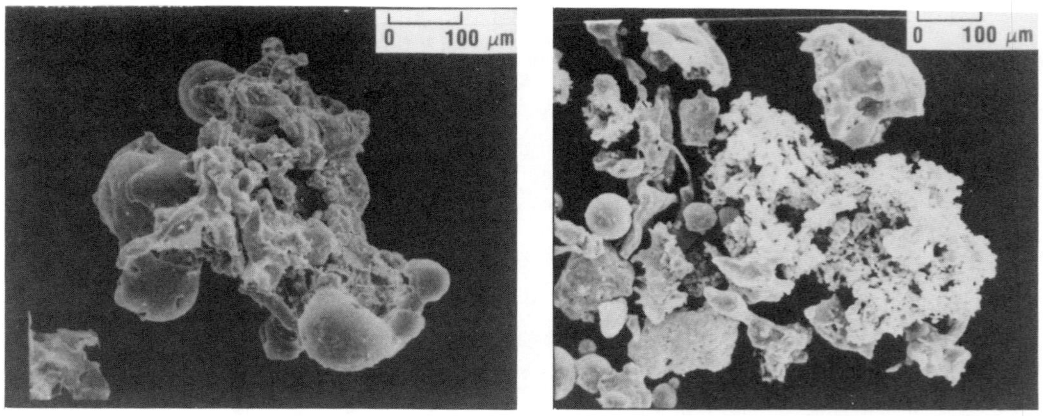

FIGURE 5. SEMs of moss-like, convolute shapes with large surface areas. Most surfaces are irregular threads to smooth, rounded globules and chunks fused together into particles, usually <20 µm in maximum dimension. Deep embayments separate globular lobes and tiny spheres that are attached to some particles. Jagged edges are rare except where ribbon-like or spongy material forms particle lobes.

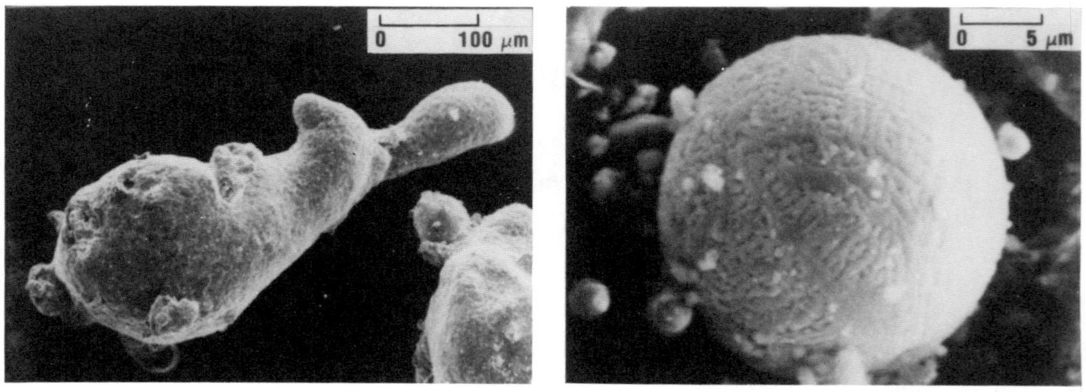

FIGURE 6. SEMs of drop-like and spheroidal particles. Surfaces show small indentations and complex patterns of interlocking crystals. Spheres often have smaller attached drops and spheres and may be, in turn, attached to a substrate of blocks and aggregate material. Many spheres have hollow centers and resemble industrial fly-ash debris.

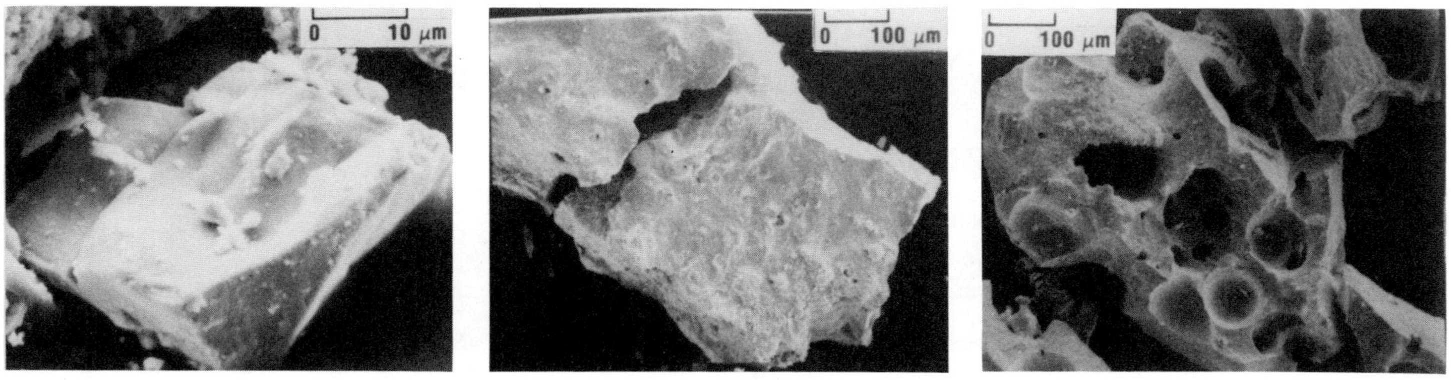

FIGURE 7. SEMs of angular, blocky particles that most resemble typical phreatomagmatic ash. Particles have one or more planar to curviplanar surfaces and may have vesicles, which are rare in experimental ash. Surfaces are often pitted but may show smooth, fluidal areas.

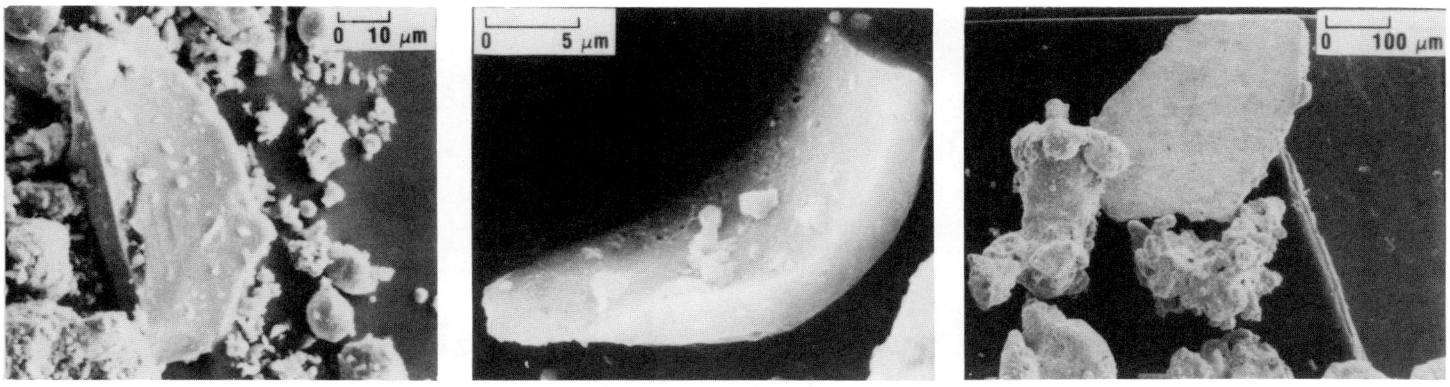

FIGURE 8. SEMs of plate-like particles that are flat, angular slabs. Where particles have curved surfaces, they are broken bubble-wall fragments.

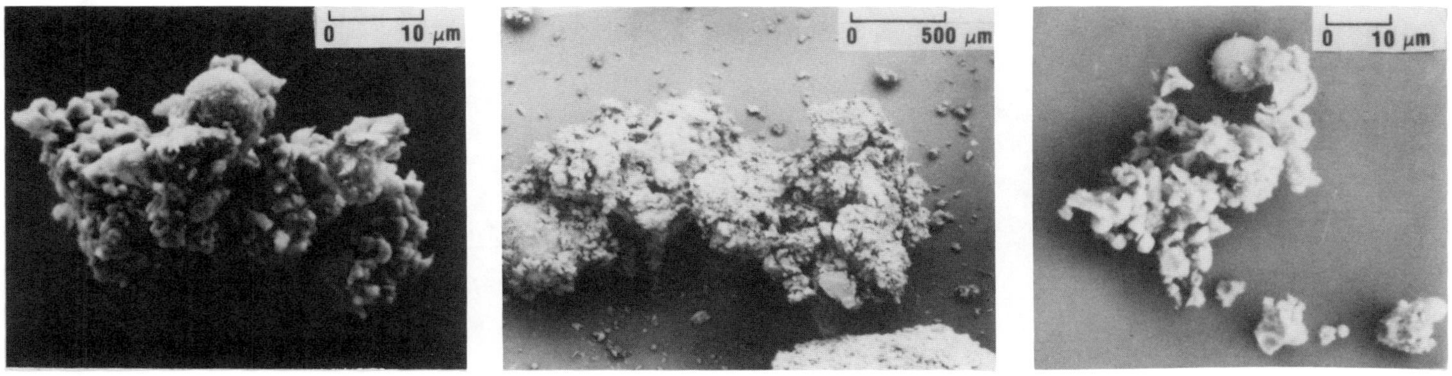

FIGURE 9. SEMs of aggregate particles common to both fine and coarse debris. These aggregates consists of droplets and spheres bonded to angular particles, often by ribbon-like, deformed melt particles. The coarse-grained particles (*middle*) strongly resemble volcanic aggregates found in wet phreatomagmatic deposits (see Cerro Colorado and Hopi Buttes sections).

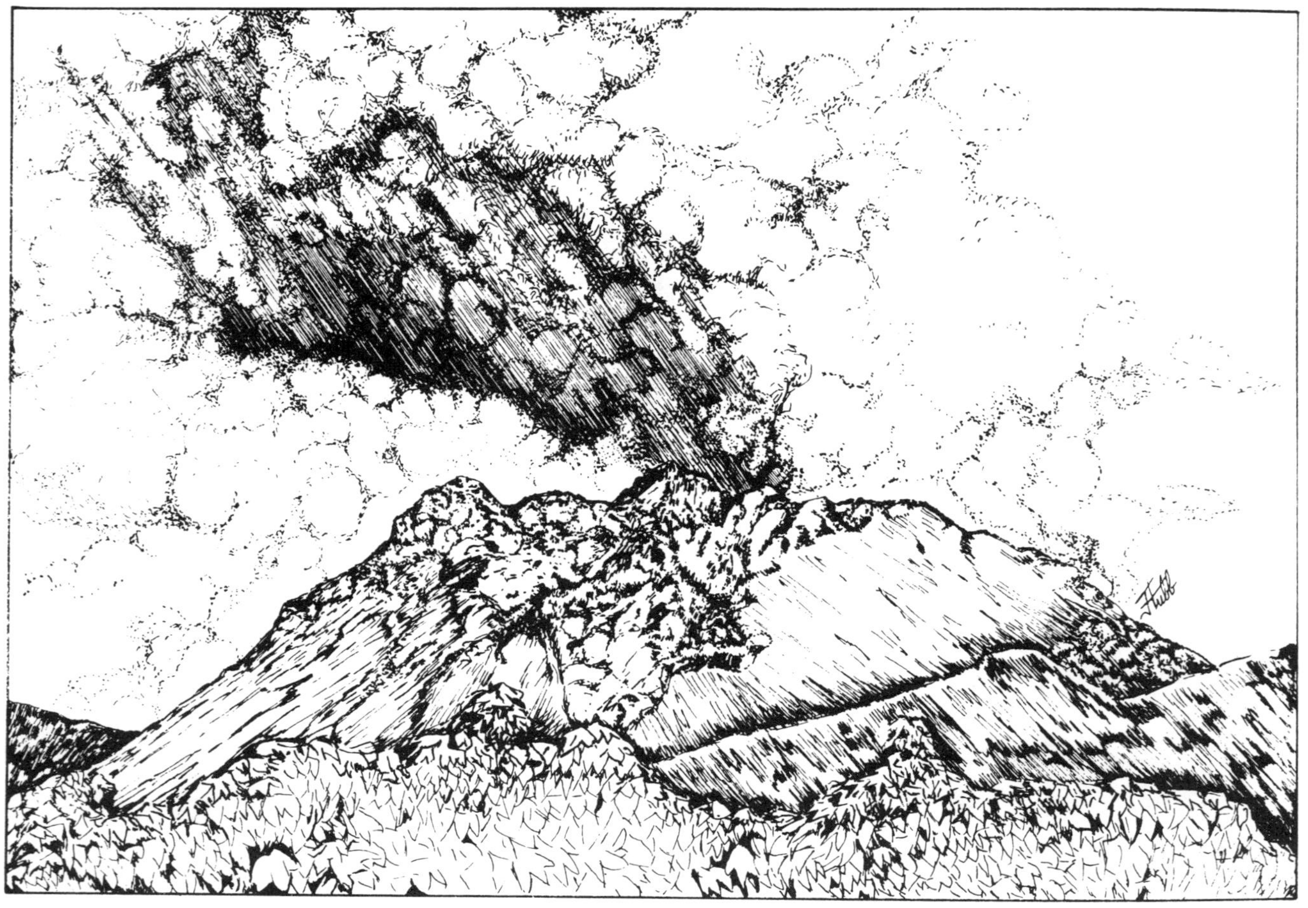

Phreatic explosions at Soufrière de Guadeloupe, 1976.

PHREATIC ERUPTIONS

Hydrothermal activity at fumeroles and hot springs can become energetic enough to erupt explosively. The volcanic ash from this activity consists of fragments from vent walls and crater fill, and it contains no juvenile ash. In some instances phreatic explosions have preceded magmatic or phreatomagmatic eruptions.

Historic Eruptions

LA SOUFRIÈRE DE GUADELOUPE, F.W.I.; 1976 ERUPTION

Introduction: La Soufrière de Guadeloupe is a steep-sided andesitic dome crossed by multiple fissures. There has been considerable explosive phreatic activity from these fissures during historic time. The last phreatic activity, during late summer and fall of 1976, caused the evacuation of thousands of residents from towns located on La Soufrière's slopes. Explosive steam eruptions from open fissures at the summit and along the southeastern flank of the dome entrained large amounts of fine-grained tephra. This material was deposited downwind, south and west of the volcano, as tephra fall and very dilute density currents (Heiken et al., 1980). These tephra formed a gray, thin, sticky deposit containing accretionary lapilli. During larger eruptions, thin streams of light gray steaming mud flowed down the flanks of the dome as lahars (Wohletz and Crowe, 1978).

During the phreatic activity, it was important to discover if fresh magmatic pyroclasts were present in the tephra because it was not possible to geophysically determine if magma had reached the surface at that time. Careful petrologic analysis of the tephra was, in this instance, of more than academic interest; it was needed, in addition to geophysical measurements, to assess the probability of a significant magmatic eruption following the period of phreatic activity.

Grain Size: Tephra from the eruption of October 2, 1976, is typical of tephra deposited throughout the fall of 1976. It is very poorly sorted and has a mean grain size of 38 μm (4.7φ).

Tephra Description: *Lithic Pyroclasts.* These are equant to subequant, subangular to subrounded dacitic and andesitic fragments that exhibit a wide range in alteration. The most common lithic tephra pyroclast consists of stubby plagioclase and orthopyroxene phenocrysts (up to 40 μm long) in a groundmass of microcrystalline (<1-μm) feldspar and pyroxene phases or colorless glass. The alteration in the lithic pyroclasts ranges from those that are nearly fresh with only a trace of hematite stain, through those laced with hematite or pyrite, to those that are mixtures of an authigenic silica phase (tridymite), pyrite, and anhydrite. This last type exhibits only relict textures that identify them as lava fragments.

Mineral pyroclasts. Plagioclase grains, with compositions of An_{50} to An_{88}, and in some cases exhibiting simple and oscillatory zoning, are present as broken, subhedral crystals. As was the case for lithic fragments, they range from fresh (slightly etched or weathered grain surfaces) to ragged skeletons. Some of the feldspars contain trains of 2- to 4-μm-long, square to circular glass inclusions.

Most of the pyroxene tephra grains are pale yellow to pale pink orthopyroxenes; some contain 20- to 200-μm-long brown glass inclusions. Small amounts of clinopyroxene are present in the tephra.

Amorphous masses of hematite and 30- to 50-μm pyrite grains are present in small amounts in many of the samples and in trace amounts in all the samples. These phases are, however, more abundant in the <38-μm fraction of the tephra. There are also traces of anhydrite and tridymite.

Glass. Glass shards are a minor component within the La Soufrière tephra. Most are angular, equant, colorless to pale brown, slightly vesicular to nonvesicular fragments ranging from andesitic to rhyolitic composition. The angular glass particles appear to be, in most cases, derived from the glassy groundmass of hyalocrystalline dome lavas.

Rare, colorless glass pyroclasts have 10 to 20% ovoid vesicles (long axes of 10 to 15 μm); some contain plagioclase phenocrysts. Nearly all are partly altered and therefore do not represent fresh juvenile pyroclasts.

Also present in the tephra are rare, orange-brown glass fragments (basaltic glass) that have palagonitic rims; these too are older, weathered, and hydrothermally altered pyroclasts.

Chemical Composition: No bulk analyses of the tephra are available. Electron microprobe analyses of the glass pyroclasts are included in Table 1.

TABLE 1. CHEMICAL ANALYSES OF GLASS PARTICLES FROM THE PHREATIC ERUPTION OF LA SOUFRIÈRE DE GUADELOUPE, 27 AUGUST 1976 (IN WT. %).*

	1	2	3
SiO_2	59.18	62.08	68.84
TiO_2	0.10	0.23	0.39
Al_2O_3	23.58	19.00	14.78
FeO	0.88	1.53	2.45
MnO	0.08	0.03	0.10
MgO	0.08	0.22	0.54
CaO	8.30	6.12	3.68
Na_2O	4.92	3.87	3.08
K_2O	0.75	1.10	1.76
Cr_2O_3	0.02	0.03	0.00
Total	97.80	94.21	95.62

*Sample 1: Small, rounded, slightly vesicular grain of colorless glass;
sample 2: slightly elongate, vesicular grain with plagioclase phenocrysts; and
sample 3: small, rounded grain; low vesicularity, but vesicles are large.

Comments on Pyroclast Formation: The tephra consist of lithic and mineral fragments that are similar in texture and composition to those of the dome lavas. Analyses of feldspars and pyroxenes from the tephra support this observation. The degree of comminution of dome lavas to fine-grained tephra ejected by the phreatic eruptions is most likely tied to the degree of hydrothermal alteration of those rocks. Explosive release of steam entrains the altered material as a relatively cool mixture of hydrothermally altered lithic and mineral fragments. Only a trace of the material was petrographically identified as older tephra that may have been derived from pyroclastic units underlying the dome lavas (the andesitic pumice mentioned by Reynal de Saint-Michel, 1966). Some of the tephra components, however, could have been derived from older, intermediate-composition lavas underlying the Soufrière dome, especially at the L'Echelle vent.

An abundance of slightly altered to nearly fresh plagioclase grains in the tephra may have resulted from the preferential alteration of a fine-grained or glassy groundmass of the lavas. This is the case, for example, in near-surface alteration of rock around hot springs at Yellowstone Park, U.S.A. (Allen and Day, 1935) and the superficial zone of acid-leached rocks at Wairakei, N.Z. (Ellis, 1967). None of the colorless pyroclastic fragments present in the Soufrière tephra are fresh enough to classify as new juvenile tephra. Incorrect identification of older, reworked vitric tephra deposited by phreatic eruptions as "fresh" tephra at La Soufrière in 1976 led some investigators to believe that the magma was, indeed, close to the surface. However, magma never did reach the surface.

Recent phreatic activity at Mount St. Helens, Washington, was followed by an explosive magmatic eruption (Sarna-Wojcicki et al., 1981). Phreatic tephra is often overlooked within the geologic record; if identified, it can be useful in determining if phreatic activity has preceded magmatic eruptions and if phreatic behavior is a typical precursor. This type of activity can occur at any volcano but appears to be most common at composite cones of intermediate composition.

REFERENCES:

Allen, E. T. and Day, Arthur L., 1935. Hot springs of the Yellowstone National Park. Carnegie Inst. Wash., Pub. 466: 515 pp.

Ellis, A. J., 1967. The chemistry of some explored geothermal systems. *In*: Barnes, H. L. (Ed.), *Geochemistry of Hydrothermal Ore Deposits*. Holt, Rinehart, and Winston, Inc., N.Y.: 465-514.

Heiken, G., Crowe, B., McGetchin, T., West, F., Eichelberger, J., Bartram, D., Peterson, R., and Wohletz, K., 1980. Phreatic eruption clouds: the activity of La Soufrière de Guadeloupe, F.W.I., August-October, 1976. Bull. Volcanol., **43**: 383-395.

de Reynal de Saint-Michel, Alain, 1966. Notice éxplicative des feuilles de Basse-Terre et des Saintes. Carte Géologique de la France, Départment de la Guadeloupe, Ministere de l'Industrie, Paris: 60 pp.

Sarna-Wojcicki, A.M., Waitt, R., Woodward, M., Shipley, S., and Rivera, J., 1981. Premagmatic ash erupted from March 27 through May 14, 1980—extent, mass, volume and composition. U.S. Geol. Surv. Prof. Paper 1250: 569-575.

Wohletz, K. H. and Crowe, B. M., 1978. Development of lahars during the 1976 activity of La Soufrière de Guadeloupe. Geol. Soc. Am. Abstracts with Programs, **10**: 154.

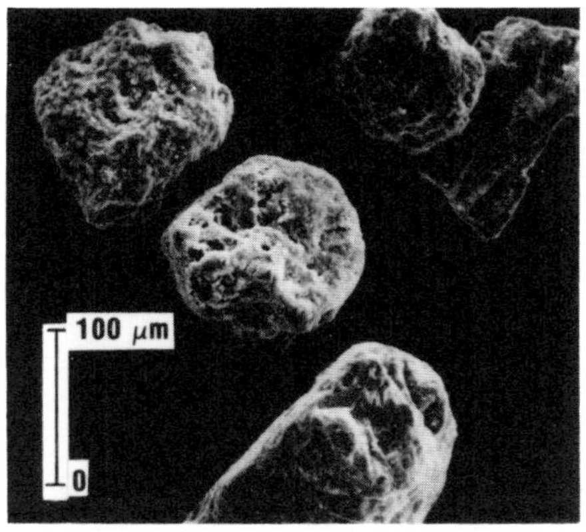

FIGURE 1. SEM of subrounded lithic fragments from the eruption of October 2, 1976. Rounding caused mainly by hydrothermal alteration of lava fragments derived from the dome.

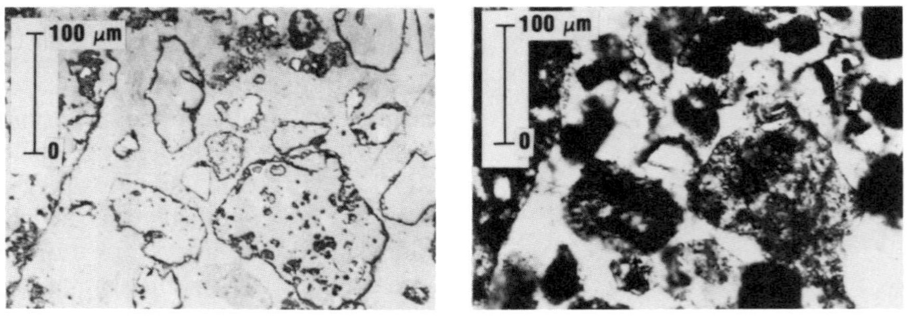

FIGURE 2. Photomicrographs (*left*—reflected light, *right*—transmitted light) of lithic and crystal pyroclasts from the eruption of August 22, 1976. Hydrothermally altered lithic pyroclasts (best examples are *lower right* and *left*) are similar in composition and texture to the dacitic lavas of the dome. Brightest phase in reflected light is pyrite.

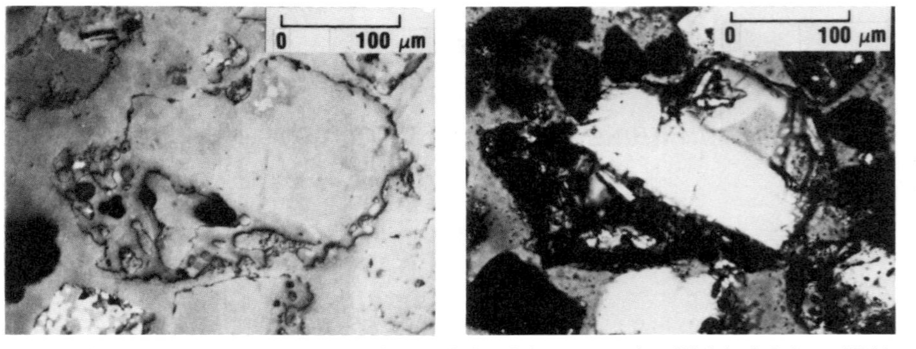

FIGURE 3. Photomicrographs (*left*—reflected light, *right*—transmitted light) of vitric and lithic pyroclasts from the eruption of September 22, 1976. *Center*—a vitric pyroclast; these are rare in the tephra of this eruption. The slightly pumiceous glass encloses a zoned plagioclase phenocryst. Traces of pyrite are indicative of some degree of hydrothermal alteration.

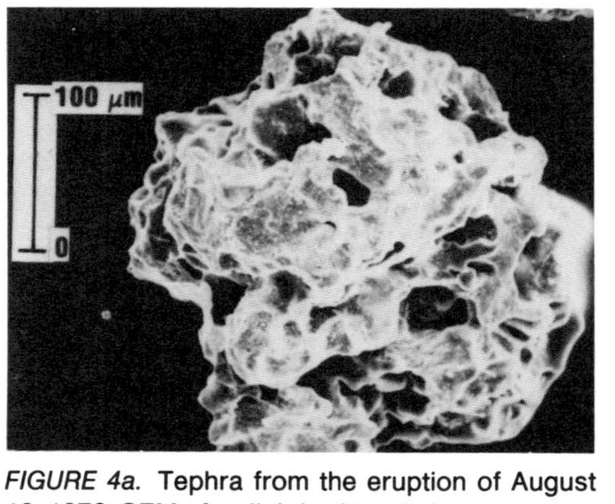

FIGURE 4a. Tephra from the eruption of August 12, 1976. SEM of a slightly altered glass pyroclast with low to medium vesicularity; vesicles are deformed ovoids. These are rare in tephra from this eruption and, at first glance, were thought to be juvenile pyroclasts; vesicular pyroclasts were described in older deposits.

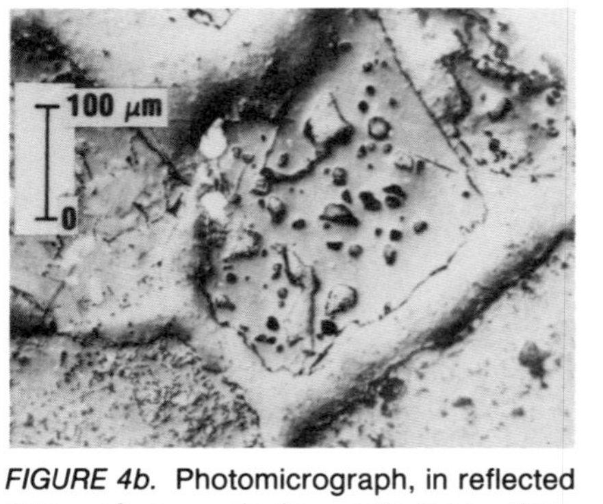

FIGURE 4b. Photomicrograph, in reflected light, of a vesicular, colorless glass pyroclast (*center*).

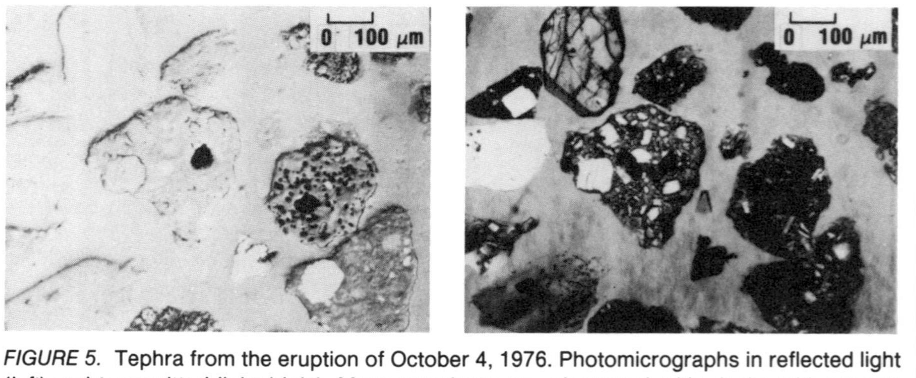

FIGURE 5. Tephra from the eruption of October 4, 1976. Photomicrographs in reflected light (*left*) and transmitted light (*right*). Most pyroclasts are microporphyritic dacite with equant, subangular to subrounded forms. Most are hydrothermally altered and contain pyrite grains. Lithic pyroclasts make up 75% of the sample. The remaining 23% consists of plagioclase, orthopyroxene, clinopyroxene, and a trace of glass.

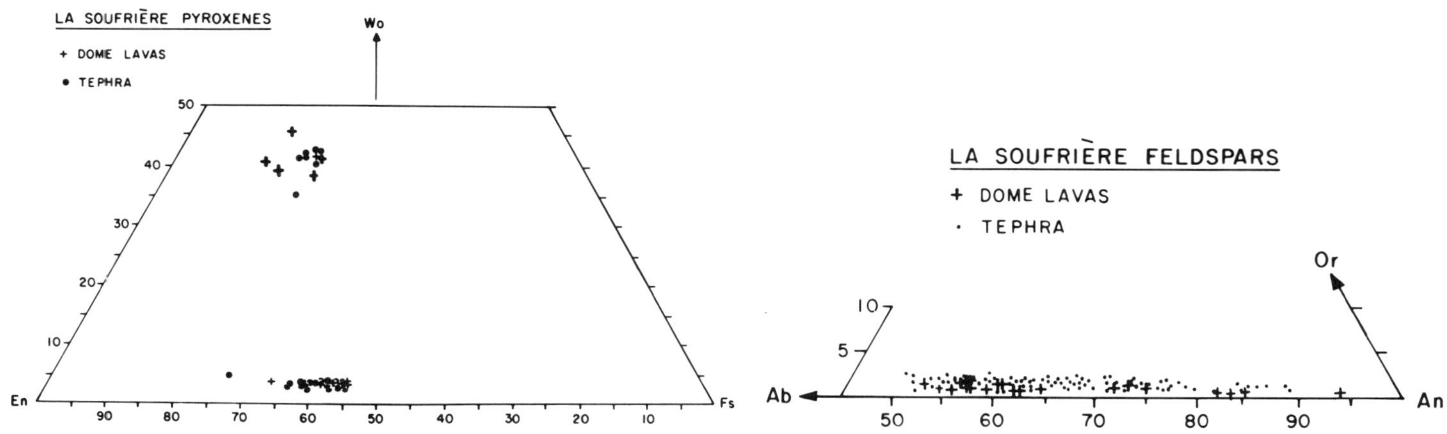

FIGURE 6. Plagioclase and pyroxene compositions from La Soufrière dome lavas and the phreatic tephra of the 1976 eruptions. This comparison supports the interpretation that all the pyroclasts were derived from hydrothermally altered dome lavas during steam explosions.

MT. BAKER, WASHINGTON; APRIL-JULY 1975

Introduction: Mount Baker, one of the northernmost volcanoes of the Cascade Range within the U.S., is a composite cone that rises above deeply dissected plutonic and metamorphic terranes. Most of its lavas and tephra are andesitic (Coombs, 1960). Over the last several hundred years, Mt. Baker has had periods of violent fumarolic activity, including what may have been a mild eruption in 1859 (Coombs, 1960).

The new fumarolic activity of 1975 was at vents located in Sherman Crater, which lies 300 m below the Summit Crater. Both craters are filled with ice. Particularly energetic fuming began during late March 1975, when steam plumes rose 200 to 300 m above the crater (Eichelberger et al., 1976). Steam issued from a 1- by 5-m fumarole in the eastern wall of Sherman Crater. Snow and ice melted by hot water ponded in the crater and formed a 100-m-diameter collapse feature (central ice plug).

Tephra from the fumarole was mostly confined to the crater and formed a continuous layer for about 100 m around the steam plume. Ash was concentrated by melting snow (Eichelberger et al., 1976). About 1-1/2 months after the activity began, the tephra layer was approximately 1 cm thick 100 m from the main fumarole.

Three and a half months after activity began, the total volume of material erupted was estimated at 500 m^3, which represents excavation within a 1- by 5- by 100-m fumarolic vent.

Tephra Description: *Lithic fragments.* The coarser tephra (>50 μm) consists of mostly equant to elongate, rounded to subrounded, granular aggregates of a 1- to 10-μm-diameter low birefringent mineral; there is 5 to 10% pyrite and a trace of hematite (Table 1). Many of the fragments are smeared with thin coats of white, very fine grained (probable) clay mineral. The granular aggregates were identified by XRD as mainly potassic analcite (J. R. Smyth, personal communication, 1975).

Many of the altered fragments retain relict textures. Fragments that appear to have been phenocrysts (feldspar?) of up to 150 μm long are set in a matrix of relict 10- to 30-μm-long tabular crystals (feldspars?). The same fragments have slightly elongate, 2- to 20-μm-long vesicles. These fragments may have been deposited as lithic or lithic-vitric ash before burial and subsequent alteration by fumarolic activity.

Rounded, granular aggregates of 1- to 10-μm analcite crystals are most common in the finer grain-size fractions (<50 μm) but still are abundant in coarser fractions. These aggregates are similar in composition to the altered lithic fragments but exhibit no relict textures. Individual, well-formed analcite crystals are present on the surface of the aggregates as octahedra of up to 10 μm in diameter. Less common spheroidal particles of similar size are composed of sulfur, and some have adhering pyrite cubes.

TABLE 1. COMPOSITION OF TEPHRA AND CLASTIC MATERIAL FROM SHERMAN CRATER, MOUNT BAKER.*

	Tephra (May 1, 1975) (%)	Clastic Material West Wall Fumaroles (%)
Granular aggregates (rock fragments and/or alteration products)	66	61
Clear glass (partially altered)	10	29
Brown glass (unaltered)	1	6
Opaque minerals	19	1
Quartz	1	2
Feldspar	3	1

*Percentages are based on grain counts of the >50-μm fraction.

Glass. Glass fragments make up a small part of the tephra; because of its fresh appearance, it is the only component of the tephra that could be interpreted as juvenile ash. Many glass fragments are colorless, equant, angular particles with low (<20%) vesicularity. In most cases, vesicles are filled with clusters of a very fine grained mineral with low birefringence (analcite). Colorless, elongate pumice grains are present in coarser grain size fractions; irregular, ovoid vesicles are up to 20 μm long. Some of the grains are stained with hematite. Altered and unaltered brown glass fragments also occur. Nearly equant fragments contain 5 to 20% ovoid to spherical vesicles that are up to 25 μm long. Some of these grains have smooth, fluidal grain surfaces.

Mineral fragments. The most common mineral type is pyrite; there are smaller amounts of alkali and plagioclase feldspars, hematite, quartz, chalcopyrite, sulfur, pyroxene, and gypsum(?).

Gas release study of the tephra: The tephra was analyzed by Everett K. Gibson, NASA Johnson Space Center, for gas content and weight loss during heating under a vacuum by the procedures outlined by Gibson and Johnson (1972). After drying at 60 to 70°C, the sample was heated at 6°C/minute to 1400°C. The 8.6% weight loss of H_20 at 350 to 500°C indicates the presence of zeolites. A 10.1% weight loss of SO_2 in the range of 650-750°C is most likely from breakdown of the pyrite observed in the sample.

Comments on Pyroclast Formation: Clastic material scraped from the west wall of Sherman Crater is similar in most aspects to the tephra. The coarser grain-size fraction (>50 μm) consists of mainly colorless, granular aggregates of analcite with thin coats of a white amorphous clay(?) mineral and equant to slightly elongate, highly altered lithic fragments. Brown, vesicular glass fragments are present in the sample and consist of broken droplets or angular, blocky fragments. Some varieties of the brown glass are pumiceous, exhibiting highly elongate "pipe" vesicles. Both unaltered and altered forms are present; vesicles in the unaltered form are filled with granular aggregates of analcite. Brown glass pyroclasts are similar to vitric ash fragments from Strombolian eruptions of basaltic or basaltic andesite lavas (Heiken, 1972). Colorless, spindle-shaped, pumiceous fragments are also present. Mineral fragments present include pyrite, plagioclase and alkali feldspars, hematite, quartz, sulfur, and possibly gypsum.

It is likely that clastic material present in the walls of the fumarole was originally deposited as lithic and lithic-vitric tuffs and was subsequently altered by fumarolic activity. More energetic fumarolic activity broke this material from the vent walls to form phreatic tephra. Glass pyroclasts within the 1975 tephra deposit represent reworked ash and not juvenile tephra.

REFERENCES:

Coombs, H. A., 1960. Active volcanoes and solfatara fields of the United States of America. *In*: *Catalogue of the active volcanoes of the world including solfatara fields.* Intern. Assoc. of Volcanol. and Chem. of the Earth's Interior, IX: 1-58.

Eichelberger, J. C., Heiken, G., Widdicombe, R., Wright, D., Keady, C., and Cobb, D., 1976. New fumarolic activity on Mt. Baker: observations during April through July, 1975. J. Volcanol. Geotherm. Res., **1**: 35-53.

Gibson, Jr., E. K. and Johnson, S. M., 1972. Thermogravimetric-quadrupole mass-spectrometric analysis of geochemical samples. Thermochim. Acta, **4**: 49-56.

Heiken, G. H., 1972. Morphology and petrography of volcanic ashes. Geol. Soc. Am. Bull., **83**: 1961-1988.

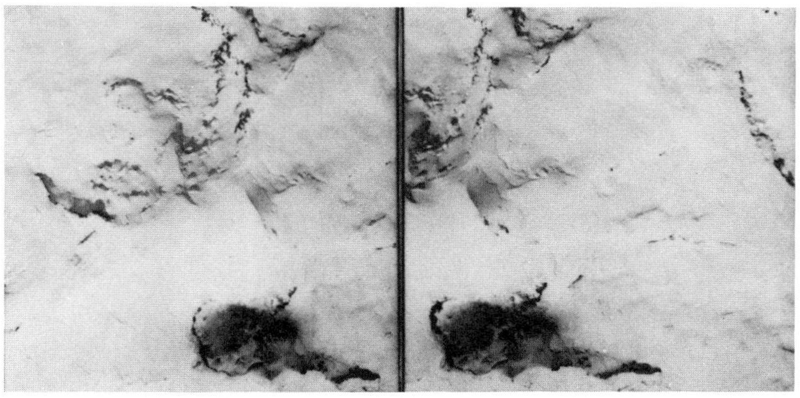

FIGURE 1. Aerial photographs (stereo pair) of Sherman Crater, Mt. Baker, taken April 11, 1975. Note the dark tephra cover within the crater and the steam plume rising from the NE crater wall.

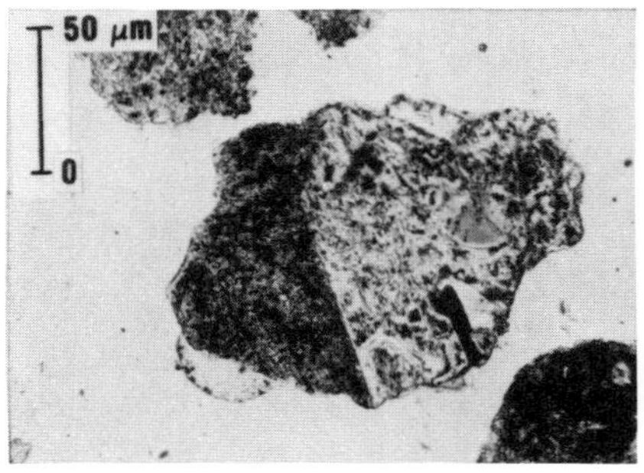

FIGURE 2a. Photomicrograph (transmitted light) of hydrothermally altered lithic pyroclasts. Most lithic fragments now consist of analcite but have faint relict textures.

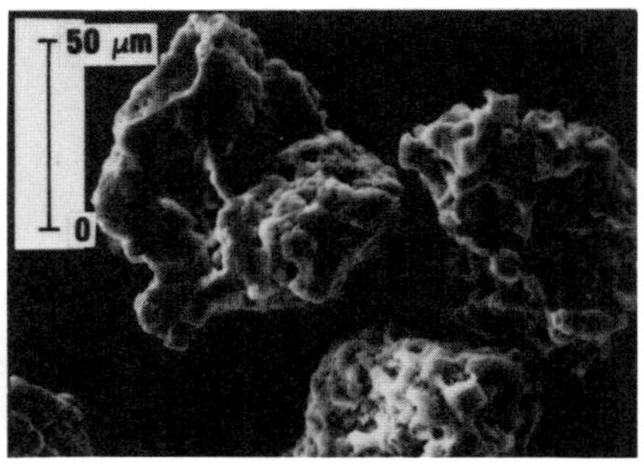

FIGURE 2b. SEM of hydrothermally altered lithic pyroclasts. Much of what may have been glassy matrix has been etched and removed.

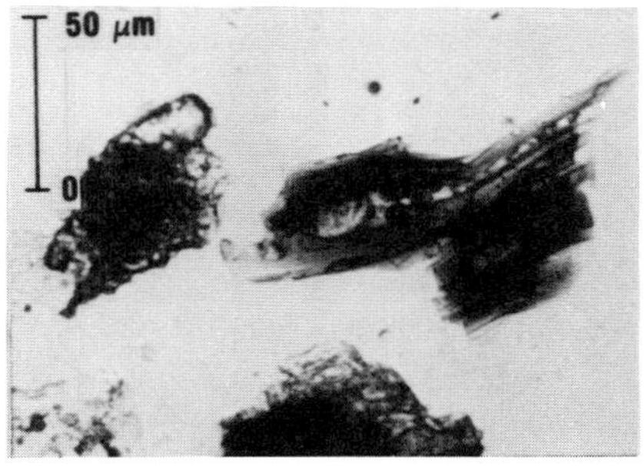

FIGURE 2c. Photomicrograph (transmitted light) of an angular vitric pyroclast that is typical of the brown-glass tephra fraction from fumarole walls and in the phreatic tephra.

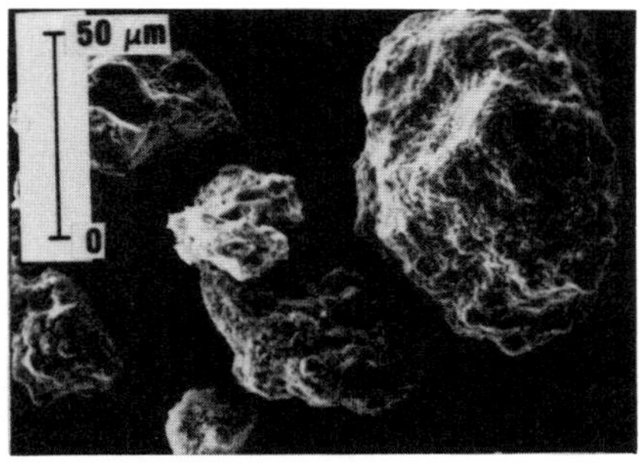

FIGURE 2d. SEM of hydrothermally altered lithic pyroclasts.

Prehistoric Eruptions

INYO CRATERS, CALIFORNIA

Introduction: The Inyo Craters consist of three explosion pits located at the southern end of the Inyo Domes chain, Long Valley caldera, eastern California (Mayo et al., 1936; Huber and Rinehart, 1967). Rinehart and Huber (1965) reported a radiocarbon age of 650 ± 200 years for the southern Inyo Crater. No juvenile ash or lava was ejected during the explosive excavation of the craters. Phreatic explosions likely resulted when rhyodacitic magma rose to a shallow depth and flashed ground water to steam (Bailey et al., 1976). Pyroclasts formed during these explosions are dominantly altered and broken pieces of the underlying basalt lavas and subordinant amounts of comminuted rhyolite.

The two largest Inyo Craters are bowl-shaped and have maximum diameters of nearly 200 m and depths of nearly 50 m. These craters are aligned along a fault trending N20W, which is the southern expression of the Inyo Dome chain (Mayo et al., 1936). The late basalts and moat rhyolites of Long Valley caldera (Bailey et al., 1976) are visible in crater walls. Overlying these lavas and exposed within crater rims is 15 to 25 m of tephra. The tephra consists of 2- to 3-m-thick, alternating layers of explosion breccia and finer grained ash fall that contains vesicles. Similar in composition and appearance to the Inyo Crater phreatic breccia and ash is the small (<100-m-diameter) phreatic tuff rampart on the south slope of Deer Mt., a rhyodacitic dome just north of the Inyo Craters.

The phreatic ash described here is from a thickly bedded ash layer in the explosion breccia at the rim of the southern Inyo Crater (sample 1) and from a massive lapilli layer at the rim of the north Inyo Crater (sample 2). These two samples are compared with a sample from the tuff rampart of Deer Mountain (sample 3). This last is an explosion breccia consisting of predominantly ash and lapilli-size clasts of lava with some blocks up to 20 cm in diameter. This phreatic crater probably has a origin similar to that of the Inyo Craters; it resulted from near-surface, explosive hydrothermal activity located along the N20W-trending fault.

Grain Size: Grain-size data for samples 1, 2, and 3 are shown in Table 1. The poor sorting (σ_φ = 237 to 2.53φ) of samples 2 and 3 indicates a pyroclastic surge emplacement, whereas the better sorting of sample 1 ($\sigma_\varphi = 1.92\varphi$) may be the result of ash fall. These phreatic tephra required some disaggregation before they were sieved. Disaggregation is often useful when analysing this material because phreatic tephra near their vents are typically lithified by the steam that condenses on particles during emplacement.

TABLE 1. GRAIN-SIZE ANALYSES.

Sample	Md_φ (mm)	σ_φ	Fine Ash (Wt. %)
1	−0.07 (1.04)	1.92	3.44
2	−1.05 (2.07)	2.37	1.78
3	0.25 (0.87)	2.53	6.90

Tephra Description: Probably the most notable characteristic of the phreatic pyroclasts is the surface aggregation of fine particles and their chemical composition, which shows the strong effects of hydrothermal alteration (Table 2). This surface aggregation and alteration gives individual grains irregular surfaces and a muddy or "dirty" appearance. Sample 1 contains pyroclasts that are rounded aggregates of smaller angular pieces. Fresh surfaces are rare, as are sharp points or edges, and vesicles are entirely absent. The fine fraction of sample 1 reveals individual blocky and angular clasts with abundant, weakly attached dust particles. Fresh surfaces are more common, as are flat or curviplanar surfaces. Clasts are commonly tabular or slab-like. Sample 2 demonstrates the surface alteration typical of phreatic ash.

Rare, vesicular basaltic clasts have a surface hydration rind that has contracted and cracked, especially within exposed vesicles. Hydration rinds tend to peel away from the substrate, like mud cracks. Sample 3 contains rhyodacitic lithic clasts that have somewhat less surface alteration than do samples 1 and 2, but these clasts still have considerable surface aggregation of fine material. Very fine grained ash is missing from sample 3, and its particles are much less slab-like than those of sample 1.

TABLE 2. CHEMICAL ANALYSES OF INYO CRATER PYROCLAST SURFACES.*

	1	3
SiO_2	67.50	55.41
TiO_2	1.59	5.38
Al_2O_3	20.08	15.02
FeO	9.71	6.45
MnO	---	4.64
MgO	0.18	3.71
CaO	0.42	1.72
Na_2O	---	1.58
K_2O	0.51	6.09

*Normalized EDS analyses of rhyolite and rhyodacitic lithic clasts.

Comments on Pyroclast Formation: Because juvenile tephra is absent in the pyroclastic deposits surrounding Inyo Craters, the explosions forming the crater are termed phreatic. Muffler et al. (1971) studied similar hydrothermal explosion craters in Yellowstone National Park. During phreatic explosions, rock overlying a magma body may be saturated with water, which fills tiny cracks and vesicle spaces. Shepherd and Sigurdsson (1982) studied the 1979 eruption of Soufrière Volcano, St. Vincent, in which the water-saturated upper part of the previously extruded lava dome was pulverized. The phreatic pulverization of lava probably occurs by hydraulic fracturing in which thermal expansion of water produces high-pressure gradients of about 1.5 MPa/C° during constant volume heating (Knapp and Knight, 1977). When the pressure of superheated pore water exceeds the strength of the rock, vapor explosion shatters the rock. The fracture density of the rock before eruption determines, in part, the size distribution of tephra. Hydraulic fracturing of the rock before eruption may increase fracture density and allow more water to infiltrate and contribute to the vapor explosion. Hydraulic fracturing followed by a large-scale vapor explosion produces fine-grained ejecta. Ejecta shapes are mainly determined by brittle fracture, and planar grain surfaces are produced. Aggregation of fine, cohesive dust and rounding during transport determines the typical shapes shown in these figures.

REFERENCES:

Bailey, R. A., Dalrymple, G. B., and Lanphere, M. A., 1976. Volcanism, structure, and geochronology of the Long Valley caldera, Mono County, California. J. Geophys. Res., **81**: 725-744.

Huber, N. K. and Rinehart, C. D., 1967. Cenozoic volcanic rocks of the Devil's Postpile quadrangle, eastern Sierra Nevada, California. U.S. Geol. Surv. Prof. Pap., 554-D: D1-D21.

Knapp, R. B. and Knight, J. E., 1977. Differential thermal expansion of pore fluids: fracture propagation and microearthquake production in a hot pluton. J. Geophys. Res., **82**: 2515-2522.

Mayo, E. B., Connant, L. C., and Chelikowsky, J. R., 1936. Southern extension of the Mono Craters, California. Am. J. Sci., **32**: 81-97.

Muffler, L. P. J., White, D. E., and Truesdale, A. H., 1971. Hydrothermal explosion craters in Yellowstone National Park. Geol. Soc. Am. Bull., **82**: 723-740.

Rinehart, C. D. and Huber, N. K., 1965. The Inyo Crater Lakes - a blast in the past. Calif. Div. Mines Geol. Mineral. Inform. Ser., **18**: 169-172.

Shepherd, J. B. and Sigurdsson, H., 1982. Mechanism of the 1979 explosive eruption of Soufrière Volcano, St. Vincent. J. Volcanol. Geotherm. Res., **13**: 119-130.

FIGURE 1. SEM of a pyroclast from sample 1 that shows rounded aggregates typical of phreatic tephra.

FIGURE 2. SEM of an aggregate from sample 1 with numerous angular particles cemented to the surface.

FIGURE 3. SEM of the fine fraction of sample 1 showing individual blocky and angular, slab-like particles with poorly attached, smaller dust grains.

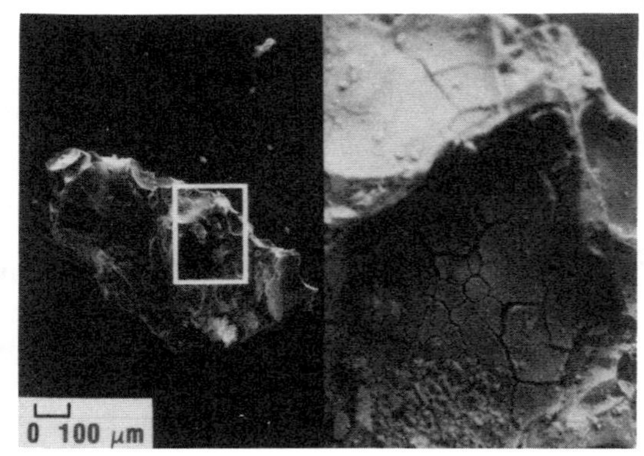

FIGURE 4. SEM of a vesicular basalt clast of sample 2 with 10X magnification in right view. Note the hydrated and cracked surface in vesicle embayment.

FIGURE 5. SEM of a sample 2 pyroclast covered with fine clay alteration products.

FIGURE 6. SEM of a rhyodacite clast of sample 3, phreatic tuff, which is a broken aggregate of finer material.

FIGURE 7. SEM of a sample 3 pyroclast showing a surface coating of fine, angular ash chips.

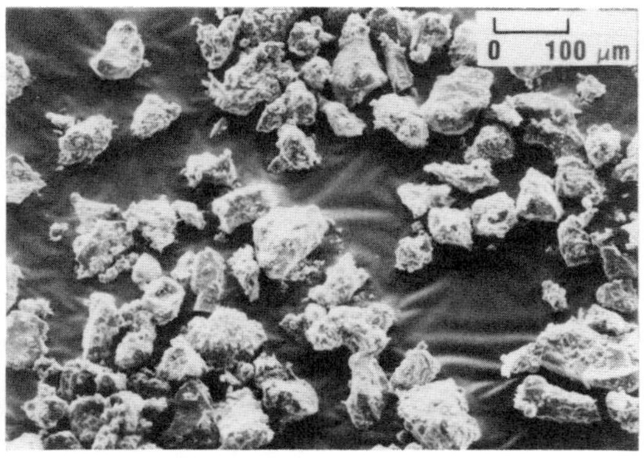

FIGURE 8. SEM of the fine fraction of sample 3. Note the equant shape and lack of fine dust, compared to particles in Fig. *3*.

Ash bed sequence, Oshima, Japan.

ERUPTION SEQUENCES

This section illustrates changes in eruption phenomena as they are reflected in ash properties from individual eruptions. Examples presented here include many of the eruption types presented in the previous section, but many of the samples exemplify the changes in eruption phenomena that can occur during an eruption, e.g., the progression from magmatic to phreatoplinian activity.

Historic Eruptions

A significant aspect of ash studies is the correlation of ash types with witnessed accounts of eruptions. Analysis of historic eruption sequences provides us with basic information on genetic aspects of tephra formation that can be used in interpretation of prehistoric eruption sequences.

LAKAGIGAR, ICELAND; 1783 ERUPTION

Introduction: The Lakagigar (Laki) eruption of 1783 may be the largest historic eruption of basaltic lava on Earth. It began on June 8, 1783, and lasted about 8 months. The eruption was from a 25-km-long line of fissures and craters and produced 12.3 km^3 of lava (Thorarinsson, 1970). The many craters include spatter cones, cinder cones, and maars; the maars are believed to have been constructed when the eruption fissure cut through lakes (Thorarinsson, 1958, 1970; S. Self and G. Larsen, verbal communication). Although the eruption is most famous for its voluminous lava flows, there was a considerable tephra component.

The tephra samples studied here were provided by G. Larsen (Nordic Volcanological Institute, Iceland) and S. Self (University of Texas, Arlington) from a site adjacent to the SW part of the Lakagigar fissure row near Laki Mountain; they are from a representative stratigraphic section in which the lowest and middle layers consist of basaltic phreatomagmatic ash and the top layer is basaltic scoria. The phreatomagmatic layers may be from only a few of the vents—perhaps from the larger maars along the line of craters. The uppermost scoria layer is from a nearby spatter and scoria rampart.

Tephra Description: The phreatomagmatic tephra layers are fine grained and consist of small, equant pyroclasts that are characterized by angular fracture surfaces and curved, smooth surfaces from vesicle walls.

The uppermost unit consists of mostly scoria lapilli. These are composed of highly vesicular basaltic glass. Many surfaces are characterized by a thin, lumpy skin with ridges and mounds developed over vesicles. Effects of surface tension are also evident in the strands of glass that formed when the outermost vesicle walls collapsed before chilling of the scoria pyroclasts. Most vesicles consist of large (30-μm to several millimeters in diameter) spheroidal to ovoid vesicles with thin vesicle walls. Large composite vesicles are present in many of the pyroclasts; they were formed when the walls of dozens of adjacent vesicles coalesced and collapsed.

Both tephra types are produced by the same magma; the marked difference is caused by interaction of magma with water.

Chemical Composition: The lava and tephra are tholeiitic basalts, which are poor in olivine but have traces of plagioclase (An_{60}-An_{64}) and clinopyroxene. For a chemical analysis, see Table 1.

TABLE 1. CHEMICAL ANALYSIS OF LAKAGIGAR TEPHRA.*

	(Wt. %)
SiO_2	50.30
TiO_2	2.11
Al_2O_3	14.16
Fe_2O_3	4.17
FeO	9.25
MnO	0.23
MgO	6.58
CaO	11.42
Na_2O	2.01
K_2O	0.53
Total	100.84

*From Thorarinsson, 1970.

Comments on Pyroclast Formation: The first two deposits (at least at this point in the line of craters) are phreatomagmatic. Quenching, mixing, and fragmentation occurred at the surface *after* complete or nearly complete vesicle development and growth. The result is a fine-grained tephra composed of vesicle-wall fragments.

During the last phase represented in this stratigraphic section, it appears that large lava fountains produced highly vesicular scoria and Pele's hair (described by Thorarinsson). The vesicles are large and equidimensional; vesicles continued to grow after eruption of molten droplets, and there was little vesicle elongation during ejection. Many of the delicate, thin-skinned lapilli broke upon hitting the ground. Lava fountains accompanied the effusion of lava flows that traveled toward the coast.

REFERENCES: Thorarinsson, S., 1958. The Öraefajökull eruption of 1362. Acta Nat. Islandica, **II**: 1-99.

Thorarinsson, S., 1970. The Lakagigar eruption of 1783. Bull. Volcanol., **33**: 910-929.

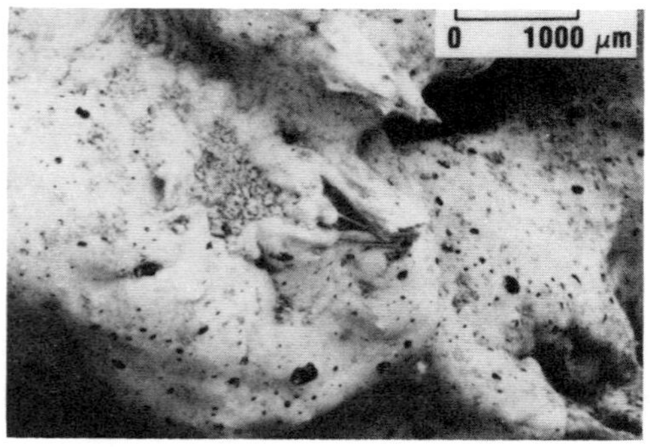

FIGURE 1a. Scoria fall unit at the surface, NW side of the Laki fissure. The surface of the scoria lapillus has a smooth, thin, glassy skin that was deformed after ejection of vesiculating melt. Many surface mounds are over vesicles; many have broken, leaving holes with ragged edges.

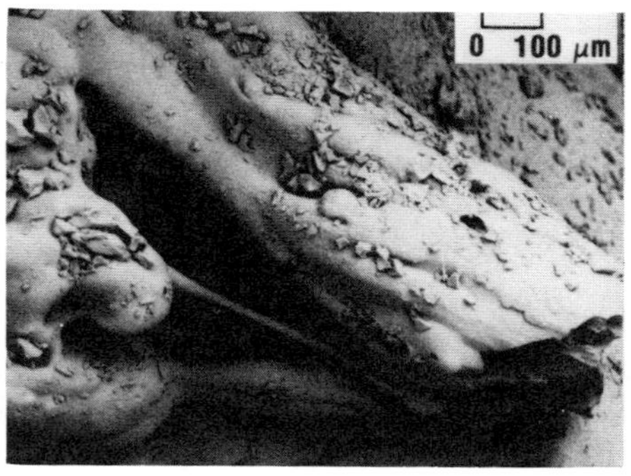

FIGURE 1b. Close-up of Fig. *1a.* The high vesicularity and thin vesicle walls are visible where the lapillus has been broken. Elongate ridges are present over elongate vesicles. Thin glass strands at the surface may have formed when still-molten broken vesicle walls were stretched.

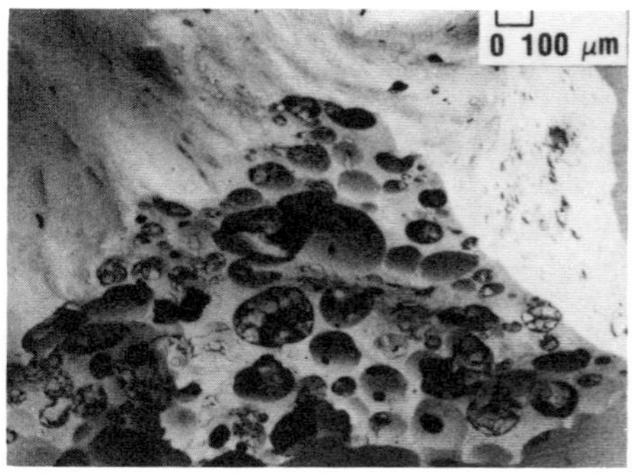

FIGURE 1c. Broken edge of a scoria lapillus shows development of spherical to ovoid vesicles within. There has been some degree of vesicle coalescence, as is noted in the vesicle walls with one or several holes.

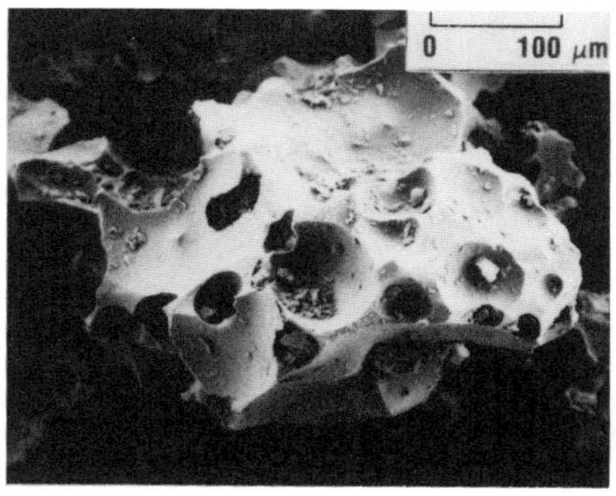

FIGURE 2a. Scoria layer at top of Laki tephra section. The broken surface of a scoria lapillus shows well-developed vesicularity and very thin vesicle walls. Larger cavities were formed during coalescence of dozens of vesicles.

FIGURE 2b. Close-up of Fig. *2a*. Basaltic glass can exist as very thin, delicate vesicle walls, such as those seen here. Thin flakes adhering to the scoria surface are only a few micrometers thick, as are the broken edges in the *center* and *top* of the photograph.

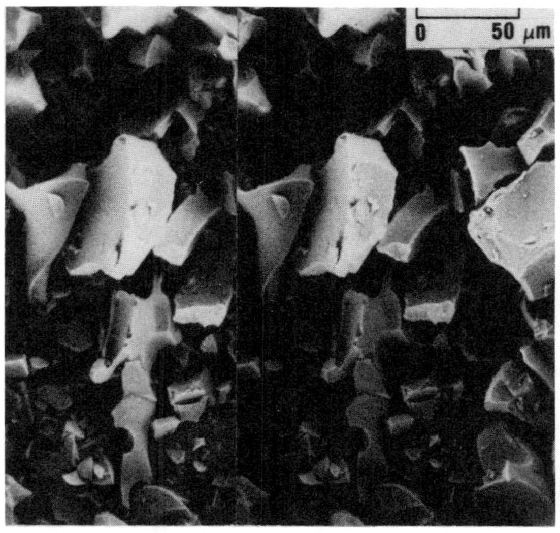

FIGURE 3a. Phreatomagmatic tephra layers at the base and middle of the Laki sequence, SW side of the fissure, near Middle (Laki) Hill. SEM of phreatomagmatic ash of basaltic composition. In contrast to the coarse scoria, this is a fine-grained tephra; most particles are less than 50 μm long. The small, angular particles have some smooth, curved surfaces that were vesicle walls before comminution. Interactions of melt with water and subsequent chilling and fragmentation occurred after vesiculation and may have been a near-surface phenomenon.

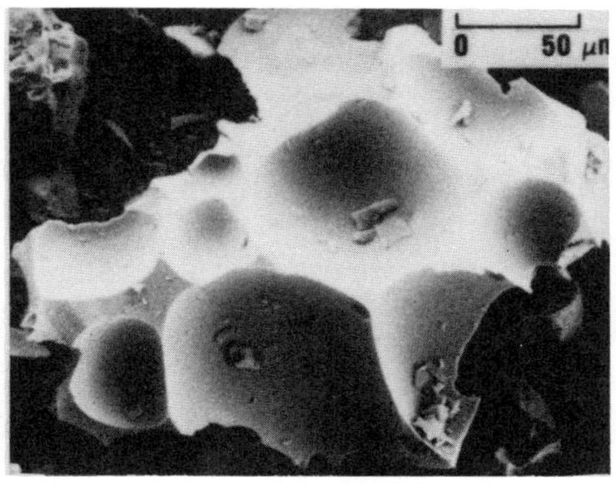

FIGURE 3b. One of the coarsest particles in this particular sample—a broken, angular scoria pyroclast. Vesicles in this pyroclast, as can be seen, were mostly spheroids and ovoids.

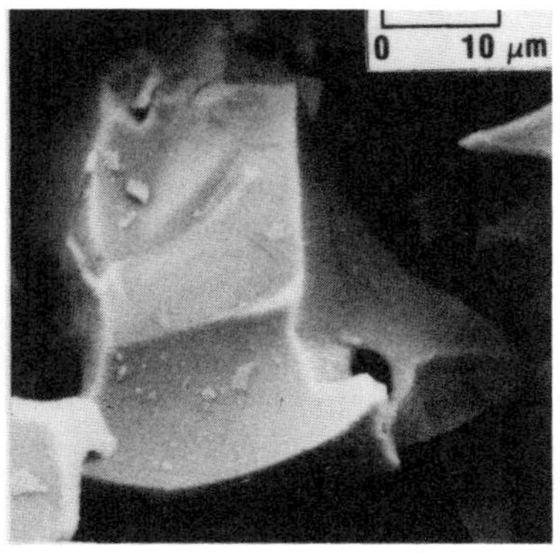

FIGURE 3c. Close-up of a typical glassy pyroclast from this tephra layer. It is a fragment from a wall between several vesicles; the former vesicle walls are curved and smooth, whereas fractures are conchoidal.

EL CHICHÓN, CHIAPAS, MEXICO; 1982 ERUPTION

Introduction: El Chichón, a small volcano located in Chiapas State, Mexico, and characterized by sub-alkaline eruption products, erupted in late March and early April 1982. The remarkably gas-rich eruption was characterized by high eruption clouds that penetrated the tropopause and were partly dispersed in the stratosphere. Volcanic ash and sulfate aerosols still remain in the stratosphere 1-1/2 years later, at the time of this writing, and have stimulated great interest within the community of atmospheric scientists. There were three main eruption phases: March 28-29, April 3, and April 4, 1982. The first phase dispersed crystal-rich ash and pumice over 37,000 km^2 (the area within the 1-mm isopach); the second enlarged the crater, producing lithic-fragment-rich ash and pumice; and the third phase spread ash to the ENE over 50,000 km^2 (1-mm isopach) (Varekamp et al., 1982; Luhr et al., 1982). Pyroclast surfaces are heavily laden with adsorbed S and Cl; Varekamp and Luhr (1982) estimate that about 2×10^6 metric tons of S and Cl were present on pyroclast surfaces. They also estimate that 0.4 to 0.43 km^3 (compacted) of tephra were erupted during the three main phases.

The samples described here were collected by J.C. Varekamp (Wesleyan University) and J.C. Luhr (University of California - Berkeley) and were given to the authors by Varekamp. SEMs of stratospheric samples were provided by D. S. McKay, NASA-Johnson Space Center, Houston, Texas.

Grain Size: Grain-size data for ash fall deposits (Table 1) were supplied by S. Carey (University of Rhode Island); detailed information on the grain-size variations may be found in a paper by Sigurdsson et al. (in press).

In most of the samples the tephra are bimodal with coarser pumice pyroclasts and very fine grained shards.

TABLE 1. GRAIN SIZE OF EL CHICHÓN TEPHRA.

Eruption	Grain Size at Two Downwind Locations: 6.5 km (μm, φ)	22 km (μm, φ)
March 29	750, 0.5	200, 2.3
April 3	125, 3.0	20, 5.7
April 3-4(1)	4000, −2.0	450, 1.2
April 3-4(2)	500, 1.0	300, 1.7

Tephra Description: In general, tephra from the El Chichón eruption (in the samples provided) are fine grained. Pumice pyroclasts are subequant to slightly elongate, and they are partly coated with shards, anhydrite crystals, and small phenocrysts. Vesicles are ovoid with long axes of 1 to 18 μm and intermediate axes of 0.5 to 10 μm. Most vesicle walls are thick relative to the vesicle size (1 to 4 μm). Coalesced vesicles are rare.

Most shards are subequant and angular; the surfaces consist of fractured vesicle walls (smooth and conchoidal fractures). Some particles have sharp, pointed edges; they too are partly coated with sublimates, H_2SO_4 droplets, smaller shards, and small phenocryst fragments. Near-vent samples are bimodal and consist of coarse-grained, angular crystal and lithic pyroclasts and fine-grained vitric pyroclasts.

Chemical Composition: A chemical analysis of trachyandesitic pumice from El Chichón is shown in Table 2.

TABLE 2. CHEMICAL COMPOSITION OF EL CHICHÓN TRACHYANDESITIC PUMICE.*

	(Wt. %)		(ppm)
SiO_2	57.5	V	165
TiO_2	0.7	Cr	---
Al_2O_3	17.7	Ni	5
FeO	5.8	Cu	40
MnO	0.2	Zn	90
MgO	2.3	Th	8
CaO	8.2	Rb	90
Na_2O	4.4	Sr	1150
K_2O	3.0	Ba	800
P_2O_5	0.4	La	35
		Ce	60
Total	100.2		

*From Varekamp and Luhr (1982).

Comments on Pyroclast Formation: There are very few highly elongate vesicles, which implies little or no deformation and elongation of vesicles during their growth and ascent in magma within the vent. Either vesicle development was suppressed by high viscosity or it occurred very near the ground surface—at the magma/atmosphere interface.

The uniformly fine grained nature of the El Chichón tephra has led to some speculation that a phreatomagmatic component was involved, but there are not enough published data to verify this hypothesis. Extensive fragmentation may have been related to the fact that the magma was very rich in volatile phases (including abundant SO_2).

REFERENCES: Luhr, J., Varekamp, J., and Prestegaard, K., 1982. The 1982 eruption of El Chichón Volcano, Chiapas, Mexico. Part II. Mineralogy and petrology of the ejecta. Geol. Soc. Am. Abs. with Programs, **14**: 551.

Sigurdsson, H., Carey, S., and Espindola, J., 1984. The 1982 eruptions of El Chichón Volcano, Mexico: Stratigraphy of pyroclastic deposits. J. Volcanol. Geotherm. Res. (in press).

Varekamp, J., Luhr, J., Prestegaard, K., and Carepa, J., 1982. The 1982 eruption of El Chichón Volcano, Chiapas, Mexico. Part I. Stratigraphy, volume, and volatile element characteristics of the ash fall deposits, Geol. Soc. Am. Abs. with Programs, **14**: 637.

Varekamp, J. and Luhr, J., 1982. The 1982 eruption of El Chichón Volcano: Nature of the eruption, ash and volatiles. Trans. Am. Geophys. Union, **63**: 1126.

ERUPTION OF MARCH 28-29, 1982:

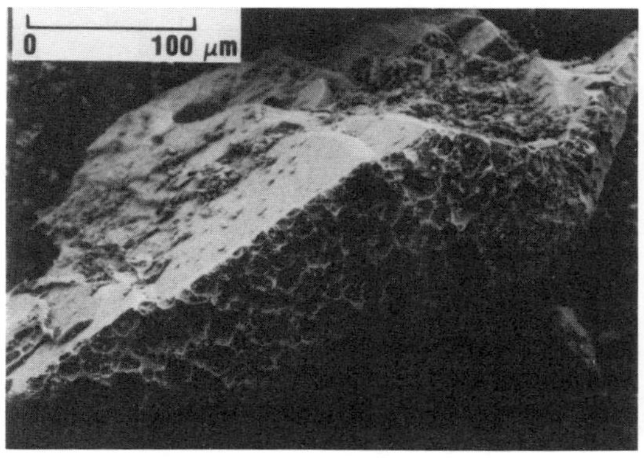

FIGURE 1a. SEMs of a sample collected 15 km E of El Chichón. Pyroxene pyroclast with a thin coating of vesicular glass (bubble-wall texture).

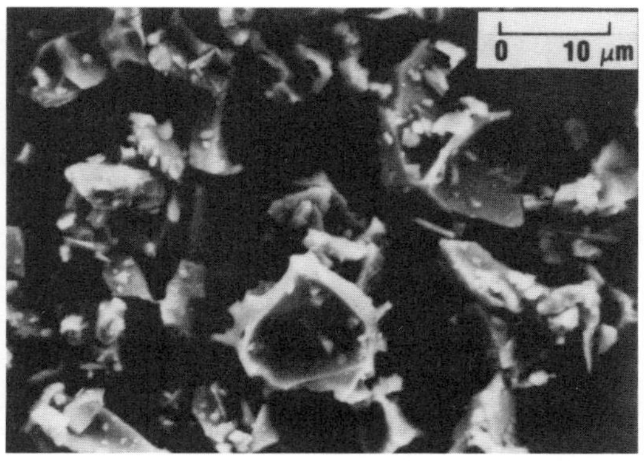

FIGURE 1b. Fine vitric ash. Most shards have been broken from a highly vesicular magma with small, slightly elongate vesicles. This particular ash is bimodal and consists of coarse-grained crystal and lithic pyroclasts (see Fig. *1a*) and very fine grained vitric pyroclasts. Some of the shards in this example are less than 1 μm long.

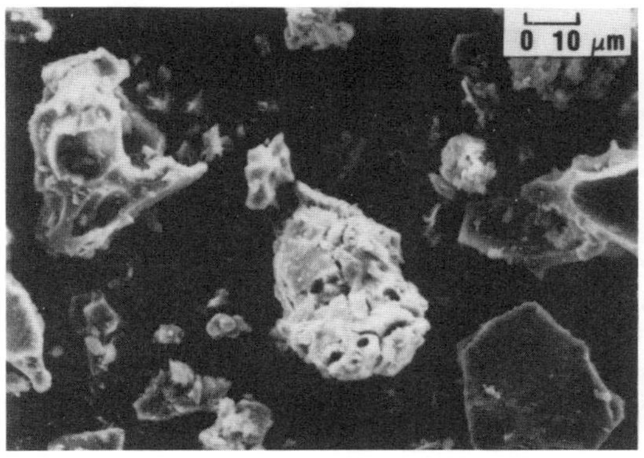

FIGURE 2a. SEMs of a sample collected 70 km SW of El Chichón. Vitric, lithic, and crystal components of fine ash. *Center* and *top right*—aggregates that appear to be hydrothermally altered lithic fragments. *Lower right*—plagioclase fragment. The remaining particles are mostly shards.

FIGURE 2b. Center and lower right—fractured pyroxene pyroclasts. Of the remaining particles, most of the larger (>10 μm) are shards.

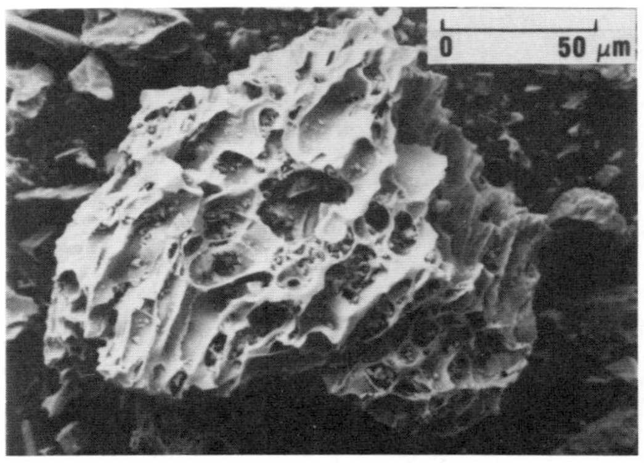

FIGURE 3a. SEMs of a sample collected 80 km NW of El Chichón. Subequant, angular pumice pyroclast. Most vesicles are ovoid and slightly elongate. Many are flattened. There is a range of vesicle sizes from 1 to about 40 μm long.

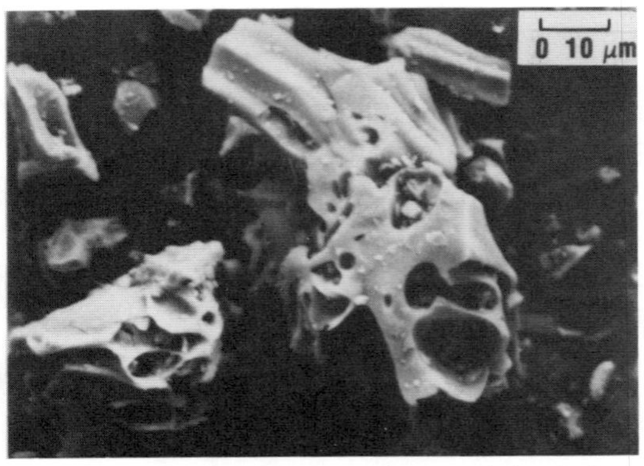

FIGURE 3b. Small pumice pyroclasts and shards from the fine ash matrix of this sample.

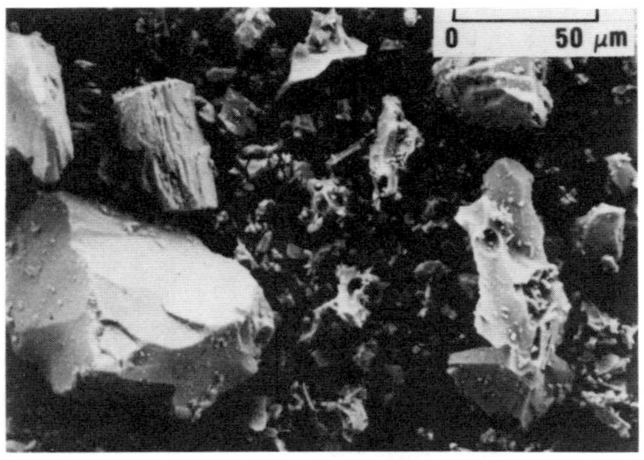

FIGURE 4. SEMs of a sample collected 125 km E of El Chichón. Even at this distance from the source, the ash is bimodal with crystal and pumice pyroclasts >40 μm long and shards <20 μm long. Most of the shards are <10 μm in size. *Bottom right*—plagioclase feldspar; *left* of the feldspar—pyroxene.

ERUPTION OF APRIL 3-4, 1982:

FIGURE 5a. SEMs of a sample collected 15 km E of El Chichón. A crystal-vitric tuff; like examples from the earlier eruption, it is bimodal with coarse crystal and lithic components and a fine-grained vitric fraction. Most of the irregular, coarse pyroclasts are finely vesicular pumices, and the blocky forms are phenocrysts. *Lower center and lower left*—plagioclase; *upper left*—Fe-Ti oxide; *upper right*—lithic fragment consisting of quartz and pyroxene phenocrysts.

FIGURE 5b. Subequant pumice pyroclast with small, slightly elongate vesicles and thick vesicle walls. It is surrounded by glass-free phenocrysts (crystal pyroclasts).

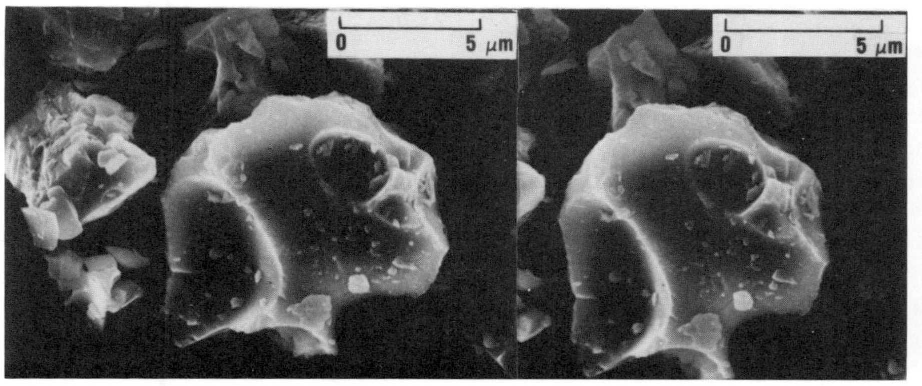

FIGURE 6a. SEMs of a sample collected 170 km E of El Chichón. Stereophotomicrograph of fine-grained vitric tephra.

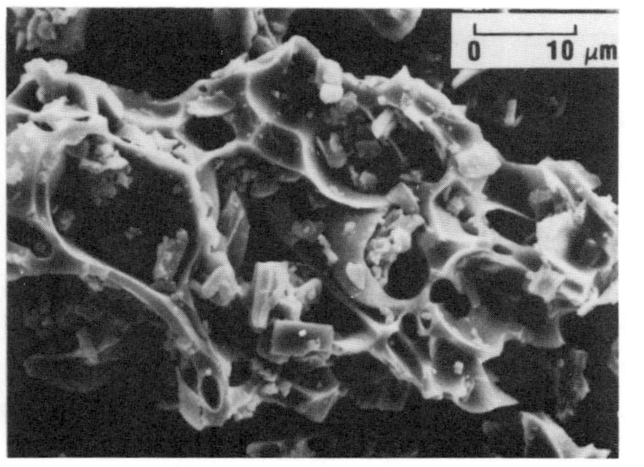

FIGURE 6b. Pumice pyroclast partly covered with fine-grained shards and tabular crystals. The tabular grains are anhydrite ($CaSO_4$) crystals.

STRATOSPHERIC SAMPLES:

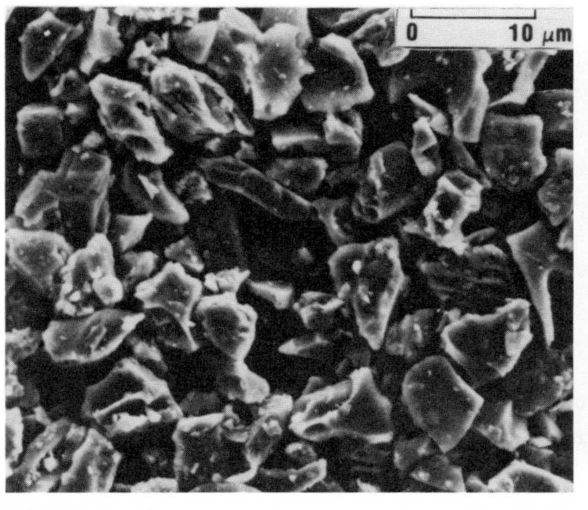

FIGURE 7a. These samples were collected by NASA aircraft in the eruption plume at an altitude of 18 km on May 7, 1982. SEMs were supplied by David S. McKay of NASA-Johnson Space Center. Equant to subequant shards, partly coated with sublimates and <1-μm shards. These are among the coarsest volcanic ash particles ever collected in an eruption plume at stratospheric heights.

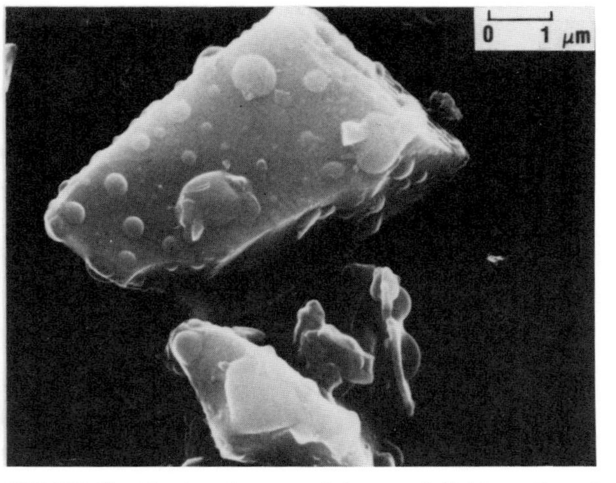

FIGURE 7b. Blocky phenocryst fragment (feldspar?) and shards. All pyroclasts are coated with droplets, which are most likely H_2SO_4 condensed on pyroclast surfaces in the eruption plume.

FIGURE 7c. Angular, blocky vitric pyroclasts. Curved surfaces are former vesicle walls.

FIGURE 7d. Fine-grained vitric ash coated with liquid droplets (most likely H_2SO_4).

MT. PELÉE, MARTINIQUE; MAY 8 AND 20, 1902 ERUPTIONS

Introduction: During the 1902 eruption of Mt. Pelée, Martinique, the town of St. Pierre was destroyed by nuées ardentes (LaCroix, 1904). Observations of this eruption led to some of the first studies of pyroclastic flows.

Recent stratigraphic and petrologic work on the deposits of the 1902 eruption have provided a clear picture of the eruption mechanism and characteristics of the Peléean sequence (Fisher et al., 1980; Fisher and Heiken, 1982). These eruptions followed periods of andesitic lava dome growth and mechanical disruption of the domes. Dense eruption columns collapsed and flowed outward through a cleft on the southwest side of the volcano. The pyroclastic flows separated into basal avalanches that left massive block and ash flow deposits and an overriding, expanding ash cloud that produced finer grained ash-cloud deposits (Fisher et al., 1980). The samples here are from the collection of R. V. Fisher.

Grain Size: Fisher and Heiken (1982) have sieved tephra samples from the Peléean deposits. Median grain sizes of the block and ash flows range from 0.43 to 0.36 mm (1.2 to 1.5 φ); ash cloud surges have median grain sizes of 0.38 to 0.1 mm (1.4 to 2.9 φ). Both types of deposits are poorly sorted.

Tephra Description: Most tephra in the devastating nuées ardentes consist of juvenile lithic fragments (43 to 81 vol.%). Most are blocky, tabular to irregular, hyalocrystalline andesite fragments consisting of 1-μm-wide, 2- to 3-μm-long feldspar and pyroxene microlites and a few feldspar phenocrysts (up to 30 μm long) in a glassy matrix (≃60%). Vesicularity is absent or very low (a few isolated, 1- to 10-μm irregular vesicles). A few pyroclasts have a diktytaxitic texture and about 10% vesicles.

There are traces of hydrothermally altered andesitic fragments in all samples; however, samples from gas pipes in block and ash flow deposits contain 18 to 35% of these fragments.

About 18 to 36 vol.% of the ash is composed of a crystal fraction; it is mostly plagioclase with smaller amounts of orthopyroxene, clinopyroxene, and Fe-Ti oxides. Most are cleanly fractured, and the forms range from splinters to euhedral crystals.

Vitric pyroclasts are present and range from splinter-shaped shards to small pumice fragments with 60 to 80% vesicularity. Pumice pyroclasts generally have thin-walled vesicles and ovoid vesicles 4 to 15 μm long. Vitric pyroclasts generally make up less than 7% of the deposits.

Chemical Composition: Table 1 shows the chemical analysis of a sample collected on deck of the ship *Roddam* in the harbor of St. Pierre on May 8, 1902.

TABLE 1. CHEMICAL COMPOSITION OF MT. PELÉE TEPHRA.*

	(Wt. %)
SiO_2	60.01
Al_2O_3	17.54
Fe_2O_3	2.88
FeO	4.30
MgO	2.26
CaO	6.80
Na_2O	3.41
K_2O	0.89
H_2O^+	0.30
H_2O^-	0.10
TiO_2	0.45
P_2O_3	0.13
MnO	0.23
Total	99.85

*Hillebrand, analyst (quoted in Diller, 1902).

Comments on Pyroclast Formation: Fisher and Heiken (1982) propose that nearly all pyroclasts in the Mt. Pelée pyroclastic flows are comminuted debris from the lava domes within the crater. Poorly vesiculated, juvenile (lithic) fragments were solidified, partly crystalline material from the dome. Vesicular glass shards and pumice are probably from molten or partly molten portions from the dome interior or from immediately below the dome. Explosive disruption and comminution of the dome could have been a combination of several processes: (1) overpressure within vesicles scattered throughout the dome during cooling and crystallization and (2) phreatomagmatic interaction of crater lake water and hot dome rocks. Fisher and Heiken (1982) note that there were heavy rains on May 7 and suggest that contact with ground water may have triggered the eruptions.

REFERENCES:

Diller, J. S., 1902. Volcanic rocks of Martinique and St. Vincent collected by Robert T. Hill and Israel C. Russell. Nat. Geogr. Mag., **13**: 285-296.

Fisher, R. V., Smith, A. L., and Roobol, M. J., 1980. Destruction of St. Pierre, Martinique by ash cloud surges, May 8 and 20, 1902. Geology, **8**: 472-476.

Fisher, R. V. and Heiken, G., 1982. Mt. Pelée, Martinique, May 8 and 20 pyroclastic flows and surges. J. Volcanol. Geotherm. Res., **13**, 339-371.

LaCroix, A., 1904. *La Montagne Pelée et ses eruptions.* Masson et Cie., Paris: 662 pp.

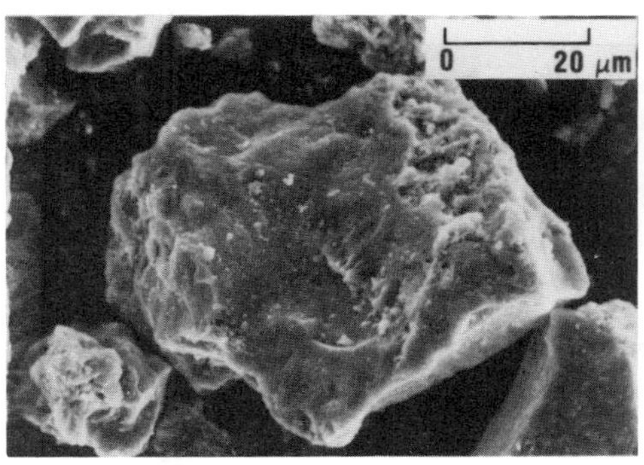

FIGURE 1a. Samples from near the base of block and ash flow deposits. Fine-grained fraction. SEM of hyalocrystalline, juvenile lithic fragments. Surfaces are irregular, with some conchoidal fractures. The equant, blocky clasts have no or rare vesicles.

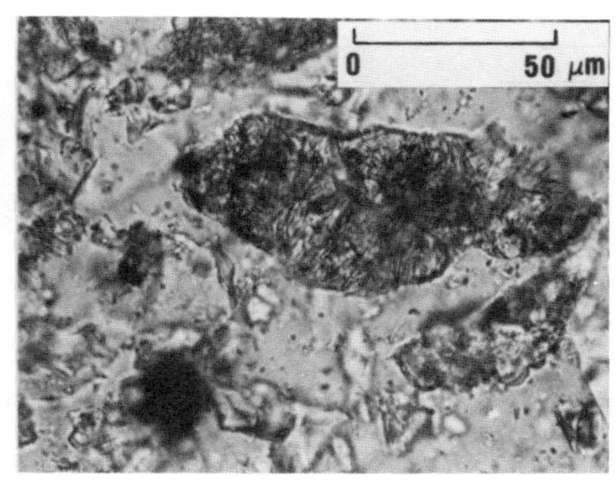

FIGURE 1b. Transmitted-light photomicrograph of mainly juvenile lithic fragments similar to those described in Fig. *2a*. Some 2- to 8-μm-wide, angular vitric shards are also visible. Microlites visible in glass groundmass are mainly feldspar.

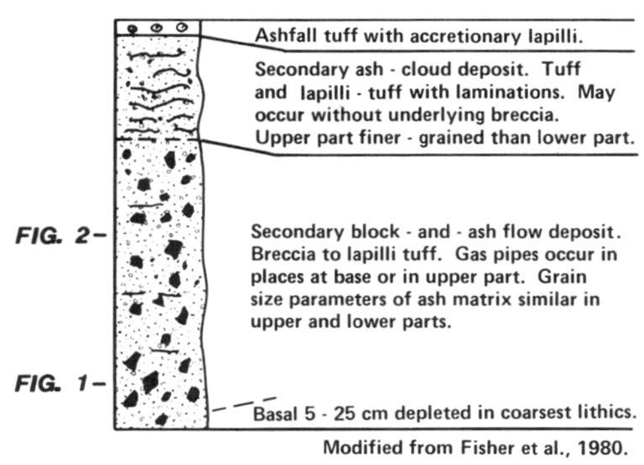

Depositional units from a Peléean ash cloud (from Fisher and Heiken, 1982).

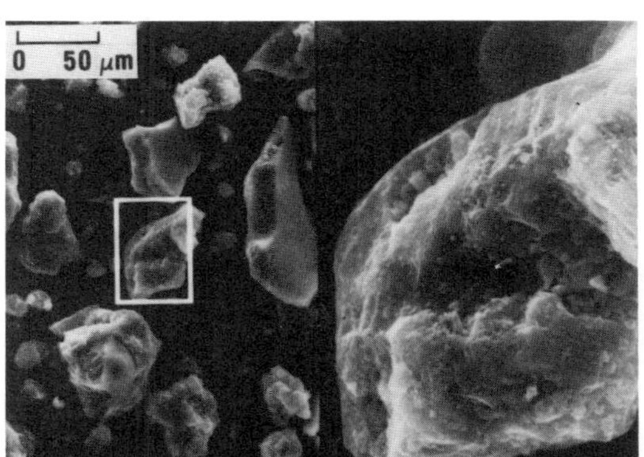

FIGURE 2a. Samples from the top of block and ash flow deposits. Fine-grained fraction. Mostly equant to slightly elongate lithic pyroclasts with rough, hackly grain surfaces. Some with a few vesicles have diktytaxitic textures. Smooth, flat pyroclasts (*right and top center*) are vitric shards. *Inset*—lithic pyroclast.

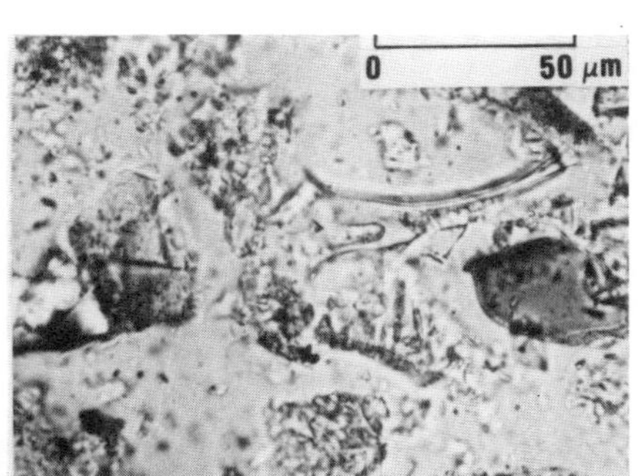

FIGURE 2b. Photomicrograph (transmitted light) of lithic ash. Mostly angular to subangular lithic and crystal pyroclasts. *Center right*—a delicate, elongate glass shard typical of the vitric fraction.

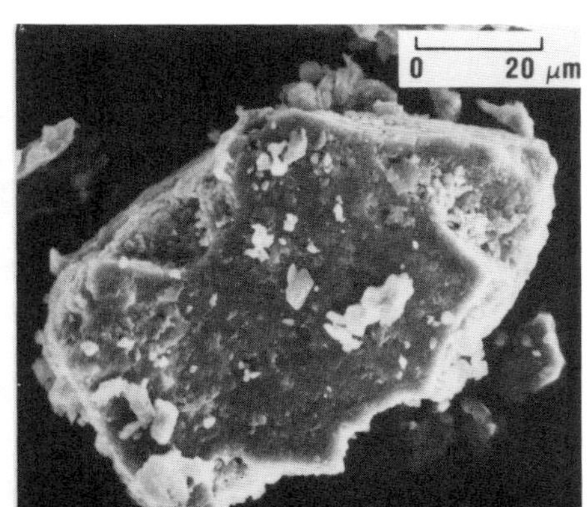

FIGURE 2c. Blocky, hyalocrystalline lithic pyroclast (juvenile). Smaller clasts on grain surface are small shards.

FIGURE 3. Sample from near the base of an ash-cloud deposit. Mostly equant, hyalocrystalline lithic pyroclasts; angular forms are bounded by rough fracture surfaces. Vesicles are rare and very small (<10 μm) when present. There are about 2% vitric shards that were broken from a pumiceous melt.

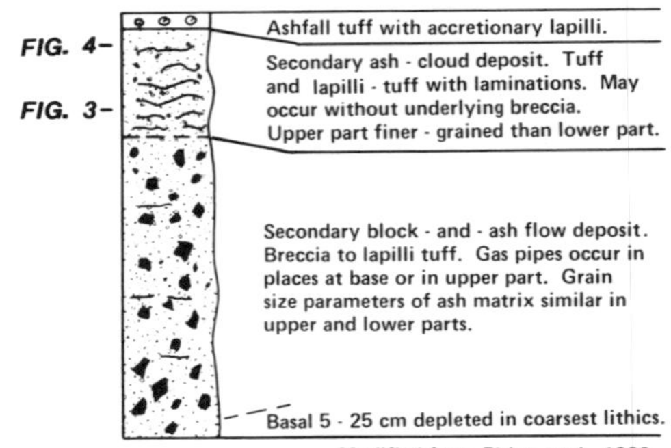

Modified from Fisher et al., 1980.

Depositional units from a Peléean ash cloud (from Fisher and Heiken, 1982).

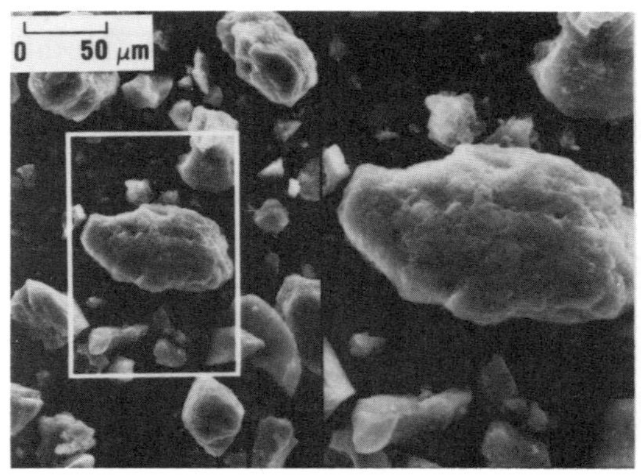

FIGURE 4a. Samples from the top of ash-cloud deposits; the upper part of this deposit is finer grained than the lower part (Fisher and Heiken, 1982). Equant to slightly elongate lithic fragments. *Inset*—clast with evidence of some rounding.

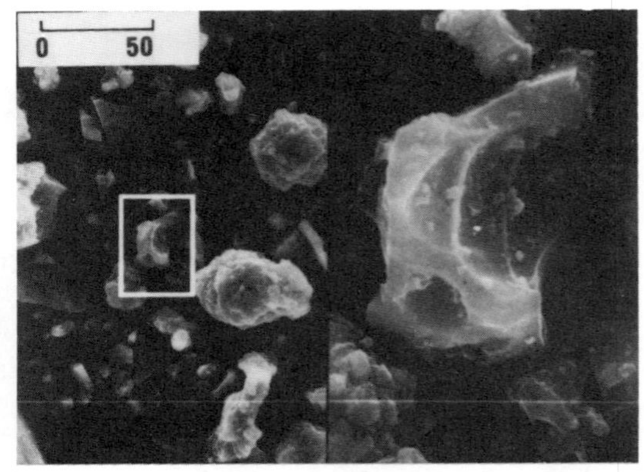

FIGURE 4b. Mostly equant lithic clasts; many of those with evidence of some rounding by abrasion. *Inset*—vesicular vitric shard. Ovoid vesicle remnants with thick vesicle walls.

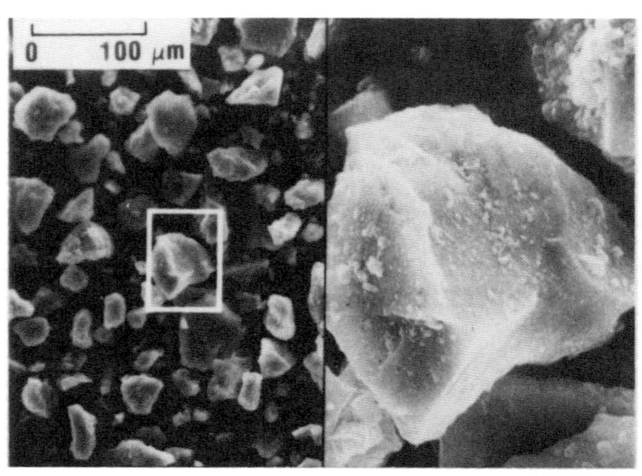

FIGURE 4c. Mostly 20- to 50-μm, equant, juvenile lithic and vitric pyroclasts. All are angular to subangular pyroclasts bounded by fracture surfaces. *Inset*—angular, juvenile lithic pyroclasts composed of hyalocrystalline andesite.

KRAKATAU, SUNDA STRAITS; 1883 ERUPTION

Introduction: Krakatau, an island volcano located in the Sunda Straits between Sumatra and Java, began erupting violently on May 20, 1883. The eruptions continued sporadically until the climactic eruption of August 27, 1883. Abundant tephra and volatiles were injected into the stratosphere and produced a worldwide haze and associated optical phenomena (Royal Society, 1888). Before 1883, the Krakatau group consisted of a 9- by 5-km island that included three overlapping volcanoes located along a fissure and two other islands that were remnants of the rim of a prehistoric caldera (Royal Society, 1883; Self and Rampino, 1981). Most of Krakatau Island was destroyed by caldera collapse during the eruption of about

20 km^3 of dacitic tephra (Self and Rampino, 1981). It has been proposed by Latter (1981) and Self and Rampino (1981) that tsunamis associated with this eruption were formed not by caldera collapse or submarine explosions, but by several cubic kilometers of pyroclastic flows entering the sea after each of several large explosions. The reader is referred to these two papers for information on the eruption sequence. Simkin and Fiske (1983) have prepared a comprehensive book on the eruption that includes eyewitness accounts from old manuscripts.

Grain Size: All the Plinian pumice fall deposits have mean grain sizes of 3 to 5 mm and are very poorly sorted (Self and Rampino, 1981). One of the pyroclastic surge deposits sampled has a 5-mm mean grain size and is also poorly sorted. The ignimbrite is finer grained, has a median of 0.2 mm, and is very poorly sorted.

Tephra Description: The Plinian pumice falls, associated with the intermittent explosive activity of May 20 to August 26, consist of mostly highly vesicular (65 to 70% vesicles) pumice pyroclasts. These are characterized by slightly elongate, flattened vesicles with thin (1- to 2-μm-thick) vesicle walls. There are many coalesced vesicles and some broken vesicle walls that have been rounded by viscous flow. Most of the finer grained pyroclasts are pumice.

Thin, possibly phreatomagmatic ash falls formed during the intermittent activity are interbedded with the pumice falls and pyroclastic surge deposits. The ash is bimodal, consisting of pumice pyroclasts in the 50- to 250-μm range and shards in the 5- to 100-μm range. Pumice vesicularities range from 40 to 70%. These are generally similar to pumices from the fall deposits but are finer grained.

An incipiently "welded" gray pumice fall that was deposited near-vent, just before eruption of the pyroclastic flows, is unique. It contains pumice pyroclasts with irregular, twisted forms whose vesicles are very well developed but curved, twisted, and flattened. Vesicle walls are 1 to 2 μm thick. Pyroclast surfaces are coated with small mounds and needle-shaped crystals (<1 μm wide); these may be sublimates. The coatings are largely responsible for cementing particles together.

Beginning at about 0530 on August 27, the first of several large explosions occurred. Self and Rampino (1981) suggest that each explosion produced a pyroclastic flow (ignimbrite). The ignimbrite sampled on Rakata Island consists of nonwelded, poorly sorted coarse ash and is a friable aggregate of pumice pyroclasts in a matrix of very fine grained (0.5- to 30-μm) shards. Many of the shards and pumice pyroclasts have fairly thick (15- to 30-μm) vesicle walls.

Chemical Composition: The 1883 ashes are dacitic; the phenocrysts, which are sparse, include plagioclase, augite, orthopyroxene and Fe-Ti oxides (Verbeek, 1886). For analyses of the pumices, see Table 1.

TABLE 1. MAJOR ELEMENT ANALYSES OF PUMICE CLASTS FROM KRAKATAU'S 1883 ERUPTION.*

	K2 (Wt. %)	K25 (Wt. %)
SiO_2	67.51	67.43
TiO_2	0.71	0.75
Al_2O_3	14.41	15.54
Fe_2O_3	4.02	3.91
MgO	1.14	0.76
MnO	0.14	0.15
CaO	2.65	2.92
Na_2O	5.26	5.16
K_2O	2.59	2.38
P_2O_5	0.10	0.09
Total	98.52	99.09

*S. Self, analyst.
K2 = gray pumice from the ignimbrite.
K25 = white pumice from the ignimbrite.

Comments on Pyroclast Formation: Nearly all the tephra were erupted during subaerial explosions—a fact noted in the field by Self and Rampino (1981). Nearly all eruption phases included pumice pyroclasts and shards that indicate complete vesicle growth; there are also high vesicularities and very thin vesicle walls. Variations in vesicle elongation and deformation indicate some variation in volatile content and/or viscosity throughout all eruption phases. The most unique pyroclasts are found in the near-vent "welded" pumice fall; twisted, deformed, very angular pumices indicate slightly lower viscosities and greater turbulence within the vesiculating magma.

The ignimbrite is strongly bimodal; it consists of pumice blocks in a poorly sorted ash matrix (Self and Rampino, 1981). The ash matrix is mostly very fine ash containing small, angular shards. It is possible that there was some magma/water interaction during this phase.

REFERENCES:

Latter, J., 1981. Tsunamis of volcanic origin: summary of causes, with particular reference to Krakatau, 1883. Bull. Volcanol., **44**: 467-490.

Royal Society of London, Krakatau Committee, 1888. *The Eruption of Krakatau and Subsequent Phenomena.* Trübner and Co., London: 494 pp.

Self, S. and Rampino, M., 1981. The 1883 eruption of Krakatau. Nature, **294**: 699-704.

Simkin, T. and Fiske, R.S., 1983. *Krakatau 1883—The volcanic eruption and its effects.* Smithsonian Inst. Press, Washington: 464 pp.

Verbeek, R. D. M., 1886. *Krakatau.* Imprimerie de l'Etat Batavia.

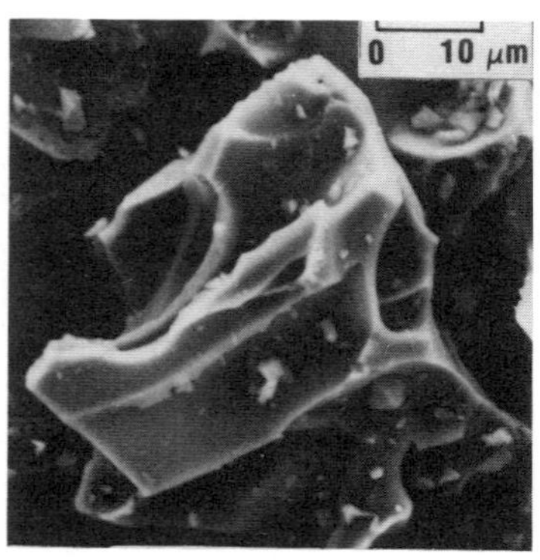

FIGURE 1a. SEM of pyroclasts from the coarsest Plinian phase. Collected on Rakata Kecil Island. Example of pyroclasts from the <62-μm fraction, which consists of mostly pumice pyroclasts in the 20- to 40-μm range and curved angular shards in the 5- to 10-μm range. Most are fragments of 1- to 6-μm-thick vesicle walls.

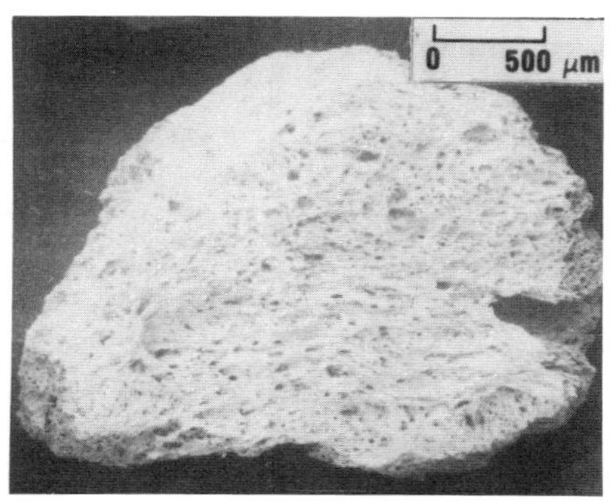

FIGURE 1b. Subrounded pumice pyroclast with about 65 to 70% vesicularity. Vesicles are well developed but were only slightly elongated by flow before eruption.

FIG. 1—

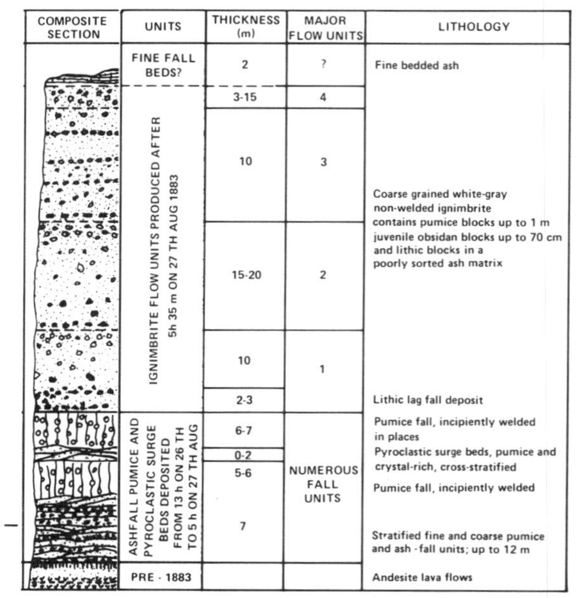

Composite section through the 1883 pyroclastic deposits of Krakatau (from Self and Rampino, 1981). All samples were collected by S. Self, University of Texas at Arlington.

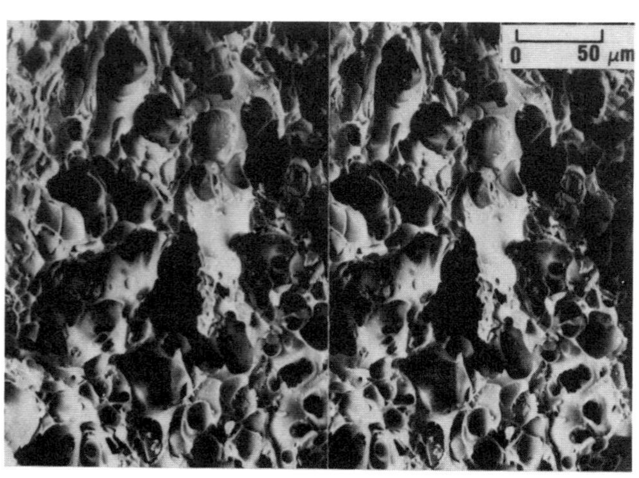

FIGURE 1c. Stereomicrographic detail of a pumice pyroclast. Most of the vesicles are slightly elongate and flattened, which produces parallel flattened ovoids. There are many coalesced vesicles with septae on vesicle walls (remnants of vesicle walls); some septae were rounded by viscous flow. Vesicle walls are mostly 1 to 2 μm thick.

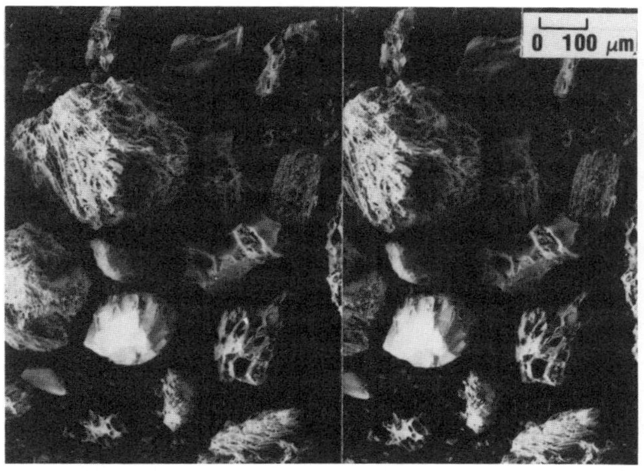

FIGURE 2a. Finest grained ash fall of phreatomagmatic origin. Collected on Rakata Kecil Island. Stereomicrograph of the tephra. The bimodality of this ash is evident; it consists of equant to slightly elongate pumice pyroclasts in the 50- to 250-μm range and glass shards in the 5- to 100-μm range. Most of the pumices are subangular to subrounded with vesicularities of less than 40 to 70%; some of this variation in vesicularity can occur within a single pyroclast. The equant pyroclast at center is a phenocryst covered with vesicular glass (bubble-wall texture).

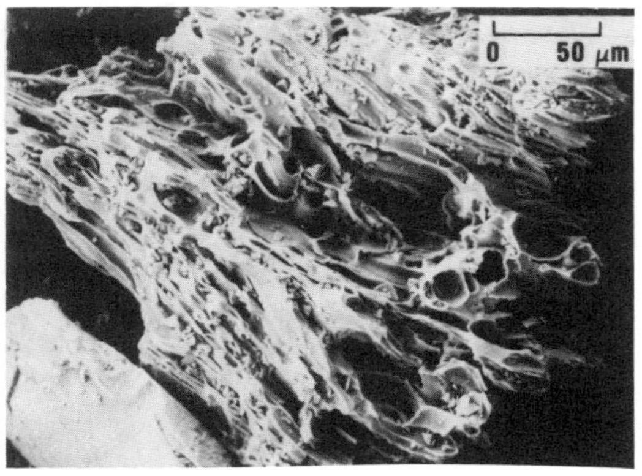

FIGURE 2b. Pumice pyroclast showing variation in vesiculation, coalescence of vesicles, and vesicle elongation. Vesicle-wall thicknesses vary from 1 to 8 μm.

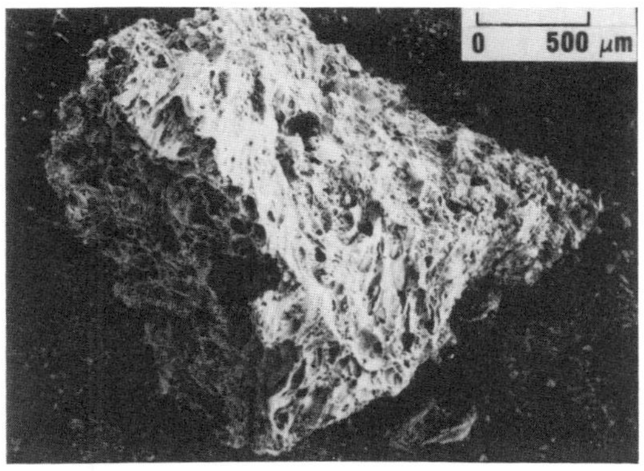

FIGURE 3a. "Welded" gray pumice fall collected on Rakata Kecil Island. Irregular, slightly elongate, ragged pumice pyroclast with a vesicularity of about 70%. Vesicles are curved, twisted, and flattened; vesicle walls are 1 to 2 μm thick. Many small particles are clinging to the pyroclast surfaces.

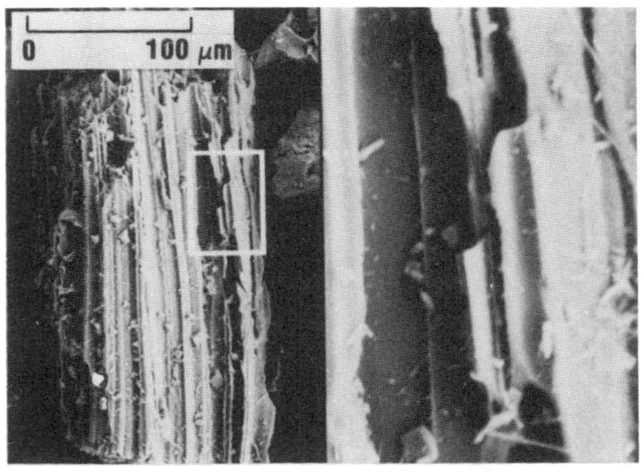

FIGURE 3b. A less common form of pumice pyroclast in this deposit has straight, parallel vesicles with thin walls. In the detailed inset, straight and slightly curved rods are visible. These are common on all pyroclasts in this deposit.

FIG. 3 —
FIG. 2 —

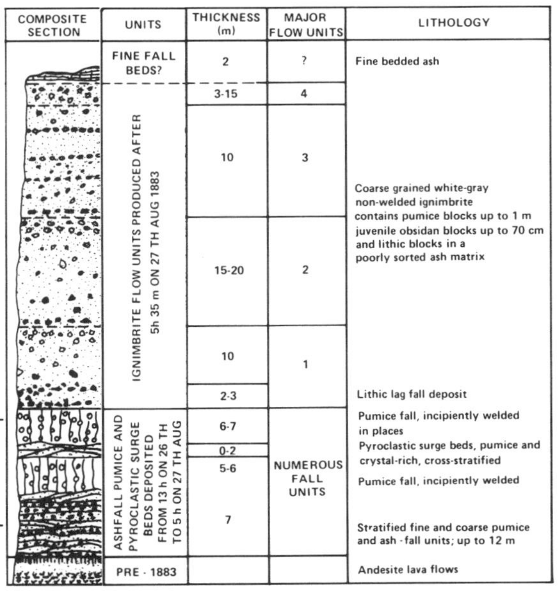

Composite section through the 1883 pyroclastic deposits of Krakatau (from Self and Rampino, 1981). All samples were collected by S. Self, University of Texas at Arlington.

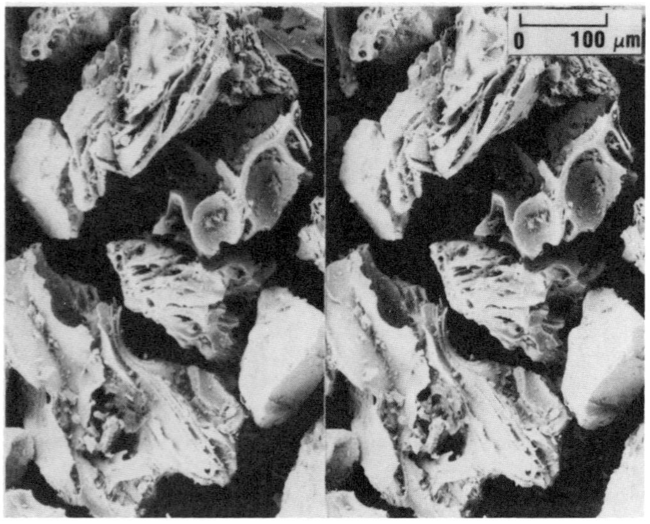

FIGURE 3c. Stereomicrograph of pyroclasts that are typical of this deposit. The most common pyroclasts are ragged, twisted, angular pumices. Vesicles are well developed (≃70% vesicularity); most are flattened ovoids separated by 2- to 20-μm-thick vesicle walls. Pyroclast surfaces are coated with 1-μm-wide, low mounds (sublimates?) and whiskers, 0.5 to 1 μm in diameter and 20 to 30 μm long. Clumps of these "whiskers" (zeolites?) bond some pyroclasts. It is probable that particle agglomeration was accomplished through cementation by secondary minerals and not through welding; perhaps it happened during release of volatiles from the cooling deposit.

FIGURE 4a. Ignimbrite collected on Rakata Island. It is an aggregate composed of flat, curved, and irregular glass shards in a fine-grained matrix. Shards as small as 0.5 μm are present, some with visible broken walls (septae).

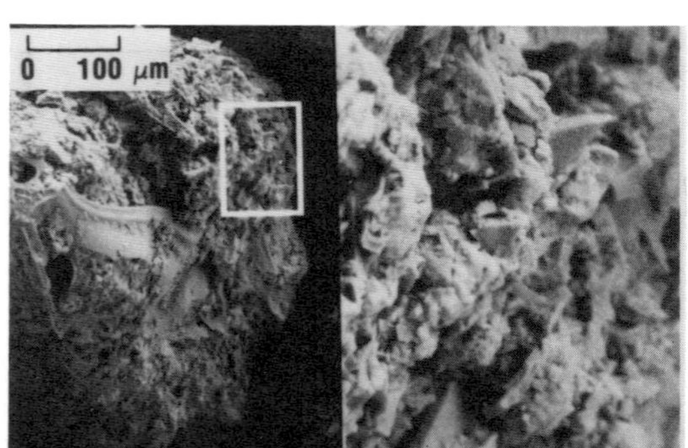

FIGURE 4b. Larger aggregate than the example in Fig. *4a* was broken and rounded by sieving of the tephra. A pumice pyroclast is visible in the aggregate and is characterized by well-developed, elongate vesicles and vesicle walls 15 to 30 μm thick. The inset is a closeup of the aggregate; there are small, curved and blocky equant shards in a very fine-grained matrix.

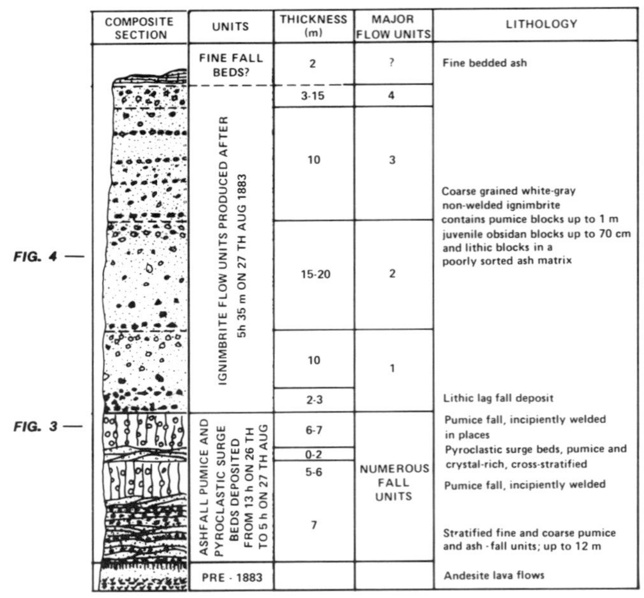

Composite section through the 1883 pyroclastic deposits of Krakatau (from Self and Rampino, 1981). All samples were collected by S. Self, University of Texas at Arlington.

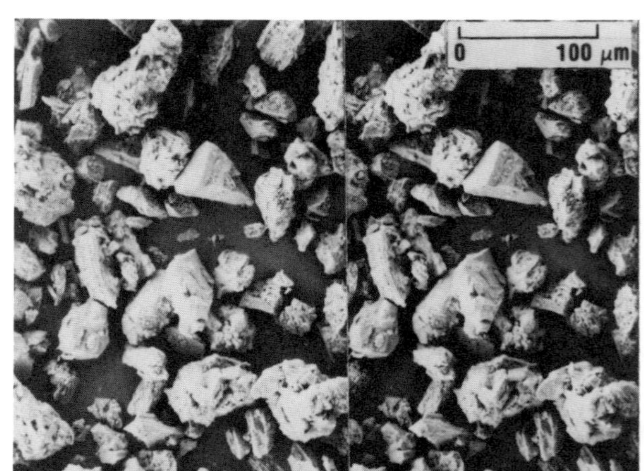

FIGURE 4c. Stereomicrograph of the <70-μm fraction. The most common particle type is that of an aggregate of fine ash (probably a relict of sieving and incomplete disaggregation). Shards are similar to those described in Figs. *4a* and *b*. Small pumice pyroclasts, equant, blocky shards, and crystal fragments are also present.

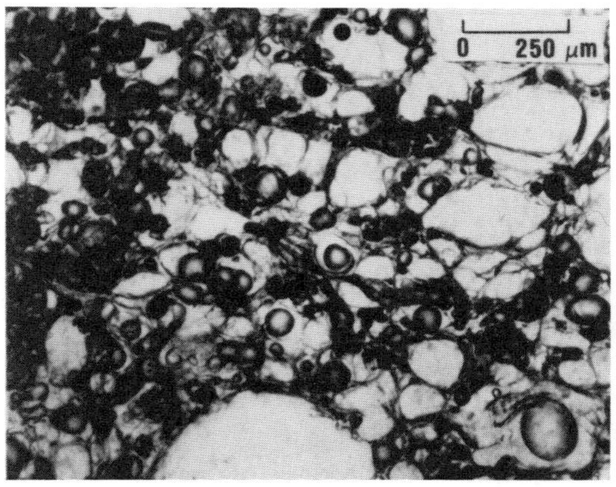

FIGURE 5a. Pumice pyroclasts from an ignimbrite collected on Rakata Island. Transmitted-light photomicrograph of a white pumice from the ignimbrite. There is a heterogeneous distribution of spherical to ovoid (deformed) vesicles in the 15- to 80-μm range. Some of the larger vesicles are in the 300- to 400-μm range. Vesicle walls, consisting of colorless glass, are 3 to 25 μm thick. There are 1% phenocrysts that consist of clinopyroxene, plagioclase, and hypersthene.

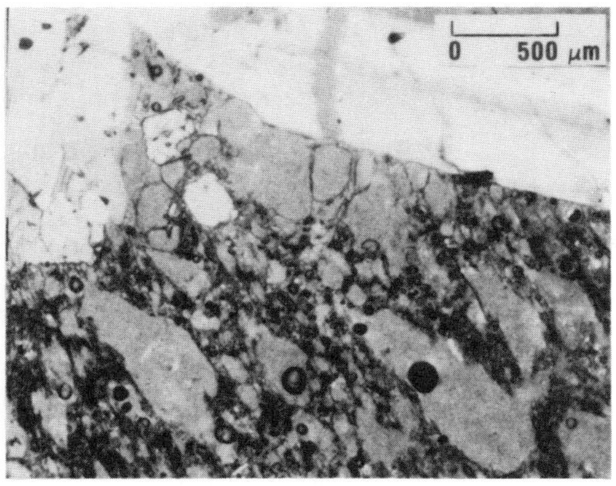

FIGURE 5b. Transmitted-light photomicrograph (partly crossed polarizers) of grey pumice from the ignimbrite. There are 5% phenocrysts and 95% colorless glass and vesicles. Phenocrysts are present as separate crystals and glomeroporphyritic clots of plagioclase, hypersthene, and Fe-Ti oxides. As can be seen here, vesicles have developed radially around the phenocrysts.

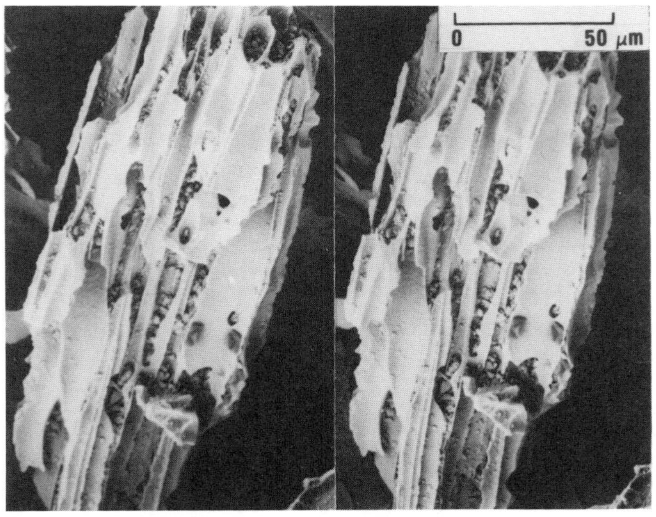

FIGURE 6a. Ash fall tephra collected on the vessel *Herschel*, in the Sunda Straits, just before the climactic phase of the eruption. These are similar to the Plinian deposits. Sample was provided by the U.S. Natural History Museum, Smithsonian Institution. Stereomicrograph of a pumice pyroclast. This pyroclast is characterized by high vesicularity and elongate vesicles. Vesicle aspect ratios range from 4:1 to 20:1.

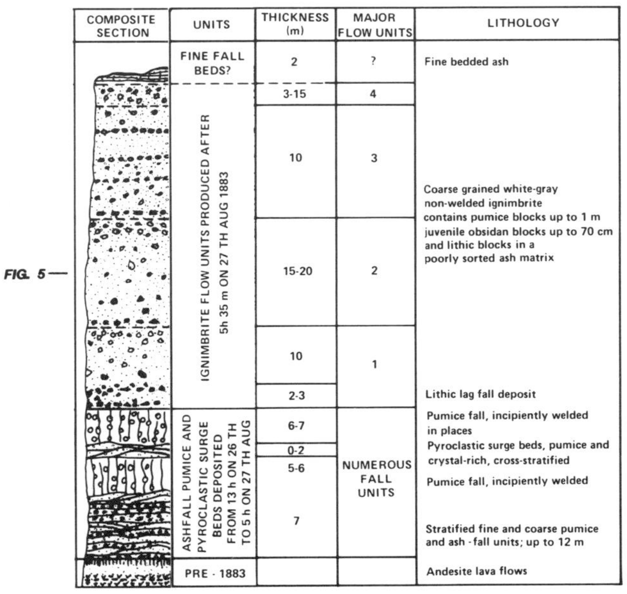

Composite section through the 1883 pyroclastic deposits of Krakatau (from Self and Rampino, 1981). All samples were collected by S. Self, University of Texas at Arlington.

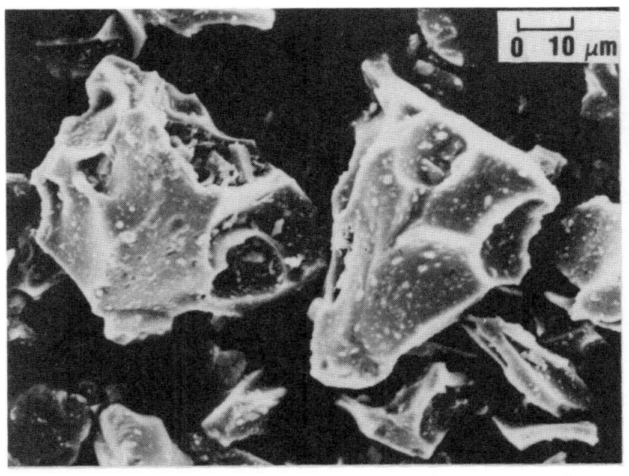

FIGURE 6b. An example of the fine-grained fraction of this tephra. It consists of mostly 10- to 70-μm-long shards and small pumice pyroclasts. Shards were broken from thick vesicle walls.

FIGURE 6c. Variety of pumice pyroclasts and thick-walled shards. Vesicles range from small ovoids to elongate, twisted tubes. Vesicle-wall thicknesses vary greatly within individual pumice pyroclasts.

MINOAN TUFF, THIRA, GREECE

Introduction: The Minoan eruption of about 1400 B.C. produced between 13 and 18 km^3 (dense rock equivalent) of rhyodacitic tephra (Watkins et al., 1978). The eruption sequence shows development from magmatic to phreatomagmatic eruption processes and back again to a mixture of both processes.

The eruption began at a vent located above sea level, where it produced a coarse-grained pumice deposit. This was followed by two later phreatomagmatic phases of the eruption from: (1) vents that propagated into an older, flooded caldera and (2) a collapsing Minoan caldera flooded by the sea. The final eruptive phase originated above sea level but may have been a mixture of magmatic and phreatomagmatic processes; during this phase, hot pyroclastic flows deposited a mixture of pumice pyroclasts and finely comminuted shards (Heiken and McCoy, 1984).

Grain Size: Samples from the Minoan tephra sequence were sieved by Bond and Sparks (1976). Plinian pumice deposits (unit 1) are poorly sorted and have a median grain size of 10 to 17 mm. Units 2 and 3 (phreatomagmatic deposits) are much finer grained and their median grain sizes range from 100 μm to 2 mm; all are poorly sorted. These deposits are all bimodal and consist of pumice bombs and lapilli in a very fine grained matrix. The ignimbrites are all very poorly sorted and have median grain sizes of 160 to 700 μm.

Tephra Description: Pumice pyroclasts (from unit 1) are mostly quite large (2- to 5-cm-long), angular to subangular forms. The pumices are characterized by bimodal vesicularity; pockets of large, coalesced ovoid vesicles are surrounded by highly elongate, tube-like vesicles. Vesicle walls are very thin. Within the upper part of unit 1 is a thin phreatomagmatic layer consisting of mostly fine-grained shards.

Deposits of units 2 and 3 are quite different. Unit 2 is characterized by plane beds and large base surge dunes, and unit 3 is massive and contains a high percentage of large lithic fragments. The tephra throughout these deposits are all similar, consisting of pumice lapilli and fragments in a matrix of fine white ash. Within the matrix, shards outnumber pumice pyroclasts 8:2. Shards are fine grained and flat to slightly curved with sharp, angular edges. There is little evidence of abrasion or edge modification on these particles. Most represent broken vesicle walls, 5 to 10 μm thick. Within the upper part of unit 3, small pumices and shards show decreasing vesicularity and thicker vesicle walls.

Within the ignimbrite there are 200- to 300-μm, subequant pumice pyroclasts; all are highly vesicular with well-developed, elongate vesicles. Some highly inflated pumices with polygonal vesicles and curved and flat shards with angular edges are also present. Most shards are in the 5- to 30-μm range.

Chemical Composition: Günther and Pichler (1973) published whole-rock chemical analyses of the Minoan tephra sequence. Analyses of the rhyodacitic tephra are shown in Table 1.

Comments on Pyroclast Formation: The most obvious change during this eruption sequence is from predominantly pumice pyroclasts in the coarse-grained Plinian deposits (shard-to-pumice ratio of 1:9) to predominantly shards in the phreatomagmatic deposits (shard-to-pumice ratio of 8:2).

All phases of the eruption demonstrate that vesicle growth was complete. The high vesicularities are characteristic of the Plinian pumices. Shards in later phases were broken from a vesiculated melt during magma/water interaction and produced thin, curved and flat, fine-grained, angular shards. Comminution was not complete, however, as a minor proportion of small pumice pyroclasts are present in all samples.

The last eruptive phase, which produced an ignimbrite, is a bimodal mixture of pumice pyroclasts and finely comminuted shards. Grain surfaces are coated with finer grained shards. It is possible that the very fine grained shards may have been produced by minor phreatomagmatic interaction during eruption of the ignimbrite.

TABLE 1. CHEMICAL ANALYSES OF PUMICE PYROCLASTS FROM THE MINOAN TUFF.*		
	Unit 1 (pumice fall) (Wt. %)	Unit 3 (phreatomagmatic) (Wt. %)
SiO_2	68.5	69.12
Al_2O_3	14.2	13.93
Fe_2O_3	1.0	0.72
FeO	1.9	1.92
MnO	0.08	0.08
MgO	0.05	0.58
CaO	2.2	2.04
Na_2O	4.5	4.23
K_2O	2.8	3.01
TiO_2	0.4	0.41
H_2O	3.9	3.7
Total	99.61	99.36

*From Günther and Pichler (1973).

REFERENCES: Bond, A. and Sparks, R., 1976. The Minoan eruption of Santorini. J. Geol. Soc. London, **132**: 1-16.

Günther, D. and Pichler, H.. 1973. Die obere und untere bimsstein-folge auf Santorin. N. Jb. Geol. Palaönt. Mh. H., **7**: 394-415.

Heiken, G. and McCoy, F., Jr., 1984. Caldera development during the Minoan eruption, Thira, Cyclades, Greece. J. Geophys. Res., **89**: 8441-8462.

Watkins, N., Sparks, R., Sigurdsson, H., Huang, T. C., Federman, A., Carey, S., and Ninkovich, D., 1978. Volume and extent of the Minoan tephra from Santorini volcano: New evidence from deep-sea sediment cores. Nature, **271**: 122-126.

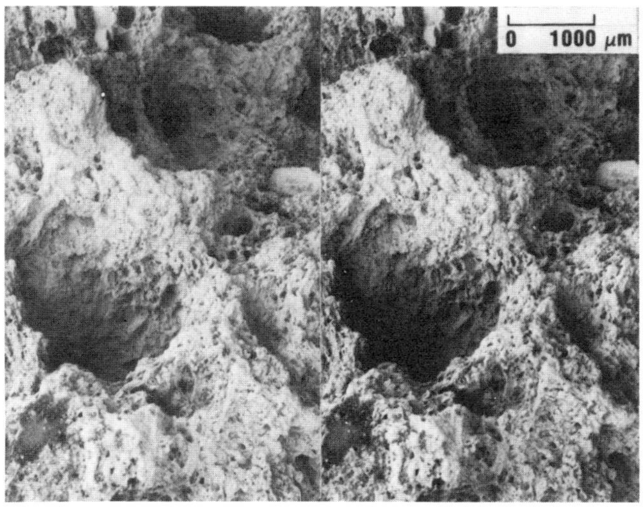

FIGURE 1a. Tephra from the Plinian pumice deposit (unit 1). SEM stereomicrograph of a typical pumice pyroclast. Lapilli-size and larger pumices are slightly elongate, blocky forms with surfaces broken by pockets of coalesced vesicles. The pockets were formed during heterogeneous vesicle growth throughout the vesiculated magma. Most of the vesicles are highly elongate and have thin vesicle walls. The tube-like vesicles are parallel to long axes of the pyroclasts; some are bent around larger cavities. The cavities, or pockets, are patches of highly inflated, coalesced vesicles. Most pockets are adjacent to pyroxene or feldspar phenocrysts.

FIG. 1 —

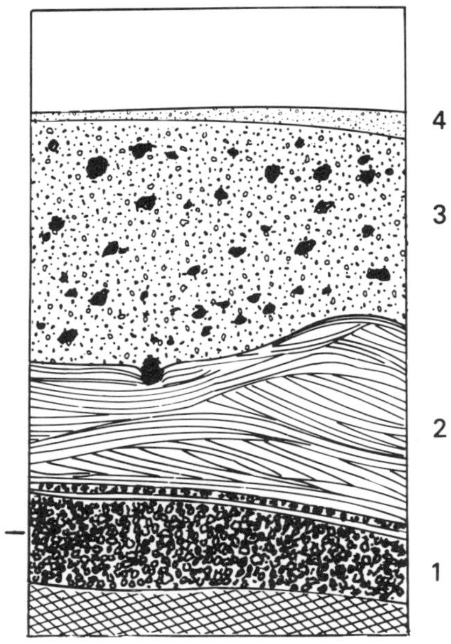

Composite section of the Minoan tephra sequence that was erupted in ≃1400 B.C. Unit 1 is a coarse-grained Plinian pumice fall; the thin bed near the top of this deposit is a fine-grained layer of phreatomagmatic origin. Unit 2 is a phreatomagmatic deposit characterized by large base surge dunes and plane beds. Unit 3 is a massive, lithic-fragment-rich phreatomagmatic deposit. Unit 4 is a nonwelded ignimbrite.

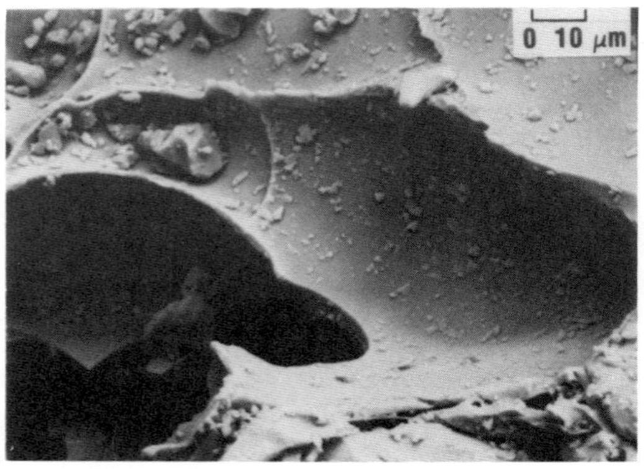

FIGURE 1b. Detail of coalesced vesicles in a pocket of the larger vesicles. Remnants of ovoid, coalesced vesicles are visible as septae or vesicle walls. Wall thicknesses vary from 1 to 5 μm.

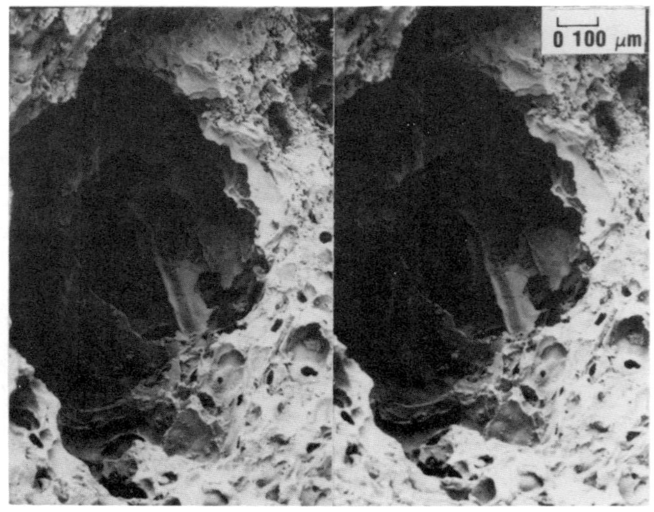

FIGURE 1c. Stereomicrograph of coalesced larger vesicles that illustrates the heterogeneity of these pumice pyroclasts.

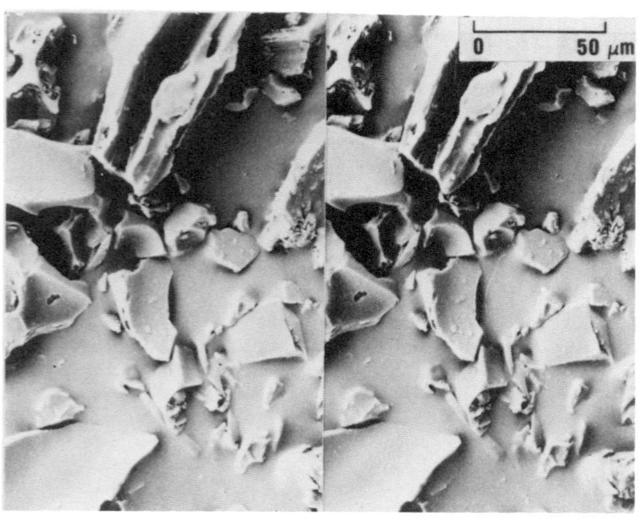

FIGURE 2a. Stereomicrograph of the tephra from a fine-grained phreatomagmatic layer within the upper part of the pumice fall. There are mostly broadly curved shards with very angular edges. Most were broken from a pumiceous magma with 50- to 70-μm-wide, flattened vesicles and 5- to 10-μm-thick vesicle walls. About 85% of the tephra consists of shards and 21% is pumice pyroclasts. The fine-grained nature (most shards are in the 20- to 40-μm range) and abundance of angular shards are typical of a highly vesicular melt that was fragmented during melt/water interactions.

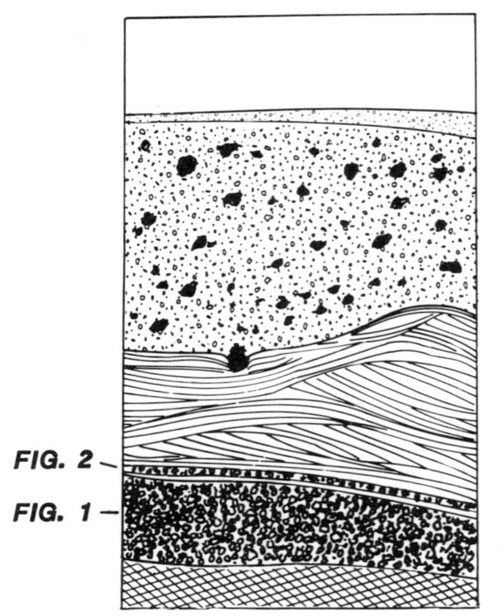

Composite section of the Minoan tephra sequence that was erupted in ≃1400 B.C. Unit 1 is a coarse-grained Plinian pumice fall; the thin bed near the top of this deposit is a fine-grained layer of phreatomagmatic origin. Unit 2 is a phreatomagmatic deposit characterized by large base surge dunes and plane beds. Unit 3 is a massive, lithic-fragment-rich phreatomagmatic deposit. Unit 4 is a nonwelded ignimbrite.

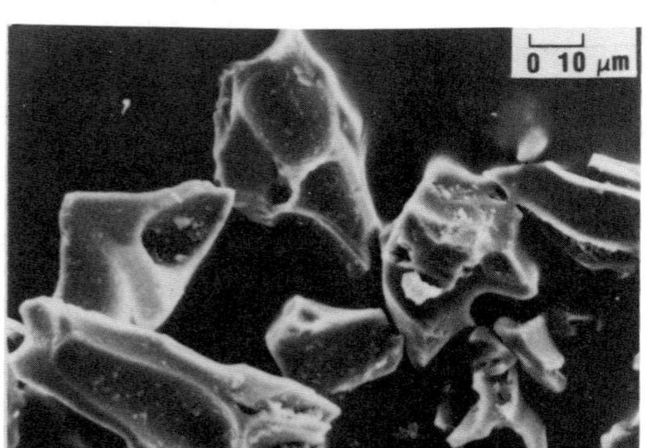

FIGURE 2b. Fine-grained shards typical of this layer.

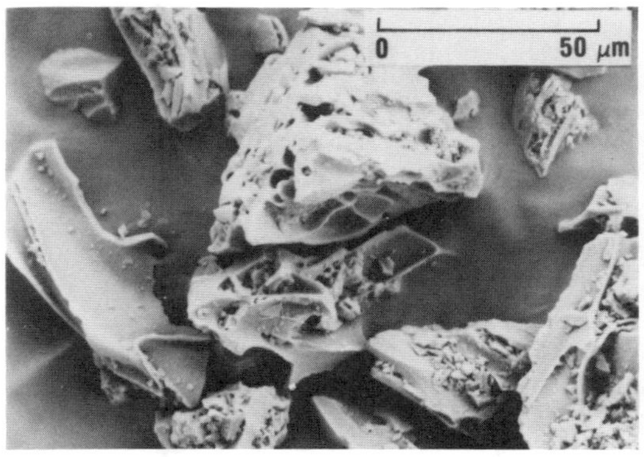

FIGURE 3a. Phreatomagmatic ash from unit 2 (not including the pumice lapilli and lithic fragments scattered through this ash matrix). The pyroclasts are very similar to those in Fig. 2. Fine-grained ash that consists of mostly angular shards (75%) and small, angular pumice pyroclasts (25%). Most shards are broken from coalesced vesicles; thin, tapered ridges are remnants of vesicle walls broken during coalescence.

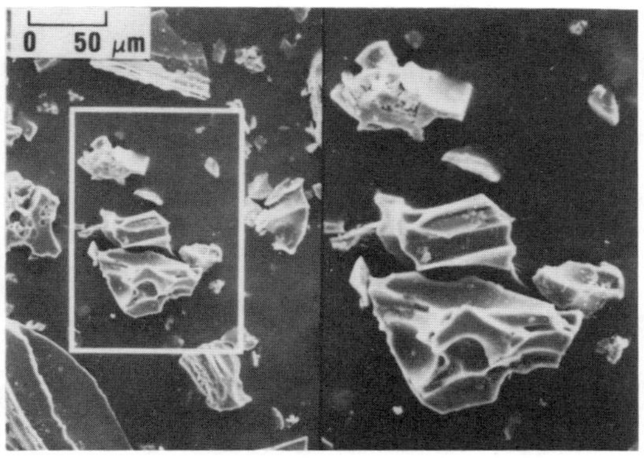

FIGURE 3b. Representative sample of shards and small pumice pyroclasts that are typical of this phreatomagmatic deposit.

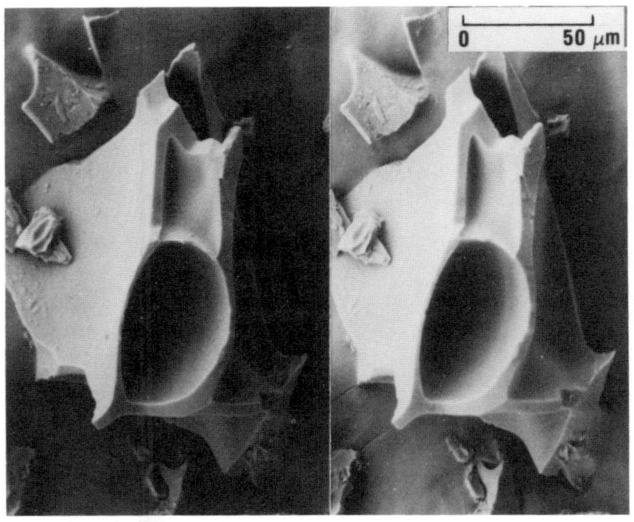

FIGURE 3c. Stereomicrograph of an angular, curved shard. Multiple vesicle remnants are visible. There is little or no abrasion of delicate edges on these pyroclasts.

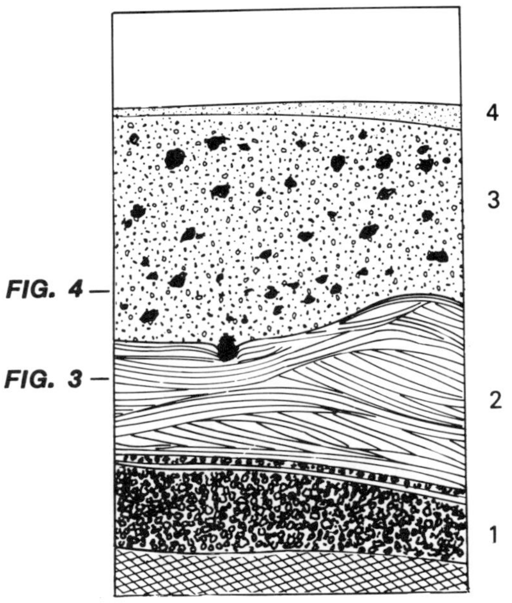

Composite section of the Minoan tephra sequence that was erupted in ≃1400 B.C. Unit 1 is a coarse-grained Plinian pumice fall; the thin bed near the top of this deposit is a fine-grained layer of phreatomagmatic origin. Unit 2 is a phreatomagmatic deposit characterized by large base surge dunes and plane beds. Unit 3 is a massive, lithic-fragment-rich phreatomagmatic deposit. Unit 4 is a nonwelded ignimbrite.

FIGURE 4a. Matrix sample from the base of unit 3. Lithic fragments and pumice bombs, which make up 5 to 40% of this unit, are not included in this analysis. There are mostly (87%) slightly curved shards with angular edges. Small pumices make up about 13% of the matrix sample studied here. Vesicle walls are 5 to 20 μm thick.

FIGURE 4b. A closer look at some of the smaller shards in Fig. *4a*.

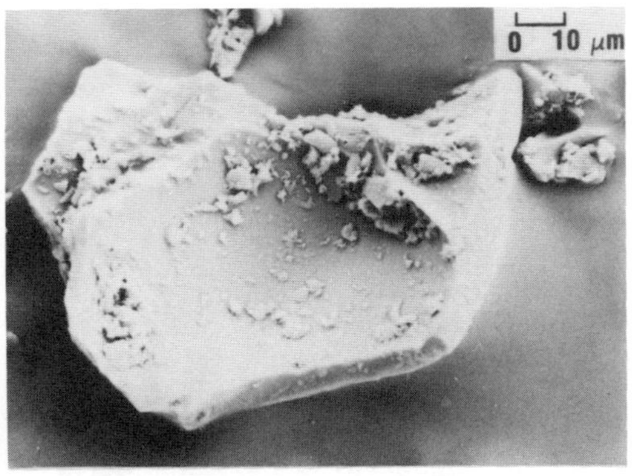

FIGURE 4c. A curved shard typical of products from finely comminuted vesiculated magma. Angular grain edges are not abraded.

FIGURE 5a. Matrix sample from the uppermost part of unit 3. Small, slightly elongate to equant shards and pumice pyroclasts. The shard-to-pumice ratio (8:2) is nearly the same as that of a sample from the base of unit 3 (Fig. *4*), although the vesicularity of the pumice pyroclasts appears to be lower. Most vesicle walls are 5 to 10 μm thick. Smaller shards are mostly equant to slightly elongate.

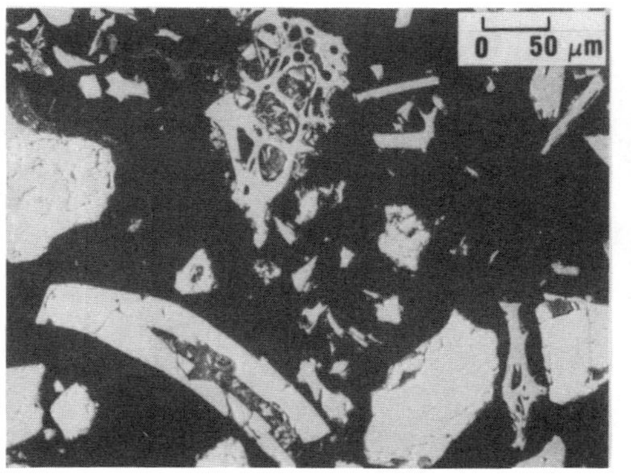

FIGURE 5b. Polished thin section of the coarser fraction from this sample. Nearly all the pyroclasts are glass; well-developed pumice (*top*) to coarse angular shards. Note the angularity of most of the smallest shards. A rounded lithic fragment (lava) is visible at the left of the photograph.

FIGURE 5c. Single angular pumice fragment. This finely comminuted pyroclast is typical of the phreatomagmatic phase of the eruption. Minor abrasion is visible only on the thinnest edges.

FIGURE 6a. Ignimbrite sample from unit 4. Bimodal distribution of pyroclasts is visible in this SEM. Subequant pumice pyroclasts are in the 200- to 300-μm range; most are highly vesicular with highly elongate vesicles and 3- to 15-μm-thick vesicle walls. There are also a few highly inflated pumices (*upper right*) with flat (polygonal) vesicle walls and some small, rounded lithic fragments (*center bottom*). Angular, curved and flat shards make up the 5- to 30-μm fraction (mostly about 20 μm long).

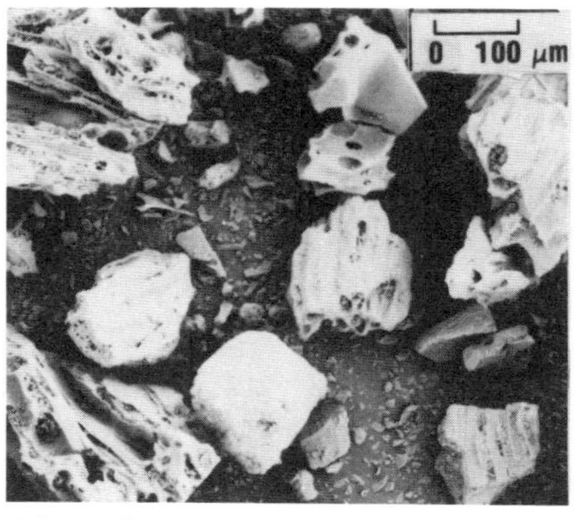

FIGURE 6b. View of the finer grained fraction, which is made up of equant, angular shards. Note that grain surfaces are coated with fine-grained particles (typical of many ignimbrites).

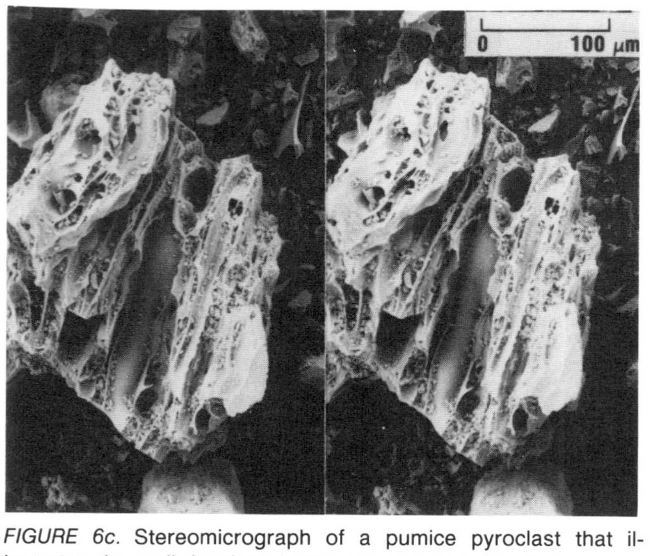

FIGURE 6c. Stereomicrograph of a pumice pyroclast that illustrates the well-developed elongate vesicles. There are some pockets (*center*) of coalesced vesicles. Note the thin vesicle walls.

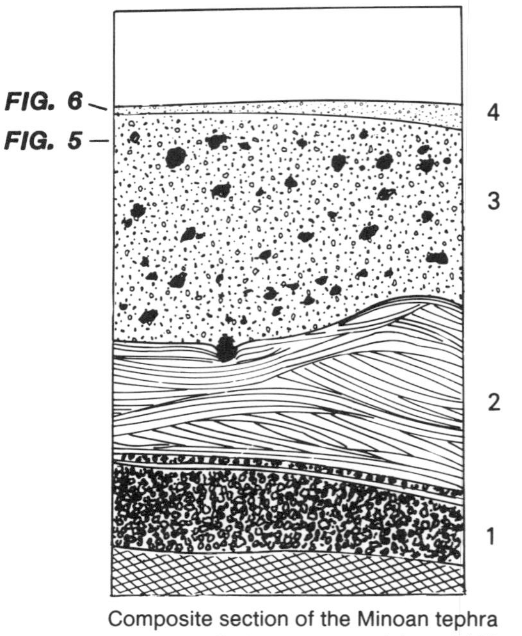

Composite section of the Minoan tephra sequence that was erupted in ≃1400 B.C. Unit 1 is a coarse-grained Plinian pumice fall; the thin bed near the top of this deposit is a fine-grained layer of phreatomagmatic origin. Unit 2 is a phreatomagmatic deposit characterized by large base surge dunes and plane beds. Unit 3 is a massive, lithic-fragment-rich phreatomagmatic deposit. Unit 4 is a nonwelded ignimbrite.

TAAL VOLCANO, PHILIPPINES; DRY PHREATOMAGMATIC ERUPTION

Introduction: Taal Volcano is located in Lake Taal, about 60 km S of Manila, Philippines. Continuous activity has been recorded there over the last several hundred years and up to recent activity in 1976 (Wolfe, 1980a). The activity of 1965 provided the first well-documented accounts of volcanic base surges, a type of pyroclastic surge (Moore et al., 1966). Surges that erupted from a vent on an island in the lake sandblasted objects as far as 6 km away. Approximately 9.0×10^4 kg of ash was ejected from the vent, of which only 10% was new basalt; the rest was older basaltic ash, cinders, and blocks from the new crater (Moore, 1967).

The stratigraphy of the 1965 deposits (Moore, 1967) supports witnessed accounts of the activity. Opening eruptions deposited ash falls of juvenile materials, including some spatter and lapilli. The second stage of eruption was marked by phreatomagmatic explosions and the excavation of a new crater when lake water gained access to the vent. Surge deposits resulting from this second stage of activity are breccias that are crudely stratified near the vent and well bedded away from the vent. The eruptions decreased in energy during the final stages and resulted in ash fall deposition and formation of a cinder cone within the earlier tuff ring.

Grain Size: Samples of Taal surge tephra from 1965 eruptions (collected by M. Sheridan) are dark gray, fine to medium ash. Sandwave beds (Table 1) are characterized by an abundant (over 35 wt.%) fine-ash fraction. Waters and Fisher (1971) give size data that show decreasing median grain size in sandwave beds farther from the vent, which possibly indicates a decrease in the surge's kinetic energy. The plotted grain-size groupings by bed form suggest different surge transport and depositional processes for each (Sheridan and Updike, 1975).

TABLE 1. GRAIN-SIZE CHARACTERISTICS OF TAAL TEPHRA.*

Sample	Bed Form*	Md_φ (mm)	σ_φ	Fine Ash (Wt. %)
1	M	0.65 (0.64)	1.42	2.53
2	S	3.10 (0.12)	2.14	36.07
3	S	2.93 (0.13)	1.99	28.42
4	M	0.51 (0.70)	1.98	4.02
5	S	2.37 (0.19)	1.57	15.39
6	S	2.08 (0.24)	1.20	8.19
7	P	0.33 (0.79)	1.55	3.01
8	M	1.12 (0.46)	2.35	12.07
9	P	−0.73 (1.66)	1.67	2.74
10	S	3.60 (0.08)	1.77	37.80
11	P	0.13 (0.91)	2.18	5.82
12	S	1.93 (0.26)	1.70	13.40
13	M	1.13 (0.46)	2.48	13.28
14	P	−0.57 (1.48)	1.28	0.33
15	M	1.35 (0.39)	1.60	6.40
16	M	1.40 (0.38)	1.40	4.71
17	M	0.53 (0.69)	1.62	2.32
18	S	1.48 (0.36)	1.36	4.88
19	M	1.06 (0.48)	1.47	3.31
20	S	2.98 (0.13)	1.83	28.80
21	S	2.67 (0.16)	1.93	23.26
22	S	2.33 (0.20)	1.96	18.33
23	M	0.96 (0.51)	2.17	9.68

*M. Maisanno and M. Sheridan, analysts. Bed forms: S = sandwave, M = massive surge, P = planar surge.

Tephra Description: The 1965 Taal ash is composed of: (1) fresh, dark gray sideromelane of basaltic andesite composition; (2) slightly altered to fresh, dark brown tachylite; and (3) plagioclase, olivine, and clinopyroxene crystals. The sideromelane is juvenile and composes up to 90% of fall deposits, whereas the tachylite is generally lithic and composes 90% of surge deposits. Crystals usually make up less than 10% of the ash.

Of the juvenile pyroclasts, three characteristic types dominate the 1965 products: (1) vesicular, (2) blocky, and (3) drop-like or vesicular ash with fluidal surfaces. The blocky ash samples are excellent examples of typical phreatomagmatic basaltic pyroclasts. The planar and curviplanar surfaces meet at nearly 90° angles and cut across vesicle walls. Vesicles, when present, are nearly spherical but may show some elongation; this shape contrasts with that of Strombolian-fall ash, which shows very irregularly shaped vesicles. Smooth, fluidal surfaces on ash are manifested as poorly vesicular, drop-like particles and, in vesicular ash, as gently curved areas that "flow" into vesicle hollows.

Lithic pyroclasts present are mostly tachylitic and exhibit diktytaxitic textures on vesicle walls. They are generally more altered than juvenile ash is, and alteration products fill the tube-like or elongated vesicles.

Alteration of juvenile pyroclasts shows two stages of development. Fine grained clays and zeolites form first in vesicle hollows because of interaction with steam that condensed there during emplacement. Further alteration is manifested by surface hydration, which forms a thin crust on the grain that often cracks after desiccation.

Fine-grained ash consists of both blocky and vesicular forms that are marked by abrasion chips and pits. Alteration products are commonly present in vesicle hollows. The late-stage growth of some small vesicles was arrested by chilling during their development. These tiny vesicles are 1 to 20 µm in diameter and represent a second stage of bubble growth; Vesicles from the first stage are 50 to 100 µm in diameter.

Comments on Pyroclast Formation: Taal magma was relatively volatile-rich, as is shown by the abundance of vesicles in both early- and late-stage ashes. These volatiles were abundant enough to tear apart the magma during their exsolution, producing Strombolian lava sprays and fountains early in the eruption, before they contacted lake water. In fact, Moore et al. (1966) described some pyroclasts formed during that stage as pumiceous. During phreatomagmatic stages, the magma was fragmented before much bubble-burst and bubble-coalescense occurred. The dominant blocky texture of the ash indicates that while it was in a nearly solid state and before it mixed with water, the magma was fragmented by propagating stress waves. However, some ash still must have been fluid when it contacted and mixed with water, as is indicated by the fluidal surfaces on vesicular and drop-like pyroclasts. Lithic ash expelled from the vent was abraded, resulting in smooth pyroclast edges. This older ash had formed mostly from a strongly crystallized magma with opaque oxide phases; therefore, it is tachylitic, unlike the juvenile material, which is glassy.

Although most vesicles attained diameters of 50 to 100 µm before bursting, a second stage of bubble nucleation and growth occurred, which is shown by the population of tiny vesicles. This second stage was probably triggered by stress waves propagating downward in the vent from a surface explosion source. This hypothesis is supported by the presence of small, linear discontinuities in glass shards, along which the tiny vesicles are aligned. Where the stress waves were tensile, they formed areas of renewed vesicle nucleation and growth in their wake. When the magma solidified, this second stage of growth was arrested before the bubbles grew to more than about 20 µm in diameter. Such secondary vesiculation areas may be preserved adjacent to a deformation plane.

A more detailed discussion of the formation of Taal ash is given by Wolfe (1980b), who described the explosive breccias of the 1976 eruptions.

REFERENCES:

Moore, J. G., 1967. Base surge in recent volcanic eruptions. Bull. Volcanol., **30**: 337-363.

Moore, J. G., Nakamura, K., and Alcaraz, A., 1966. The 1965 eruption of Taal Volcano. Science, **151**: 955-960.

Sheridan, M. F. and Updike, R. G., 1975. Sugarloaf Mountain tephra—a Pleistocene rhyolitic deposit of base-surge origin. Geol. Soc. Am. Bull., **56**: 571-581.

Waters, A. C. and Fisher, R. V., 1971. Base surges and their deposits: Capelinhos and Taal Volcanoes. J. Geophy. Res., **76**: 5596-5614.

Wolfe, J. A., 1980a. Phreatomagmatic eruption of Taal Volcano in 1976. Trans. Am. Geophys. Union., EOS, **61(7)**: cover photo.

Wolfe, J. A., 1980b. Fluidization versus phreatomagmatic explosions in breccia pipes. Econ. Geol., **75(7)**: 1105-1109.

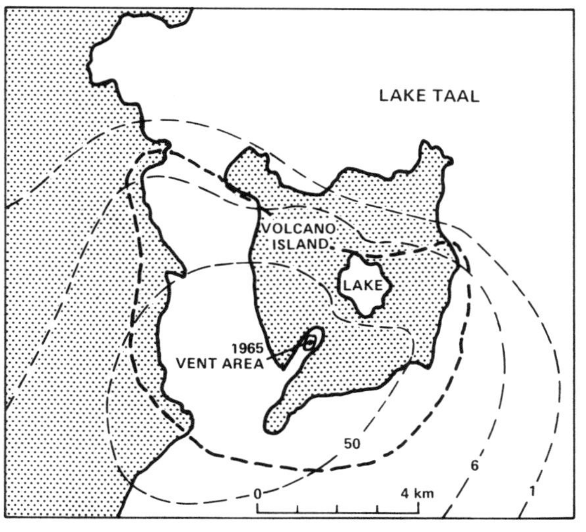

FIGURE 1. Map of Taal ash deposits adapted from Moore (1967). Dashed lines show 50-, 6-, and 1-cm isopachs of ash fall deposits from the 1965 eruption; the heavy dashed line shows the extent of base surge.

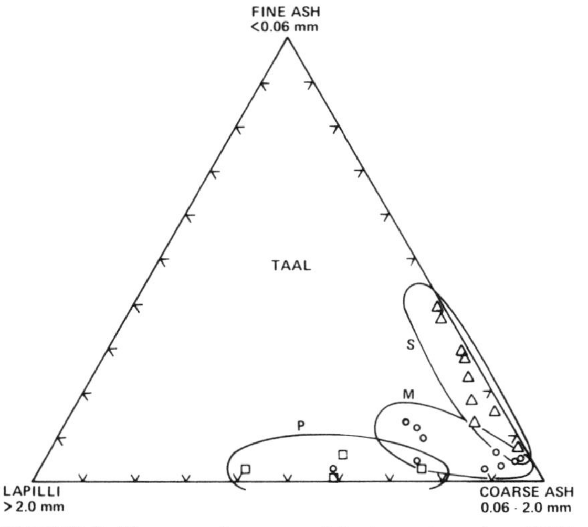

FIGURE 2. Ternary diagram of tephra sizes for 1965 Taal deposits. Size components are fine ash (<0.063 mm), ash (0.063 to 2.00 mm), and lapilli (>2.00 mm). Median sizes of samples are strongly correlated to bed form: S = sandwave, M = massive, and P = planar pyroclastic surge deposits.

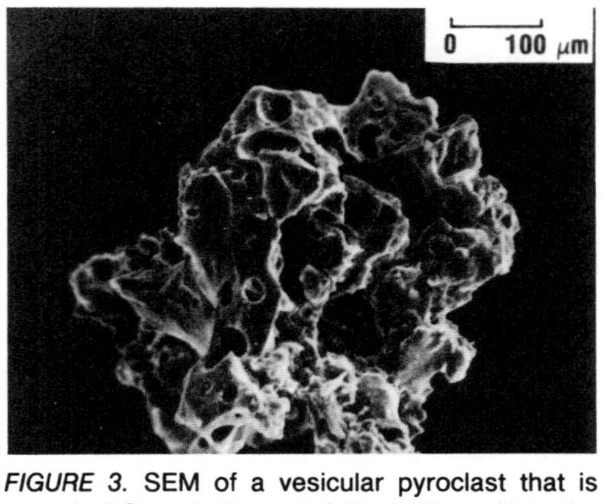

FIGURE 3. SEM of a vesicular pyroclast that is typical of Strombolian ash fall eruptions. Note the irregular shape of vesicles and their porous, spongy appearance.

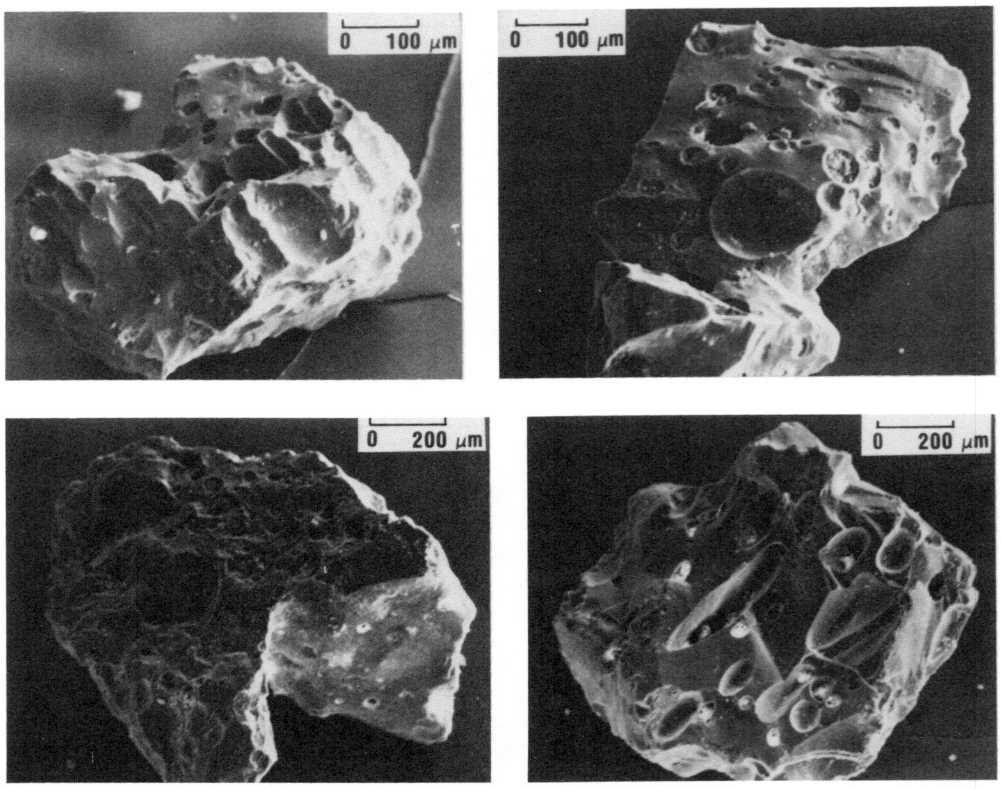

FIGURE 4. SEMs of blocky pyroclasts from surge deposits. Particles are poorly vesicular with vesicles ranging from spherical to ovoid shapes. Edges were chipped during transport, and surfaces are generally fresh.

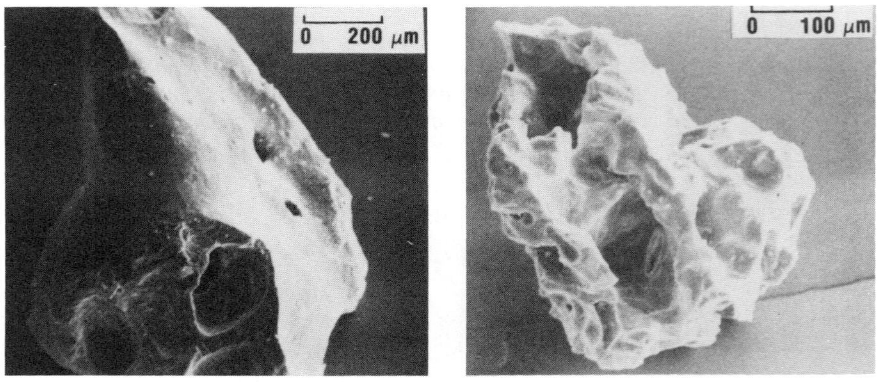

FIGURE 5. SEMs of drop-like and vesicular pyroclasts with fluid-formed surfaces. These shapes are found in deposits from the last stages of Surtseyan eruption. The shapes indicate that the magma was still partly fluid after it was fragmented by contact with water.

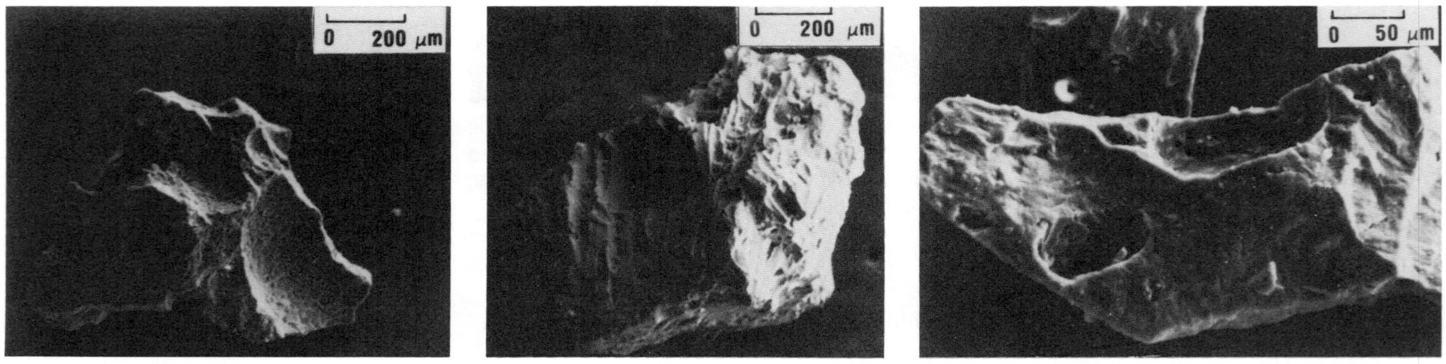

FIGURE 6. SEMs of lithic tachylite ash from early eruptions. These particles make up nearly 90% of the 1965 deposit. Phenocrysts produce the textured surface patterns that are apparent on the left and right pictures. *Center*—many older lithic ash pyroclasts contain elongated vesicles formed by plastic deformation during flow of lava within an older vent.

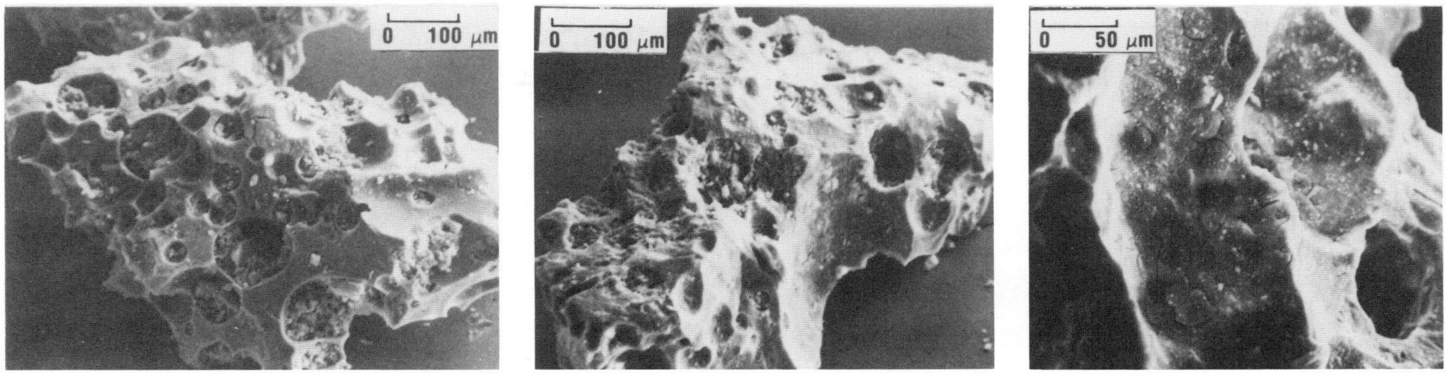

FIGURE 7. SEMs of altered ash from Taal. Clay and zeolite particles fill the vesicles of partly altered ash (*left picture*), and hydrated skin covers the particle surfaces in more strongly altered ash (*center and left pictures*). Note the cracks formed in hydrated skin.

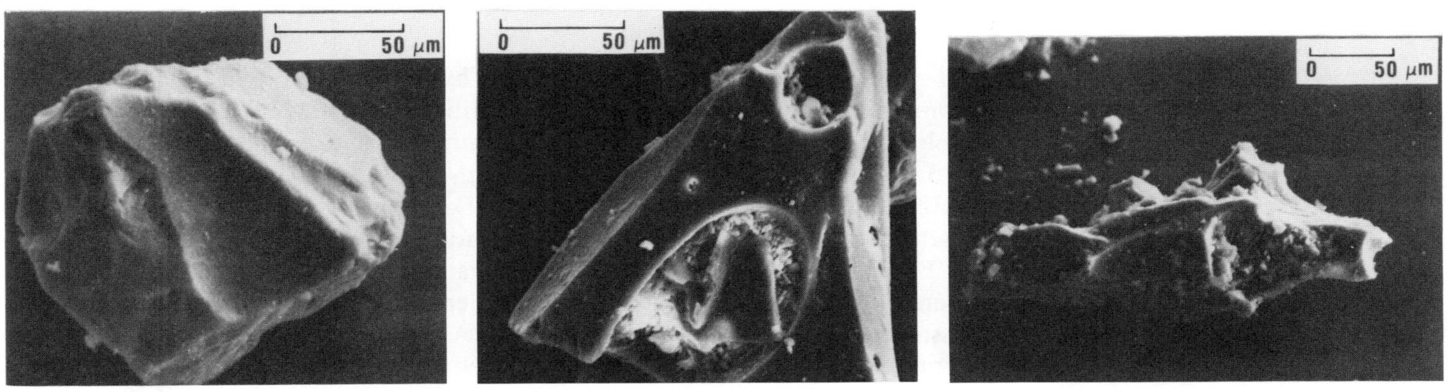

FIGURE 8. SEMs of fine-fraction sideromelane particles. Angular blocky shape is well developed (*left picture*), vesicles are filled with fine adhering dust (*center*), and slight alteration has developed within the adhering dust (*right*).

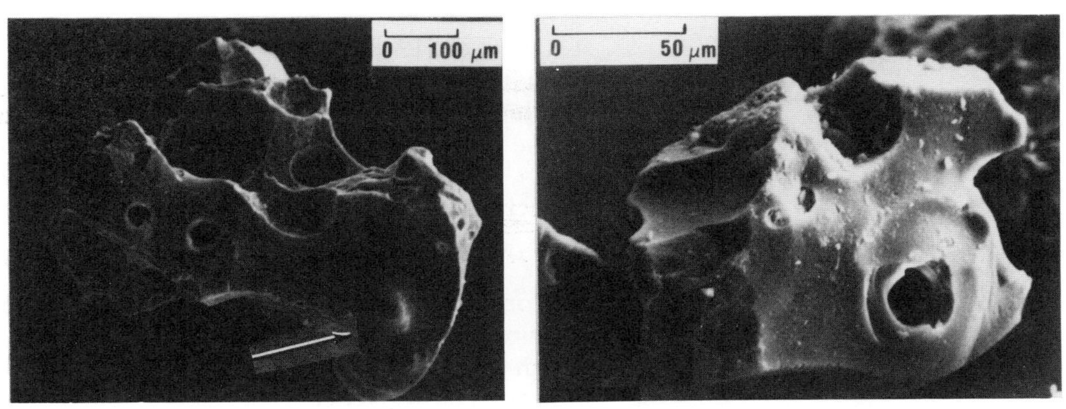

FIGURE 9. SEMs of pyroclasts that show arrested vesiculation. Left view (*arrow*) shows vesicle bulge over a bubble that was frozen as it expanded through the grain surface after fragmentation. Right view shows a vesicle that "popped" after fragmentation; the curved bubble walls are preserved.

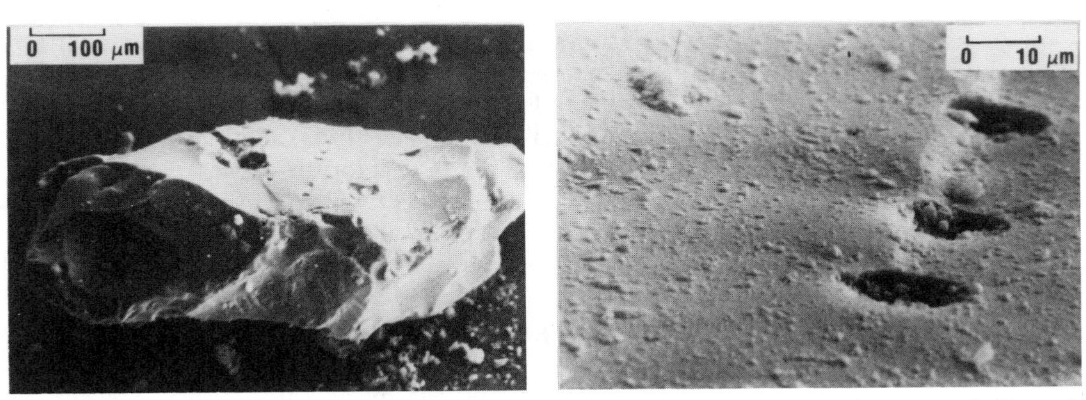

FIGURE 10. SEMs of a line of secondary vesicles formed in a pyroclast. Vesicle growth was arrested by rapid solidification after late-stage nucleation. A closer view (*right*) shows bubbles aligned along a surface irregularity. This texture may have been formed by a propagating stress wave that caused plastic deformation of the glass, and it was preserved as a result of quick solidification.

Prehistoric Eruptions

By studying old, well-dissected volcanoes, we can gain perspective on the many variations in eruption phenomena that may occur during a single period of activity. Deep erosion allows us to examine complete eruption sequences that sometimes are not visible at young vents. Using tephra characteristics that were defined on the basis of samples from observed eruptions, it is possible to reconstruct eruption histories of prehistoric volcanoes. Examples of this approach to tephra studies are presented in this section.

BANDELIER TUFF, JEMEZ MOUNTAINS VOLCANIC FIELD, NEW MEXICO

Introduction: The Bandelier Tuff, originally described by Smith (1938) and Griggs (1964), is composed of two major sequences of rhyolitic tuff that flank the Jemez Mountains volcanic field, New Mexico. The bulk volume of the tuff is about 600 km^3. It was erupted during two cycles, 1.4 and 1.1 m.y. ago and is associated with a major caldera complex. Much of the pioneer work on pyroclastic flow deposits by Ross and Smith (1961) was based upon their studies of the Bandelier Tuff.

Each of the two cycles began with a Plinian eruption of pumice and ash falls that were followed by pyroclastic flow sequences. Pyroclastic surge deposits are associated with the flows (Fisher, 1979). A facies model for ignimbrite-forming eruptions (Wright et al., 1981) was based chiefly on studies of the Bandelier Tuff.

Massive pyroclastic flow sequences, emplaced at temperatures of 600 to 800°C (Smith and Bailey, 1966), exhibit many secondary textures and mineralogical changes caused by welding, compaction, vapor-phase, and fumarolic alteration. For detailed descriptions of these petrologic changes, see Ross and Smith (1961) or Crowe et al. (1978).

Individual pyroclasts within this eruption sequence are often not preserved as a result of vapor-phase alteration. The examples used here are from a distal section of the tuff where there is little or no welding. These samples were collected for a study by Crowe et al. (1978) and are mainly from units where individual pyroclasts or relict textures are still preserved.

Grain Size: Unconsolidated samples from nonwelded zones within the upper member (Tshirege Member) are all poorly or very poorly sorted, with mean grain sizes ranging from 61 to 412 μm. All samples are within the pyroclastic flow field described by Walker (1971).

Tephra Description: The variety of pyroclast types is similar for both upper and lower members of the Bandelier Tuff. Pumice pyroclasts are most common, and there are fewer shards, crystals, and lithic clasts. Sanidine and quartz phenocrysts are "clean" or are encased by a thin, vesicular glass coating. Vesicles are flattened ovoids that are parallel to phenocryst surfaces. These are similar to the bubble-wall texture described by Fisher (1963) and Meyer (1972).

There are at least three varieties of pumice pyroclasts:

(1) Those with thin, elongate vesicles whose aspect ratios are >50:1. These are mostly small, elongate pyroclasts with a "silky" appearance. The vesicles have been flattened (they are ovoid to lens-like in cross-section) and have thin vesicle walls.

(2) Pyroclasts with large ovoid vesicles that range from 100 to 500 μm in length and have low aspect ratios (30:1 to 1:1). The vesicle walls are very thin. The pyroclasts are equant to slightly elongate and are mostly angular.

(3) Dense pumice pyroclasts, characterized by ovoid to spherical vesicles that are 5 to 80 μm long (or in diameter if spherical). These vesicle walls are thick relative to those of the two other pumice types.

Most shards are fragments of comminuted pumice pyroclasts that range from thin, curved plates to Y-shaped forms derived from the mutual intersection of three vesicles. Ash-size lithic fragments include a variety of types but are mostly equant, subrounded lava clasts of intermediate composition; some have been hydrothermally altered.

Chemical Composition: The tephra are rhyolitic with only minor variations in major elements. For variations in trace elements, see Smith (1979) and Smith and Bailey (1966). Analyses of glassy pyroclasts are shown in Table 1.

TABLE 1. ELECTRON MICROPROBE ANALYSES OF GLASS COMPOSITIONS, LOWER (OTOWI) MEMBER, BANDELIER TUFF.*

Sample	Basal Pumice Fall	Lower, Pyroclastic Flow	Center, Pyroclastic Flow	Top, Pyroclastic Flow	Top, Pyroclastic Flow
SiO_2	73.78	74.03	73.59	73.92	74.94
TiO_2	0.10	0.08	0.10	0.11	0.07
Al_2O_3	11.74	11.92	12.05	11.73	12.05
FeO**	1.15	1.06	1.11	1.09	1.14
MgO	0.05	0.04	0.04	0.05	0.03
CaO	0.30	0.29	0.29	0.29	0.25
Na_2O	3.38	2.82	3.23	2.95	3.29
K_2O	4.42	4.68	4.65	4.89	4.70
F	0.1	0.09	0.06	0.04	0.05
Total	95.02	95.01	95.12	95.07	96.52
No. of analyses	6	6	6	5	6

*From Crowe et al., 1978.
**Total iron as FeO.

Comments on Pyroclast Formation: Meyer (1972) used the amount of glass remaining on phenocrysts as an indicator of eruption energy; the presence of fewer phenocrysts with glass "crusts" indicates increased explosivity. Within the lower member of the Bandelier Tuff, this general observation holds. For example, within the base of the Plinian pumice fall, only a few per cent of the crystal pyroclasts have glassy coatings, whereas in the top and throughout the pyroclastic flow, 25 to 30% of all pyroclasts are phenocrysts with glass coatings. With the exception of the lowest tephra fallout unit, the same variation exists in the upper member.

Pumice pyroclasts with thin, elongate vesicles, which indicate considerable laminar flow in the vent of a vesiculating melt before eruption, are abundant (25%) in the lowest tephra fall units of each member and decrease to 10 to 13% throughout overlying deposits. Dense, finely vesicular pyroclasts dominate (60%) only the lowest tephra fall in the lower member and decrease to 5 to 25% in overlying pyroclastic flows. Pumice pyroclasts with large, ovoid vesicles, indicative of lower viscosity and vesiculation and fragmentation near the magma/atmosphere interface, are present in varying proportions in all samples.

REFERENCES:

Crowe, B. M., Linn, G. W., Heiken, G., and Bevier, M. L., 1978. Stratigraphy of the Bandelier Tuff in the Pajarito Plateau. Los Alamos Scientific Laboratory Report LA-7225-MS: 57 pp.

Fisher, R. V., 1963. Bubble-wall texture and its significance. J. Sed. Petrol., **33**: 224-235.

Fisher, R.V., 1979. Model for pyroclastic surges and pyroclastic flows. J. Volcanol. Geotherm Res., **6**: 305-318.

Griggs, R. L., 1964. Geology and ground water resources of the Los Alamos area, New Mexico. U.S. Geol. Surv. Water Supply Paper 1753: 107 pp.

Meyer, J. D., 1972. Glass crust on intratelluric phenocrysts in volcanic ash as a measure of eruptive violence. Bull. Volcanol., **35**: 358-368.

Ross, C. S. and Smith, R. L., 1961. Ash-flow tuffs: their origin, geologic relations and identification. U.S. Geol. Surv. Prof. Paper 366: 81 pp.

Smith, H. T. U., 1938. Tertiary geology of the Abiquiu Quadrangle, New Mexico. J. Geol., **46**: 933-965.

Smith, R. L., 1979. Ash-flow magmatism. Geol. Soc. Am. Spec. Paper 180: 5-27.

Smith, R.L. and Bailey, R.A., 1966. The Bandelier Tuff: a study of ash-flow eruption cycles from zoned magma chambers. Bull. Volcanol., **29**: 83-104.

Walker, G. P. L., 1971. Grain-size characteristics of pyroclastic deposits. J. Geol. **79**: 696-714.

Wright, J. V., Self, S., and Fisher, R. V., 1981. Towards a facies model for ignimbrite-forming eruptions. *In*: S. Self and S. Sparks (Eds.), *Tephra Studies*, Reidel Pub. Co., Dordrecht: 433-439.

FIGURE 1. Pumice pyroclasts from the lower member. Most are angular, relatively dense pumices with small vesicles and thick vesicle walls.

FIGURE 2. Tephra sample from the upper half of the tephra fall bed, lower member. *Center*—a quartz(?) phenocryst coated with vesicular glass; vesicles are broad ovoids flattened parallel to crystal faces. The remaining pyroclasts are composed of a mixture of curved vesicles and fractured mineral fragments.

FIG. 4
FIG. 3
FIG. 2
FIG. 1

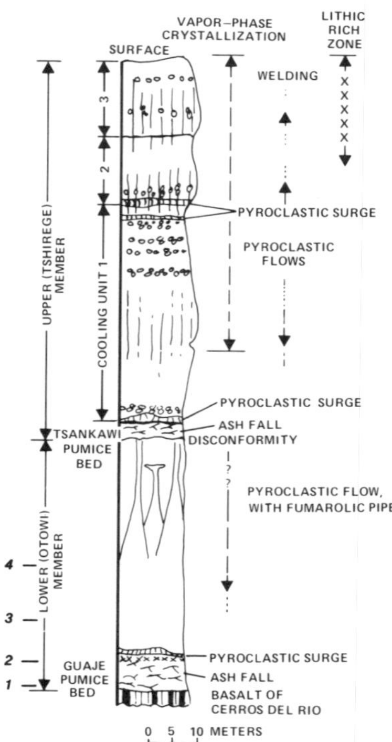

Stratigraphic section of the Bandelier Tuff.

FIGURE 3. Sample from near the base of the pyroclastic flow, lower member. Pumice pyroclasts of all three types described in the text are present. *Center*—a pyroclast with large, flattened ovoid vesicles. *Bottom*—a pumice with straight, elongate, narrow vesicles. All pumice pyroclasts are coated with fine-grained shard fragments.

FIGURE 4. Disaggregated tephra from the middle of the pyroclastic flow, lower member. Equant to slightly elongate phenocrysts (with and without glass coatings) and a variety of pumice pyroclasts are visible. Pumices include those with thin, elongate vesicles (elongate, angular pyroclasts) and dense pumices with small vesicles (*upper right*).

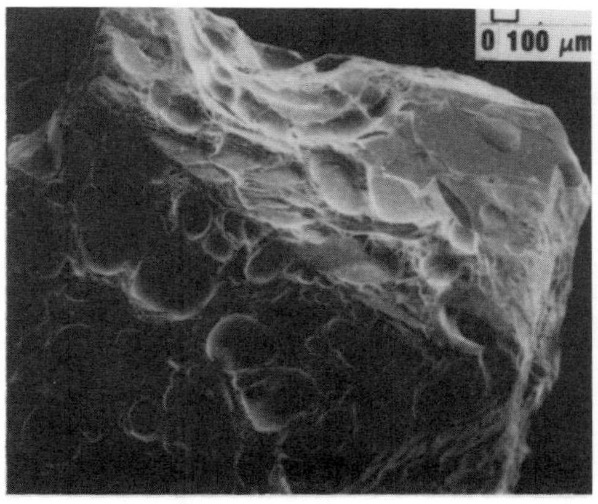

FIGURE 5a. Samples from thin but widespread tephra fall beds at the base of the upper member of the Bandelier Tuff. Large phenocryst coated with a thin layer of vesicular glass. Note that vesicles are flattened parallel to crystal faces.

FIGURE 5b. Stereomicrograph of ash from a relatively fine grained tephra fall bed. It is composed chiefly of shards derived from a vesicular melt with well-developed, elongate flattened vesicles. Note the large, curved vesicle walls (now shard surfaces). There are a few small pumice pyroclasts with narrow, tube-shaped vesicles.

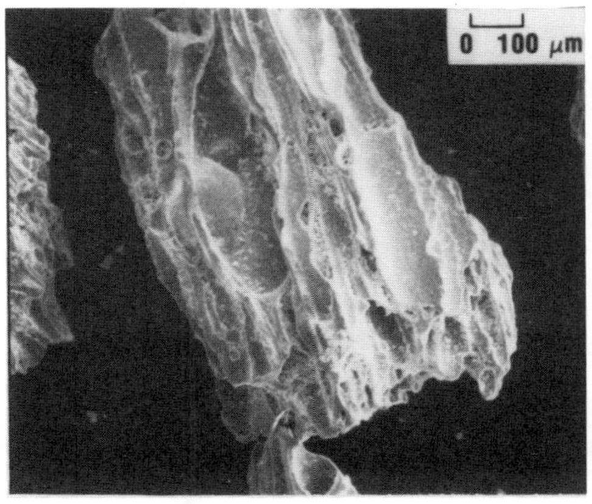

FIGURE 5c. Pumice pyroclast with well-developed, elongate ovoid vesicles; most have been flattened.

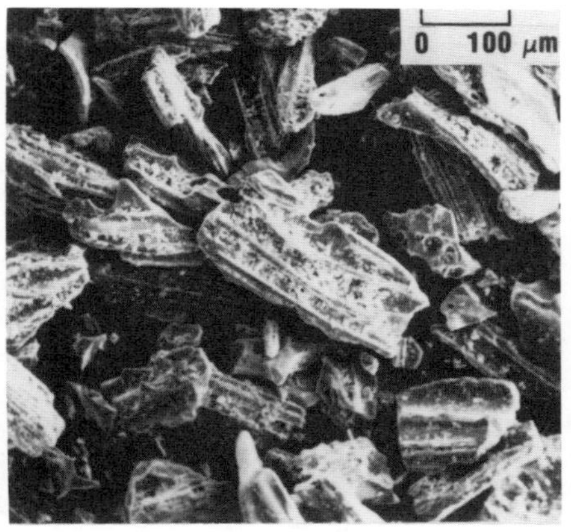

FIGURE 5d. These pyroclasts are mostly small pumices with narrow, elongate parallel vesicles; the shapes of these pumices reflect the nature of the vesicles.

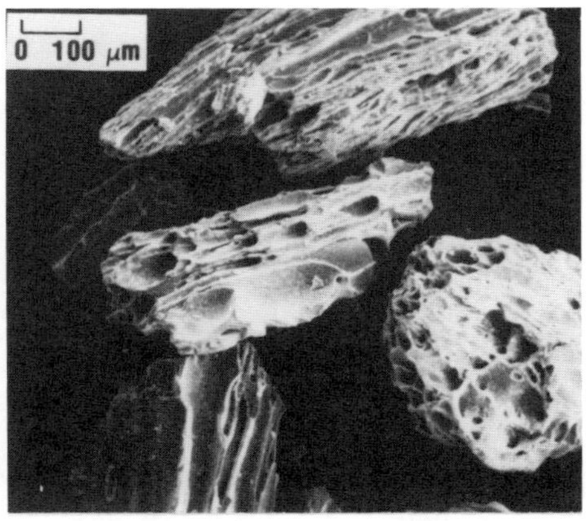

FIGURE 5e. Angular to subangular pumice pyroclasts with mostly large, well-developed ovoid vesicles (*lowest 4 particles*) and thin vesicle walls. The uppermost pumice contains both elongate ovoid vesicles and thin, tube-like vesicles with high aspect ratios.

FIGURE 6. Tephra from a ground surge bed at the base of the lowest pyroclastic flow, upper member. There are mostly equant to slightly elongate shards, each with curved or parallel ridges remaining from adjacent vesicles. There are also a few needle- or wedge-shaped shards.

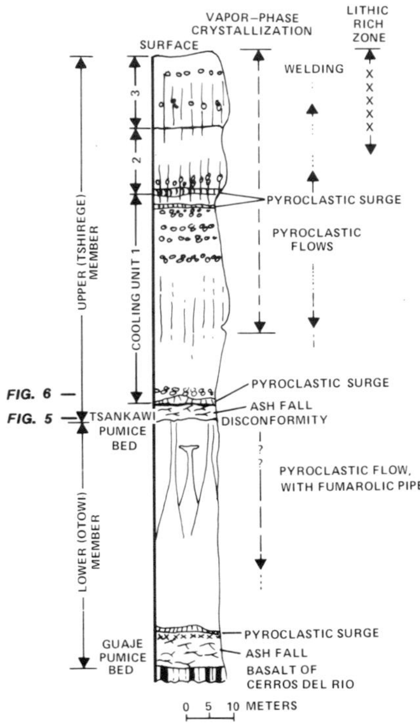

Stratigraphic section of the Bandelier Tuff.

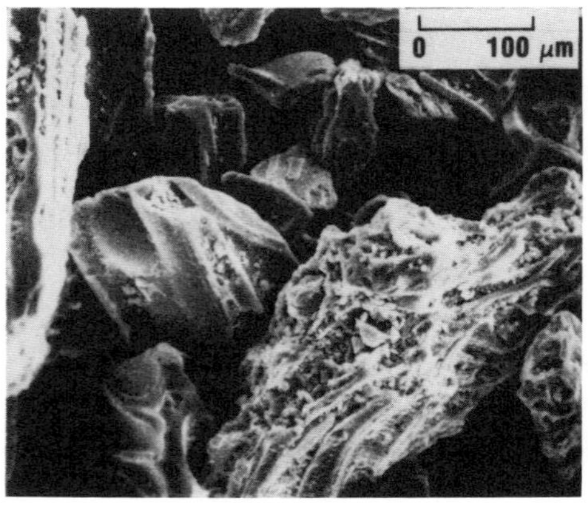

FIGURE 7a. Sample from near the base of the lowest pyroclastic flow, upper member. Mixture of elongate pumice pyroclasts and nearly equant shards. Note the fine-grained shards clinging to pyroclast surfaces—a feature that is typical of pyroclastic flow deposits.

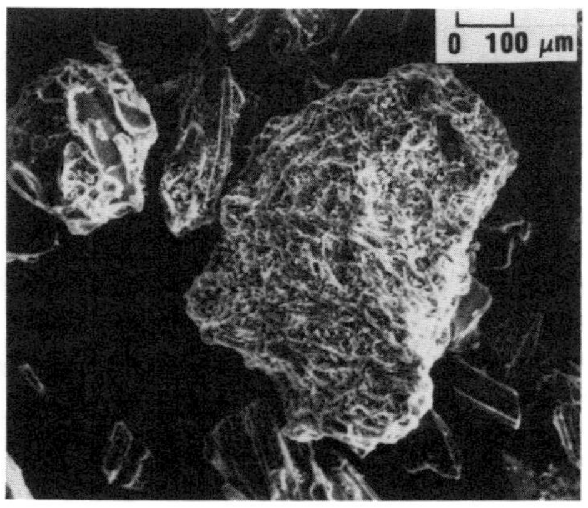

FIGURE 7b. Large, subequant pumice pyroclast (*center*) with finely vesicular, thick-walled vesicles. *Lower left*—quartz phenocrysts coated with vesiculated glass (bubble-wall texture described by Fisher, 1963).

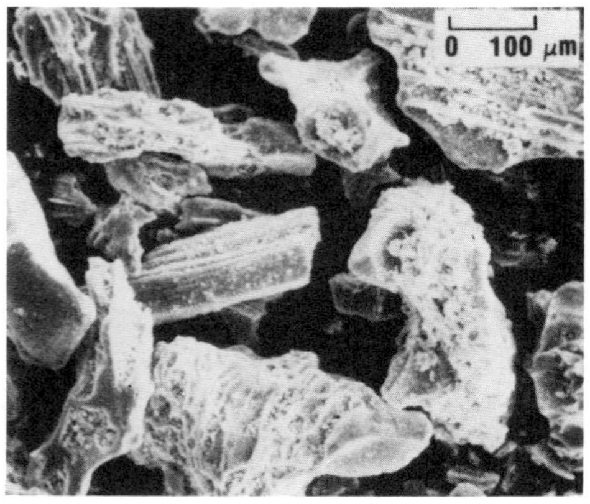

FIGURE 8a. From the lower half of the lowest pyroclastic flow, upper member. Small pumice pyroclasts and shards partly coated with fine-grained shards.

FIGURE 8b. Mostly elongate pumice pyroclasts with thin, tube-like vesicles. All pyroclasts are coated with fine-grained shards and alteration products.

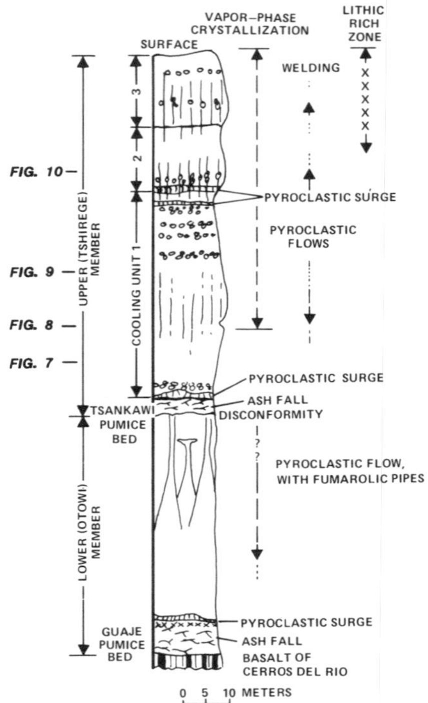

Stratigraphic section of the Bandelier Tuff.

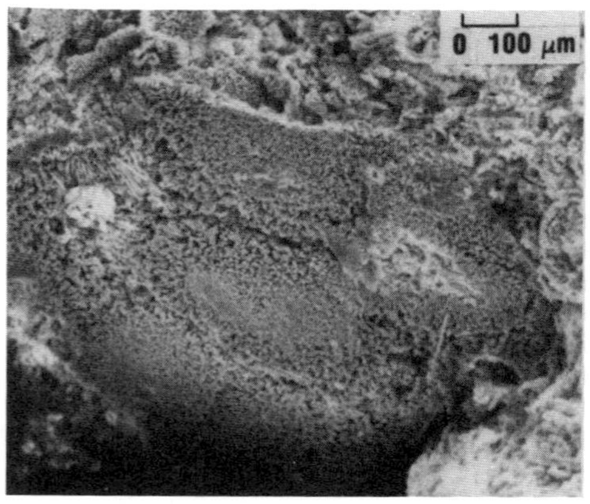

FIGURE 9a. Consolidated, cemented sample from cooling unit 1, upper member. This partially welded tuff is in the vapor-phase zone, where all glass pyroclasts have been replaced by authigenic alkali feldspar and silica. The flat surface is the "mold" of a shard—a relict form or "fossil" pyroclast. Relicts of smaller glass shards are visible to the left of the large relict pyroclast.

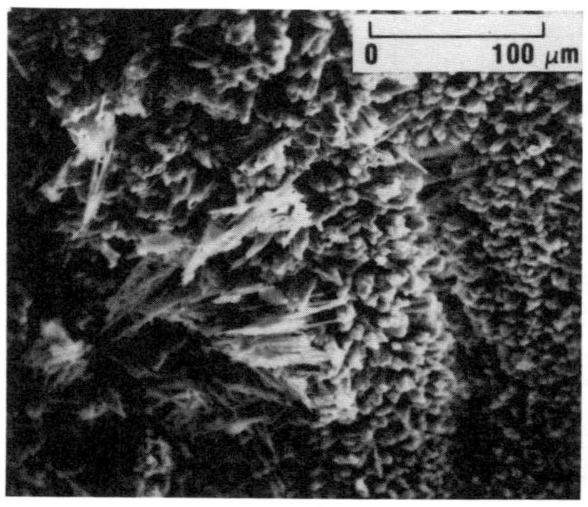

FIGURE 9b. Close-up of the relict shard surface in Fig. *9a*, which shows parallel intergrowths of alkali feldspar and cristobalite.

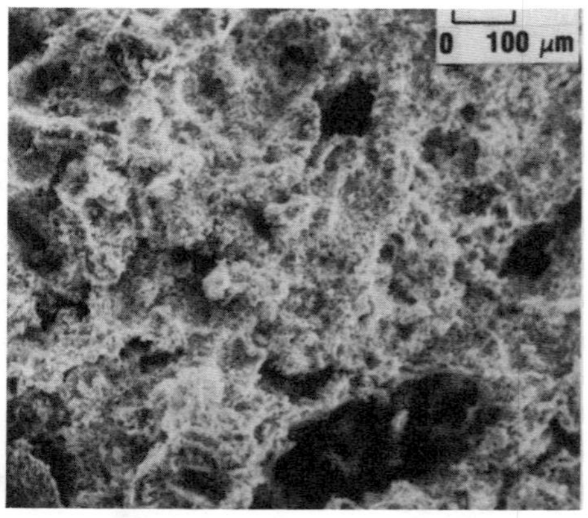

FIGURE 10. Sample from cooling unit 2, upper member. Vapor-phase alteration of the tuff has replaced vitric pyroclasts and filled void space with alkali feldspar, cristobalite, and tridymite. Grain boundaries and relict textures have been completely destroyed in this unit; there is nothing left to indicate the types of pyroclasts involved in this phase of the eruption.

HOPI BUTTES, ARIZONA; VULCANIAN ERUPTION OF DIATREMES

Introduction: The geology of the Hopi Buttes has been described by Williams (1936), Hack (1942), Shoemaker et al. (1962), and Sutton (1974). This area in northeastern Arizona is primarily underlain by the Pliocene Bidahochi Formation, which includes six members that are composed mostly of lake sediments. The fifth member (White Cone Member) is mostly volcanic flows and tuffs up to 100 m thick; it was derived from 200 to 300 vents that erupted within the 600-km^2 area during Pliocene time. Most of the vents are diatremes exposed at various depths by erosion. Vents that show little erosion are maar craters, whereas those that have been eroded are breccia dikes and necks. Some of the maars are capped by lava. Lavas of the Hopi area are of monchiquitic affinity (sodium-amphibole-bearing lamprophyre), and the tuffs have been termed limburgites (extrusive lamprophyre that contains glass). The structure and composition of these diatremes is very similar to those of the Schwabian Alb, Germany (Cloos, 1941).

Most of the Hopi vents erupted along a lake shore—a setting that promoted steam explosions during initial phases of eruption. Williams (1936) noted that the angular and subangular chips of black limburgite and monchiquite in the pyroclastic deposits indicated that lavas were solid or nearly so before their eruption. Hence, the eruptions were a Vulcanian type, erupting both pyroclasts and large volumes of pulverized sandstone and shale. Shoemaker et al. (1962) explained the diatreme formation by magmatic explosions that followed exsolution of magmatic volatiles from a volatile-rich magma. Evidence discussed in this section strongly supports the idea that phreatomagmatic eruptions occurred, during which meteoric water caused fragmentation of the magma.

The pyroclast samples are from two vents, Bidahochi Butte and Coliseum Diatreme.

Grain Size: Both tuff rings studied are composed of well-indurated, bedded pyroclasts that are lapilli-bearing tuffs to fine-grained tuffs. They are moderately sorted except for lens-shaped concentrations of pebbles. Table 1 lists median sizes, sorting, and fine-ash abundances for the studied samples. The samples from Bidahochi Butte are somewhat finer grained and have more abundant fine ash than those of Coliseum Diatreme; however, they are slightly less sorted. This distinction is manifested by the smaller degree of lithification at Bidahochi Butte.

TABLE 1. GRAIN-SIZE CHARACTERISTICS OF HOPI BUTTES TEPHRA.*

Sample	Bed Form*	Md_φ (mm)	σ_φ	Fine Ash (Wt. %)
1	F	1.45 (0.37)	1.72	6.39
2	M	2.03 (0.24)	1.54	11.65
3	S	0.16 (0.89)	1.28	0.34
4	M	0.68 (0.62)	1.56	4.45

*Bed forms: F = fall, M = massive surge, S = sandwave surge.

Tephra Description: Williams (1936) noted that Hopi Butte pyroclasts are mostly angular. There is a marked paucity of vesicular fragments in these samples. Bidahochi Butte ash fall sample 1 consists of abundant aggregated particles. Most of the small particles are prismatic richterite crystals. The fine fraction of this sample contains curved splinters and chips of relatively fresh glass with little fine, adhering dust. In contrast to sample 1, sample 2 is from a massive surge bed and shows abundant flat surfaces of blocky particles. Where present, vesicles are cut by these flat surfaces. The alteration of grain surfaces to clay and zeolite is minimal, and a few aggregates are evident in the sample. The finer grained particles of sample 2 have a mossy appearance, high surface area, and abundant small dust particles clinging to surfaces. No abrasion or edge modification features are observed in this sample.

Sample 3, from Coliseum Diatreme, is from a coarse sandwave surge bed. It consists of very angular grains with broken, flat surfaces and nearly spherical, drop-like particles. Some of these spherical particles have adhered to the surfaces of larger grains. Hydration cracks and crusty surface overgrowths caused by alteration are abundant, and fine particles show abundant tiny crystals adhering to glass surfaces. Sample 4 is from a massive surge bed and is made up of subangular to subrounded grains with smooth surfaces that are partly covered by crusty alteration. Botryoidal surfaces formed from the aggregation of small, drop-like particles. Abrasion features such as v-shaped scratches and chipped edges are common and contribute to grain rounding. The fine fraction contains many angular, blocky particles of glass and crystals.

Chemical Composition: Analyses listed in Table 2 show the differentiation trends of the monchiquite to trachybasalt and syenite-like compositions, as well as the effects of surface alteration after emplacement in wet pyroclastic surges. The limburgite tuffs were derived from a silica-undersaturated melt that differentiated to a more silica-rich composition during extrusion. Abundant crystals of richterite and analcime occur in the tuffs and lavas.

TABLE 2. GRAIN SURFACE CHEMISTRY OF HOPI BUTTES TEPHRA.*

	A	1	2	3	4
SiO_2	47.62	61.03	55.6	66.98	73.57
TiO_2	8.45	6.51	7.21	5.66	4.32
Al_2O_3	8.38	9.78	9.40	7.76	7.58
FeO	11.96	8.15	9.36	7.31	4.33
MnO	1.67	1.50	1.53	2.09	1.53
MgO	6.02	4.05	4.88	2.46	2.39
CaO	13.14	6.02	8.66	4.43	3.00
Na_2O	1.03	0.95	1.03	0.75	0.75
K_2O	1.73	2.02	2.27	2.57	2.52
Glass		79.73	87.01	88.06	
Crystal		15.86	9.45	7.46	
Lithic		4.41	3.54	4.48	

*Normalized standardless EDS analyses. A = analysis of freshest grain from fall sample 2, a trachybasalt similar to that provided by Williams (1936).

Comments on Pyroclast Formation: The absence of vesicles and the ubiquity of blocky pyroclasts with planar surfaces is strong evidence for phreatomagmatic eruptions. Drop-like and spherical particles indicate that some magma erupted while partly fluid. The occurrence of surge deposits (indicative of wet and cohesive tephra and particle aggregates) supports the contention that phreatomagmatic eruptions occurred when magma interacted with ground or lake water. Vulcanian eruptions consist of expulsions of steam and particles derived from previously solidified lava; they are often interspersed with phreatomagmatic bursts of fresh lava (Frazzetta et al., 1983). Furthermore, drop-like particles produced during phreatomagmatic eruptions indicate very fluid lava and strong, turbulent mixing of water with magma before eruption. At Vesuvius, this mixing occurs at depths as great as 5 km (Sheridan et al., 1981) and thus may be responsible for the admixture of lithic fragments from underlying Mesozoic strata.

REFERENCES:

Frazzetta, G., La Volpe, L., and Sheridan, M. F., 1983. Evolution of the Fossa Cone, Vulcano. *In*: M. F. Sheridan and F. Barberi (Eds.), *Explosive Volcanism*, J. Volcanol. Geotherm. Res., **17**: 329-361.

Cloos, H., 1941. Bau and Tätigkeit von Tuffschloten—Untersuchungen an dem schwäbischen Vulkan. Geol. Rundschau, **32(H6-8)**: 709-800.

Hack, J. I., 1942. Sedimentation and volcanism in the Hopi Buttes, Arizona. Geol. Soc. Am. Bull., **53**: 335-372.

Sheridan, M. F., Barberi, F., Rosi, M., and Santacroce, R., 1981. A model for the Plinian eruptions of Vesuvius. Nature, **289**: 282-285.

Shoemaker, E. M., Reach, E. H., and Byers, F. M., Jr., 1962. Diatremes and uranium deposits in the Hopi Buttes, Arizona. *In: Petrologic Studies—A Volume to Honor A. F. Buddington*, Geol. Soc. Am. Mem., Buddington Vol.: 317-355.

Sutton, R. L., 1974. The geology of Hopi Buttes, Arizona. *In:* Guidebook to northern Arizona. Geol. Soc. Amer. Rocky Mt. Section Meeting, 1974.

Williams, H., 1936. Pliocene volcanoes of the Navajo-Hopi country. Geol. Soc. Am. Bull., **47**: 111-172.

FIGURE 1a. SEMs of sample 1, an irregularly shaped aggregate particle of elongate richterite crystals. These crystals were winnowed from erupted ash falls because of their contrasting density with glass particles. They adhered to each other while tumbling around in a steam-rich eruption cloud.

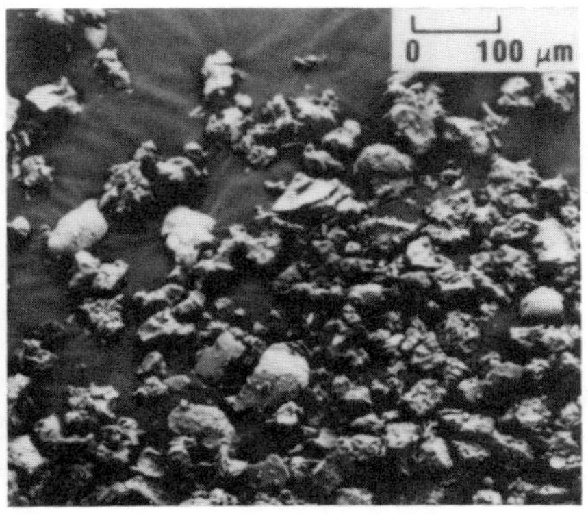

FIGURE 1b. Fine-fraction particles are crescent-shaped and curved splinters of glass with fresh surfaces.

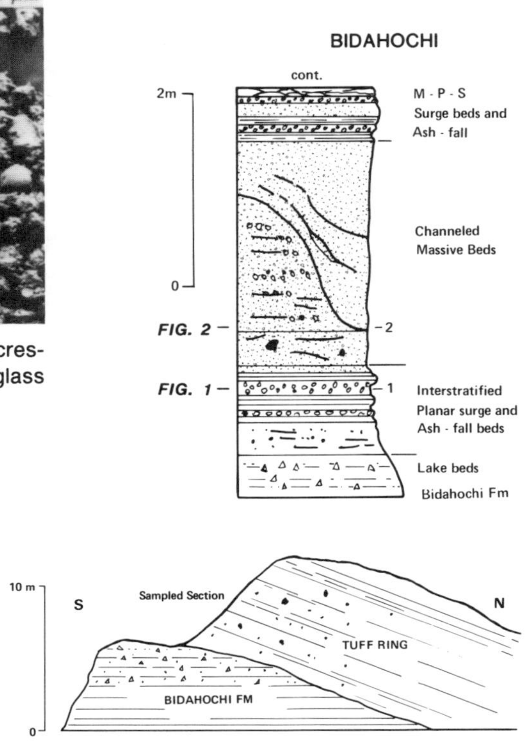

Stratigraphic columnar section and cross section of the S tuff rampart of Bidahochi Butte. The Bidahochi Formation is dominantly flat-lying lake sediments and is overlain by massive, consolidated tuff beds dipping 12 to 15° towards the vent. The tuff consists of ash and lapilli fall beds interlayered with dominantly massive, channeled pyroclastic surge tephra.

FIGURE 2a. SEMs of sample 2, a blocky pyroclast with flat surfaces, patchy altered overgrowths, and angular, chipped corners.

FIGURE 2b. Fine-fraction particles are mossy aggregates of tiny crystals, dust, and angular glass fragments.

FIGURE 3a. SEMs of sample 3, an analcime crystal with surface accumulation of glass droplets that have been altered to clay. Note the crystal faces, abraded chips, and hydrated glass cracks.

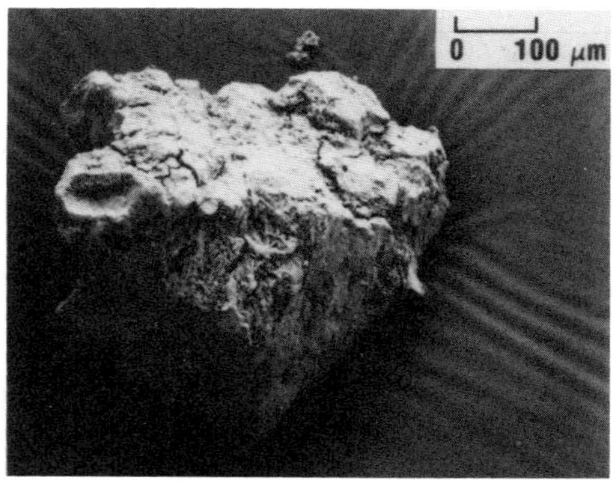

FIGURE 3b. Blocky, altered pyroclast with cracked, altered skin and clay minerals attached to embayed areas.

COLISEUM DIATREME

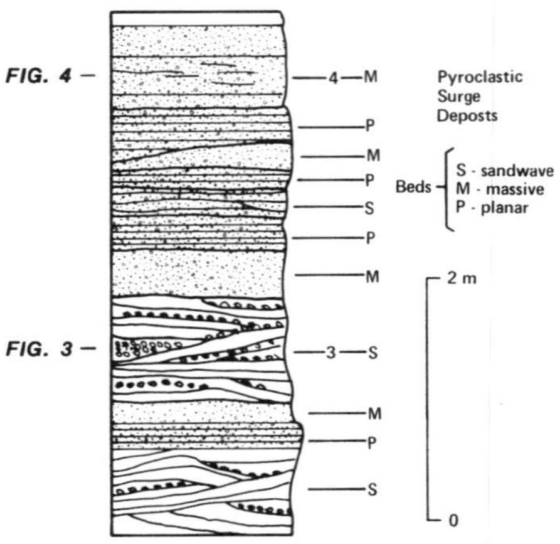

Stratigraphic columnar section of a part of the Coliseum Diatreme tuff rampart. This well-lithified tuff is well bedded and primarily consists of surge beds with interspersed ash fall layers.

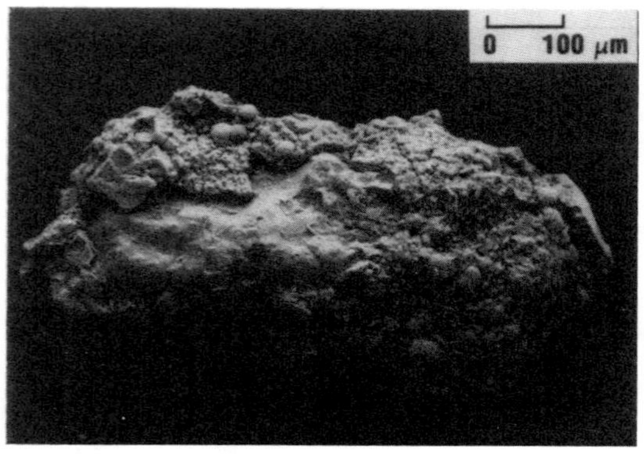

FIGURE 4a. SEMs of sample 4. This glassy, rounded pyroclast shows surface agglutination of drop-like particles and crystalites. The sample is highly altered.

FIGURE 4b. Angular chips and blocks of the fine fraction include numerous, elongate richterite crystals.

BISHOP TUFF, CALIFORNIA; PLINIAN AND PHREATOPLINIAN ERUPTIONS

Introduction: The Bishop Tuff, which crops out over 900 km^2 in eastern California, was first described by Gilbert (1938). This calc-alkaline rhyolitic tuff was erupted from the Long Valley Caldera 0.71 m.y. ago (Bailey et al., 1976), and Bishop ash beds have been found as far away as Nebraska 1800 km to the E (Izett, 1981). The total volume of the Bishop tuff is calculated to be at least 500 km^3, 100 km^3 of which is a pumiceous tephra fall overlain by 150 km^3 of welded ash flow tuff that is exposed as outflow units around the caldera (Bailey et al., 1976). Descriptions of the Bishop Tuff include a paper by Bateman (1965), maps by Crowder and Sheridan (1972) and Huber and Rinehart (1967), and mineralogic and petrologic studies by Sheridan (1965) and Hildreth (1979). The samples and their SEM photos described in this report were donated by M. Sheridan, who collected them from the basal ash fall and lower ash flow sections exposed in the Owen's River Gorge near Tom's Place, California.

Tephra Description: The sample of the Plinian lapilli fall and the two samples of unconsolidated ash flow are representative of early eruptions of the Bishop Tuff. The chemical and petrologic zonation throughout the total thickness of the Bishop Tuff are well documented in articles by Sheridan (1965), Ragan and Sheridan (1972), and Hildreth (1979). These zonation features contribute to ash textures that vary upward in the sequence; however, some textures are difficult to identify because of welding and crystallization of the tuff.

Glass. Pyroclasts from the pyroclastic flow deposit are only poorly vesicular and consist of dominantly blocky and equant grains. Plates and curved splinters from broken pieces of larger bubble walls are subordinate. Hydration cracks are evident on a few grains, and fine dust fills vesicles of pumiceous particles.

Fine-ash samples of the fall deposit are delicate, prong-like shards of vesicle walls with thicknesses of 30 to 50 μm. Vesicle surfaces are spherical to ellipsoidal and have dimensions >50 μm. The fine ash shows clean surfaces with no hydration or adhering fine dust. Coarser grained ash shows elongate and curved vesicles of from 0.1 to >1.0 mm long. Fine dust and alteration products stick to vesicle surfaces. The overall grain shape is blocky to subrounded.

Crystals. Crystalline pyroclasts of the Bishop lapilli fall and ash flow have been described by de Rita et al. (in press). The edges of quartz and feldspar grains from the pyroclastic flow deposit have been rounded by transport abrasion. Numerous small chips, v-shaped pits, and a frosted appearance distinguish these grain edges from those in fall deposits. The dish-shaped concavities and smooth, planar faces of cleavages are well developed.

Ash and lapilli fall crystals consist of nearly equal amounts of quartz and feldspar; they are irregularly shaped grains with delicate edges and have clean surfaces. No abrasion features are apparent, although the original crystal faces are rare. The grain surfaces have small, dish-shaped fractures and are coated with solution and precipitation surfaces. The feldspars show small cleavage surfaces, whereas quartz grains have large, conchoidal fractures.

Comments on Pyroclast Formation: The initial eruptions of the Bishop Tuff produced ash and lapilli fall, surge, and pyroclastic flow deposits. The delicate, prong-like, vesicular shapes of glass and pumice in fall products indicate magmatic eruptions where decompression of a volatile-rich magma caused violent vesiculation and fragmentation. The initial Plinian eruption probably widened a vent to the point that the eruption column collapsed (Wilson et al., 1980), first in pulses that formed pyroclastic surge deposits (Fisher, 1979), and then in a continuous fashion that produced pyroclastic flows. The dominantly blocky, angular glass and pumice fragments of the flow deposit, which are less vesicular than those of fall deposits, most likely attained their shape by large-scale interaction of meteoric water with vesiculating magma in the vent (Self and Sparks, 1978). This phreatoplinian eruption of early Bishop products may have been initiated as the vent widened and allowed water access to the magma. This process is probable because a Pleistocene lake may have occupied the Long Valley Caldera area before eruption of the Bishop Tuff. Later Bishop eruption products dominantly consist of welded tuff, and they make up most of the total erupted volume of the Bishop Tuff. Because these tuffs are compacted and welded, samples were not investigated and their origin by phreatoplinian or Plinian mechanisms is not known.

REFERENCES:

Bailey, R. A., Dalrymple, G. B., and Lanphere, M. A., 1976. Volcanism, structure, and geochronology of Long Valley Caldera, Mono County, California. J. Geophys. Res., **81**: 725-744.

Bateman, P. C., 1965. Geology and tungsten mineralization of the Bishop district, California. U.S. Geol. Sur. Prof. Paper 470: 465-482.

Crowder, D. F. and Sheridan, M. F., 1972. Geologic map of the White Mountain Peak quadrangle, Mono County, California. U.S. Geol. Sur. Map GQ-1012.

Fisher, R. V., 1979. Models for pyroclastic surges. J. Volcanol. Geotherm. Res., **6**: 305-318.

Gilbert, C. M., 1938. Welded tuff in eastern California, Geol. Soc. Am. Bull., **49**: 1829-1862.

Hildreth, W., 1979. The Bishop Tuff: Evidence for the origin of compositional zonation in silicic magma chambers. Geol. Soc. Am. Spec. Paper 180: 43-75.

Huber, N. K. and Rinehart, C. D., 1965. Geologic map of the Devil's Postpile quadrangle, California. U.S. Geol. Surv. Prof. Paper 554-D: D1-D21.

Izett, G. A., 1981. Volcanic ash beds: Recorders of upper Cenozoic silicic pyroclastic volcanism in the western United States. J. Geophys. Res., **86**: 10200-10222.

Ragan, D. M. and Sheridan, M. F., 1972. Compaction of the Bishop Tuff, California. Geol. Soc. Am. Bull., **83**: 95-106.

de Rita, D., Sheridan, M. F., and Marshall, J. R., in press. SEM surface textural analysis of phenocrysts from pyroclastic deposits in the Sabatini volcanic field, Latium, Italy. *In*: W. B. Whalley and D. H. Krinsley (Eds.), *Scanning Electron Microscopy in Geology*, Geoabstracts, Norwich, England.

Self, S. and Sparks, R. S. J., 1978. Characteristics of widespread pyroclastic deposits formed by the interaction of silicic magma and water. Bull. Volcanol., **41**: 196-212.

Sheridan, M. F., 1965. The mineralogy and petrology of the Bishop Tuff, Ph.D. dissertation, Stanford University, Palo Alto, California: 165 pp.

Wilson, L., Sparks, R. S. J., and Walker, G. P. L., 1980. Explosive volcanic eruptions-IV, the control of magma properties and conduit geometry on eruption column behaviour. Geophys. J. Roy. Astron. Soc., **63**: 117-148.

FIGURE 1. SEMs of vitric pyroclasts from the Bishop Tuff pyroclastic flows. Several size fractions are shown (left to right): 250 to 500 μm and 125 to 250 μm (*top*), 63 to 125 μm and <63 μm (*bottom*). Note the mixture of blocky, nonvesiculated glass and pumiceous particles in the coarser size fractions. The fine fraction shows very few vesicle surfaces. Plate-like and angular shards are characteristic of phreatoplinian fine ash.

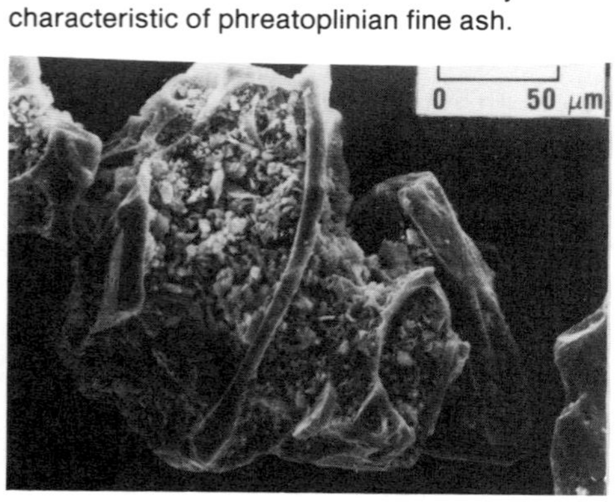

FIGURE 2. SEM of a vesicular ash particle from the flow deposits shows vesicles completely filled with the fine, adhering dust that is characteristic of hydrovolcanism.

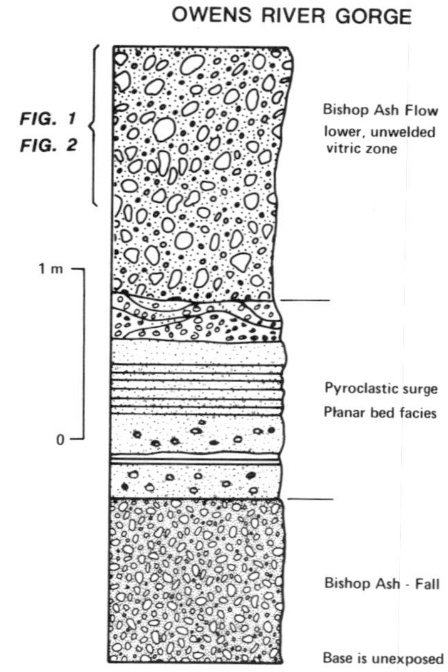

Stratigraphic columnar section of the lower part of the Bishop Tuff near Tom's Place, California. Samples studied were taken from the pyroclastic fall and pyroclastic flow units.

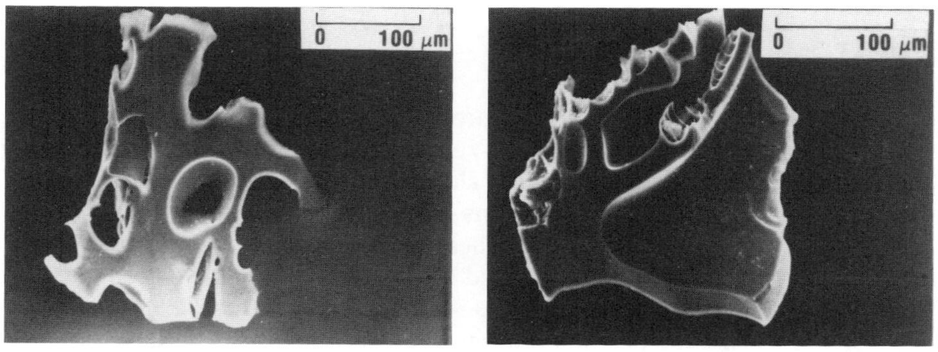

FIGURE 3. SEMs of vesicular, prong-like shards of fine ash in Bishop fall deposits. Note the delicate shapes and sharp edges determined by bubble walls; compare with fine ash in Fig. *1* (*lower right picture*).

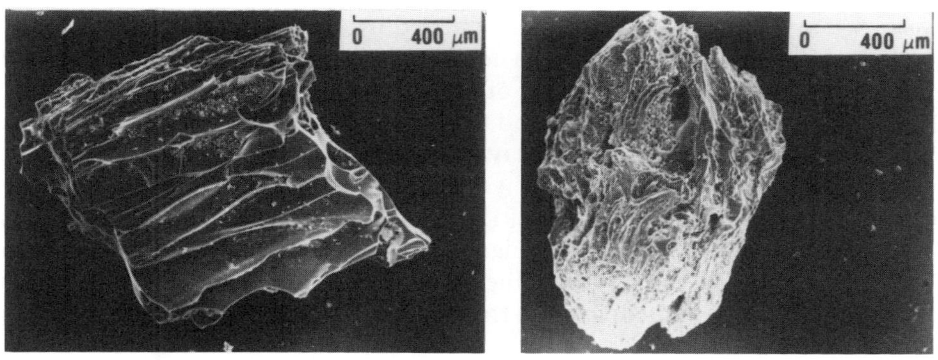

FIGURE 4. SEMs of pumiceous ash particles that dominate the fall deposit materials and are typified by high vesicularity, elongated vesicles, and some fine dust adhering to vesicle surfaces. Note that these pumices are more vesicular than those of the flow products shown in Fig. *1*.

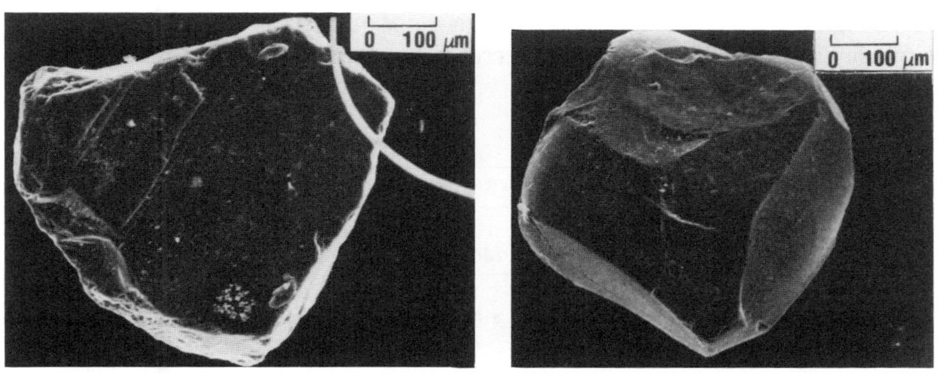

FIGURE 5. SEMs of crystals from the flow deposit show that they are rounded and have abundant abrasion chips and indentations along edges. A feldspar is shown on the *left* and quartz on the *right*. Note that large conchoidal fractures on the quartz have smoothed edges and that the cleavage of the feldspar is well developed.

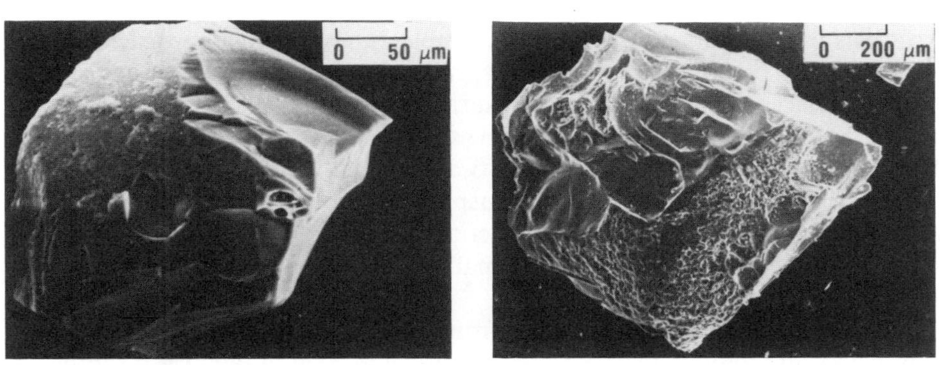

FIGURE 6. SEM of a quartz (*left*) and a feldspar (*right*) crystal from the fall deposit. Note the sharp edges of conchoidal fractures and dish-shaped fractures. Cleavage is only partly developed on the feldspar. Edges of both grains show little modification from percussion.

LOWER BISHOP TUFF
OWENS RIVER GORGE

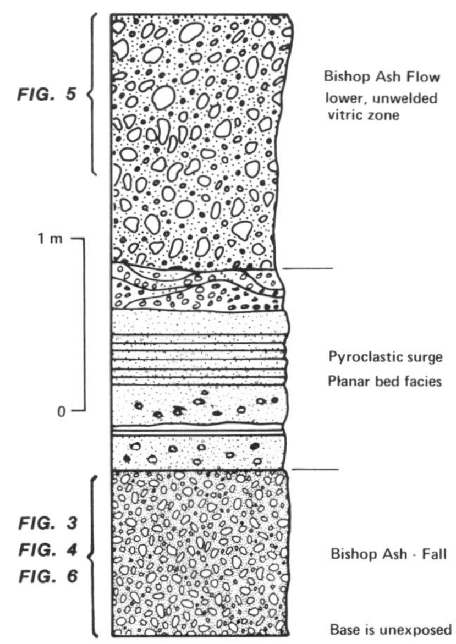

Stratigraphic columnar section of the lower part of the Bishop Tuff near Tom's Place, California. Samples studied were taken from the pyroclastic fall and pyroclastic flow units.

OBSIDIAN DOME, INYO DOMES, CALIFORNIA; SUB-PLINIAN AND PHREATOMAGMATIC ERUPTION

Introduction: Obsidian Dome is one of the five rhyolitic to rhyodacitic lava domes that are the youngest volcanic features of the Long Valley caldera, eastern California (Bailey et al., 1976). These domes are aligned along a N-trending fracture on the NW margin of the caldera, and their eruptions have been bracketed between 720 ± 60 and 1040 ± 250 y.b.p. (Wood, 1977). These vents have been studied by Mayo et al. (1936), and eruptions of the associated Mono Crater chain have been dated by Dalrymple (1967) and Friedman (1968).

The domes show evidence of a polygenetic history. The initial sub-Plinian to phreatomagmatic eruptions were caused by venting of a volatile-rich magma with an increasing hydromagmatic component as ground water gained access to the vent. These eruption phases were followed by passive lava extrusion. Our diagrammatic cross section illustrates a typical crater-rim profile. Initial explosions created the shallow crater surrounded by a tuff ring rampart, and final extrusions filled the crater with a lava dome. Samples described below were taken from the crater rim on the SW side of Obsidian Dome (the north Glass Creek Dome mentioned by Wood, 1977) and the tuff rampart of north Deadman Creek Dome (Wood, 1977). The erupted lavas are composed of a fluidal mixture of pumiceous, coarsely porphyritic rhyodacite and dark, aphyric to sparsely porphyritic rhyolitic obsidian (Lajoie, 1968; Jack and Carmichael, 1968). There is evidence that suggests mixing of Long Valley magmas with those of the Mono Craters. An increasing proportion of the rhyolitic obsidian in the lavas of the Inyo Domes from S to N is noted, and the rhyolite resembles that of rhyolite of Mono Craters.

Grain Size: Grain-size analyses for three samples taken from the Obsidian Dome tuff ring are shown in Table 1. Note the decrease in median diameter from the sample of early sub-Plinian fall deposit (sample 1) through those of the later phreatoplinian surge eruptions (samples 2 and 3). This decrease in grain size is accompanied by an increase in fine ash, which is evident as fine, adhering dust on larger pyroclasts.

TABLE 1. GRAIN-SIZE ANALYSIS OF OBSIDIAN DOME TEPHRA.*

Sample	Bed Form*	Md_ϕ (mm)	σ_ϕ	Fine Ash (Wt. %)
1	F	−2.11 (4.32)	1.74	1.77
2	P	1.68 (0.31)	1.52	7.80
3	M	2.10 (0.23)	1.75	5.68

*Bed forms: F = fall, P = planar surge, and M = massive surge.

Tephra Description: Sample 1 is a pumiceous fall deposit of highly vesicular rhyodacitic glass fragments. Pyroclasts have a jagged outline that consists of delicate, sharp points. Pyroclasts from a planar surge deposit (sample 2) have elongated vesicles. Grain shapes are not determined by vesicle surfaces but are blocky with curviplanar fracture surfaces. Alteration of vesicle surfaces is moderate and vesicle walls are broken. Sample 3 is from a rhyolitic vesiculated tuff that was deposited as a massive surge bed. Individual pyroclasts are weakly vesicular and are blocky in shape. Surfaces were smoothed by transport abrasion and are strongly altered; each has a hydration rind. The fine fraction of sample 3 is composed of broken bubble walls, chips, and flakes. These small pieces are covered with adhering dust particles 1 to 10 μm in diameter.

Dominantly rhyodacitic pyroclasts from north Deadman Creek Dome show both blocky and vesicular shapes. Vesicles are elongated in the mixed sample (as described for tephra 1; Wood, 1977) and are characteristic of rhyodacitic clasts formed during the sub-Plinian eruptions. In contrast to the vesicular rhyodacite, obsidian pyroclasts are nonvesicular (for example, those from Devil's Punch Bowl crater on the S end of the Mono Crater chain near its intersection with the Inyo chain).

Chemical Composition: Representative EDS chemical analyses are shown for Obsidian Dome samples in Table 2. The change in composition going up section from sample 1 to sample 3 is reflected as a increase in SiO_2 from 65 to 70 wt.%. There are two explanations for this change: (1) a progressive increase in the rhyolitic component of the glass during the eruption (glass analyses of rhyodacite from south Inyo Dome show SiO_2 at an average 62 wt.%, and analyses of Devil's Punch Bowl rhyolitic obsidian show SiO_2 at 76 wt.%); or (2) progressive pyroclast alteration caused by phreatomagmatic eruptions in which vaporized water condensed on pyroclasts and hydrated their surfaces. Both explanations are supported by field and microscopic evidence. Strong surface alteration is evident in sample 3, as observed by SEM, and its origin as a wet surge deposit suggests that surface hydration contributed to the observed chemical composition. However, the decrease in crystal content in sample 3 suggests an increasing rhyolitic component in the tephra during eruption.

TABLE 2. SURFACE CHEMICAL COMPOSITION, OBSIDIAN DOME.*

	1 (Wt. %)	2 (Wt. %)	3 (Wt. %)
SiO_2	65.56	68.58	70.02
TiO_2	2.33	0.57	0.66
Al_2O_3	14.44	17.68	16.34
FeO	3.15	2.32	1.54
MnO	1.85	0.02	---
MgO	1.60	0.37	0.39
CaO	3.03	2.25	2.30
Na_2O	1.37	0.91	0.94
K_2O	6.68	7.29	7.72
Glass	74.0	83.0	88.0
Crystal	19.0	15.0	9.0
Lithic	7.0	2.0	3.0

*Normalized standardless EDS analyses of grain surfaces.

Comments on Pyroclast Formation: Phreatomagmatic and phreatoplinian eruptions formed tuff rings around the Inyo Domes. Magmatic volatiles were a strong contributing factor in earlier eruptions. The volatile-rich top of the magma column decompressed as it vented, and that action initiated strong vesiculation and fragmentation of the melt. These early eruptions deposited mostly pumice falls. During this phase of sub-Plinian activity, rapid bubble growth was likely aided by propagation of expansion waves from the surface downward into the vent. These stress waves could have both shattered the viscous melt and promoted the formation of vesicular pyroclasts. As volatiles in the magma were depleted, there was fragmentation during interaction with ground water. Abundant water gained access to the vent when the local aquifer was disrupted by the initial magmatic bursts. Stress waves resulting from rapid water vaporization exceeded the strength of the melt as they propagated through vent. The blocky, nonvesicular pyroclasts formed were ejected from the vent as pyroclastic surges. In view of these ash formation mechanisms (Wohletz, 1983), sample 1 is probably magmatic fragmentation, sample 2 indicates a transitional phase of phreatomagmatic activity, and sample 3 is wholly phreatomagmatic. Variations in volatile content caused by mixing of porphyritic, volatile-rich, rhyodacitic and aphyric, dry rhyolite caused fluctuations in the degree of magmatic fragmentation. Magma mixing has been suggested as a trigger for silicic explosive eruptions (Sparks et al., 1977). The lack of crystals in the rhyolitic magma suggests that it was most likely hotter than the rhyodacite. If the rhyodacite was saturated with volatiles, an increase in temperature resulting from mixing with the hotter rhyolite would have caused supersaturation. Rapid dissolution of water from supersaturation then initiated vesiculation and magmatic eruption.

REFERENCES:

Bailey, R. A., Dalrymple, G. B., and Lanphere, M. A., 1976. Volcanism, structure, and geochronology of the Long Valley caldera, Mono County, California. J. Geophys. Res., **81**: 725-744.

Dalrymple, G. B., 1967. Potassium-argon ages of recent rhyolites of the Mono and Inyo Craters, California. Earth Planet. Sci. Letters, **3**: 289-298.

Friedman, I., 1968. Hydration rind dates rhyolite flows. Science, **159**: 878-880.

Jack, R. N. and Carmichael, I. S. E., 1968. The chemical 'fingerprinting' of acid volcanic rocks. Calif. Div. Mines Geol. Spec. Report 100: 17-32.

Lajoie, K. R., 1968. Late Quaternary stratigraphy and geologic history of Mono Basin. Ph.D. dissertation, Univ. of Calif., Berkeley: 271 pp.

Mayo, E. B., Conant, L. C., and Chelikowsky, J. R., 1936. Southern extension of the Mono Craters, California. Am. J. Sci., **32**: 81-97.

Sparks, R. S. J., Sigurdsson, H., and Wilson, L., 1977. Magma mixing: a mechanism for triggering acid explosive eruptions. Nature, **267**: 315-318.

Wohletz, K. H., 1983. Mechanisms of hydrovolcanic pyroclast formation: size, scanning electron microscopy, and experiments. J. Volcanol. Geotherm. Res., **17**: 31-64.

Wood, S. H., 1977. Distribution, correlation, and radiocarbon dating of late Holocene tephra, Mono and Inyo Craters, eastern California. Geol. Soc. Am. Bull., **88**: 89-95.

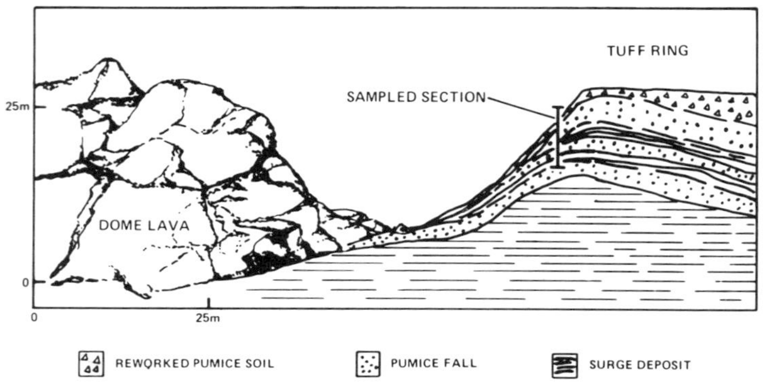

FIGURE 1. Schematic cross section of the rim section of a lava dome surrounded by a tuff ring. The tuff ring forms a collar of pyroclastic material that is often draped over an earlier crater rim. The section sampled at Obsidian Dome is located on the inner slope of a tuff ring that forms a moat around the dome.

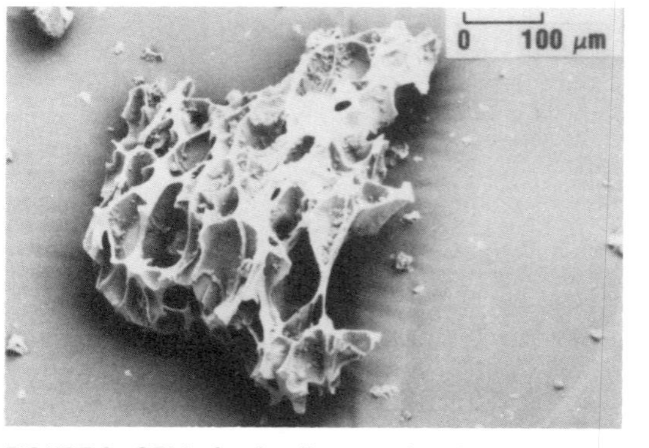

FIGURE 2. SEM of a rhyolite pyroclast from sample 1 that shows the strongly vesicular, delicate shape.

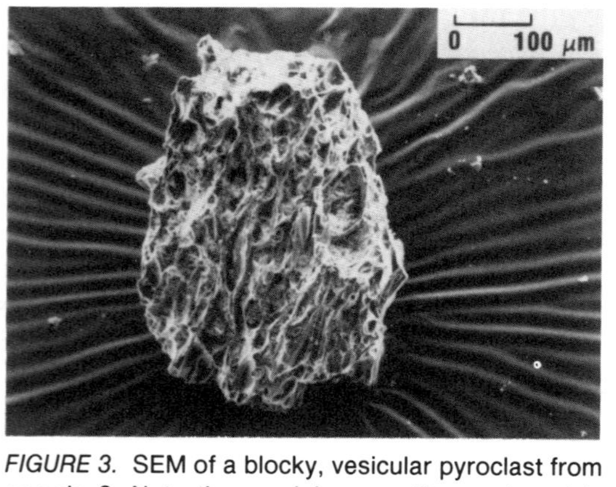

FIGURE 3. SEM of a blocky, vesicular pyroclast from sample 2. Note the curviplanar outline and vesicle fillings.

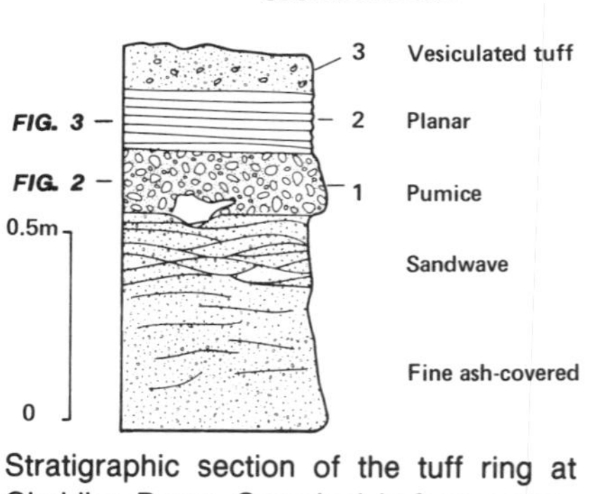

Stratigraphic section of the tuff ring at Obsidian Dome. Sample 1 is from a sub-Plinian pumice fall deposit, and samples 2 and 3 are from phreatoplinian deposits composed of planar and vesiculated massive pyroclastic surge beds.

FIGURE 4. SEM of a blocky pyroclast from sample 3 that shows rounded abraded edges and altered surface.

FIGURE 5. SEM of the fine fraction from sample 3. Note the vesicle chips and blocky particles covered with adhering dust.

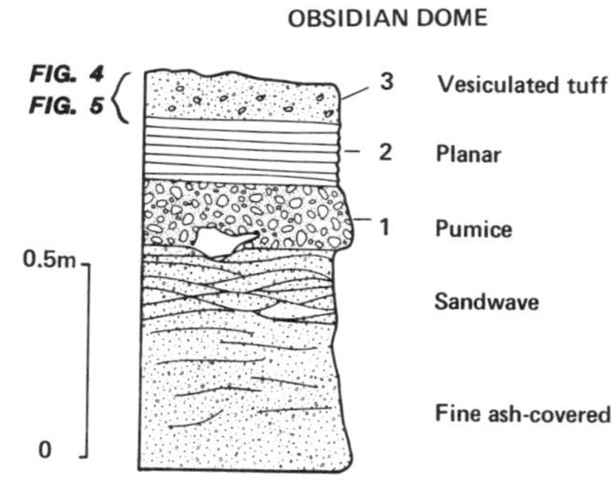

Stratigraphic section of the tuff ring at Obsidian Dome. Sample 1 is from a sub-Plinian pumice fall deposit, and samples 2 and 3 are from phreatoplinian deposits composed of planar and vesiculated massive pyroclastic surge beds.

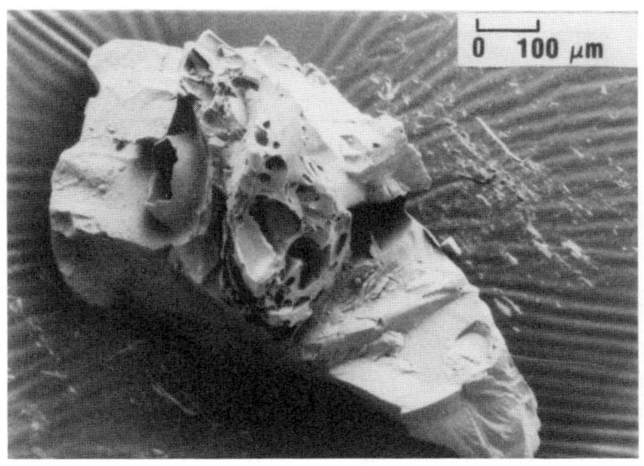

FIGURE 6. SEM of a pyroclast from north Deadman Creek Dome that shows both blocky and vesicular surfaces.

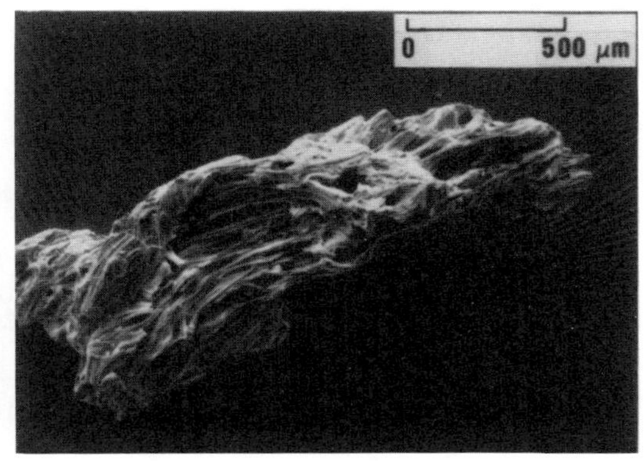

FIGURE 7. SEM of a blocky pyroclast with elongated vesicles that formed during fluid mixing of vesicular and nonvesicular lavas at north Deadman Creek Dome.

FIGURE 8. SEM of the fine fraction from north Deadman Creek Dome; it shows dominantly broken bubble-wall chips and splinters.

FIGURE 9. SEM of the fine fraction of Devil's Punch Bowl ash. Note the blocky, nonvesicular nature of this rhyolitic ash.

CERRO COLORADO, SONORA, MEXICO; WET PHREATOMAGMATIC ERUPTION

Introduction: Cerro Colorado is a tuff cone located in the Pinacate volcanic field near Crater Elegante, Sonora, Mexico. It has been studied by Jahns (1959); its stratigraphy is discussed by Gutmann and Sheridan (1978) and Wohletz and Sheridan (1983); and preliminary SEM analyses of tephra are shown by de Rita et al. (in press). Eruptions that built the steep-sided cone have been tentatively dated at 10,000 years (Shakel and Harris, 1972). The tuff cone deposits consist of highly palagonitized and indurated pyroclastic surge and fall beds that contain abundant accretionary lapilli; these features are indicative of wet eruptions in which steam condensed on particle surfaces and resulted in wet deposits. The cone encircles a crater nearly 1 km in diameter, and it rises over 100 m above the crater floor. The rim stratigraphy consists of a 10-m-thick basal unit of thinly bedded, slightly palagonitized tephra that overlies lake beds. This basal unit is, in turn, overlain by massive, thickly bedded, highly palagonitized tuff that makes up the bulk of the cone.

Grain Size: Tephra grain-size characteristics are given in Table 1 and show a gradual fining of median grain diameter upward in the section. The tephra are moderately sorted, and accretionary lapilli account for the lack of abundant fine ash, which is noted especially in samples of fall and planar bedded ash.

TABLE 1. GRAIN-SIZE CHARACTERISTICS OF CERRO COLORADO TEPHRA.

Sample	Bed Form*	Md_ϕ (mm)	σ_ϕ	Fine Ash (Wt. %)
1	S	1.40 (0.38)	1.64	7.38
2	M	1.96 (0.26)	1.80	10.53
3	F	0.38 (0.77)	1.64	1.07
4	M	1.39 (0.38)	1.74	8.91
5	P	0.44 (0.74)	1.52	0.91

*Bed forms: S = sandwave, M = massive, P = planar, and F = fall.

Tephra Description: Table 2 lists textural features of the samples described below. This tephra sequence is characterized by ash that has been extensively hydrated as a result of palagonitization during and after emplacement. Subangular to subrounded grains in the surge deposits show edge modification that resulted from abrasion of the original curviplanar, blocky pyroclast shapes. The sample of lapilli fall tuff shows few chipped and abraded edges. All samples except sample 1 show surface alteration coatings and desiccation cracks on hydrated skins and palagonitic surface material. Sample 1 from the basal, thinly bedded, poorly consolidated, dry surge deposit shows pyroclasts that are relatively fresh.

TABLE 2. DESCRIPTIONS OF TEPHRA FROM CERRO COLORADO.

Sample	Bed Form*	Grain	Edges	Alteration	Fine Fraction (Wt. %)	Distinguishing Features
1	S	blocky, drop-like	subrounded, smoothed, chipped	crusty hydration rind	blocky with abundant adhering dust (7%)	adhering dust on coarse grains
2	M	accretionary, blocky	rounded, dish-shaped fractures	high; granular material covering surface	blocky cores with accreted fine dust (11%)	glass enclosing crystals, accretionary lapilli
3	F	vesicular	subangular, few chipped surfaces	a few vesicle fillings, fresh surfaces	angular plates and blocks (1%)	solution pits
4	M	blocky vesicular	subrounded, grooves and chips	moderate to high; granular encrustation	angular, blocky, and drop-like (9%)	crystal molds in glass
5	P	blocky	subrounded, grooves, dish-shaped fractures	moderate; hydration cracks and skin, fresh areas	plate-like (1%)	glass enclosing crystals

*Bed forms: S = sandwave, M = massive, P = planar, and F = fall.

Chemical Composition: The EDS analyses of pyroclast surfaces are shown in Table 3. Sample A is an analysis of a fresh, basaltic fall deposit from nearby Crater Elegante for comparison. Note that Colorado samples all show greater SiO_2 values and smaller values for most of the other oxides. These differences arise from extensive palagonitization that affects most pyroclast surfaces—even those in the relatively fresh example, sample 1. Ash constituent analyses of glass, crystal, and lithic abundances show little variation throughout the eruption sequence, with the exception of the fall deposit, sample 3, in which crystals are separated from glass and lithics (possibly during emplacement).

TABLE 3. GRAIN SURFACE CHEMISTRY OF CERRO COLORADO TEPHRA.*

	A	1	2	3	4	5
SiO_2	50.45	69.67	62.55	69.55	69.88	71.03
TiO_2	4.26	2.97	3.56	2.63	2.68	2.43
Al_2O_3	16.85	10.78	13.54	11.58	12.48	11.58
FeO	6.70	4.33	6.57	8.17	5.21	5.94
MnO	2.21	1.96	1.66	---	0.83	---
MgO	3.84	2.49	2.99	1.67	1.63	1.62
CaO	8.31	3.64	4.92	5.45	5.24	5.22
Na_2O	2.36	1.14	1.28	---	0.88	0.62
K_2O	2.84	3.02	2.78	1.95	1.68	1.58
Glass	86.00	87.00	86.00	92.50	88.00	82.00
Crystal	11.00	12.00	14.00	4.50	10.00	17.00
Lithic	3.50	1.00	1.00	2.00	1.50	1.00

*Normalized standardless EDS analyses. Sample A is a fresh pyroclast from nearby Crater Elegante.

Comments on Pyroclast Formation: The following discussion illustrates probable eruption and emplacement variations that occurred throughout the course of activity and that resulted in the observed sequence of textural features. A more detailed version of this interpretation is available in Wohletz and Sheridan (1983).

Opening eruptions were high-energy explosions that produced fine-grained tephra and dry steam. These eruptions produced pyroclastic surges in which steam separated from pyroclasts before their emplacement. Hence, particles were deposited with relatively dry surfaces. The generally slight surface alteration of these pyroclasts contrasts with that of pyroclasts emplaced in wet deposits and is evidence that hydration of pyroclast surfaces occurs during and shortly after emplacement because of the wetness of the deposit. Blocky pyroclasts were smoothed by abrasion during emplacement. These pyroclasts are covered with fine adhering dust, which also attests to high-energy explosions that were capable of producing abundant fine-grained particles.

Later eruptions were wetter; they produced grains that stuck together because of the cohesion of wet surfaces. The strong alteration of these particles is possibly related to the presence of steam droplets that condensed on pyroclast surfaces before and during emplacement. Glass that clings to the crystals likely indicates a less energetic disruption of the magma than that which produced the glass-free crystal surfaces common in earlier eruptions.

Phreatomagmatic activity was punctuated by explosive pulses in which there was little magma/water interaction. Sample 3 is from an eruption caused by vesiculation. Grains in the fine fraction of this sample show little hydration, which suggests a relatively dry environment after emplacement. Those of the coarse fraction, however, have cracked, hydrated skins, which indicates that some water was probably present in the deposit after emplacement.

Samples 4 and 5 are from planar and massive bedded deposits emplaced by surges during wet eruptions; the massive surge deposits are characteristic of tuff cones. Sample 4 originated from a more energetic eruption than that which produced sample 5; it is finer grained, and phenocryst surfaces are completely separated from glass coatings. Other evidence for a more energetic eruption is seen in the blocky and drop-like fine fraction of sample 4, which indicates explosive mixing on a fine scale (see section in this book on the experimental production of ash). Sample 5 fine fraction pyroclasts formed more as a result of larger pyroclasts breaking than by the energetic mixing with water. Larger clasts in these samples often form the nuclei of accretionary lapilli—another feature that typically indicates wet emplacement of surges at tuff cones.

REFERENCES:

Gutmann, J. T. and Sheridan, M. F., 1978. Geology of the Pinacate volcanic field: State of Arizona Bureau of Geology and Mineral Technology, Special Paper No. 2: 47-59.

Jahns, R. H., 1959. Collapse depressions of the Pinacate volcanic field, Sonora, Mexico. Arizona Geol. Soc. Southern Arizona Guidebook II: 165-184.

de Rita, D., Sheridan, M. F., and Marshall, J., in press. SEM surface textural analysis of phenocrysts from pyroclastic deposits at Baccano and Sacrofano volcanoes, Latium, Italy. *In*: W. B. Whalley and D. H. Krinsley (Eds.), *Scanning Electron Microscopy in Geology*, Geo Abstracts, Inc., Norwich, England.

Shakel, D. W. and Harris, K. M., 1972. Potsherd evidence for minimum age of Cerro Colorado Crater, Pinacate volcanic field, northwestern Sonora, Mexico, Geol. Soc. Am. Abs. with Programs **4**: 408.

Wohletz, K. H. and Sheridan, M. F., 1983. Hydrovolcanic explosions II: evolution of basaltic tuff rings and tuff cones. Am. J. Sci., **283**: 385-413.

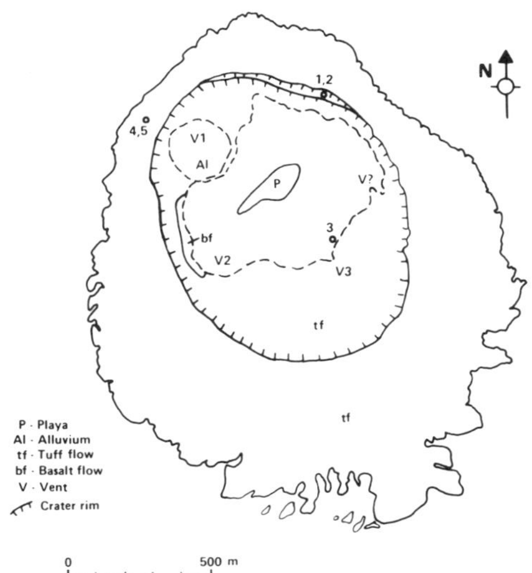

FIGURE 1. Geologic sketch map of Cerro Colorado with map units shown by key (adapted from Jahns, 1959). Samples were taken from locations 2 and 3.

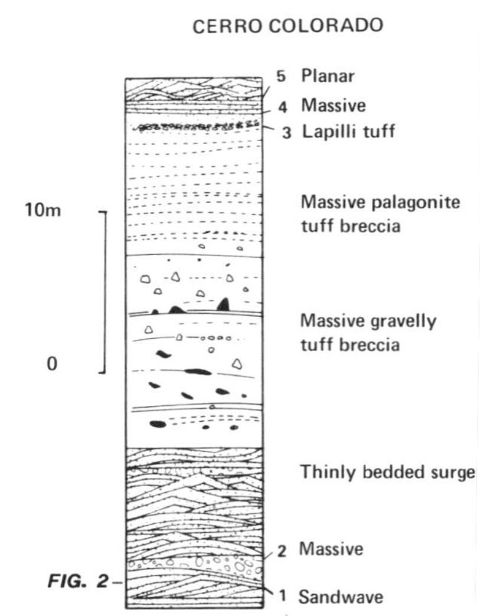

FIG. 2

Composite stratigraphic section of the rim of Cerro Colorado that shows a thinly bedded basal unit, thickly bedded upper unit, sample locations, and deposit characteristics.

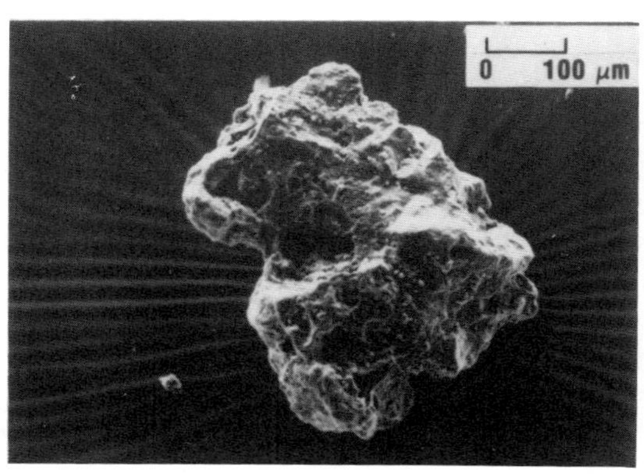

FIGURE 2a. SEMs of sample 1. Coarse-fraction pyroclast with blocky shape and several vesicle surfaces. It has relatively fresh-appearing surfaces and fine, adhering dust particles.

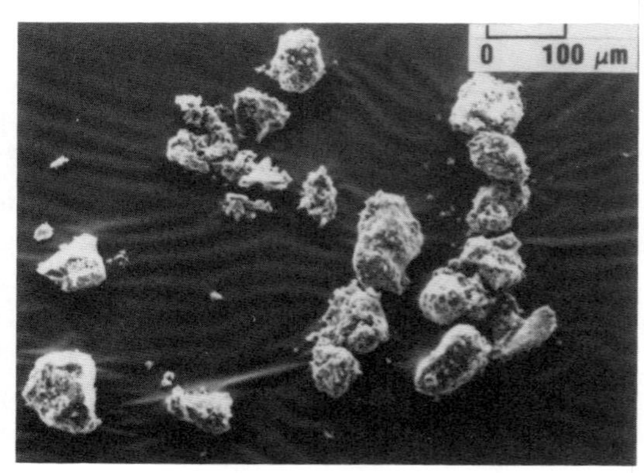

FIGURE 2b. Fine fraction consisting of blocky pyroclasts with abundant adhering dust.

FIGURE 3a. SEMs of sample 2, a glassy pyroclast with crystal molds. Note the smooth phenocryst surfaces and crusty glass coating with an uneven surface.

FIGURE 3b. Accretionary ash grain of hydrated particles bonded tightly by fine particles and altered glass. Note the rounded grain shape.

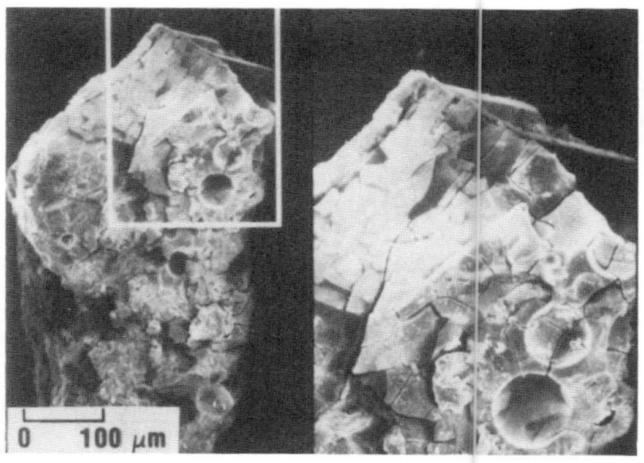

FIGURE 4a. SEMs of sample 3, a blocky, vesicular pyroclast with hydrated glass skin showing desiccation cracks. The right hand view is a 5X enlargement of the cracked surface that shows vestiges of a crystal mold.

FIGURE 4b. Fine-fraction particles are angular plates and blocks with little adhering dust.

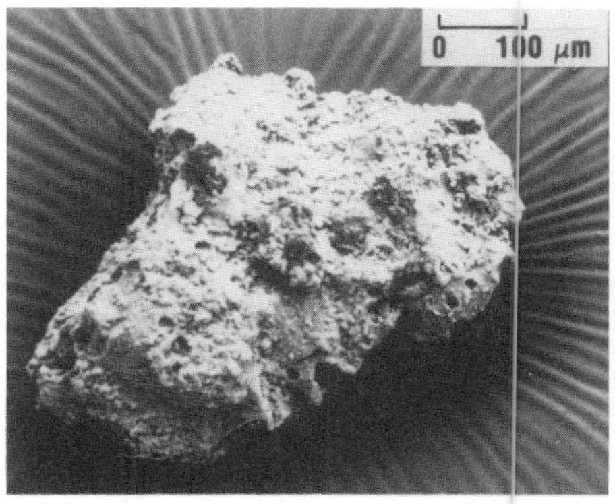

FIGURE 5a. SEMs of sample 4, a coarse-fraction pyroclast with highly altered surface and surface encrustation of clay materials. This grain is highly rounded and shows few vesicles.

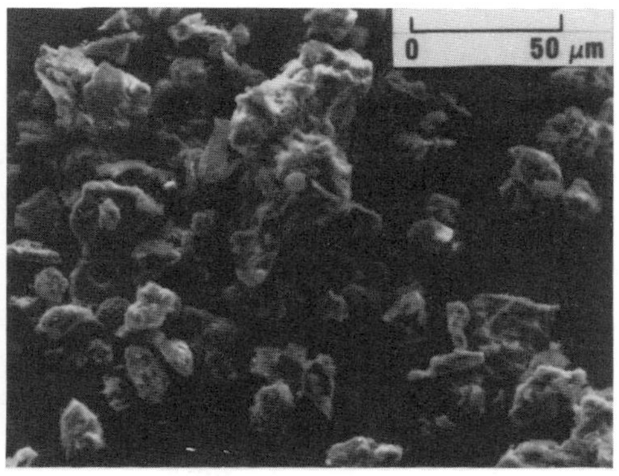

FIGURE 5b. Fine-fraction particles are irregularly shaped and angular. Grains show highly irregular shape.

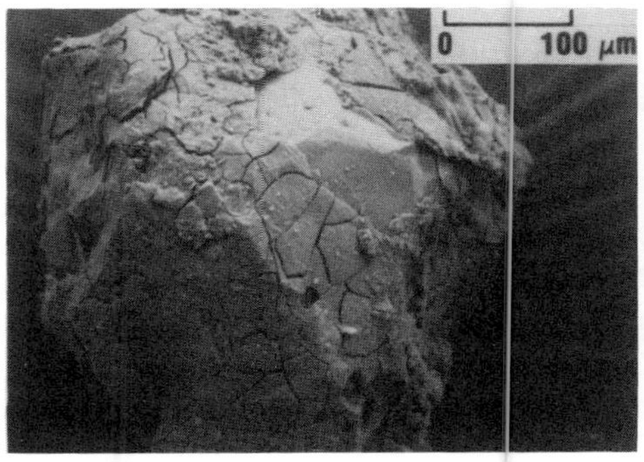

FIGURE 6a. SEMs of sample 5, a crystal enclosed in hydrated glass coating. Note the crystal faces and desiccation cracks.

FIGURE 6b. The fine fraction is composed of plate-like fragments with sharp, angular edges.

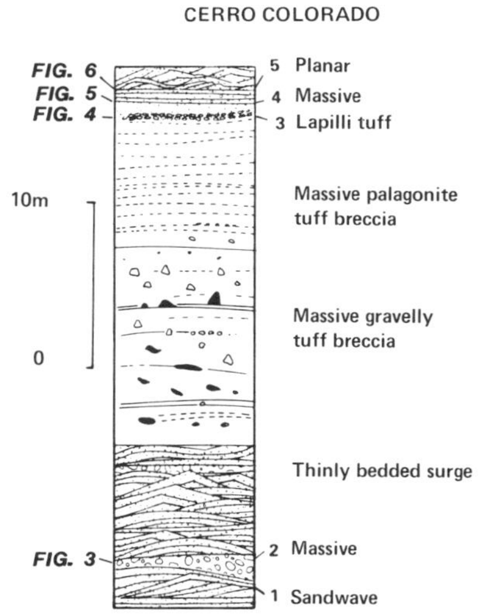

Composite stratigraphic section of the rim of Cerro Colorado that shows a thinly bedded basal unit, thickly bedded upper unit, sample locations, and deposit characteristics.

CRATER ELEGANTE, SONORA, MEXICO; DRY PHREATOMAGMATIC ERUPTION

Introduction: Crater Elegante is in the Pinacate volcanic field, northern Sonora, Mexico (Jahns, 1959). This maar crater, 1.6 km in diameter and 0.24 km in depth, was studied in detail by Gutmann (1976). The crater is surrounded by a 50-m-thick tuff ring. The tuff ring is composed mostly of pyroclastic surge deposits that were mapped and described by Wohletz and Sheridan (1979; 1983). The earliest, vent-opening pyroclastic eruptions at Crater Elegante produced Strombolian fall deposits and were followed by phreatomagmatic eruptions; this sequence is visible in crater-rim stratigraphic sections. Samples from a typical stratigraphic section are discussed below.

Grain Size: Samples are moderately sorted and are fine-ash to ash size with grain-size distributions that are typical of surge deposits (Table 1). An ash fall sample (1) has a median size of 7.88 mm, whereas surge samples range from 0.12 to 0.88 mm. Sandwave beds are the finest grained, and planar beds are the coarsest grained of the surge deposits. The abundance of fine ash (Md < 0.06 mm) is a measure of the fine fragmentation caused where the basalt lava mixed explosively with water during eruption.

TABLE 1. GRAIN-SIZE CHARACTERISTICS, CRATER ELEGANTE.

Sample	Bed Form*	Md_φ (mm)	σ_φ	Fine Ash (Wt. %)
1	AF	−2.98 (7.88)	2.07	1.97
2	S	2.56 (0.17)	1.81	14.54
3	S	1.81 (0.29)	2.06	11.83
4	M	3.02 (0.12)	1.77	25.12
5	P	0.17 (0.88)	1.55	0.78
6	P	0.72 (0.61)	1.54	1.86
7	M	2.71 (0.15)	1.85	13.36
8	S	1.76 (0.29)	1.35	4.53

*Bed forms: AF = ash fall, S = sandwave, M = massive, P = planar.

Tephra Description: The tephra samples of Crater Elegante illustrate a spectrum of SEM textures that are typical of basaltic phreatomagmatic ash. These textural features are summarized in Table 2 and are classified under overall grain shape, edge modification features produced by transport abrasion, alteration caused by postdepositional diagenetic processes, fine-fraction characteristics, and distinguishing textural features. The typical basaltic phreatomagmatic tephra (samples 2 through 8) consist of angular and blocky pyroclasts with curviplanar surfaces; they contrast with tephra from Strombolian deposits (sample 1) that are characterized by vesicular ash, surfaces bounded by bubble walls, and a lack of very fine ash particles.

Above the basal Strombolian fall deposit (sample 1), tephra were emplaced during phreatomagmatic eruptions; ash particles are weakly vesicular and show broken surfaces, surface alteration, rounding during transport abrasion, and abundant dust-size material in the fine fraction. Features distinctive of ash from this deposit include jagged edges and moderately fresh surfaces for pyroclasts from sandwave beds, rounded particles with highly altered surfaces from massive beds, and blocky, angular shapes with moderately fresh surfaces from planar beds.

Chemical Composition: The EDS analyses of grain surfaces are given in Table 3. Analysis of sample 1 is an average of several grain surfaces and represents typical unaltered hawaiite basalt compositions of the Crater Elegante area (Gutmann and Sheridan, 1978). Pyroclast surfaces remained mostly unaltered during Strombolian eruptions in which little water interacted with the melt. Samples 2 through 8 have EDS analyses showing strong chemical changes. These changes reflect alteration of grain surfaces by the abundant water in phreatomagmatic eruptions that was emplaced as steam with the tephra.

TABLE 2. DESCRIPTIONS OF CRATER ELEGANTE TEPHRA.

Sample	Bed Form*	Grain Shape	Modification	Alteration	Fine Fraction (Wt. %)	Distinguishing Features
1	F	vesicular	none	none	broken, clean surfaces (2%)	delicate shape preserved
2	SW	blocky, planar surfaces; few vesicles	rounded, large chips	slight to moderate	moss-like (15%)	abundant adhering micron-size surface material
3	SW	blocky, planar surfaces; few vesicles	rounded, large chips	moderate, hydration cracks	moss-like (12%)	adhering material, but less than sample 2
4	M	blocky accretionary masses	rounded, smooth	high, totally altered to palagonite	blocky, little adhering material (25%)	uneven surfaces of many small blocks and plates stuck together
5	P	blocky and vesicular	subangular, many small chips on edges	moderate; some fresh surfaces	broken, angular chips with fine adhering grains (1%)	abrasion features well preserved, many fracture surfaces
6	P	blocky with few vesicle surfaces	subrounded, chipped surface	moderate to high, vesicles filled with alteration material	broken, angular chips with fine adhering grains (2%)	abrasion features well preserved, many fracture surfaces
7	M	blocky and vesicular	subangular, small chips on edges	moderate vesicle fillings	blocky, abundant adhering material (13%)	alteration masking abrasion features
8	SW	blocky with few vesicles	subangular, smoothed edges	slight to moderate with some fresh surfaces	clean, blocky, nonvesicular (5%)	hydration rind in vesicles

*Bed forms: F = fall, SW = sandwave, M = massive, and P = planar.

TABLE 3. GRAIN SURFACE CHEMISTRY OF CRATER ELEGANTE TEPHRA.*

	1	2	3	4	5	6	7	8
SiO_2	50.45	62.69	63.78	58.59	67.07	60.81	52.35	62.61
TiO_2	4.26	3.29	3.67	3.46	3.28	3.50	3.58	2.36
Al_2O_3	16.85	14.41	16.60	19.61	14.36	17.18	21.66	15.89
FeO	6.70	9.37	2.93	3.67	4.21	5.50	5.62	6.01
MnO	2.21	0.06	1.11	1.60	0.44	0.89	1.37	0.62
MgO	3.80	1.70	2.72	3.28	2.50	3.74	4.94	2.68
CaO	8.31	6.83	5.05	5.06	4.68	3.70	5.12	6.68
Na_2O	2.36	0.11	1.50	1.37	0.93	1.36	1.88	1.23
K_2O	2.84	2.03	2.64	3.36	2.55	3.24	3.50	1.93
Glass	86.0	22.0	88.0	89.0	80.0	87.0	89.0	67.0
Crystal	11.0	12.0	7.0	9.0	13.0	9.0	9.0	26.0
Lithic	3.5	67.0	5.0	2.0	7.0	4.0	2.0	7.0

*Normalized standardless EDS analyses.

Comments on Pyroclast Formation: Phreatomagmatic ashes owe their overall shapes to fragmentation of the magma by contact with water. The thermal shock and vaporization pressure waves break up the melt into blocky, irregular, and sometimes drop-like and moss-like shapes. These shapes are strongly dependent upon the amount of water interacting with melt, the chemical composition, and physical characteristics of the melt, such as viscosity and yield strength. Subsequent dispersal of the ash in base surges promotes mechanical abrasion of the surfaces and particle rounding. Crater Elegante tephra were deposited as dry surge deposits in which much of the steam had segregated from the ash before condensation. In this fashion, ash particles were emplaced with relatively dry surfaces; subsequent alteration of the glass is minor, except in areas where there is abundant ground water. These dry deposits can be contrasted to wet surge deposits discussed in the section on Cerro Colorado.

REFERENCES: Gutmann, J. T., 1976. Geology of Crater Elegante, Sonora, Mexico, Geol. Soc. Am. Bull. **87**: 1718-1729.

Gutmann, J. T. and Sheridan, M. F., 1978. Geology of the Pinacate volcanic field. State of Arizona Bureau of Geology and Mineral Technology Spec. Paper No. 2: 47-59.

Jahns, R. H. 1959. Collapse depressions of the Pinacate volcanic field, Sonora, Mexico. Arizona Geol. Soc. Southern Arizona Guidebook II: 165-184.

Wohletz, K. H. and Sheridan, M. F., 1979. A model of pyroclastic surge. Geol. Soc. Am. Spec. Paper 180: 177-193.

Wohletz, K. H. and Sheridan, M. F., 1983. Hydrovolcanic explosions II: evolution of basaltic tuff rings and tuff cones. Am. J. Sci., **283**: 385-413.

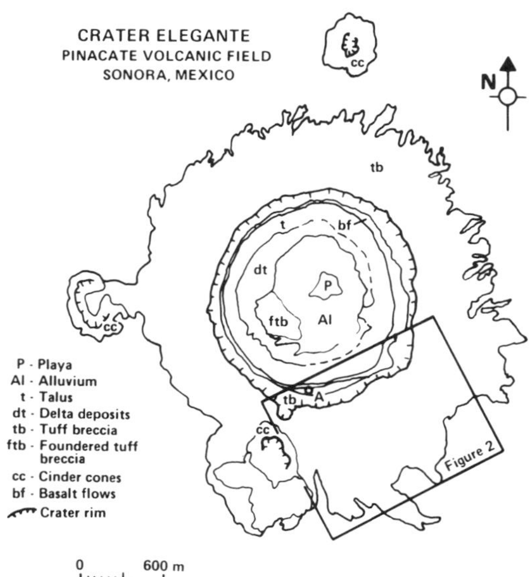

FIGURE 1. Geologic map of Crater Elegante showing the relationship of the surge deposit (tb) to the crater. Stratigraphic section and samples described in this chapter are from the S rim at location A.

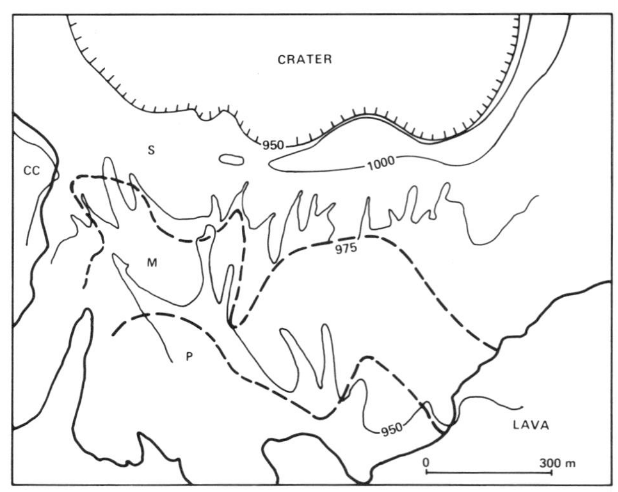

FIGURE 2. Pyroclastic surge facies map of the S rim of Crater Elegante (from Wohletz and Sheridan, 1979), which shows the extent of depositional facies characterized by dominantly sandwave (S), massive (M), and planar (P) bed forms.

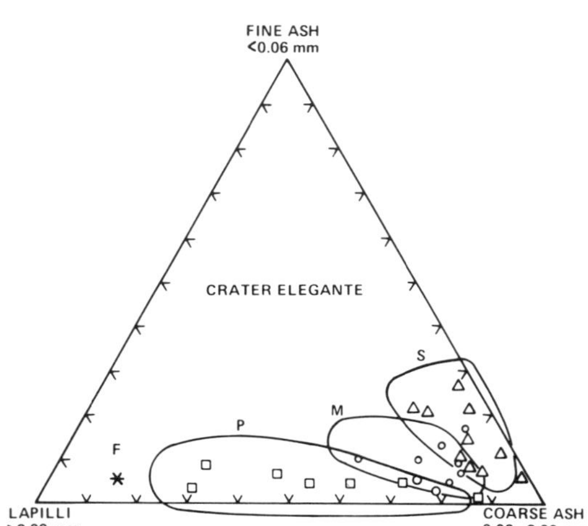

FIGURE 3. Triangular plot of sample median diameters by bed form: S = sandwave, M = massive, and P = planar surge beds.

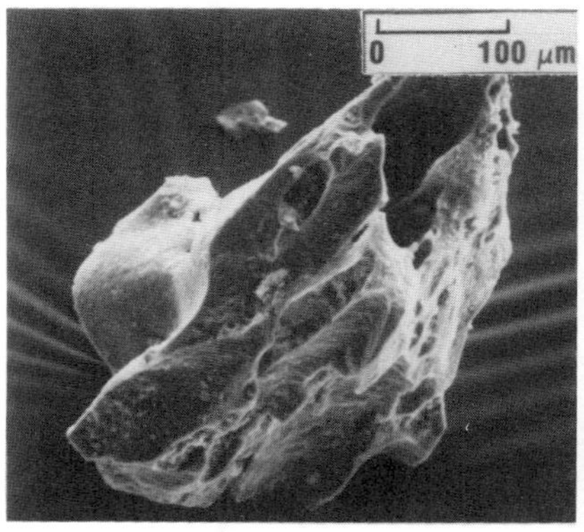

FIGURE 4. SEM of a characteristic pyroclast from sample 1 of the Strombolian fall deposit. The typical vesicular shape with fresh surfaces was formed by fragmentation during vesicle bursts. This texture indicates little interaction with meteoric water during eruption and marks a dominantly magmatic pyroclast.

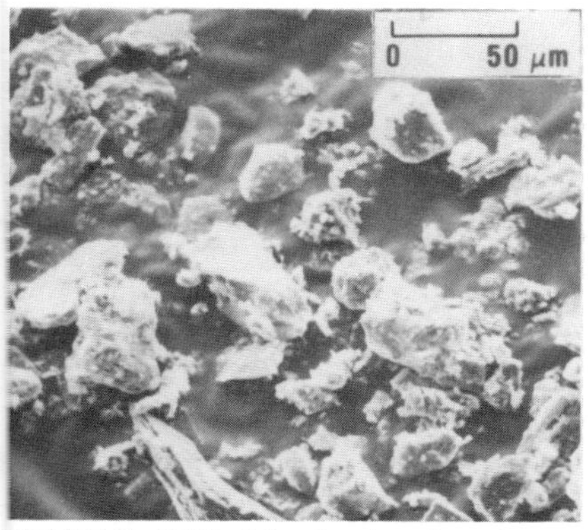

FIGURE 5. SEM of the fine fraction of sample 1, which accounts for less than 2 Wt.% of the total sample. Note the absence of very fine material adhering to grain surfaces. Blocky shapes with few vesicle surfaces are characteristic of particle sizes that are smaller than average vesicle diameters.

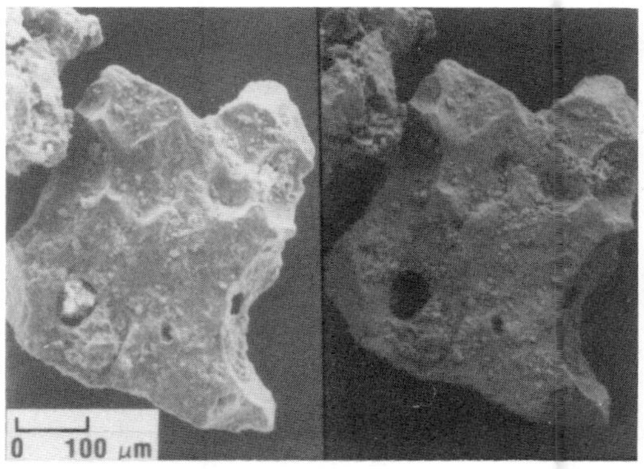

FIGURE 6. SEM of basaltic pyroclast from sample 2 that shows a blocky shape with few vesicle surfaces (typical of phreatomagmatic "Surtseyan" ash). The left view is a secondary electron image and the right view is a back-scattered image. Jagged edges result from abrasion during emplacement by pyroclastic surges. Fine, adhering dust on this grain is evidence of strong explosions, which tend to form abundant fine ash. Note that this deposit contains dominantly lithic material that is characteristic of deposition during the opening stages of phreatomagmatic eruptions (Table 2).

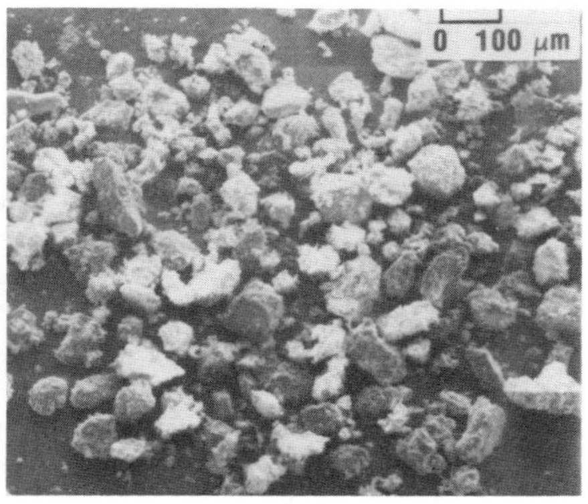

FIGURE 7. SEM of a pyroclast from sample 3; note the rounded and frosted surfaces.

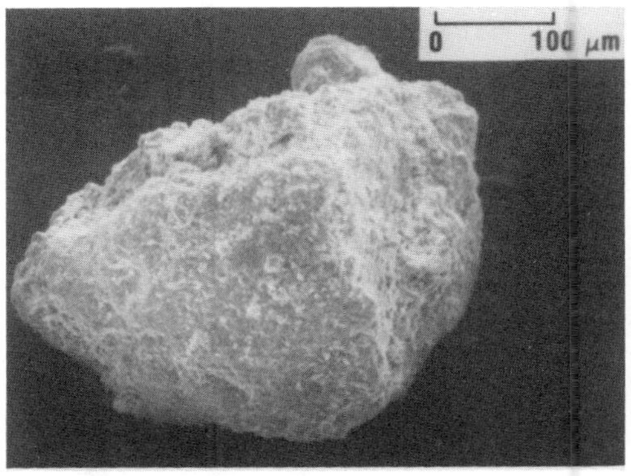

FIGURE 8. SEM of the fine fraction of sample 3, which has irregular, moss-like surfaces and abundant, fine, adhering dust.

FIGURE 9. SEM of a pyroclast from sample 4, a massive-bedded, pyroclastic surge deposit. The aggregation of small particles and strong surface alteration are evidence of emplacement in a wet surge cloud.

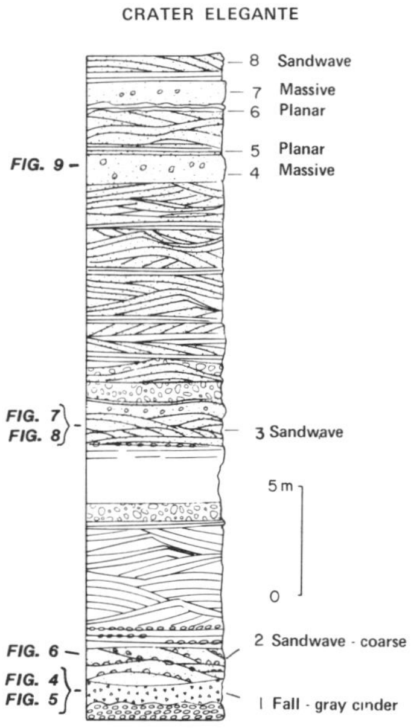

Rim stratigraphic section measured at location A (Fig. 1) that shows the sampled beds, their bed forms, and their grain-size characteristics. Sample 1 is at the base of the tuff ring deposit and is a Strombolian fall deposit. Sample 2 marks the beginning of phreatomagmatic activity that persisted throughout the emplacement of the rest of this stratigraphic section.

FIGURE 10. SEM of a pyroclast from a planar bedded surge deposit (sample 5). Both a secondary electron image (*left view*) and a back-scattered image (*right view*) are shown. Note the blocky shape with curviplanar surfaces showing dish-shaped fractures.

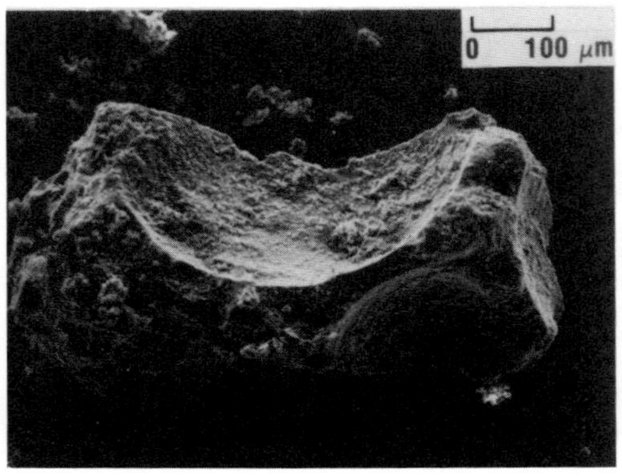

FIGURE 11. SEM of of blocky, vesicular pyroclast from a planar surge bed. Note the angular surfaces and slight incrustation of alteration products on vesicle surfaces.

FIGURE 12. SEM of a pyroclast from sample 7, a massive bedded surge deposit. The grain is well rounded and covered with palagonite alteration.

FIGURE 13. SEM of a blocky pyroclast from sample 8. Numerous pits and angular fractures that cover the grain surface have resulted from transport abrasion. Note the relatively fresh surface.

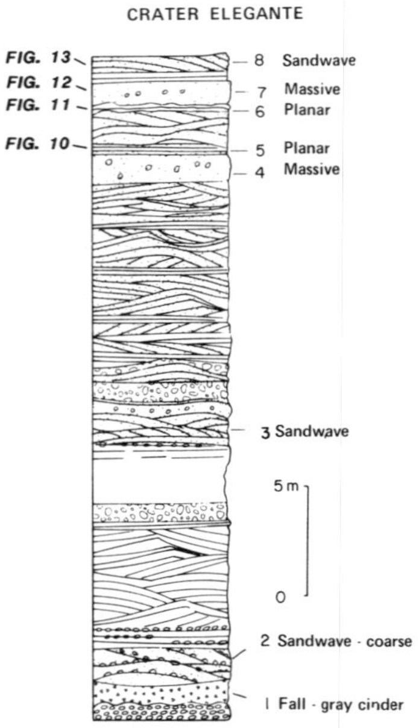

Rim stratigraphic section measured at location A (Fig. 1) that shows the sampled beds, their bed forms, and their grain-size characteristics. Sample 1 is at the base of the tuff ring deposit and is a Strombolian fall deposit. Sample 2 marks the beginning of phreatomagmatic activity that persisted throughout the emplacement of the rest of this stratigraphic section.

Eruption plume, Ngauruhoe, New Zealand.

ERUPTION PLUMES AND DISTAL ASH FALLS

Because gravity segregates the coarser grained and denser pyroclasts near the volcano, atmospheric scientists and tephrochronologists usually see only the finest grained volcanic ash. The ash examples used here are from observed historic eruptions and well-documented prehistoric eruptions. The samples were collected in eruption plumes or from ash beds located over 500 km from their sources.

Eruption Plumes

The examples here were collected by aircraft within dispersing eruption plumes in the troposphere and stratosphere.

VOLCÁN PACAYA AND VOLCÁN FUEGO, GUATEMALA; FEBRUARY 1978 ERUPTION

Introduction: In 1978, a U.S. research expedition to Guatemala used aircraft to sample volcanic eruption plumes from Volcán Pacaya and Volcán Fuego. Samplers were mounted on the aircraft, which flew through dispersing eruption plumes.

Volcán Fuego, characterized by violent, short-lived eruptions, was in a mild Vulcanian phase during the sampling period. It produced plumes of vapor and pyroclasts that rose to elevations of 4000 to 5000 m before being swept downwind. Fuego was erupting olivine-bearing, high-alumina basalts at that time.

During the sampling expedition, Volcán Pacaya was in a state of intense vapor emission that began in November 1977 and continued through May 1978. Lava was visible in the crater; recent lavas there are olivine-bearing, high-alumina basalts.

Both examples used here were provided by W. I. Rose, Jr.; the plates are from a paper by Rose et al. (1980).

Grain Size: The Fuego samples have a trimodal distribution with peaks at 0.4 to 0.8, 3, and 12 μm. Because the Pacaya plume consisted of liquid droplets only, no size data are available.

Tephra Descriptions: Glassy pyroclasts in the Fuego eruption plume have irregular, lumpy forms similar to the much larger pyroclasts that are typical of Vulcanian and some Strombolian eruptions. Angular cleavage fragments of plagioclase are also present. Most of the plumes from Fuego and Pacaya consist of impure acid droplets (mostly H_2SO_4).

Chemical Composition: Table 1 presents a whole-rock chemical analysis of Fuego basalt.

TABLE 1. WHOLE-ROCK CHEMICAL ANALYSIS OF 1974 FUEGO BASALT.*	
	(Wt. %)
SiO_2	50.3
Al_2O_3	19.4
FeO_3	10.4
MgO	5.0
CaO	9.0
Na_2O	3.4
K_2O	0.8
TiO_2	0.9
Total	99.4

*From A. Anderson, University of Chicago, unpublished data.

Comments on Pyroclast Formation: Mild but gas-rich eruptions of high-alumina basalt similar to those observed by the sampling team produce glassy pyroclasts with few vesicles. Grain surfaces are usually relatively smooth areas formed by surface tension effects or rough, hackly surfaces formed by brittle fracture. Heterogeneities in the erupting magma are evident in the mixture of glassy and tachylitic pyroclasts. There may be some degree of interaction between magma and ground water during these explosive eruptions, but this is very difficult to verify. The abundance of sulfuric acid droplets in the plume is evidence for the $S0_2$-rich magma at Fuego.

REFERENCES: Cadle, R. D., Lazrus, A. L., Huebert, B. J., Heidt, L. E., Rose, W. I., Jr., Woods, D. C., Chuan, R. L., Stoiber, R. E., Smith, D. B., and Zielinski, R. A., 1979. Atmospheric implications of studies of Central American volcanic eruption clouds. J. Geophys. Res., **84**: 6961-6968.

Rose, W. I., Jr., Chuan, R. L., Cadle, R. D., and Woods, D. C., 1980. Small particles in eruption clouds. Am. J. Sci., **280**: 671-696.

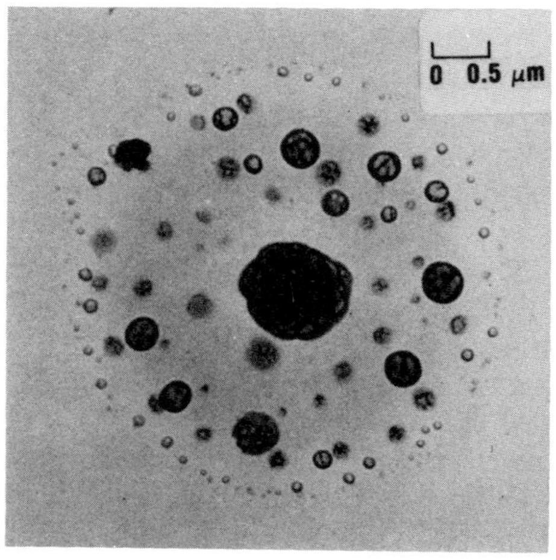

FIGURE 1a. Transmission electron micrographs of conventional impactor samples from the Fuego eruption plume of February 11, 1978 (from Rose et al., 1980). The impactor is a sampling tool used on an aircraft to sample particulates in the atmosphere. Probably an H_2SO_4 droplet with impurities, which is surrounded by spatter formed on impact with the collector.

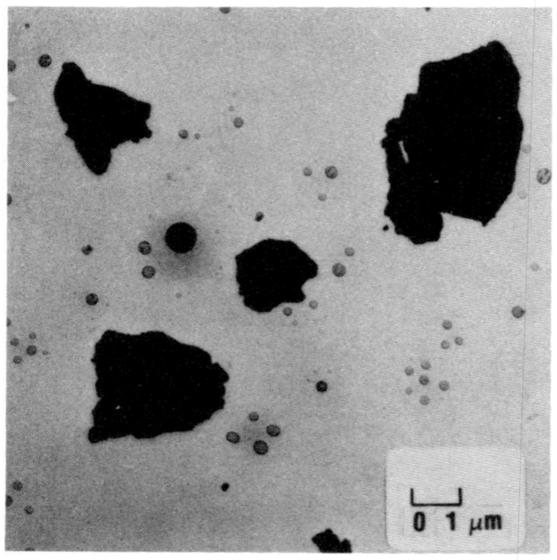

FIGURE 1b. Angular silicate pyroclasts and acid droplets. In another sample (see Cadle et al., 1979), similar particles were a mixture of plagioclase and glass.

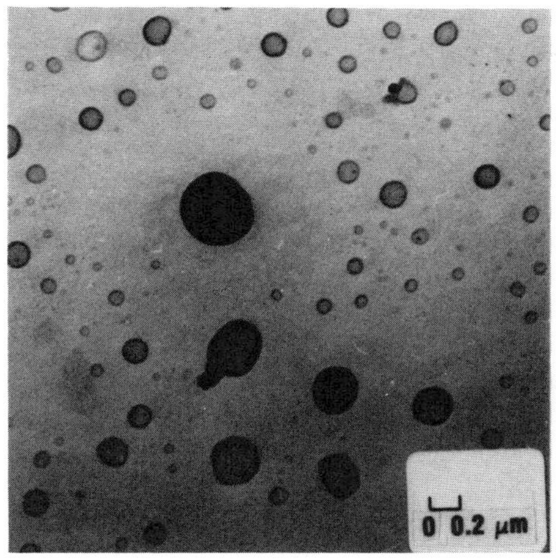

FIGURE 1c. Acid droplets, the most common component of dilute volcanic eruption plumes.

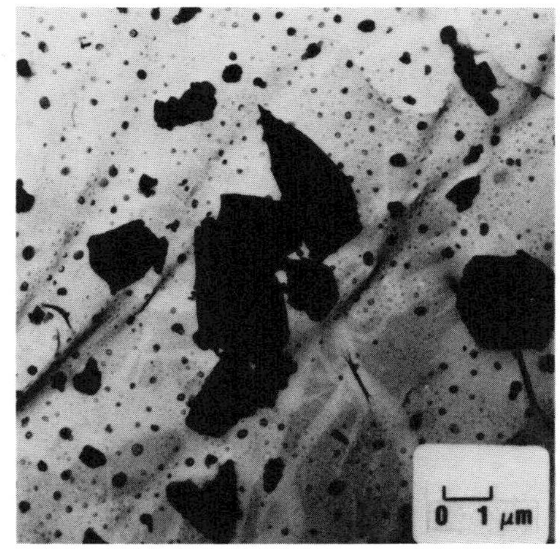

FIGURE 1d. Angular silicate particles and much smaller acid droplets. Those pyroclasts with well-developed cleavage are probably plagioclase and those with lumpy, botryoidal surfaces are most likely glass pyroclasts.

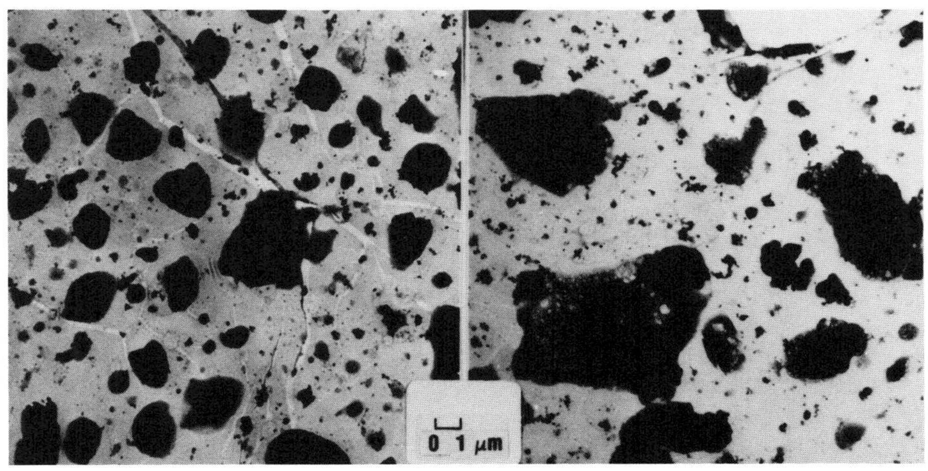

FIGURE 2. Transmission electron micrographs of impactor samples from the eruption plume of Pacaya Volcano, February 1978. Both samples illustrated here consist of impure acid droplets. Rose et al. (1980) note that silicate pyroclasts were rare in this eruption plume.

SANTIAGUITO, GUATEMALA; RECENT ERUPTIONS

Introduction: Santiaguito, located in western Guatemela, was the third volcano sampled by the 1978 U.S. research expedition. Santiaguito volcano consists of a group of dacitic lava domes that have been growing continuously since 1922; since April 1975, there have been daily explosive eruptions from the Caliente vent on the east side of the volcano. During the period of investigation, several large, ash-laden clouds rose to more than 5 km above sea level (2.5 km above the summit of the volcano) (Rose et al., 1980). The information for this example and the photographs were generously supplied by W. I. Rose, Jr., Michigan Technological University. Samples were collected with a 10-stage, quartz crystal, microbalance cascade impactor; these 10 stages sample 2 g/cm^3-density-equivalent particles of 25 to 0.05 µm in size.

Grain Size: Grain-size distributions within the cloud of February 25, 1978, are bimodal, with peaks at 0.12 to 0.4 and 3 to 12 µm. The sample was collected between 5.5 and 11 km downwind from the vent. These samples represent ~2% by weight of the ash erupted; most of the ash falls to the ground near the volcano and is not carried along in the plume.

Tephra Description: Tephra sampled with conventional and cascade impactors consist of glass pyroclasts of rhyolitic composition, plagioclase, magnetite, anhydrite, and sylvite. These pyroclasts are partly or completely coated with sulfuric acid. There are also numerous sulfuric acid droplets. There are aggregates present that consist of all the types of pyroclasts described above.

Glass pyroclasts are blocky, angular, equant to slightly elongate forms. Pyroclasts are indistinct due to the sulfuric acid coatings. Some vesicular pyroclasts exhibit irregular, equant to elongate vesicles and vesicle walls that range from <1 to 5 μm thick.

On lower stages of the impactor, there are mostly <1-μm angular fragments of plagioclase and magnetite, which are encrusted with liquid coatings (acids and Fe, K, Na, and Al salts).

Chemical Composition: Table 1 presents chemical composition data for Santiaguito dacite.

TABLE 1. CHEMICAL COMPOSITION OF SANTIAGUITO DACITE.*

	Glass (Wt. %)	Whole Rock (Wt. %)
SiO_2	72.8	63.45
Al_2O_3	13.5	16.94
FeO	1.3	4.57
MgO	0.4	1.87
CaO	1.1	5.03
Na_2O	4.2	4.91
K_2O	2.8	1.59
TiO_2	0.3	0.47
P_2O_5	0.3	0.22
MnO	0.0	0.14
H_2O	---	0.29
Total	96.7	99.53

*From Rose et al. (1980).

Comments on Pyroclast Formation: Rose et al. (1980) note that the vesiculated melt was fragmented by some other mechanism than bursting of vesicles; several possibilities are outlined, including particle collisions in the vent, explosion of overpressured glass inclusions in crystals, pressure-wave mechanisms, and melt/ground water interaction. They note that the >1-μm pyroclast population is dominantly glass, whereas the <1-μm pyroclasts are mostly phenocryst fragments. It is possible that phenocrysts with perfect cleavages are more easily fragmented. Liquid has accumulated on nearly all the silicate particles, which demonstrates the condensation and scavenging of dilute acid droplets by pyroclasts in the eruption plumes.

REFERENCE:

Rose, W. I., Jr., Chuan, R. L., Cadle, R. D., and Woods, D. C., 1980. Small particles in volcanic eruption clouds. Am. J. Sci., **280**: 671-696.

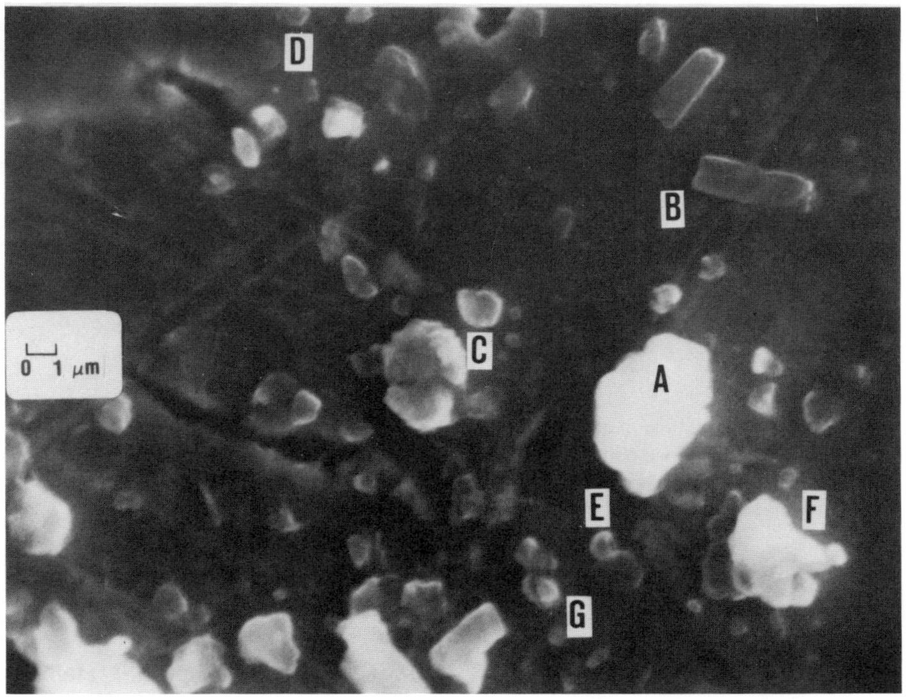

FIGURE 1. SEM of eruption cloud particles of February 22, 1978. A and F are rhyolitic glass pyroclasts; B is a tabular $CaSO_4$ particle; C a plagioclase fragment; and D, E, and G are acid droplets (probably H_2SO_4).

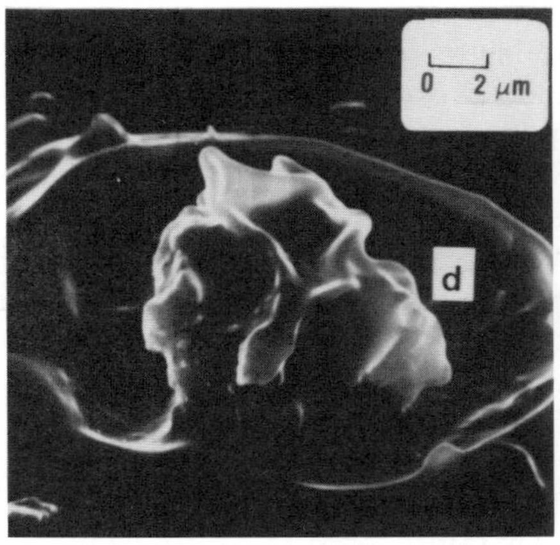

FIGURE 2a. Particles from the February 28, 1978 eruption. SEM of a vesicular shard of rhyolitic glass that is coated and surrounded by a ring of dilute sulfuric acid (spatter formed when droplet or coated pyroclast hit the impactor surface).

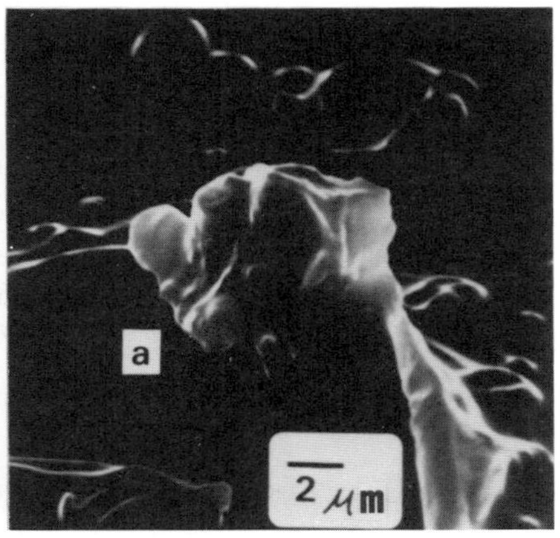

FIGURE 2b. Angular shard, which may be glass of rhyolitic composition, coated with $AlCl_3$.

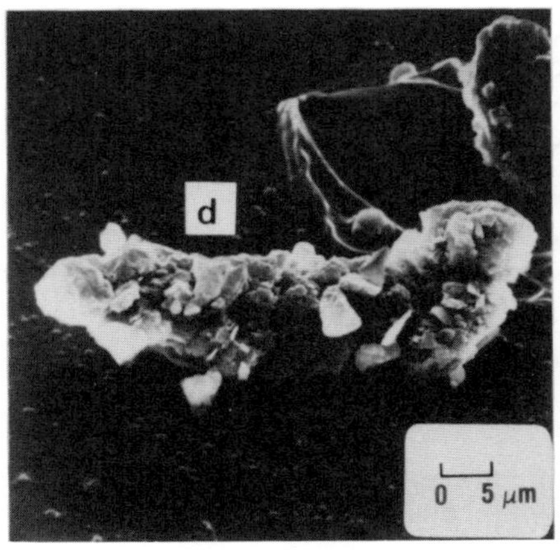

FIGURE 3a. Particles from the eruption of February 28, 1978, collected on stage 2 of the cascade impactor. These "clusters" (agglomerates) of silicate particles, including magnetite and $CaSO_4$, are bonded by acids.

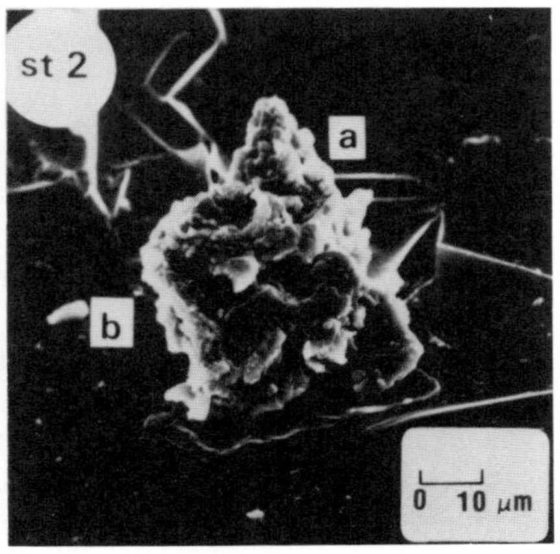

FIGURE 3b. Another "cluster." Analysis of particle *A* indicates a mixture of plagioclase and groundmass; *B* is a nonsilicate crystalline particle.

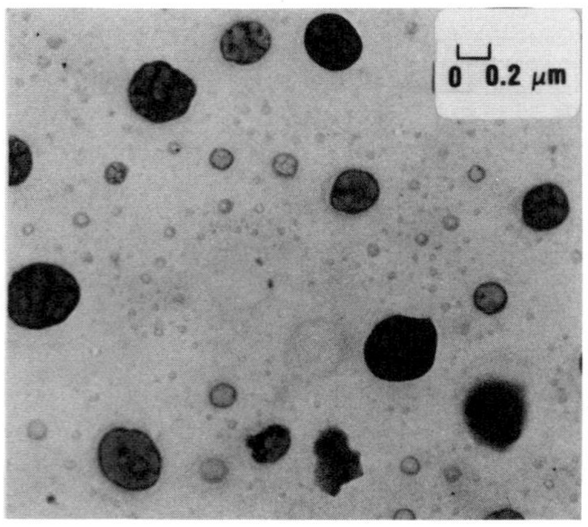

FIGURE 4a. Finer grained (0.4- to 0.2-μm) particles from the eruption cloud of February 28, 1978. Transmission electron micrograph of acid droplets collected on a conventional impactor.

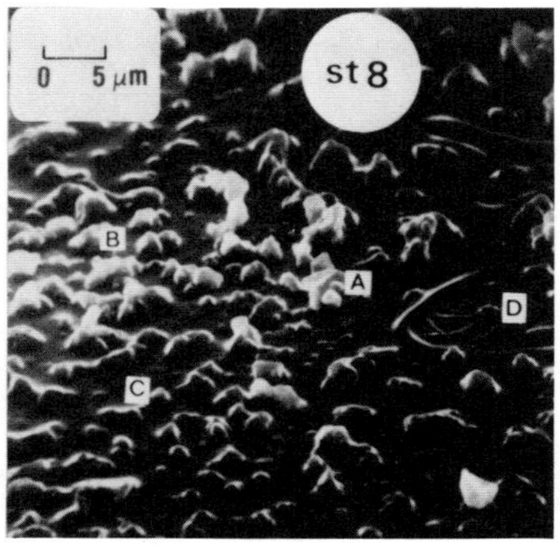

FIGURE 4b. Liquid-coated silicates. Fragments *A* and *C* are plagioclase; *B* and *D* are rhyolitic glass. The ring around *D* may be liquid spatter.

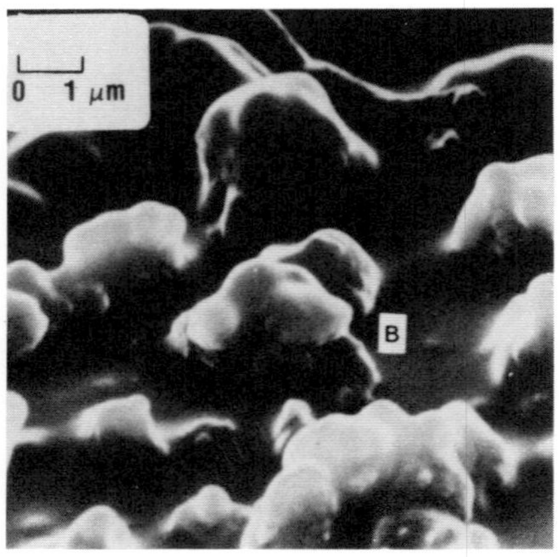

FIGURE 4c. Close-up of particle *B*, showing smooth surfaces due to liquid coatings.

ST. HELENS, WASHINGTON; MAY 18, 1980 ERUPTION

Introduction: The 1980 eruption of Mount St. Helens was the first whose eruption clouds were studied at the volcano and from aircraft, balloons, and spacecraft. The eruption brought together workers from many disciplines, including volcanologists, atmospheric physicists and chemists, meteorologists, and life scientists. Eruption plumes from the May 18 eruption were followed around the earth through satellite observations, lidar (laser radar), and twilight observations. *In situ* sampling was done at lower elevations by aircraft and in the stratosphere by high-altitude aircraft and balloons. For comprehensive syntheses of this information, see publications edited by Lipman and Mullineaux (1981), Newell and Deepak (1982), and Deepak (1982). As the plume moved E and SE across the U.S. only days after that eruption, aircraft were being prepared for sampling missions. Samples described here were collected in the stratosphere on May 20 (altitude of 16.8 km over SW Kansas) and May 21 (altitude of 18.3 km over S-central Wyoming and W-central Colorado). Detailed information on these flights is available in a paper by Sedlacek et al. (1982).

Particle Size: Silicate particles adhering to filters were measured by SEM. Samples from the May 20 flight (altitudes of 15.2 to 16.8 km) range from 0.5 to 10 μm and have a median size of 1.5 μm. Those collected on May 21 (altitudes of 15.2 to 18.3 km) range from 0.5 to 4 μm and have medians of 1.0 and 1.5 μm. Estimated concentrations in the central part of the plume (with an assumed median grain size of 2.0 μm) were 15 particles per cm^3 (May 20) and 1.5 particles per cm^3 (May 21).

Tephra Description: Silicate particles in the stratospheric plume from the May 18 eruption exhibit very little variety. The most common pyroclast types are slightly elongate, angular glass shards. Most are blocky, but some are irregular, scalloped forms; curved surfaces are vesicle-wall fragments. Most are partly coated with smaller glass fragments or liquid droplets.

Farlow et al. (1982) observed, in addition to glass shards, smaller amounts of fine-grained plagioclase, hornblende, pyroxenes, biotite, and chromitite fragments in the stratospheric plume a few days after the eruption. However, these particular clasts were not observed in the samples used here.

Chemical Composition: Chemical compositions of St. Helens ash are shown in Table 1. The variations in silicate and surface chemistry of Mount St. Helens ash are discussed in books edited by Newell and Deepak (1982) and Lipman and Mullineaux (1981).

TABLE 1. CHEMICAL COMPOSITION OF MOUNT ST. HELENS MAY 18, 1980 ASH.*

	1 (Wt. %)	2 (Wt. %)
SiO_2	67.00	71.40
Al_2O_3	16.10	14.60
TiO_2	0.64	0.37
Fe_2O_3	4.50	2.40
CaO	4.10	2.60
MgO	1.50	0.53
K_2O	1.70	2.00
Na_2O	4.60	4.30
P_2O_5	0.13	0.99
Total	101.07	99.19

*Samples are: (1) Bulk ash collected in Missoula, Montana, and analyzed by Rose and Hoffman (1980) by XRF. (2) Glass shard analyzed by Nelson et al. (1980) with an electron microprobe.

Comments on Pyroclast Formation: Interaction of ground and surface water after vesiculation of the dacitic magma is probably responsible for the very fine grained ash produced during the St. Helens eruption. Other lines of evidence for phreatomagmatic interaction are outlined in Rose et al. (1982). Sulfuric acid droplets, although relatively low in this eruption plume, were still much more voluminous than silicate particles.

REFERENCES:

Deepak, A. (Ed.), 1982. Atmospheric effects and potential climatic impact of the 1980 eruptions of Mount St. Helens. NASA Conf. Pub. 2240: 314 pp.

Farlow, N. H., Snetsinger, K. G., Oberbeck, V. R., Ferry, G. V., Polkowski, G., and Hayes, D. M., 1982. Time variations of aerosols in the stratosphere following Mount St. Helens eruptions. *In*: A. Deepak (Ed.), Atmospheric effects and potential climatic impact of the 1980 eruptions of Mount St. Helens, NASA Conf. Pub. 2240: 55-82.

Lipman, P. W. and Mullineaux, D. R. (Eds.), 1981. The 1980 eruptions of Mount St. Helens, Washington. U.S. Geol. Surv. Prof. Paper 1250: 844 pp.

Newell, R. E. and Deepak, A. (Eds.), 1982. Mount St. Helens eruptions of 1980. NASA Spec. Pub. SP-458: 119 pp.

Rose, W. I., Jr., Harris, D. M., Heiken, G., Sarna-Wojcicki, A., and Self, S., 1982. Volcanological description of the 18 May 1980 eruption of Mount St. Helens. *In*: R. Newell and A. Deepak (Eds.), Mt. St. Helens eruptions of 1980, NASA Spec. Pub. SP-458: 1-36.

Sedlacek, W. A., Heiken, G. H., Mroz, E. J., Gladney, E. S., Perrin, D. R., Leifer, R., Fisenne, I., Hinchliffe, L., and Chuan, R., 1982. Physical and chemical characteristics of Mount St. Helens airborne debris. *In*: A. Deepak (Ed.), Atmospheric effects and potential climatic impact of the 1980 eruptions of Mt. St. Helens, NASA Conf. Pub. 2240: 83-107.

FIGURE 1a. These silicate pyroclasts were collected on a filter at an altitude of 16.8 km over SW Kansas, on May 20, 1980. View of filter fibers that are coated with glassy pyroclasts and sulfuric acid droplets.

FIGURE 1b. The angular glass shard in this image is one of the largest from this sampling mission. Curved surfaces are vesicle-wall fragments. Vesicle walls visible in this pyroclast are 0.2 to 2 μm thick. Small, angular, and subrounded particles adhering to the shard are also glass fragments.

FIGURE 1c. Irregular glass shard with smaller shards adhering to the pyroclast surface. Only one remnant is visible on this blocky pyroclast. The substrate is a filter fiber.

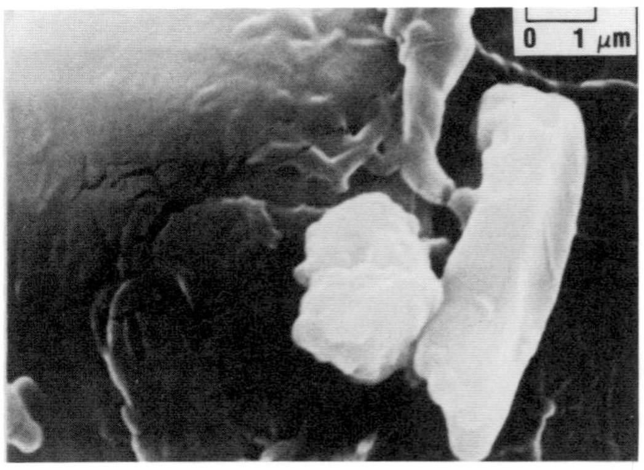

FIGURE 1d. Glass shards; because of liquid droplets adhering to grain surfaces, both have a rounded, lumpy appearance.

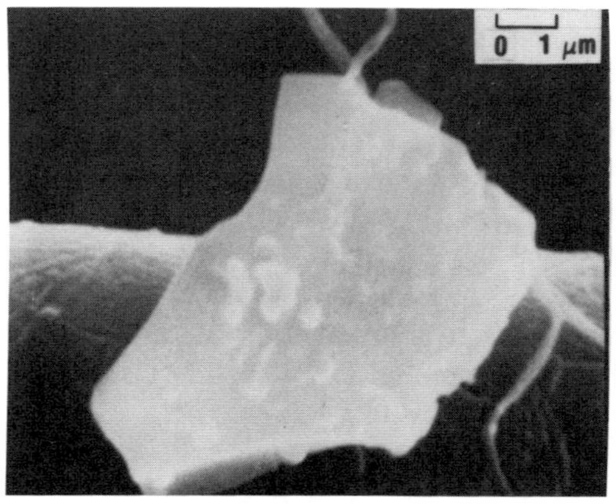

FIGURE 2. Glass shard collected at an altitude of 18.3 km over S-central and W-central Colorado on May 21, 1980. The angular, subequant pyroclast is partly coated with smaller shard fragments.

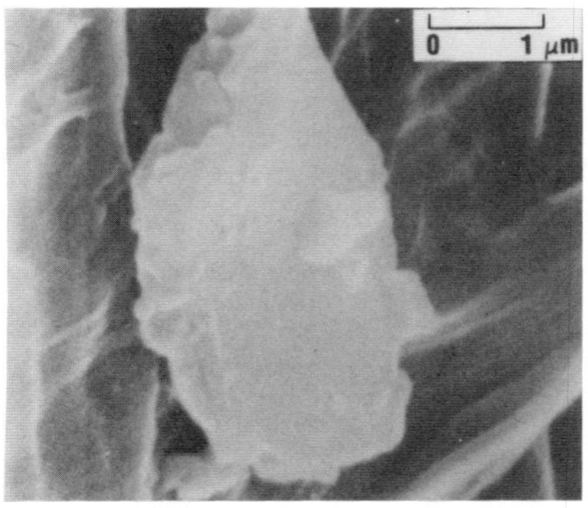

FIGURE 3. Small shard collected at an altitude of 9.5 km over central Kansas on May 20, 1980.

Distal Ash Falls

Plumes from larger eruptions can deposit ash fall layers over thousands to millions of square kilometers and these layers are convenient stratigraphic markers. Older ash beds provide information on the magnitude of large prehistoric eruptions. The "fingerprinting" and stratigraphic utility of ash beds has grown into the discipline of *tephrochronology*. For reviews of this subject, see Wilcox (1965), a bibliography by Westgate and Gold (1974), and papers in a volume edited by Self and Sparks (1981).

BANDELIER TUFF; BLANCO, TEXAS

Introduction: This example of a distal ash fall is a tephra layer correlated with the lower member of the Bandelier Tuff. It was collected near Mt. Blanco, Texas, 528 km SE of the Jemez volcanic field, New Mexico, which is the source of the Bandelier Tuff. Izett et al. (1972) made a correlation between this ash bed and the lower member of the Bandelier Tuff, proposing that it was related to the Plinian phase pumice fall of the lower member, whereas Wright et al. (1981) suggest that at least part is a co-ignimbrite ash fall deposit associated with pyroclastic flows of the lower member. Although it is not a definitive means of correlation, the shard shapes suggest that there is a closer affinity between this ash bed and the fine-grained fraction of the Plinian pumice fall. For a description of the Bandelier Tuff, see the example in Eruption Sequences.

Sample was provided by J. V. Wright of the British Petroleum Co.

Grain Size: Pyroclasts in this sample range from 1 to 100 μm.

Tephra Descriptions: The sample consists of nearly all glass shards and small pumice pyroclasts. They appear to have been derived from a physically heterogeneous magma and are of two pyroclast types.

(1) Slightly curved, plate-shaped shards represent wall fragments from large, flattened vesicles and clusters of coalesced vesicles. Reconstruction of these vesicles indicates that they were flattened, elongate ovoids with maximum intermediate axes of 50 to 130 μm. Thin ridges preserved on smooth shard surfaces represent remnants of vesicle walls (septae) that burst while still fluid. Surviving vesicle walls are 3 to 20 μm thick.

(2) Small pumice pyroclasts are characterized by highly elongate vesicles; in cross section, the vesicles are flattened ovoids with intermediate axes of 3 to 20 μm. Fracture surfaces across vesicle walls are both smooth and conchoidal. There is very little abrasion of these delicate shards.

Chemical Composition: Table 1 presents an electron microprobe analysis of the glass pyroclasts from a near-source Plinian pumice fall deposit of the Bandelier Tuff (from Crowe et al., 1978).

TABLE 1. COMPOSITION OF THE LOWER MEMBER OF THE BANDELIER TUFF.*

	(Wt. %)
SiO_2	73.78
TiO_2	0.10
Al_2O_3	11.74
FeO	1.15
MgO	0.05
CaO	0.30
Na_2O	3.38
K_2O	4.42
F	0.10
Total	95.02

*From Crowe et al., 1978.

Comments on Pyroclast Formation: See the comments on pyroclast formation within the section on the Bandelier Tuff sequence. Considering the age (1.4 m.y.) of this ash deposit, it is interesting to note that there is little alteration or chemical etching of glass shards and small pumices.

REFERENCES: Crowe, B. M., Linn, G. W., Heiken, G., and Bevier, M. L., 1978. Stratigraphy of the Bandelier Tuff in the Pajarito Plateau. Los Alamos Scientific Laboratory Report LA-7225-MS: 57 pp.

Izett, G. A., Wilcox, R. E., and Borchardt, G. A., 1972. Correlation of a volcanic ash bed in Pleistocene deposits near Mount Blanco, Texas, with the Guaje pumice bed of the Jemez Mountains, New Mexico. Quat. Res., **2**: 554-578.

Self, S. and Sparks, R. (Eds.), 1981. *Tephra Studies.* Reidel Pub. Co., Dordrecht: 481 pp.

Westgate, J. and Gold, C., 1974. *World Bibliography and Index of Quaternary Tephrochronology.* U. of Alberta, Edmonton: 528 pp.

Wilcox, R., 1965. Volcanic ash chronology, *In*: H. E. Wright, Jr., and D. C. Frey (Eds.), *The Quaternary of the United States.* Princeton U. Press, Princeton: 807-816.

Wright, J., Self, S., and Fisher, R. V., 1981. Towards a facies model for ignimbrite-forming eruptions. *In*: S. Self and R. Sparks, (Eds.), *Tephra Studies.* Reidel Pub. Co., Dordrecht: 433-439.

FIGURE 1a. SEMs of an unsieved ash sample collected near Mount Blanco, Texas, 528 km SE of its source in the Jemez Mountains, New Mexico. Slightly curved, smooth, plate-like shards dominate this view; some are crossed by thin septae that are remnants of vesicle walls that burst while still fluid.

FIGURE 1b. Similar to Fig. *1a*. At the base of the image is a vitric pyroclast with a conchoidal fracture surface. All the smaller fragments are glass shards as well.

FIGURE 1c. Blocky, angular shards and small pumice pyroclasts. The pyroclasts were produced by fragmentation of a brittle, solid vesicular magma. Most fracture surfaces are normal to elongate vesicles. Fracture surfaces are both smooth and conchoidal. Smooth, slightly curved to semicircular surfaces were vesicle walls.

FIGURE 1d. This image illustrates the two types of pyroclasts in this sample: slightly curved, plate-like shards broken from large vesicles and small pumice pyroclasts with thin, elongate vesicles.

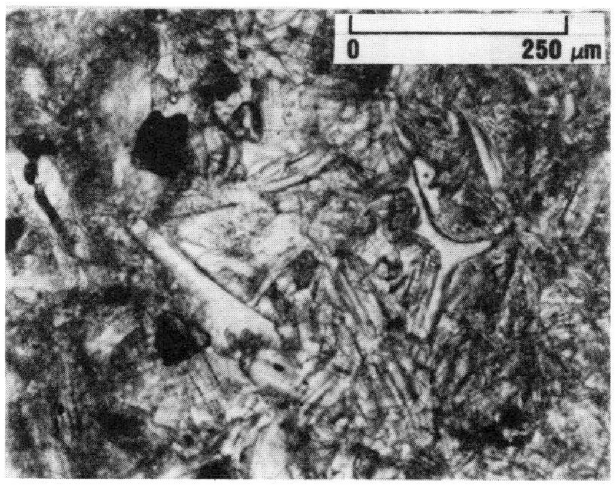

FIGURE 2a. Plane-light photomicrographs of the sample shown in Fig. *1*. Nearly all the ash consists of small, colorless shards. Note slightly curved, plate-like, and Y-shaped shards.

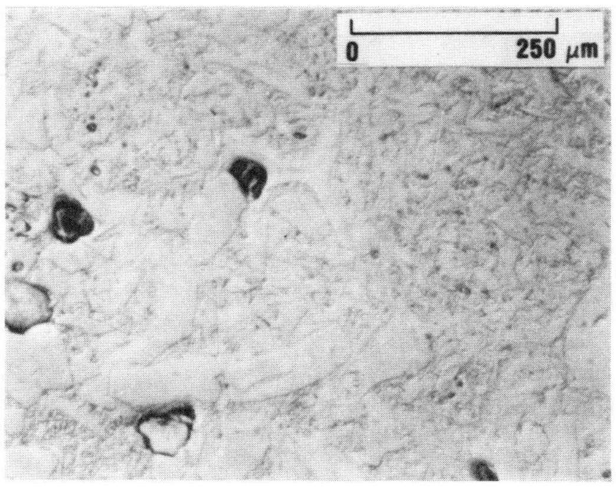

FIGURE 2b. Reflected-light photomicrograph of the sample seen in Fig. *2a*.

MAZAMA TEPHRA; STEPTOE CANYON, WASHINGTON

Introduction: Crater Lake, located in S-central Oregon, is one of the better known calderas of the Cascade Range. A large eruption that occurred about 7,000 years ago (^{14}C dates, as cited by Kittleman, 1973; Bacon, 1983) erupted 51 to 59 km^3 of rhyodacite to mafic andesite as Plinian pumice falls, pyroclastic flows, and preclimactic lava flows (Williams and Goles, 1968; Kittleman, 1973; Bacon, 1983). The extensive ash fall associated with the climactic phase covered much of the Pacific Northwest. Caldera collapse associated with the climactic phase formed the crater now occupied by Crater Lake.

Measurable ash fall deposits cover at least 900,000 km^2 (Fryxell, 1965), making this deposit an important stratigraphic marker that can be used by stratigraphers and archeologists. This sample of Mazama tephra was collected by R. Fryxell in Steptoe Canyon, Whitman County, Washington (600 km NE of Crater Lake).

Grain Size: Thorough analyses of grain-size variations within the Mazama tephra are given by Fisher (1964) and Kittleman (1973). Median grain size decreases from 23 mm (-4.5φ), at a distance of 5 km, to 350 μm (1.5φ), at a distance of 190 km. This deposit is moderately sorted near the source (5 km) and well sorted at a distance of 100 km.

Tephra Descriptions: An "average" sample, as described by Kittleman (1973), is a dacitic crystal ash (up to 60 Wt.% crystals). The distal ash fall sample described here was collected 600 km from its source; it is a vitric ash that consists of colorless flat and curved shards and small pumice pyroclasts (up to 100 μm long). Shards range in size from 1 to 100 μm long and have a variety of shapes, including Y-shapes, curved plates, and flat plates.

Chemical Composition: Table 1 gives an average composition of Mazama ash.

TABLE 1. AVERAGE COMPOSITION OF MAZAMA ASH.*

	(Wt. %)	
SiO_2	72.59	±0.27
Al_2O_3	14.42	±0.16
FeO	2.08	±0.06
CaO	1.71	±0.09
MgO	0.54	±0.08
Na_2O	5.15	±0.16
K_2O	2.70	±0.06
TiO_2	0.48	±0.02
Total	99.67	

*Average of 58 microprobe analyses from Westgate et al., 1969.

Comments on Pyroclast Formation: This vitric ash appears to have been derived from a heterogeneous, highly vesicular magma. Shards and pumice pyroclasts broken from a pumiceous melt have highly elongate, narrow vesicles and large, flattened, irregular vesicles. Much of the fine-grained ash may have been derived partly through a phreatomagmatic interaction of magma and surface water, but there are no data available at this time to confirm this interpretation.

REFERENCES:

Bacon, C., 1983. Eruptive history of Mount Mazama and Crater Lake caldera, Cascade Range, U.S.A.. J. Volcanol. Geotherm. Res., **18**: 57-115.

Fisher, R., 1964. Maximum size, median diameter, and sorting of tephra. J. Geophys. Res., **69**: 341-355.

Fryxell, R., 1965. Mazama and Glacier Peak volcanic ash layers; relative ages. Science, **147**: 1288-1290.

Kittleman, L., 1973. Mineralogy, correlation and grain-size distributions of Mazama tephra and other postglacial pyroclastic layers, Pacific Northwest. Geol. Soc. Am. Bull., **84**: 2957-2980.

Westgate, J., Smith, D., and Nichols, H., 1969. Late Quaternary pyroclastic layers in the Edmonton area, Alberta. *In*: Pedology and Quaternary Research, Symp. Proc. University of Alberta: 179-186.

Williams, H. and Goles, G., 1968. Volume of the Mazama ash-fall and the origin of Crater Lake Caldera. *In*: Dole, H. (Ed.), Andesite Conference Guidebook. Ore. Dept. Geol. and Min. Indus., Bull. **62**: 37-41.

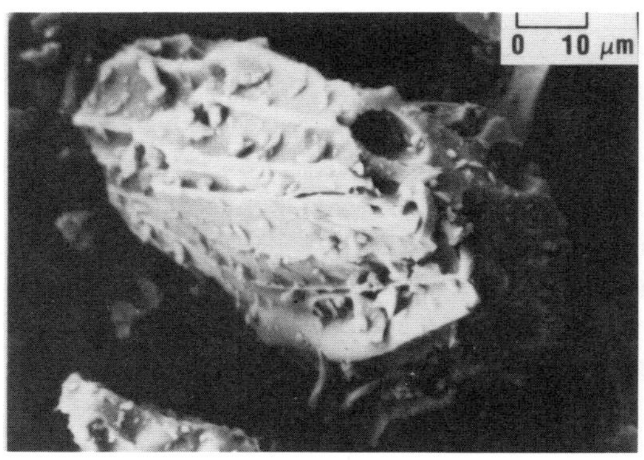

FIGURE 1a. Pumice pyroclasts from the Mazama tephra. Pumice pyroclast with highly elongate vesicles. They are slightly flattened ovals in cross section. Vesicle walls range in thickness from <1 to 10 μm. Small fragments clinging to the pumice pyroclasts are shards.

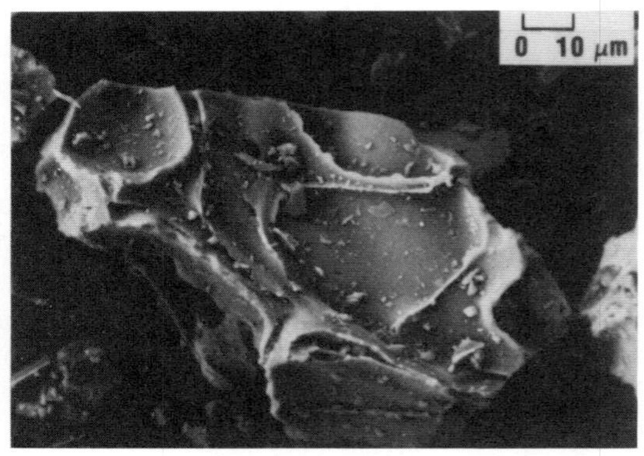

FIGURE 1b. A pumice pyroclast characterized by well-developed, flattened vesicles; many flat or slightly curved shards may have been derived from pumices similar to this one.

FIGURE 2a. Shards from the Mazama tephra. There are a variety of shards visible in this SEM photo. They range from 1 to 70 μm and vary from broad, flat plates to complex Y-shaped forms. The variety of shapes reflects heterogeneities in the size, shape, and degree of vesicle coalescence in the magma. The "lump" visible on a large plate near the center of the photo is most likely over a small phenocryst.

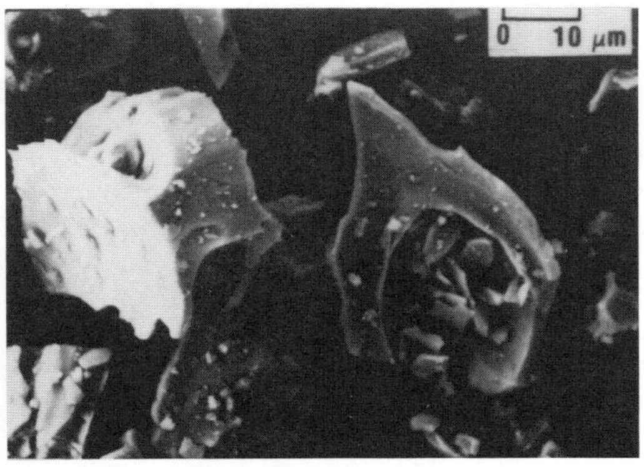

FIGURE 2b. Angular, blocky shards that are partly coated with aggregates of very fine-grained, thin shards.

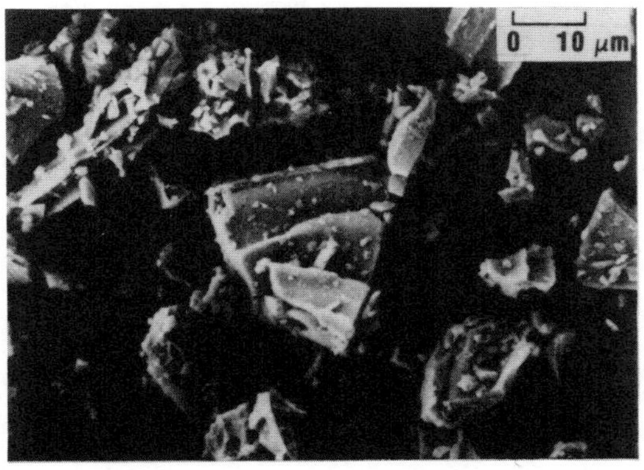

FIGURE 3. Aggregates or clusters of very fine grained vitric ash.

This weathered tuff from St. Vincent, West Indies, shows a 600-µm-long weathered pyroxene in a clay matrix.

WEATHERED AND METAMORPHOSED TUFFS AND RELICT PYROCLASTIC TEXTURES

Volcanic ash deposits form an important part of the geologic record. Little is done to identify weathered ash deposits and ashy soils. Most pre-Tertiary ash deposits are not recognized and consequently are classified as sedimentary rocks or lava flows. Only recently, investigators in Canada have begun to recognize the diversity of pyroclastic deposits that are preserved in Archean-age volcanic terranes. It is possible to identify relict pyroclastic textures in metamorphosed and deeply weathered tuffs, and several examples are provided here.

Introduction

The following section discusses the features of ash that result from secondary processes of hydration and palagonitization. The examples are drawn from other sections in this book. Instructive literature concerning alteration of volcanic ash is directed towards distinguishing a volcanic origin of clays and specific examples of ash diagenesis (Wise and Weaver, 1979; Wise et al., 1973). In this section we have attempted to present a general treatment of the topic.

ALTERATION OF VOLCANIC ASH: AN OVERVIEW OF HYDRATION AND PALAGONITIZATION STUDIES DONE BY SEM

Introduction: The processes of hydration and palagonitization (see the discussion by Fisher and Schmincke, 1984) involve physical and chemical changes that are intermediate between those of weathering (diagenesis) and metamorphism. Hydration of glass occurs when silicic pyroclasts or lavas are altered by water or steam (Friedman et al., 1966) and may be compared with the gradual alteration of basaltic glass during the process of palagonitization. Palagonite is formed by the alteration of basaltic glass to smectite clays and zeolites (Peacock, 1926; Peacock and Fuller, 1928; Honnorez, 1962; Hay and Iijima, 1968). Some investigators, including Nayudu (1964) and Bonatti (1965), suggest that palagonitization involves hydration of glass. The difference between the alteration of basaltic glass and that of silicic glass is not well understood, but experimental studies (Marshall, 1961; Hawkins and Rustim, 1963) show that basaltic glasses are more reactive to aqueous solutions than silicic glasses are. This reactivity may be a reflection of the greater concentration of silicon/aluminum polymer networks in silicic glasses.

Hydrated obsidian (perlite) is valuable for dating silicic rocks because it can be easily distinguished from fresh glass with an optical microscope. Because hydration "fronts" move into glass from an initially fresh, exposed surface at a constant rate, knowledge of local climatic conditions makes it possible to calibrate hydration rind thickness to age (Friedman and Smith, 1960). The alteration of silicic ash deposits of tuff cones and tuff rings that are commonly associated with lava domes is poorly understood; these deposits have not been studied as thoroughly as their basaltic counterparts.

Palagonite tuffs were first recognized over a century ago (von Walterhausen, 1845); they are most commonly associated with pillow lavas (Moore, 1966) and basaltic ash deposits of tuff rings and tuff cones such as Surtsey (Jakobsson, 1978). Tuff cone and tuff ring deposits are generally intensely altered around the craters by posteruptive hydrothermal activity. The widely dispersed tuffs from caldera-forming eruptions and Strombolian ash fall blankets generally show only minor diagenetic and weathering effects unless they were deposited in a tropical environment.

The descriptions below are based upon SEM observations of hydration and palagonitization textures and upon EDS analyses of the surface chemical compositions of many altered pyroclastic samples discussed in other sections of this book. For most cases, sample disaggregation was necessary to separate particles cemented by alteration products. The samples were prepared for SEM examination by a thorough cleansing with acetone and ultrasound, which removed loosely adhering material from grains and exposed individual grain surfaces.

Grain Size: Strongly palagonitized or hydrated ash that still contains individual particles is generally too consolidated to be accurately sieved. Less altered ash, however, can be disaggregated and sieved for a rough estimate of size distribution. Regardless of the degree of hydration or palagonitization, the grain size of altered pyroclastic deposits appears to deviate very little from that of unaltered equivalents. It appears that hydration and palagonitization do not affect grain size significantly. In general, within deposits from an individual eruption, the degree of palagonitization increases slightly with increasing grain size, which indicates that porosity and permeability are important controls of alteration. This relationship does not hold true, however, for massive and planar-bedded surge deposits of tuff rings, where the degree of palagonitization increases slightly with decreasing grain size and probably is a function of total grain-surface area. These trends of alteration versus grain size are shown in Fig. 1. The trends are not strictly conclusive for all pyroclastic deposits because the data base is limited to 50 sample analyses of basaltic tuff cones and tuff rings, including Zuni Salt Lake, Koko Crater, Kilbourne Hole, Taal Volcano, and Surtsey.

Alteration Description: Peacock (1926) described two main varieties of palagonite: "gel-palagonite," a banded isotropic material that encloses cores of unaltered glass and "fibrous palagonite," which is birefringent and more strongly colored, contains opaque material, and sometimes shows microtubes penetrating into the glass. Although these terms are difficult to apply to SEM images of altered grain surfaces, two distinct palagonite textures are evident on grains and exhibit various combinations of phases that correspond to several stages of alteration. These stages of alteration noted below are associated with hydration and palagonitization but not with solution and precipitation of secondary phases.

Cracks. The first stage of basaltic glass alteration is the formation of a skin on grain surfaces. This skin, or rind, is generally lighter or more reflective than fresh glass surfaces and first develops in vesicle hollows. With progressive alteration, the rind will cover the entire grain surface. Palagonite rinds are usually broken by straight or slightly curved cracks (Noble, 1968). These cracks strongly resemble desiccation cracks in mud but are distinguishable from those formed by eruptive fragmentation and grain collisions during transport; both of the latter crack types tend to project through the grain.

In silicic glasses, the cracks tend to be curved and circular, forming what has been called the "onion skin" texture of perlite (Ross and Smith, 1955). Perlitic cracks are formed by the secondary hydration of glass (Noble, 1968) and develop more rapidly in phenocryst-bearing glasses than in aphyric glasses.

Close observation of cracks in ashes of both basaltic and silicic compositions shows that they are one to several micrometers wide and they separate thin polygonal blocks, which curl at the edges. The curled-back skin is only several micrometers thick and may overlie more altered skin layers beneath the surface. Cracks appear to grow in stages and they intersect other cracks in a polygonal fashion. The first cracks appear in vesicles and then are visible over the entire grain. Finally, they form pull-aparts of the grain skin and skin-plates with curled edges. Crack orientation may be controlled by the presence of microlites such as feldspar laths. The cracked glass skin probably represents a hydration front that progressively moves into the grain (Moore, 1966).

Cracks probably result from volume changes that take place during hydration. It is not clear why they resemble desiccation features, but this resemblance suggests that hydrated skin gradually dehydrates after it separates from fresh, underlying glass. Perhaps gradual crystallization of the hydrated skin accounts for a second volume change.

Microcrystalline Material. In the second stage of glass alteration, microcrystalline material forms, initially in vesicle hollows and then over the entire grain surface. The material may be superimposed over a hydrated skin or occur alone.

Because microcrystalline material is so typical of samples taken from the most highly altered basaltic deposits, it seems likely that it represents an advanced stage of alteration. In addition, formation of microcrystalline material corresponds to the greatest degree of chemical change documented in glasses, as we discuss in the next section. In the most advanced stages of alteration, both cracked skin and microcrystalline material cover the entire grain surface, and solution of the altered glass takes place, which results in a wormy appearance and a variety of relict textures.

Alteration stages that are concurrent with the hydration of silicic pyroclasts can be seen in the microcrystalline material on grain surfaces. Secondary phases appear first in vesicle hollows and then cover progressively more of the grain surface.

At high magnification (>20,000X), the euhedral forms of microcrystalline material may be observed. Rhomb-shaped chabazite and prismatic and spherulitic phillipsite are two common zeolites associated with palagonite. Fibrous to fan-like crystals of the natrolite-thompsonite series are less common. Clay minerals (mostly smectites) are usually too small (<0.01 μm) to distinguish easily, even at very high magnifications (greater than 100,000X).

Solution and Precipitation. In addition to the more common aspects of alteration (hydration-palagonitization-crystallization), solution and precipitation often affect glassy pyroclasts. These processes are common where circulating ground water encounters lithic materials such as sandstone or limestone and tephra. Commonly, solution forms tiny pits and frosted coatings on grain surfaces, whereas opal coatings or carbonate fillings are precipitated between grains.

Chemical Composition:

Bunsen (1851) first recognized the chemical differences between fresh basalt and palagonite. In general, alteration (especially palagonitization) involves addition of H_2O—as much as 30 Wt.%. Hydration of glass has an effect on the total elemental abundances within individual pyroclasts. A number of studies document the chemical changes that occur during alteration; they all find that the relative mobility of each major element and trace element is different (for example, Hay and Iijima, 1968; Furnes, 1975; Allen et al., 1981). Relative gain or loss of elements during alteration varies with the degree of hydration.

To illustrate the effects of alteration on the chemical composition of tephra, averaged, normalized EDS analyses of fresh and altered grain surfaces were compared. Samples were taken from individual beds in stratigraphic sequences where the relative degree of alteration in bed forms could be assessed in the field. Two basaltic volcanoes (Crater Elegante tuff ring and Cerro Colorado tuff cone) and two rhyolitic tuff rings (Panum Crater and Obsidian Dome) were studied. The tabulated analyses of these samples are given in their respective sections of this book. Figure 14 illustrates the gain and loss of major elements in samples from these stratigraphic sections. Note that the analyses were not recalculated for an assumed "immobile" element; however, they were calculated on a water-free basis. For each of the two basaltic examples, a series of analyses is shown in which the first is from a lower fall deposit that represents a fresh composition and subsequent analyses represent altered compositions. The effect of chemical alteration increases upwards in individual deposits and is thought to be related to water involved in hydrovolcanic explosions; higher in the sequence, steam has condensed on ash particles during emplacement. Wet eruption deposits (massive beds) show the greatest alteration, and dry eruption deposits (sandwave beds) the least. The two rhyolitic examples demonstrate the reduced chemical effect of alteration by hydration upon silicic glasses. Again, the degree of alteration is related to bed form.

Comments on Alteration Mechanisms: Silicic glass clearly shows the effects of hydration. It is not well defined on SEM images but is clearly evident by optical inspection of thin sections. Although basaltic glass becomes hydrated, it simultaneously changes to palagonite, which is unstable and gradually crystallizes to form smectites and zeolites.

There is still some debate as to whether palagonitization occurs at ambient temperatures or high temperatures during and shortly after eruptions (Peacock, 1926; Jonsson, 1961; Fuller, 1931; Bonatti, 1965). Both options are likely. Field evidence supports the hypothesis that the rate of palagonitization depends on temperature and bulk water content of the deposit. If water is present during eruption and emplacement, glass becomes hydrated at accelerated rates. Where magmatic fall deposits (dry eruptions) are interstratified with phreatomagmatic surge deposits (wet eruptions), the phreatomagmatic deposits are generally altered, whereas the magmatic deposits are fresh. On the other hand, alteration zones may be independent of vent location and bedding. These zones appear to be caused by water-table fluctuations after emplacement. The rate and type of alteration is affected by Eh-pH conditions, vent locations, deposition (subaerial or subaqueous), and temperature.

In summary, three mechanisms of ash alteration are recognizable. (1) Alteration that is caused by glass reaction with high-temperature water during eruption and deposition shows a bed-form dependence. Successive layers of ash show varying amounts of alteration that may be correlated to mode of emplacement (such as fall versus surge) and/or moisture in the eruption plume. (2) Postemplacement hydrothermal activity near the vent may progressively alter all near-vent deposits. This alteration is not controlled by either bedding planes or water-table horizons, so individual bedding sets can be traced from their fresh outcrops to the highly altered outcrops at near-vent locations. (3) Slow alteration occurs when ground water percolates through porous and permeable deposits. This alteration shows a horizontal and topographic control, and it may cut across bedding planes regardless of grain size.

Of the above mechanisms, bed-form control of alteration may be the most ambiguous, albeit the most instructive about eruption conditions. This ambiguity results from the nonlinear effects that porosity, grain size, and permeability have upon aqueous movement. For example, consider the following complications: smaller grain sizes promote alteration as a result of increased surface area—however, permeability usually decreases with decreasing grain size and sorting—but fine-grained deposits may retain longer the water that is emplaced with them.

Fall deposits from a magmatic burst often alter by mechanisms 2 and 3, above. Therefore, coarse-grained, well-sorted fall deposits (emplaced during dry eruptions) often show more alteration than do fine-grained deposits. In contrast, steam-rich eruptions produce relatively fine-grained tephra, which may be emplaced with wet steam that condenses on ash surfaces or dry steam that separates from the ash before condensation. Ash grains are subjected to hot fluids from the time of their formation until they finally cool to ambient temperatures within the deposit (mechanism 1). Under these circumstances, alteration may progress quickly and it is most rapid for the finest grain sizes. This relationship is especially true for vesicular, massive surge beds, which often contain accretionary lapilli and are composed of very fine ash. However, many sandwave surge beds are emplaced when steam separates from the ash before condensation, and the resulting bed remains relatively unaltered.

REFERENCES: Allen, C. C., Gooding, J. L., Jercinovic, M., and Keil, K., 1981. Altered basaltic glass: A terrestrial analog of the soil of Mars. Icarus, **45**: 347-369.

Bonatti, E., 1965. Palagonite, hyaloclastite and alteration of volcanic glass in the ocean. Bull Volcanol., **28**: 1-15.

Bunsen, R., 1851. Ueber die prozesse der vulkanischen gesteinbildungen islands. Prog. Ann., **83**: 197-272.

Fisher, R. V. and Schmicke, H.-U., 1984. *Pyroclastic rocks.* Springer-Verlag, New York: 472 pp.

Friedman, I. and Smith, R. L., 1960. A new dating method using obsidian - Pt. 1, the development of the method. Am. Antiq., **25(4)**: 476-522.

Friedman, I. I., Smith, R. L., and Long, W. D., 1966. Hydration of natural glass and formation of perlite. Geol. Soc. Am. Bull., **77(3)**: 323-327.

Furnes, H., 1975. Experimental palagonitization of basaltic glasses of varied composition. Contrib. Min. Petrol., **50**: 105-113.

Fuller, R. E., 1931. The aqueous chilling of basaltic lava on the Columbia River Plateau. Am. J. Sci., **21**: 281-300.

Hawkins, D. B. and Rustim, R., 1963. Experimental hydrothermal studies on rock alteration and clay mineral formation. Geochim. Cosmochim. Acta, **27**: 1047-1054.

Hay, R. L. and Iijima, A., 1968. Nature and origin of palagonite tuff of the Honolulu group on Oahu. Geol. Soc. Am., Mem., **116**: 331-376.

Honnorez, J., 1962. Sur l'origine des hyaloclastites. Bull. Soc. Belge Geol., **70**: 407-412.

Jakobsson, S. P., 1978. Environmental factors controlling the palagonitization of Surtsey tephra, Iceland. Geol. Soc. Denmark Bull., **27**: 91-105.

Jonsson, J., 1961. Some observations on the occurrence of sideromelane and palagonite. Bull. Geol. Inst. Uppsala, **40**: 81-86.

Marshall, R. R., 1961. Devitrification of natural glass. Geol. Soc. Am. Bull., **21**: 273-291.

Moore, J. G., 1966. Rate of palagonitization of submarine basalt adjacent to Hawaii. U.S. Geol. Sur. Prof. Paper, 550-D: D163-D171.

Nayudu, Y. R., 1964. Palagonite tuffs (hyaloclastites) and the products of post-eruptive processes. Bull. Volcanol., **27**: 392-410.

Noble, D. P., 1968. Stress-corrosion failure and the hydration of glassy silicic rocks. Am. Mineral., **53**: 1756-1759.

Peacock, M. A., 1926. The volcano glacial palagonite formation of Iceland. Geol. Mag., **63**: 385-399.

Peacock, M. A. and Fuller, R. D., 1928. Chlorophaeite, sideromelane and palagonite from the Columbia River Plateau. Am. Mineral, **13**: 360-382.

Ross, C. S. and Smith, R. L., 1955. Water and other volatiles in volcanic glasses. Am. Mineral., **40**: 1071-1089.

von Waltershausen, W. S., 1845. Über die submarinen Ausbrüch in der tertiären Formation des Val di Noto im verwandten Erscheinungen am Ätna. Gött. Stud., **1**: 371-431.

Wise, S. W. and Weaver, F. M.,1979. Volcanic ash: Examples of devitrification and early diagenesis. *In*: *Scanning Electron Microscopy/1979/I.* SEM Inc., AMF O'Hare, Illinois: 511-518.

Wise, S. W., Weaver, F. M., and Gûven, N., 1973. Early silica diagenesis in volcanic and sedimentary rocks: Devitrification and replacement phenomena. *In*: C. J. Arceneaux (Ed.), *Proceedings, 31st Ann. Meet. of the Electron Microscopy Soc. Am.*. Claitor's Publishing Division, Baton Rouge, Louisiana: 206-207.

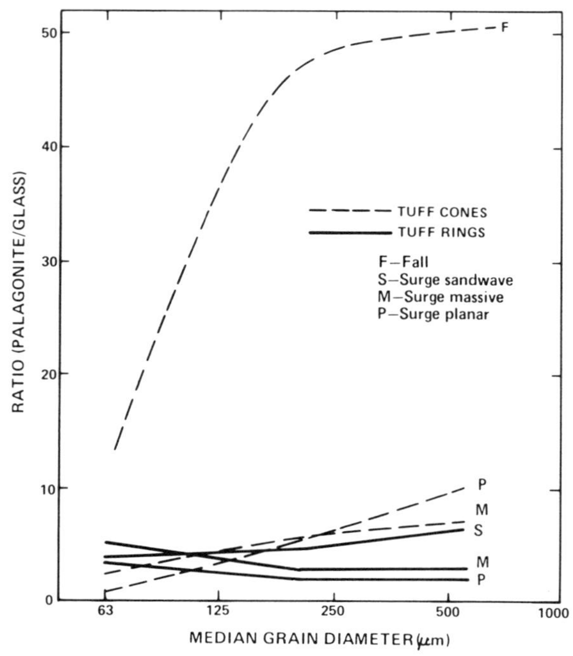

FIGURE 1. Plot of palagonite-to-glass-volume ratio versus median grain diameter for samples of basaltic tephra. Results of 50 sample analyses from tuff rings and tuff cones are grouped by bed form to show dependency of palagonitization on eruption and emplacement mechanisms. Overall palagonitization increases with grain size as a probable result of increased porosity and permeability. Samples from tuff rings, however, show a slight increase in palagonitization with decreasing grain sizes, probably because of increased grain-surface area for fine deposits emplaced in wet conditions.

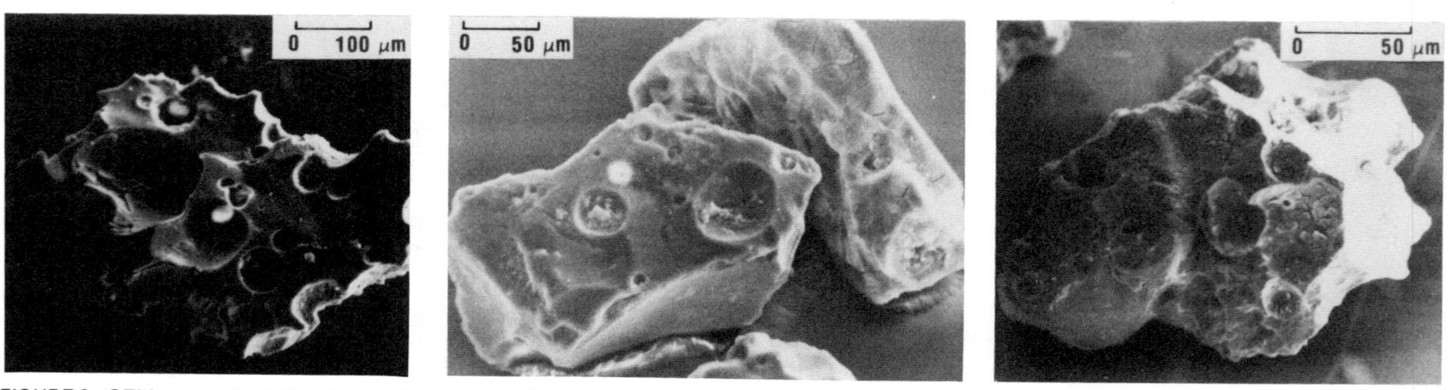

FIGURE 2. SEMs show the initial stages of alteration of basaltic pyroclasts. Alteration is indicated by a cracked skin forming first in vesicle hollows and then progressively covering the entire grain (as in Figs. *3* and *4*).

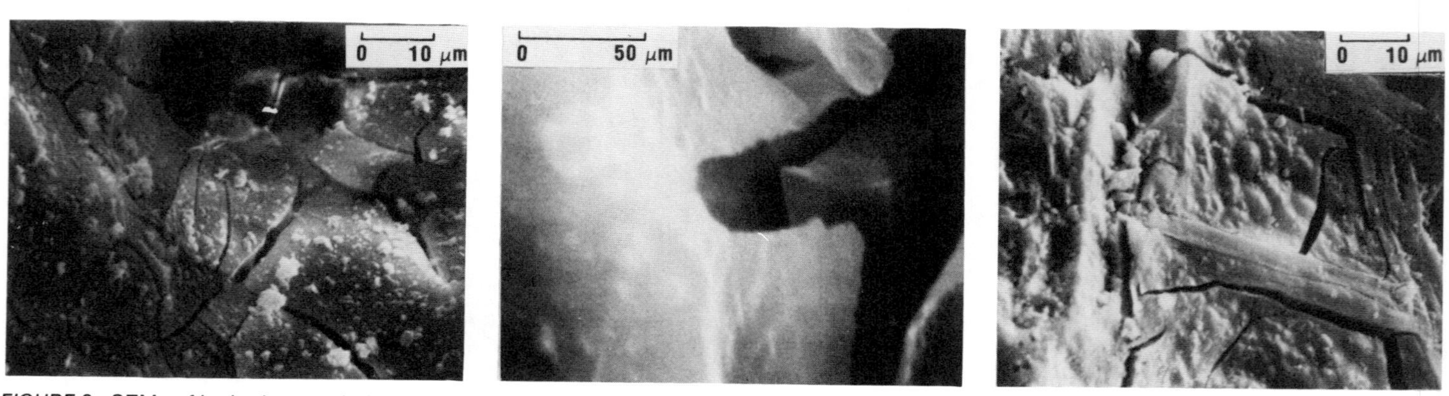

FIGURE 3. SEMs of hydration-crack detail. Note the polygonal structure in left view, the formation of two skin layers in the middle view, and microlitic control of crack orientation (feldspar laths) in right view.

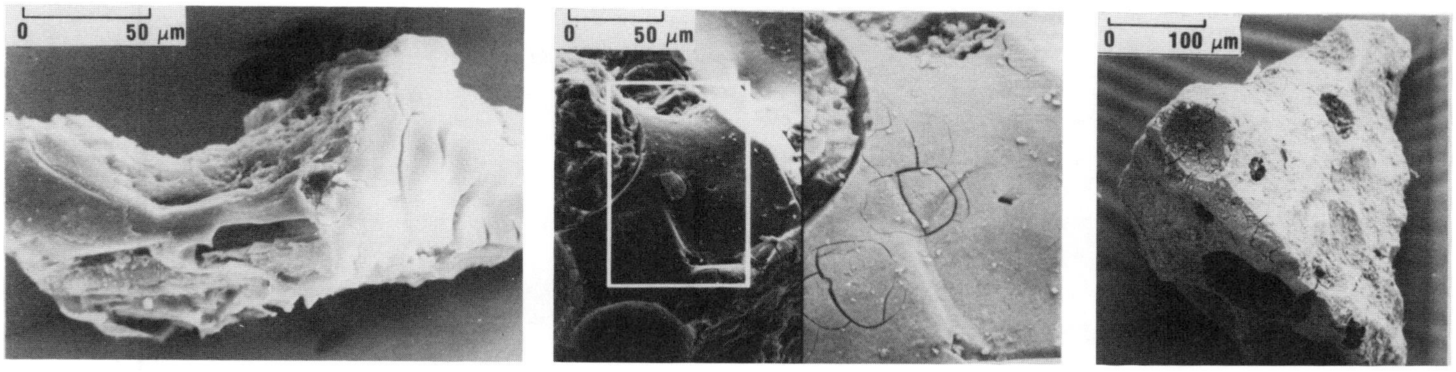

FIGURE 4. SEMs showing completed development of hydrated or palagonitized skin; it gives the grain a bright, lustrous appearance in secondary-electron images (left picture). In the middle picture, the left half (secondary-electron image) shows a bright, hydrated area surrounded by cracks, which, when viewed under 2X magnification (right half), shows a detailed crack structure with a back-scattered electron image. The right picture of this set is a back-scattered image of a hydrated grain. Note the dull, earthy appearance of the altered glass in this view.

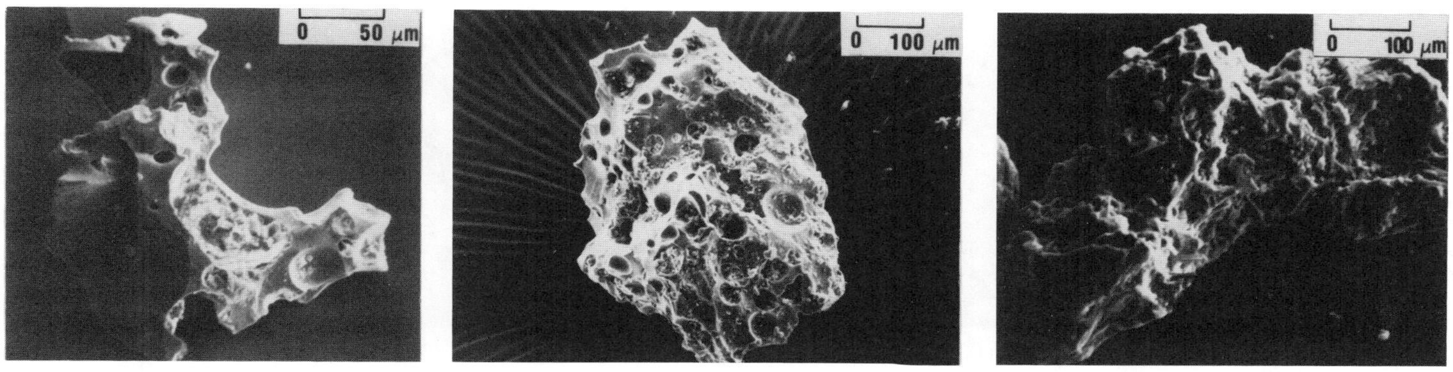

FIGURE 5. SEMs show stages of partial to extensive (left to right) alteration of pyroclasts by microcrystalline secondary minerals filling vesicles and partly covering the grain surface.

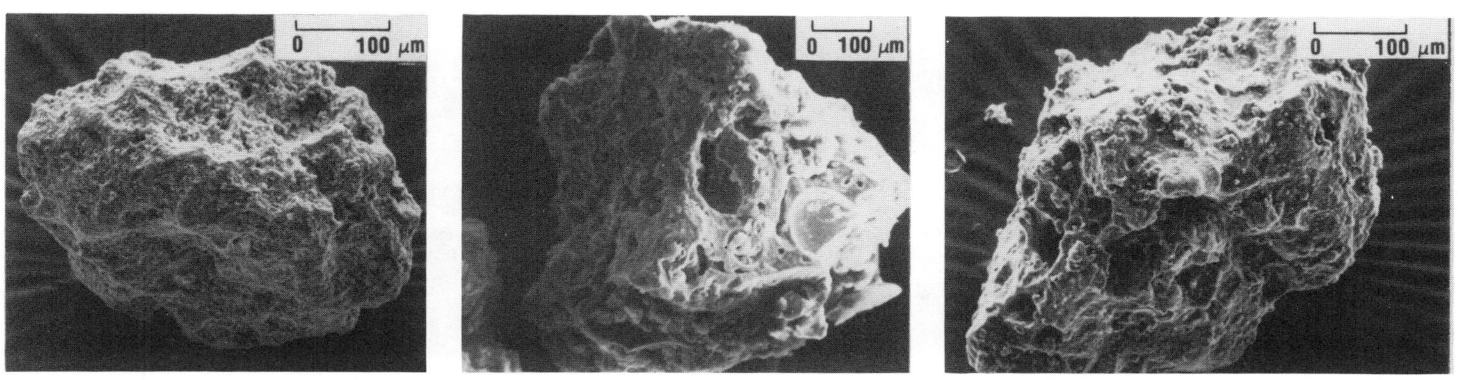

FIGURE 6. SEMs of strongly altered glassy pyroclasts; surface is rough, with smooth, glassy surfaces completely altered to microcrystalline minerals and palagonite.

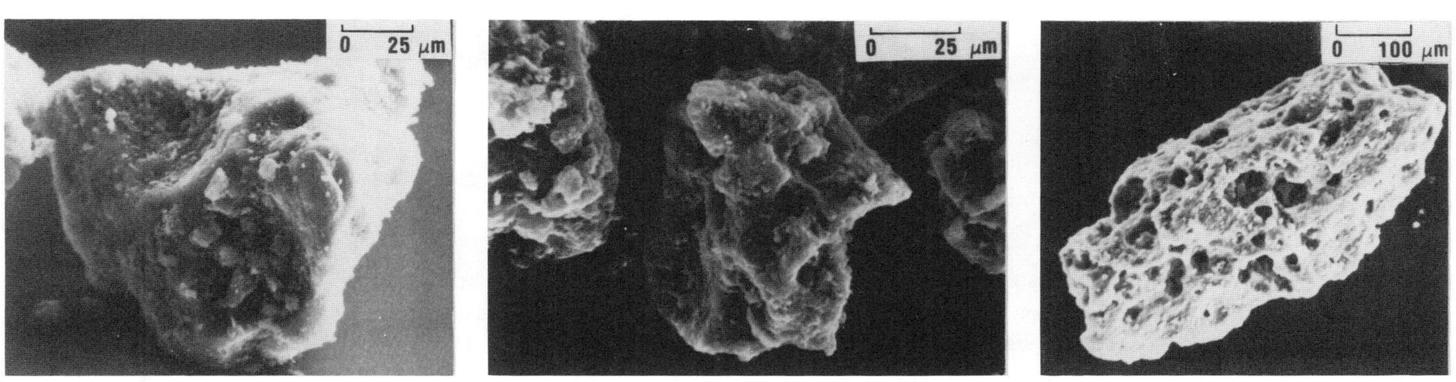

FIGURE 7. SEMs of grains that are strongly altered; they appear light gray to white in both SEM (shown) and light microscope (not shown) views. Strong hydration of the glass gives shards a sugary appearance.

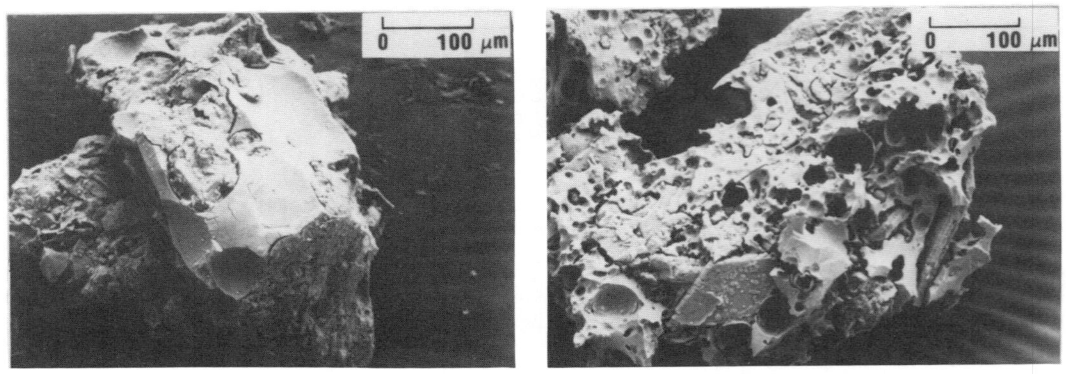

FIGURE 8. SEMs show that complete palagonitization of ash particles is often followed by solution of altered material; solid grains (left view) have a wormy appearance (compare with descriptions of the weathered Pahala ash).

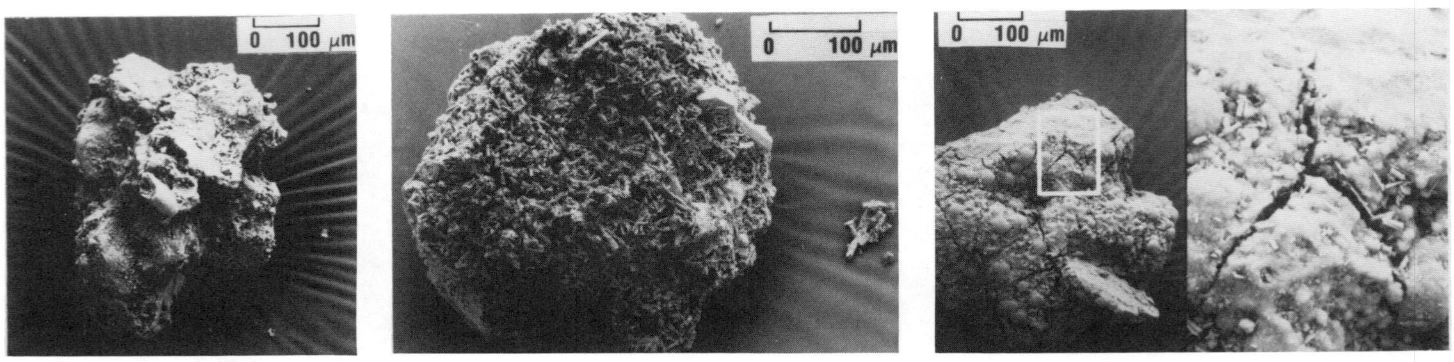

FIGURE 9. SEMs of basaltic ash from strongly lithified, tuff cone, palagonite breccias of the Hopi Buttes. These consist of aggregates of clay-like and microcrystalline material. Vesicles are no longer visible, the matrix is cracked, and tiny crystals form a network on grain surfaces where hydrated glass has been dissolved.

FIGURE 10. Photos a through d are SEMs of the alteration stages of silicic glass, from a fresh pyroclast on the upper left to altered ones on the lower right. The pyroclast on the lower right is from the green tuff, Pantelleria, and shows hydration typical of these pyroclasts.

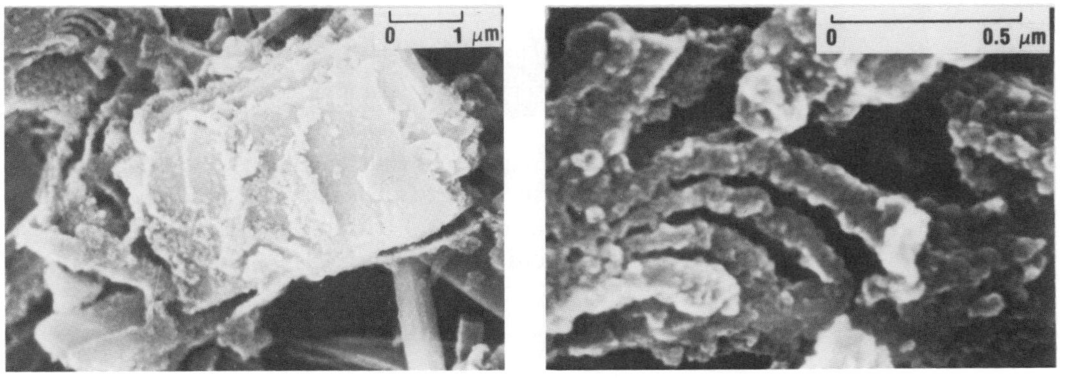

FIGURE 11. SEM high-resolution images of secondary minerals, showing a rhomb-like zeolite covered by tiny spheres that are probably smectite clays.

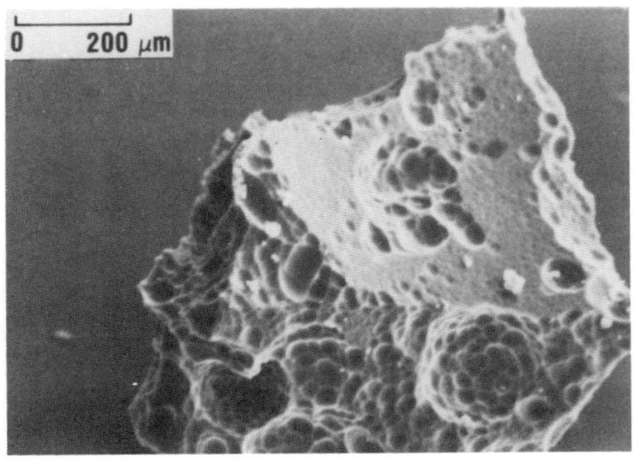

FIGURE 12. SEMs of numerous pits and rough, frosted surfaces formed during solution and precipitation. Note the original vesicle hollows that have been pitted by solution. Surfaces between vesicles have a crusty coat of precipitated material that is probably opal.

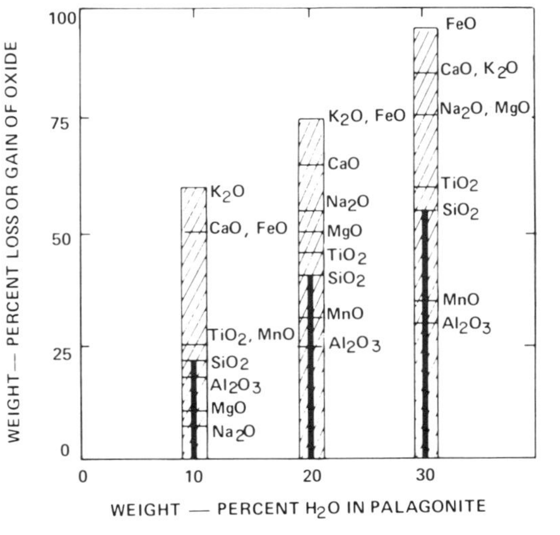

FIGURE 13. Plot of weight percentage loss or gain of major elements in palagonites with various water contents (adapted from Furnes, 1975).

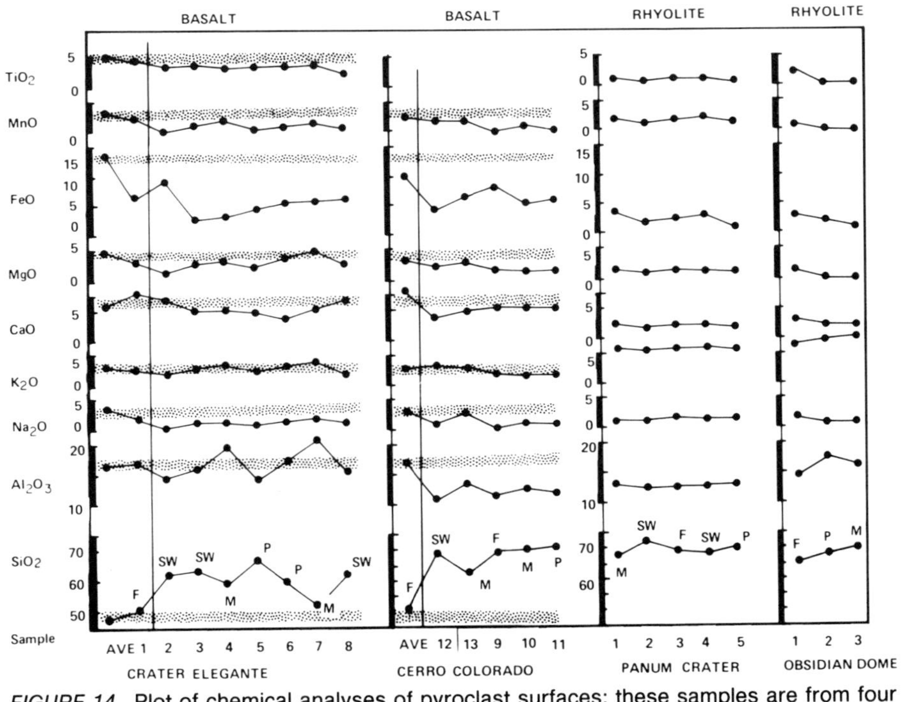

FIGURE 14. Plot of chemical analyses of pyroclast surfaces; these samples are from four stratigraphic sections where alteration varies from fresh ash to highly altered ash. The samples are numbered in ascending stratigraphic order; samples 1 and AVE (average Crater Elegante and Cerro Colorado) are fresh basalts. Note the variation in surface chemistry as a function of alteration, which varies from dry, magmatic fall deposits (F) to wet, hydromagmatic surge deposits: SW = sandwave, M = massive, and P = planar bed forms.

Weathered Tuffs

Weathered volcanic ash is often considered to be an important component in fertile soils. Many of the rich agricultural areas are located on volcanic islands and in continental arcs such as Central America and other regions of the Circum-pacific "ring of fire."

PAHALA ASH, HAWAII—AN EXAMPLE OF WEATHERED BASALTIC TEPHRA

Introduction: The Pahala ash is a widespread, weathered basaltic tephra unit on the island of Hawaii (Hay and Jones, 1972). Present rainfall over the Pahala ash ranges from 25 to 635 cm/year. The Pahala ash consists of deposits from many eruptions between 10,000 and 17,000 years ago (Hay and Jones, 1972); all are of basaltic composition. Most pyroclasts in the Pahala ash are (or were) vitric pumices, droplets, and shards. Most examples used here were probably erupted from Kilauea Volcano, although examples from South Point may have been erupted from the south rift of Mauna Loa Volcano (Macdonald and Abbott, 1970). For this study, deposits ranging in thickness from 1.5 to 3 m were sampled during a traverse across southern Hawaii.

Weathering of sideromelane produces palagonite and clay. Palagonite is a general term for a very fine grained, intergrown mixture of smectite clays, iron oxides, carbonate, and zeolites; it replaces sideromelane and forms a cement between grains. In the examples used here, the increased alteration of glass to palagonite can be correlated with increasing rainfall. All samples were provided by R. L. Hay, University of Illinois.

Grain Size: Because the deposits are weathered, only a subjective evaluation of grain size is available. Field descriptions indicate that particles range from clay size (<4 μm) to lapilli size (2 to 64 mm).

Tephra Description: Samples from sites 1 and 2 (Figs. 1 and 2) consist of fine-grained, flat to blocky pyroclasts 1 to 20 μm long that are bonded into clusters by montmorillonitic clay. These clusters, plus shards and phenocrysts coated with fine-grained particles and clay, form rounded aggregates that make up yellowish-orange, clayey deposits (Hay and Jones, 1972).

Weathered tephra from sampling sites 3 through 5 appear to have originally consisted of lapilli-size basaltic pumice (present as droplets and broken droplets) and shards. (Refer to the section on Hawaiian eruptions for examples of fresh pyroclasts.) These pyroclasts range in vesicularity from 50 to 70% by volume; most vesicles are 10- to 160-μm ovoids or spheres. There are some curved, plate-like shards that may have been derived from much larger vesicles.

Palagonitic alteration products increase from site 3 to site 5. At site 3, the outer surfaces of pumice pyroclasts have been altered to single layers of palagonite that are 1 to 2 μm thick; this surface is rough and lumpy. Where broken away, the glass at the palagonite/glass contact is smooth. At site 4, multiple layers of palagonite have formed on the outer surfaces and vesicle walls of the pyroclasts and have pseudomorphed the original grain shape. Some palagonite surfaces are lumpy, others are botryoidal, and still others are covered with a network of desiccation cracks. Some retain their cores of basaltic glass. Palagonite rinds are 20 to 50 μm thick on the outer surfaces of pyroclasts and 5 to 10 μm thick on interior vesicles. At site 5, little if any glass is present; most has been dissolved, leaving only the palagonite. If the original vesicles are regarded as molds, only the clayey palagonite casts remain, the "mold" having been dissolved. These spherical and ovoid linings remain; they resemble cracked fish eggs. Pumice pyroclasts with elongate vesicles have undergone a similar process and only stacks of tubular palagonite linings remain.

Chemical Composition: Fresh basaltic glass and palagonite rinds were separated by Hay and Jones (1972) for chemical analyses. This example of the Pahala ash is from near site 4, Kilauea Caldera (Table 1).

TABLE 1. CHEMICAL COMPOSITION OF PAHALA ASH.*

	Glass from Palagonitic Pumice Lapilli	Palagonite Rind
SiO_2	49.31	29.43
TiO_2	2.57	3.68
Al_2O_3	12.58	17.12
Fe_2O_3	1.65	16.29
FeO	10.51	1.82
MnO	0.17	0.08
MgO	10.81	1.38
CaO	10.29	1.70
Na_2O	1.81	0.22
K_2O	0.36	0.11
P_2O_5	0.20	0.23
H_2O	0.36	14.46
H_2O^-	0.21	13.59
Total	100.20	100.13

*K. Aoki, analyst (from Hay and Jones, 1972).

Pyroclast Formation: Little comment is required here, as pyroclast formation in basaltic eruptions is described in other sections. The Pahala ash, however, is a composite unit that is made up of tephra from many eruptions. Among the samples described here, it appears that those from sites 3 to 5 were the products of lava fountains. Samples from sites 1 and 2 are very poorly preserved but may have been from littoral or phreatomagmatic activity; the tephra deposits there are very fine grained, and the samples were collected from near the coast. Palagonitic rinds are preserved and pseudomorph the original pyroclasts. Where no glass remains, only the castings are visible as thin palagonitic linings.

REFERENCES: Hay, R. L. and Jones, B. F., 1972. Weathering of basaltic tephra on the Island of Hawaii. Geol. Soc. Am. Bull., **83**: 317-332.

Macdonald, G. A. and Abbott, A. T., 1970. *Volcanoes in the Sea.* Univ. of Hawaii Press, Honolulu: 441 pp.

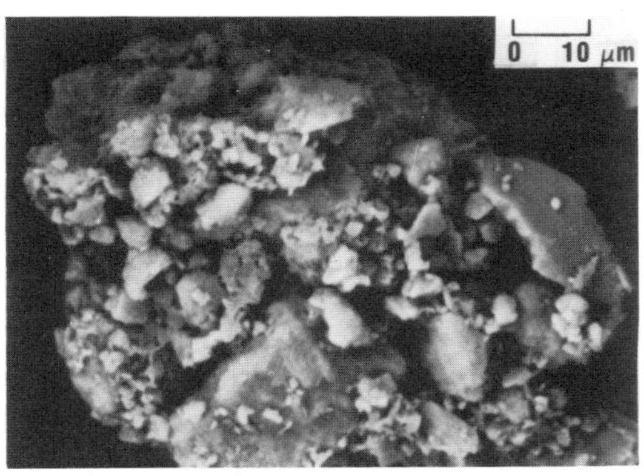

FIGURE 1. Clayey, yellowish-brown ash from South Point. These are mostly grainy pellets that consist of aggregates of fine ash and montmorillonite with and without cores consisting of coarse ash.

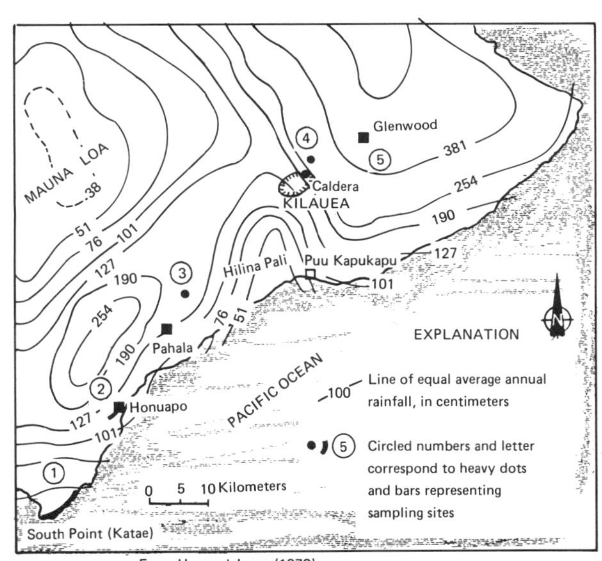

From Hay and Jones (1972)

Map of Southeastern Hawaii shows sampling sites, Mauna Loa and Kilauea volcanoes, and lines of equal average rainfall. Circled numbers 1 through 5 correlate to samples shown in Figs. 1 through 5.

FIGURE 2. Clayey, yellowish-brown ash from Honuapo. Close-up of a rounded pellet (similar to those in Fig. *1*) that consists of flat and blocky 1- to 20-μm-long vitric pyroclasts. The aggregate is probably bonded by montmorillonitic clay. Although it is difficult to see, some shards have pitted surfaces.

FIGURE 3a. Ash and lapilli from 7.5 km NE of Pahala are mostly glass, but some palagonitic rims are visible on vitric pyroclasts. Stereophotomicrograph of coarse basaltic ash. Smooth, angular surfaces are basaltic glass, and rough, lumpy surfaces are palagonitic rinds on glassy pyroclasts.

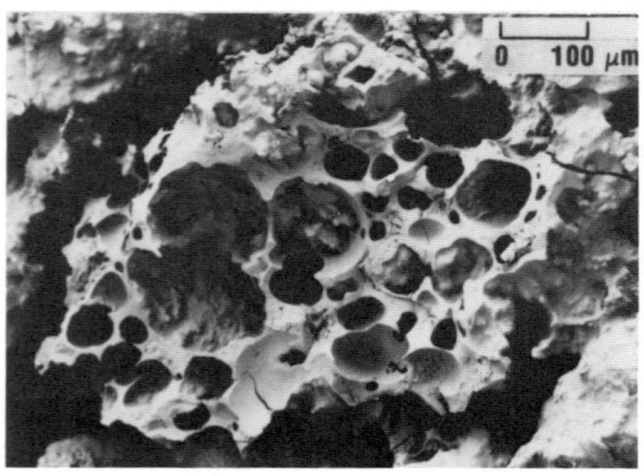

FIGURE 3b. Basaltic pumice with a thin palagonitic rind. Rough surfaces are palagonite and smooth surfaces are glass. The pyroclast is bonded to adjacent pyroclasts by palagonite.

FIGURE 3c. Basaltic pyroclasts with thin (1- to 2-μm) rinds of palagonite. The palagonite has spalled off a few grain surfaces, exposing the smooth underlying glass surface.

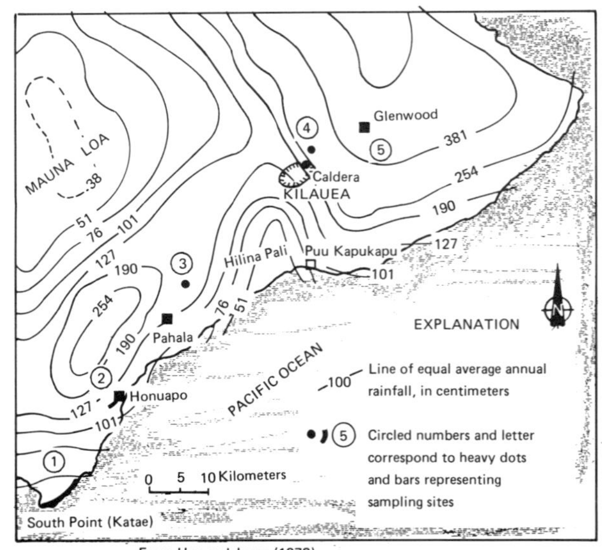

From Hay and Jones (1972)

Map of Southeastern Hawaii shows sampling sites, Mauna Loa and Kilauea volcanoes, and lines of equal average rainfall. Circled numbers 1 through 5 correlate to samples shown in Figs. 1 through 5.

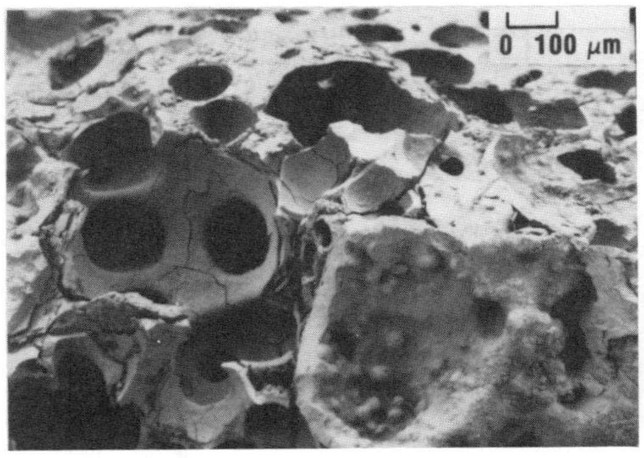

FIGURE 4a. Reddish-brown, clayey ash and lapilli that were collected 2.4 km NNE of Kilauea Caldera. Surface of a basaltic pumice lapillus. Outer grain surfaces have been altered; a 20- to 30-μm palagonite rind pseudomorphs the original pyroclast. Inner surfaces (vesicle walls) have a thinner (5- to 10-μm) palagonite rind; it is relatively smooth and crossed by a network of desiccation cracks.

FIGURE 4b. Box-like, multi-layered palagonite rinds are all that is left of most of these pyroclasts. The agglomeration of pyroclasts is bonded by palagonite cement. Palagonite surfaces vary from those crossed with desiccation cracks to those with a botryoidal appearance.

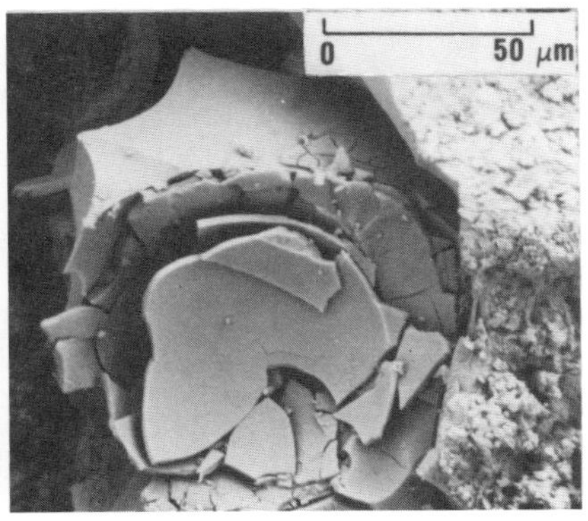

FIGURE 4c. Vitric pyroclast remnant with a 5- to 10-μm-thick void space separating it from a well-developed palagonite rind. Some alteration of the pyroclast has produced exfoliation features and desiccation cracks on the surface.

FIGURE 5. Reddish-brown clay and palagonitic lapilli tuff from a location 1 km SE of Glenwood. The round forms that look like fish eggs are palagonitic shells from the interior vesicles of a scoria fragment; the glass surrounding these vesicles has been dissolved. These vesicle "casts" are only 1 to 3 μm thick. The tubular objects (*left*) are palagonite casts of elongate vesicles from another scoria pyroclast. Holes that connected adjacent vesicles are visible as circular apertures in these delicate shells.

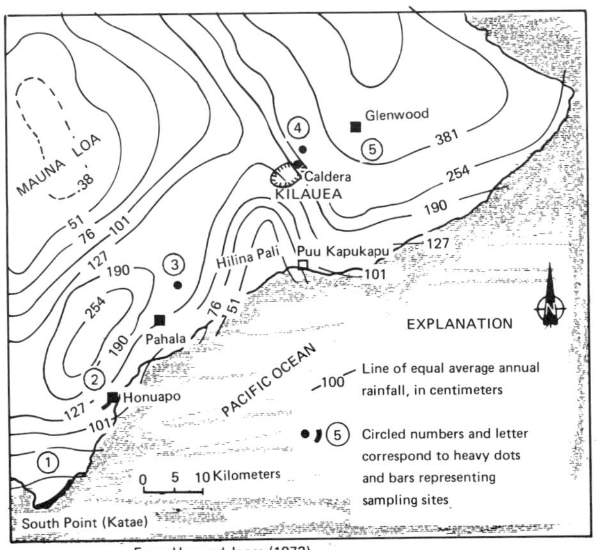

From Hay and Jones (1972)

Map of Southeastern Hawaii shows sampling sites, Mauna Loa and Kilauea volcanoes, and lines of equal average rainfall. Circled numbers 1 through 5 correlate to samples shown in Figs. 1 through 5.

ST. VINCENT, WEST INDIES; WEATHERED VOLCANIC ASH OF ANDESITIC COMPOSITION

Introduction: Tephra deposits of late Pleistocene age cover much of the island of St. Vincent, Lesser Antilles; there are many lateral facies variations that are based upon depositional mode and grain size as well as facies based upon rainfall and degree of weathering. Most of these ash deposits were erupted from the Soufriere, a large composite cone located at the northern end of the island. Ash fall deposits drape the rugged countryside, and pyroclastic flow deposits are present in some valleys (Hay, 1959). Two papers have been published on the weathering of volcanic ash in the Lesser Antilles (Hay, 1959; 1960). The first discussed a sequence of late Pleistocene ash deposits of andesitic composition, and the second concerned soil formation on a 4000-year-old pyroclastic flow sequence. R. L. Hay of the University of Illinois provided the samples used for this example of ash weathering and relict textures.

One sequence of pyroclastic deposits, dated at 4090 ± 50 y.b.p., forms a fan on the east side of the Soufriere (Hay, 1960). These deposits, which underlie deposits of the 1902 eruption, are as much as 23 m thick at the coast. Soil has developed on the 4000-year-old ash deposit to depths of 1.3 to 2.6 m. Samples from unaltered ash at a depth of 3.4 m and samples in the soil at depths of 1.3 and 0.3 m are used here to demonstrate the weathering of andesitic ash. This type of ash provides the base for widespread, fertile soils in many tropical subduction zones.

Grain Size: The late Pleistocene ash falls of St. Vincent have median grain sizes of ~80 to 200 μm (Hay, 1959); these sizes are approximate, however, as they have been altered to clay and are very difficult to sieve.

Tephra Description: *Late Pleistocene Ash.* These examples from the eastern coastal facies were described in the field by Hay (1959) as a moderately indurated, yellowish-brown, poorly sorted medium- and coarse-grained ash. The original ash appears to have been mostly porphyritic andesite pumice with phenocrysts of plagioclase, augite, hypersthene, and olivine (Table 1). Some lithic-rich ashes are also present (Fig. 3c). All glass phases and parts of the plagioclase phenocrysts have been altered to halloysite, but relict textures are very clear; the original vesicularity of pumices and the relation of vesicles to phenocrysts are visible in all samples. Ovoid vesicles of all sizes are visible (Figs. 3a, b, and d).

TABLE 1. CHEMICAL COMPOSITION OF LATE PLEISTOCENE ASH: CHIBARABU PT., ST. VINCENT.*

	(Wt. %)
SiO_2	53.03
Al_2O_3	18.78
FeO	5.49
Fe_2O_3	3.76
TiO_2	1.07
MnO	0.12
CaO	8.72
MgO	4.02
K_2O	0.44
Na_2O	3.14
P_2O_5	0.05
H_2O^+	1.45
Total	100.07

*W. H. Herdsman, analyst.

Pyroxene phenocrysts are deeply etched and are characterized by multiple needle-like terminations. Olivine is etched but retains its original shape. Plagioclase cores are, in some examples, altered to clay. Within hyalocrystalline lithic fragments, matrix minerals are altered to clay, phenocrysts are etched, and glass phases are dissolved (Fig. 3c). Pore space is filled with halloysite (Hay, 1959).

4000-Year-Old Ash. This is a crystal-rich ash (crystals form 60 to 80%) consisting of plagioclase, hypersthene, augite, olivine, magnetite, andesitic pumice, and andesitic lithic fragments (Hay, 1960). Most phenocrysts are not weathered (near the top of the soil the feldspars are only slightly etched), but vitric pyroclasts from the top of the soil horizon are totally altered to halloysite. Relict forms of vitric pyroclasts are difficult to identify in the most thoroughly weathered zone.

Comments on the Weathering of Ash: These examples from St. Vincent illustrate the physical changes that occur in pyroclasts during weathering. As shown, vitric pyroclasts are replaced by clay, but the weathering process does not always destroy pumice or shard shapes. In most of the samples studied here, relict shards and pumices are easily recognized; size, shape, and distribution of vesicles can be measured. The features of pyroclasts from ashy soils and deeply weathered tephra deposits can be described: this is especially useful when unraveling the history of volcanoes located in the tropics.

REFERENCES: Hay, R. L., 1959. Origin and weathering of late Pleistocene ash deposits on St. Vincent, B.W.I. J. Geol., **67**: 65-87.

Hay, R. L., 1960. Rate of formation and mineral alteration in a 4000-year-old volcanic ash soil on St. Vincent, B.W.I. Am. J. Sci., **258**: 354-368.

FIGURES 1 AND 2. Andesitic crystal-lithic ash from a 4000-year-old pyroclastic flow deposit along the Rabaka River, on the slopes of Soufriere of St. Vincent. Variations in the degree of weathering of a massive pyroclastic flow sequence are demonstrated here. Samples were collected 3.4, 1.3, and 0.3 m below the surface of a 4000-year-old soil (Figs. 1 and 2 *a*, *b*, and *c*, respectively).

FIGURE 1a. Mostly lithic fragments and phenocrysts. Slightly weathered pyroclasts are very well preserved and have only a thin coating of fine-grained particles. *Center*—vesicular porphyritic lithic fragment with no alteration of pyroxene and plagioclase phenocrysts. *Top*—Pumice fragment. Other pyroclasts are phenocrysts and lithic fragments.

FIGURE 1b. Center—well-preserved, unaltered plagioclase phenocryst. Subrounded, porphyritic andesite fragments are also visible.

FIGURE 1c. Center—unaltered pyroxene fragment; irregular surfaces are remnants of glass that are attached to the phenocrysts but are now altered to halloysite. *Top*—partly etched plagioclase with irregular, wormy holes. *Lower left*—porphyritic scoria pyroclast that is altered to halloysite. There are also andesitic lithic fragments that have been altered, in part, to halloysite.

FIGURE 2a. Pumice pyroclasts (juvenile). A fresh, unaltered vitric pyroclast with low vesicularity but well-developed vesicles (deformed ovoids). Finer grained clasts clinging to the surfaces are typical of pyroclastic flow deposits.

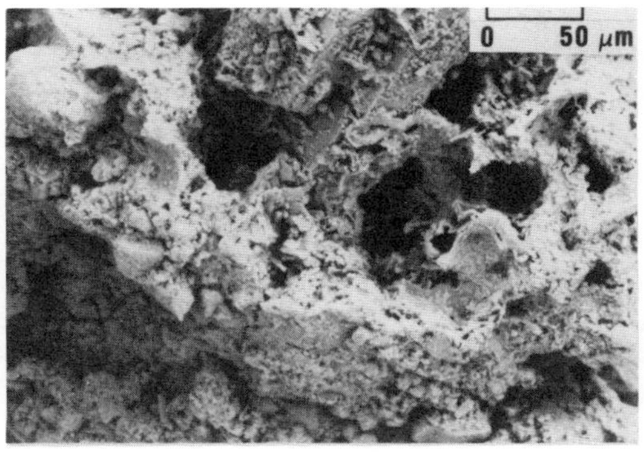

FIGURE 2b. Vitric pumice pyroclast. The glass has been altered to halloysite but relict textures remain, including vesicles that surround a phenocryst (*center*).

FIGURE 2c. Aggregate of plagioclase and pyroxene phenocrysts and halloysite. Individual aggregates might be relict pumice pyroclasts, but there are few relict textures remaining to confirm this hypothesis. It is possible that ovoid holes in the aggregate were vesicles. Pyroxene phenocrysts exhibit little or no alteration.

FIGURE 3a. Late Pleistocene andesitic ash from the east coast of St. Vincent. Stereophotomicrograph of a relict pumice pyroclast. The vitric portions of the pyroclast have been totally replaced by clay, but the original form remains. Most vesicles are 5- to 120-μm-long, irregular ovoids except where adjacent to phenocrysts (*left*), where they are as much as 200 μm long. Center of the plagioclase phenocryst has been partly replaced by halloysite.

FIGURE 3b. Surface of a pumice pyroclast where much of the original vesicularity remains. Glass has been replaced by halloysite. Notice the desiccation cracks in the clay.

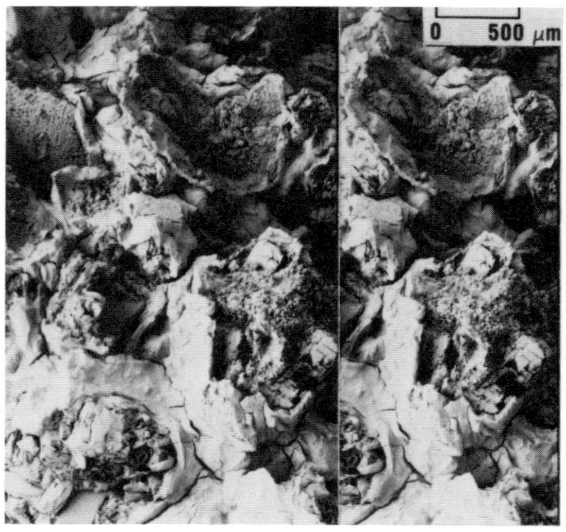

FIGURE 3c. Stereophotomicrograph of a lithic-fragment-rich andesitic tuff. Relict lithic fragments are visible as subrounded "aggregates;" these consist of etched phenocrysts and void space (perhaps originally glass between groundmass minerals, now dissolved away), surrounded by a network of groundmass minerals that is replaced by clay. Each relict grain is coated with irregular layers of clay that filled the void spaces between pyroclasts.

FIGURE 3d. Pyroxene phenocryst in a pumice pyroclast. Although the glass has been replaced by clay, vesicles are very well preserved in this example. The pyroxene has been etched both along a twin plane and at random (note the honeycomb weathering in lower left corner of the phenocryst).

FIGURE 3e. Etched pyroxene with needle-like terminations that are typical of weathered pyroxenes.

RELICT TEXTURES IN BENTONITE

Introduction: *Bentonite* is a term first used for clay-like material from the Benton Formation (Cretaceous) of Wyoming. Hewitt (1917) and Wherry (1917) established that these clays are the alteration products of volcanic ash. There have been dozens of bentonite studies over the past 60 years, including a comprehensive account of the Mowry bentonites by Slaughter and Early (1965). Bentonite beds have been useful as a source of clay for drilling muds and as marker beds for stratigraphic correlation over large areas. Slaughter and Early (1965) wrote the following field description: "a typical bentonite is a relatively thin and distinctly light-colored unindurated rock composed dominantly of clay." The major component is montmorillonite with minor relict phenocryst phases, zeolites, and authigenic silica phases.

Several SEM studies of bentonite have revealed well-preserved relict textures of pumice pyroclasts and shards (Khoury and Eberl, 1979; Wise and Weaver, 1979). The example used here to illustrate relict pyroclastic textures in bentonites is from the upper Tesuque Formation (Pliocene) of the Santa Fe Group and was collected by J. Stix near Ojo Caliente, New Mexico.

Tephra Description: The mold-cast relationship (described in the section on Pahala Ash) is well demonstrated in this example. The most apparent pyroclasts are small pumices in which long, thin parallel rods are casts of elongate vesicles. The curved, plate-like surfaces are casts of large shards. Alteration of vitric pyroclasts leaves a "rind" that is preserved, whereas the glassy core is dissolved, leaving voids.

Chemical Composition: No chemical analyses are available for the sample shown in Figs. 1 through 3. However, analyses of bentonite deposits near Santa Rita, New Mexico, and Otay, California, were published by Slaughter and Early (1965) (Table 1). Variations from bentonite to bentonite reflect the original magma composition and alteration environment.

TABLE 1. CHEMICAL ANALYSES OF TWO BENTONITES.

	Santa Rita, New Mexico (Wt. %)	Otay, California* (Wt. %)
SiO_2	58.10	57.80
Al_2O_3	17.15	16.25
Fe_2O_3	1.02	1.10
TiO_2	0.14	0.16
CaO	0.15	0.06
MgO	4.33	6.56
Na_2O	2.80	3.87
K_2O	0.14	0.14
Ignition loss	16.11	14.10
Total	100.04	100.04

*From Osthaus (1956).

Comments on Weathering Alteration of Ash: Most relict pyroclasts in this and other described examples appear to have been rhyolitic or rhyodacitic pumice pyroclasts and shards. The widespread distribution of bentonitic deposits is consistent with derivation from silicic tephra layers deposited during large-volume eruptions.

The alteration of silicic glass to bentonite (mixture of clays, silica phases, and zeolites) is described by Slaughter and Early (1965) and Grim and Güven (1978).

REFERENCES:

Grim, R. and Güven, N., 1978. *Bentonites; Geology, Mineralogy, Properties and Uses*. Elsevier, Amsterdam: 256 pp.

Hewitt, D., 1917. The origin of bentonite. J. Wash. Acad. Sci., 7: 196-198.

Khoary, H. and Eberl, D., 1979. Bubble-wall shards altered to montmorillonite. Clays Clay Miner., **27**: 291-292.

Osthaus, B., 1956. Kinetic studies on montmorillonite and nontronite by the acid dissolution technique. 4th Nat'l. Conf. on Clays and Clay Min., Proc.: 301-321.

Slaughter, M. and Early, J., 1965. Mineralogy and geological significance of the Mowry Bentonites, Wyoming. Geol. Soc. Am. Spec. Paper 83: 116 pp.

Wherry, E., 1917. Clay derived from volcanic dust in the Pierre of South Dakota. J. Wash. Acad. Sci., **7**: 576-583.

Wise, Jr., S. and Weaver, F., 1979. Volcanic ash: examples of devitrification and early diagenesis. Scanning Electron Microsc.: 511-518.

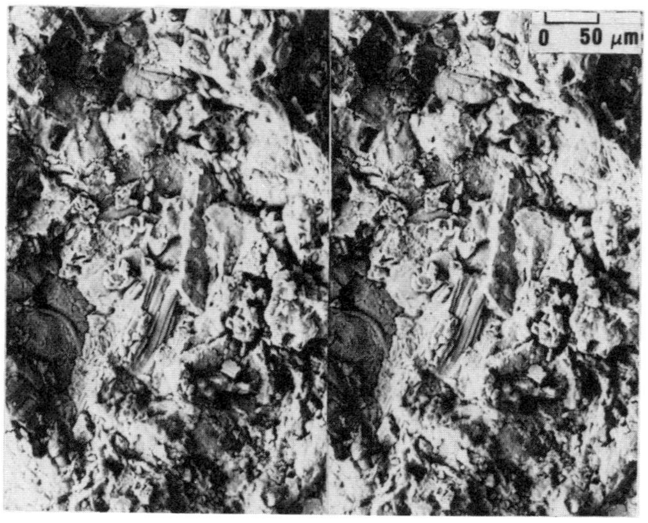

FIGURE 1. Stereophotomicrograph of bentonite from the Santa Fe Formation, Ojo Caliente, New Mexico. Casts of relict pyroclasts in clay (the pyroclasts themselves were the molds). The most apparent pyroclast is a 160-μm-long pumice located in the photo center; the long, thin, parallel rods are casts of elongate vesicles. Curved, plate-like surfaces are casts of large shards (these are less obvious).

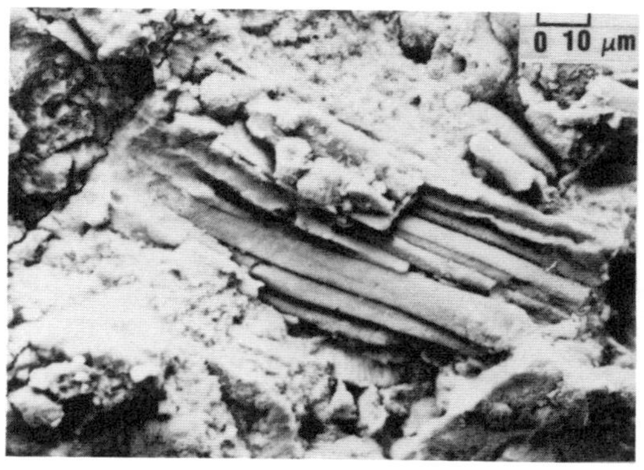

FIGURE 2a. Relict pumice pyroclasts. This is a closer view of the relict pumice pyroclast of Fig. 1. Clays have filled and pseudomorphed elongated vesicles and the original glass of the pyroclast was dissolved. The pseudomorph is that of a 90- by 35-μm, angular pumice fragment.

FIGURE 2b. Relict pumice pyroclast similar to the one in Fig. *2a.*, but broken, perhaps in sample preparation.

FIGURE 3a. Relict shards. The curved form visible here (*center*) is the casting of a shard (the glass has dissolved). Fuzzy, irregular clumps on the right wall are most likely montmorillonitic clays.

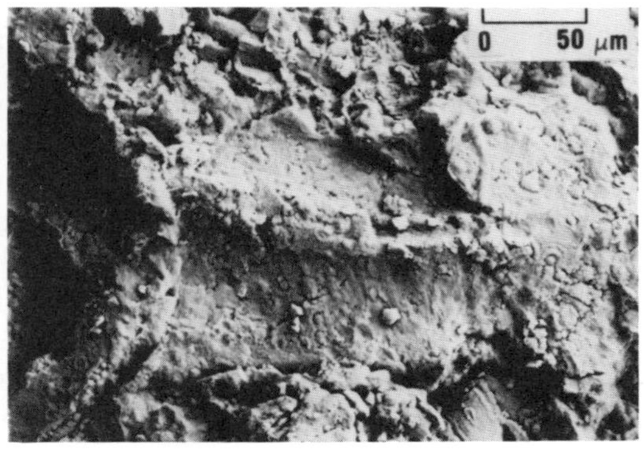

FIGURE 3b. Relict texture consisting of a "casting" of one or several shards (one multi-walled shard or several closely spaced shards).

FIGURE 3c. A more three-dimensional view of a shard casting. The original shard may have been tri-cornered and about 60 μm long. The 4-μm-thick "casting" may represent alteration products on the shard surface; the glass core probably dissolved after a clay "rind" formed.

Metamorphosed Tuffs

Metamorphosed volcanic ash deposits may be difficult to recognize. However, examples of "metatuffs" have been found in rocks from zeolite facies to amphibolite facies in Precambrian terranes. For example, rocks bearing a strong textural resemblance to pyroclastic flows have been found in Precambrian greenschists of central Arizona (Thorpe and Burt, 1978) and in the Precambrian Grenville metamorphic sequence of northern New York (Carl and van Diver, 1975). Knowledge of original pyroclast compositions and morphologies permits detailed analysis of mineral paragenesis and mechanical textural variations that accompany regional metamorphism. The two examples below illustrate common features associated with low-grade metatuffs of ultramafic and silicic compositions.

ZEOLITE FACIES IN SILICIC TUFFS; TIMBER MOUNTAIN—OASIS VALLEY VOLCANIC FIELD, NEVADA

Introduction: Pyroclasts in pyroclastic flow deposits (ash flow tuffs) may undergo many physical and chemical changes after deposition. In the well-insulated interior of a pyroclastic flow, individual pyroclasts may be flattened by the weight of the overlying deposit and, subsequently, weld together. Blocky pumice pyroclasts are flattened into pancake-shape disks. The slow release of volatiles allows hot gases to interact with glass pyroclasts in the upper parts of the flow; the glass is devitrified and often leaves only friable pyroclast relicts. After the flow has cooled, interaction of ground water with remaining vitric pyroclasts will form authigenic feldspars, silica phases, smectite clays, and zeolites. Within these rocks, the search for relict pyroclastic textures is often difficult, but identifying those textures may help in characterizing the tuffs.

For a comprehensive review of pyroclastic flow deposits, see Ross and Smith (1961) and Lipman et al. (1966). A mineralogical approach to the study of alteration products in tuffs is provided by a volume on zeolite minerals edited by Mumpton (1978).

The examples used here are from the Timber Mountain-Oasis Valley volcanic field of Nevada; these 9.5- to 16-million-year-old tuff deposits cover 11,000 km^2 and have a total volume of over 5000 km^3 (Byers et al., 1976). Detailed analyses of authigenic minerals in these deposits have been made in support of the U.S. Nuclear Waste Isolation Program (e.g., Sykes et al., 1979).

Chemical Composition: Most of the examples here are tuffs that have quartz latitic to rhyolitic compositions.

Tephra Description: In pyroclastic flow deposits affected by zeolite-facies metamorphism (diagenesis), relict textures are best seen in thin sections by transmitted light (plane) at low magnification. Outlines of pyroclasts and shards are often visible because of <1-μm hematite grains located on former pyroclast surfaces and on vesicle walls. In many cases, cores of large pyroclasts have been dissolved, leaving open vugs, and authigenic phases (silica, K-feldspar, and zeolites) may line or fill these voids. Finer grained phases are frequently replaced by finely crystalline, intergrown silica phases, authigenic feldspars, smectite clays, and zeolites. Relict textures are often completely destroyed by spherulite growth.

Phenocrysts and lithic fragments in these tuffs are rarely affected by postdepositional vapor-phase alteration and interaction with ground water.

REFERENCES:

Byers, Jr., F., Carr, W., Orkild, P., Quinlivan, W., and Sargent, K., 1976. Volcanic suites and related cauldrons of Timber Mountain-Oasis Valley Caldera complex, Southern Nevada. U.S. Geol. Surv. Prof. Paper 919: 70 pp.

Carl, J. D. and van Diver, B. B., 1975. Precambrian Grenville alaskite bodies as ash-flow tuffs, northwest Adirondacks, New York. Geol. Soc. Am. Bull., **86**: 1691-1707.

Lipman, P., Christiansen, R., and O'Conner, J., 1966. A compositionally zoned ash-flow sheet in Southern Nevada. U.S. Geol. Surv. Prof. Paper 524-F: 47 pp.

Mumpton, F. (Ed.), 1978. *Natural Zeolites*. Pergamon, Oxford: 546 pp.

Ross, C. and Smith, R., 1961. Ash-flow tuffs: their origin, geologic relations and identification. U.S. Geol. Surv. Prof. Paper 366: 81 pp.

Sykes, M., Heiken, G., and Smyth, J., 1979. Mineralogy and petrology of tuffs from the UE25a1 drill site, Yucca Mountain, Nevada. Los Alamos Scientific Laboratory report LA-8139-MS: 76 pp.

Thorpe, D. G. and Burt, D. M., 1978. Precambrian metavolcanic rocks of the Squaw Peak area, Maricopa County, Arizona. *In*: Burt, D. M. and Péwé, T. L. (Eds.), *Guidebook to the geology of Central Arizona*, Ariz. Bur. Geol. Mineral. Tech. Spec. Pap. No. 2: 101-106.

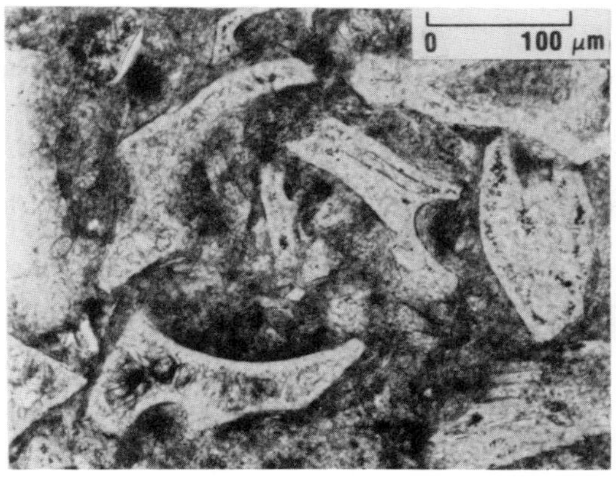

FIGURE 1a. Photomicrographs (plane light) of samples from altered pyroclastic flow and fall units. Nonwelded vitric tuff from the Prow Pass Member of the Crater Flat tuff. Curved outlines are former rhyolitic shard surfaces. Shapes are preserved by fibrous zeolite phases. Shard interiors were dissolved; the open cavities are now lined with euhedral, tabular clinoptilolite crystals. Dark, fine-grained phases between relict shards are zeolites and traces of <1-μm hematite grains.

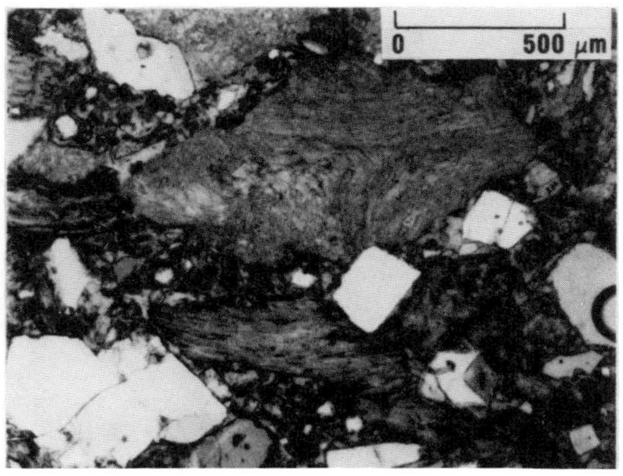

FIGURE 1b. Photomicrograph (plane light) of a vitric tuff from the "Bedded Tuff" unit of the Paintbrush Tuff. A relict pumice pyroclast is surrounded by a matrix of pale brown shards. All have been replaced by finely crystalline clinoptilolite. Outlines of pyroclasts and vesicles are visible because of chains of <1-μm hematite grains located at pyroclast surfaces.

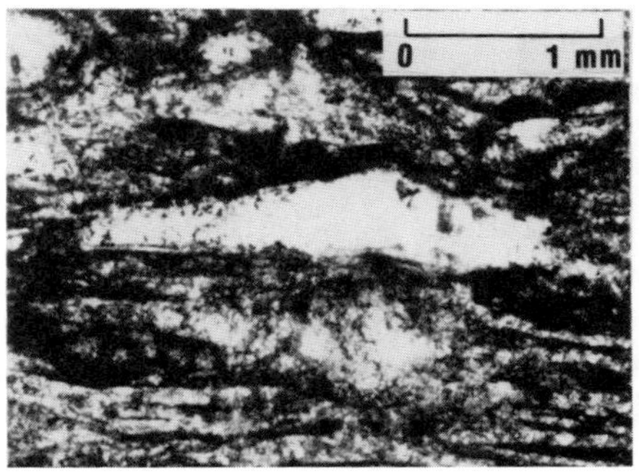

FIGURE 1c. Welded tuff from the upper part of the Topopah Spring Member of the Paintbrush Tuff. Deformation and welding of pumice pyroclasts and shards is evident, but shard shapes are very poorly preserved. As in Fig. *1b*, relict pyroclast shapes are outlined by chains of fine-grained hematite. The flattened, welded pyroclasts have been replaced by finely crystalline authigenic silica and K-feldspar.

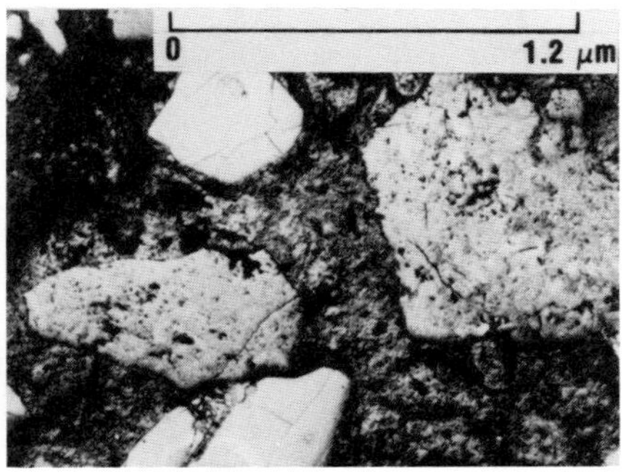

FIGURE 2. Reflected-light photomicrograph of nonwelded vitric tuff from the "Bedded Tuff" unit of the Paintbrush Tuff. Sanidine, quartz, and plagioclase phenocrysts and lithic fragments are visible (*left-center*). Lithic fragments are mostly older welded tuffs. Vitric pyroclasts have been replaced by pale brown, finely crystalline phases (the low-relief matrix); in this case the authigenic phases are mostly clinoptilolite.

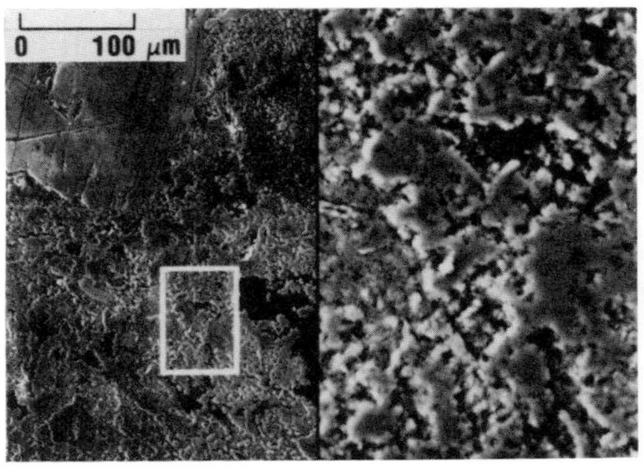

FIGURE 3. SEM of a polished thin section of nonwelded tuff from the Tram Member of the Crater Flat Tuff. Pumice and shard relicts are nearly impossible to see in this manner but are easily seen in transmitted light. Only large phenocrysts and the altered vitric matrix are visible here. The matrix consists of very finely crystalline phases (*inset*) that have replaced vitric pyroclasts, including smectites (low-relief phases) and authigenic alkali-feldspars and silica (high-relief phases).

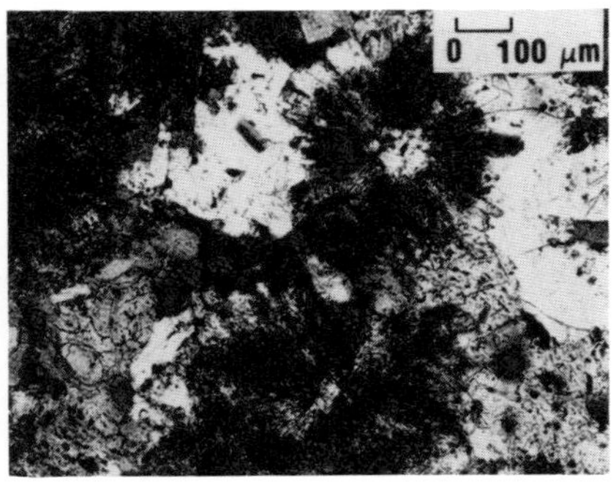

FIGURE 4. Transmitted-light photomicrograph (crossed polarizers) of a densely welded vitric tuff from the Topopah Spring Member of the Paintbrush Tuff. This is the interior of an irregularly zoned pumice pyroclast. Much of the original texture has been obscured by spherulites that consist of K-feldspar and quartz.

FIGURE 5a. SEM of intergrown, authigenic silica and K-feldspar adjacent to a vug. These phases replace vitric pyroclasts as a finely crystalline matrix or spherulites and, in many cases, obscure relict textures. This example is from a vitric-crystal welded tuff (Prow Pass Member of the Crater Flat Tuff).

FIGURE 5b. Granular and fibrous phases (alkali feldspar and quartz) that have replaced vitric pyroclasts in a welded vitric-crystal tuff (Prow Pass Member of the Crater Flat Tuff).

FIGURE 5c. Although rims of relict pyroclasts may be visible, the cores are often dissolved, leaving a void space. The void space allows growth of relatively coarse zeolites. In this example from the Topopah Spring Member of the Paintbrush Tuff, the well-developed authigenic crystals are clinoptilolite.

META-TUFF OF ULTRAMAFIC COMPOSITION; SPINIFEX RIDGE, QUEBEC

Introduction: An Archean-age pyroclastic deposit is associated with ultramafic (komatiitic) lava flows along Spinifex Ridge, Lamotte Township, Quebec (Gélinas et al., 1977). This deposit consists of well-bedded, graded pyroclasts located in channels between lava flows. A genetic interpretation of these pyroclasts has been difficult. There are several unanswered questions concerning the Spinifex Ridge deposit: (1) Did it have a phreatomagmatic or magmatic origin? (2) What volatiles were present in the magma to drive the eruption? and (3) What was the nature of the explosive volcanic activity? Gélinas et al. (1977) propose that water was required for explosive activity, but experimental petrologic evidence (Brooks and Hart, 1974; Green et al., 1975) requires that ultramafic lavas of this composition were erupted without significant water present.

The samples were supplied by L. Gélinas.

Grain Size: Pyroclasts range from 300 μm to 5 mm long, and the approximate mean size is 1.5 mm.

Tephra Description: Excellent relict textures remain in this meta-tuff, which allows description of the pyroclasts and interpretation of their genesis. The pyroclasts consist of spheres (19%), ovoids (45%), and compound ovoids (those with more than one grain annealed to another). Approximately 80% of the droplets are broken fragments of spheres and ovoids; 44% of the droplets studied are partly crystalline and contain what appear to have been dendritic to skeletal quench-texture pyroxenes or olivines. Some of the pyroclasts have been compressed or sheared, but many are not deformed at all.

Chemical Composition: A recalculated, volatile-free analysis of the meta-tuffs from Spinifex Ridge was published by Gélinas et al. (1977) (Table 1).

TABLE 1. CHEMICAL COMPOSITION, SPINIFEX RIDGE, QUEBEC.*

	(Wt. %)
SiO_2	48.58
Al_2O_3	6.28
TiO_2	0.33
FeO	11.22
MgO	30.77
CaO	5.66
Na_2O	0.08
K_2O	0.02
Total	102.94

*From Gélinas et al. (1977).

Comments on Pyroclast Formation: The genesis of this deposit has had several interpretations. Gélinas et al. (1977) proposed that water was needed to drive an explosive eruption, but experimental petrologic evidence for komatiitic magmas requires genesis and eruption of these magmas *without* significant water (Brooks and Hart, 1974; Green et al., 1975). Gélinas also suggested that these tuffs are of hyaloclastic origin.

There is yet another explanation. Heiken (1979) found that pyroclastic rocks from the Earth's moon have compositions similar to those of Spinifex Ridge komatiites. Samples collected at the Apollo 15 landing site are the most similar; they consist of fresh, unaltered glassy pyroclasts and thus are much easier to study than those from Quebec. These lunar pyroclasts are mostly vesicle-free glass droplets containing quench-crystal olivines and pyroxenes. The well-bedded, graded deposit at Spinifex Ridge is also similar to a deposit sampled by coring at the Apollo 17 landing site. Grading may have been related to variation in the intensity or dispersion angle of a lava fountain. Vesicle-poor or vesicle-free glass droplets can form in lava fountains of low-viscosity, ultramafic magmas. Water is not necessary for explosive lava-fountaining of these magmas; in the lunar case, CO was the chief volatile phase (Meyer et al., 1975; Sato, 1977).

REFERENCES:

Brooks, C. and Hart, S. R., 1974. On the significance of komatiite. Geology, **2**: 107-110.

Gélinas, L., LaJoie, J., and Brooks, C., 1977. The origin and significance of Archean ultramafic volcaniclastics from Spinifex Ridge, Lamotte Township, Quebec. *In*: Baragar, W., Coleman, L., and Hall, J. (Eds.), *Volcanic Regimes in Canada*. Geol. Assoc. Canada, Special Paper 16: 297-309.

Green, D. H., Nicholls, I. A., Viljoen, M., and Viljoen, R., 1975. Experimental demonstration of the existence of peridotite liquids in earliest Archean magmatism. Geology, **3**: 11-14.

Heiken, G., 1979. Archean ultramafic pyroclastic deposits: the use of lunar deposits to solve some terrestrial problems (abs.). Tenth Lunar and Planet. Sci. Conf., Houston, Texas.

Meyer, Jr., C., McKay, D., Anderson, D., and Butler, Jr., P., 1975. The source of sublimates on the Apollo 15 green and Apollo 17 orange glass samples. *In: Proceedings of the 6th Lunar Science Conference*. Pergamon Press, New York: 1673-1699.

Sato, M., 1977. The driving mechanism of lunar pyroclastic eruptions inferred from the oxygen fugacity behavior of Apollo 17 orange glass. Trans. Am. Geophys. Union, **58**: 425.

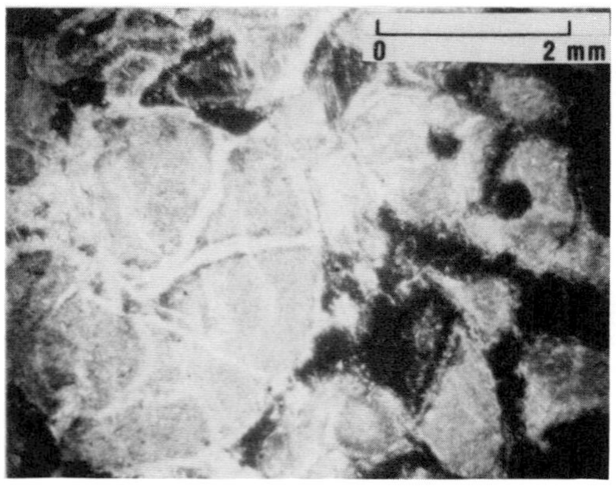

FIGURE 1a. Relict spheres and sphere fragments. All pyroclasts have been replaced by chlorite and are surrounded by a fine-grained talc matrix. Cracks in the large sphere are filled with talc. Photomicrograph taken in partly polarized light.

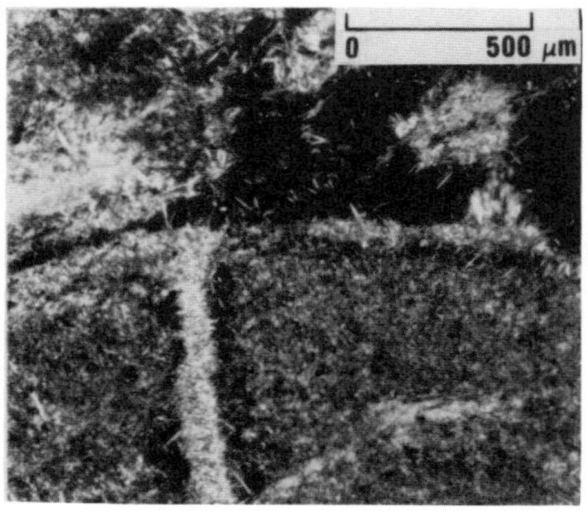

FIGURE 1b. Edge of sphere in Fig. 1.

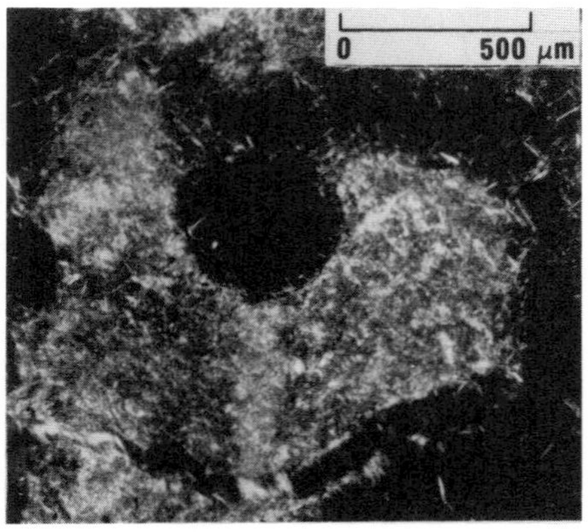

FIGURE 1c. Sphere fragment with vesicle.

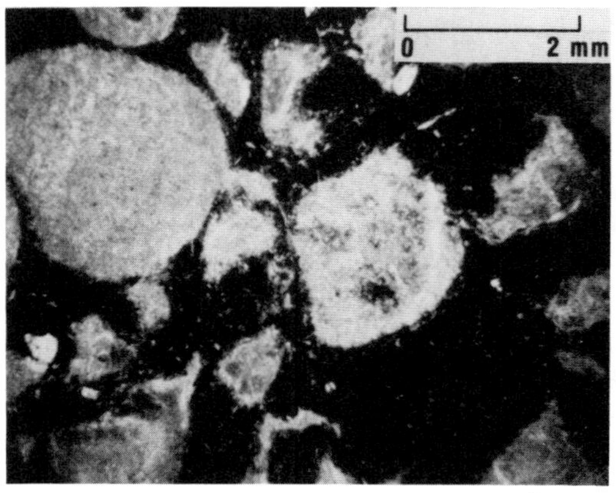

FIGURE 2. Relict spheres and sphere fragments; pseudomorphed by chlorite.

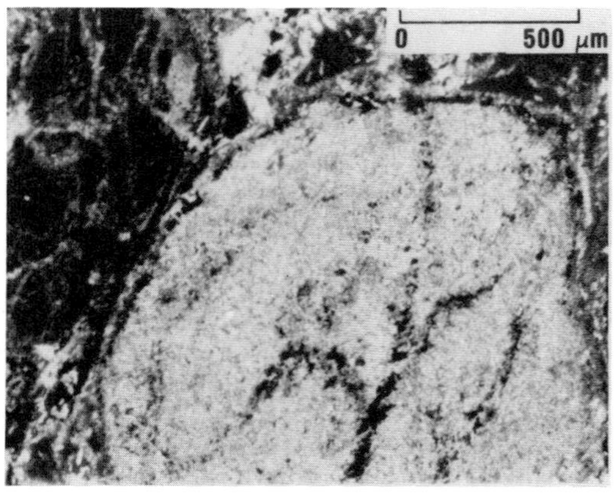

FIGURE 3. Relict droplet that has been deformed slightly by shearing.

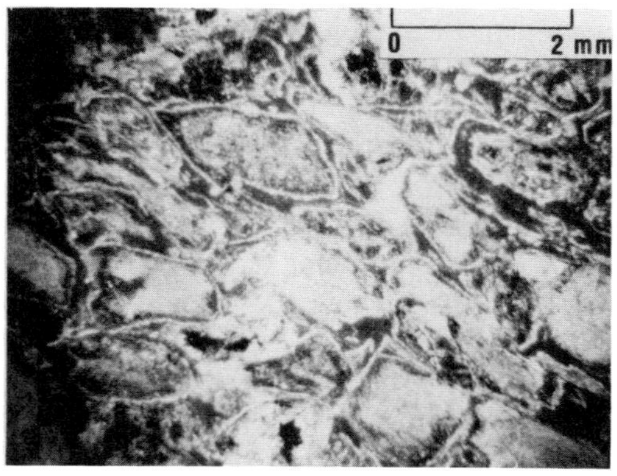

FIGURE 4a. Relict droplets and droplet fragments similar to those in Fig. 1 but deformed by shear.

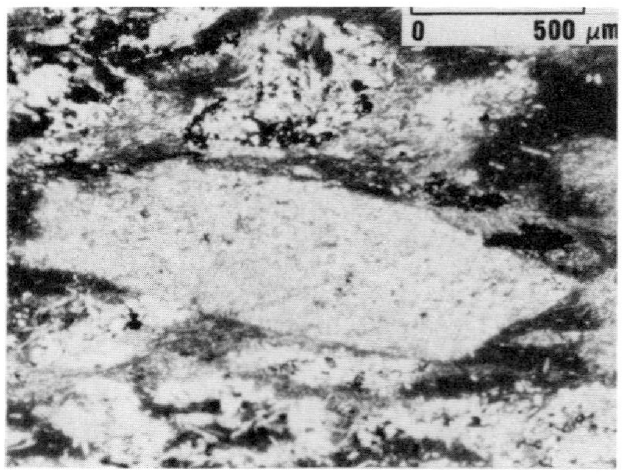

FIGURE 4b. Closer view of deformed tuff similar to that illustrated in Fig. *4a*. Relict pyroclast textures are no longer preserved, and thus identification of this unit as a tuff is very difficult.

APPENDIX: CLASSIFICATION OF COMMON IGNEOUS ROCKS

The creation of igneous rock categories, complete with names, is artificial; in reality, igneous rocks grade into one another chemically, texturally, and mineralogically.

There is a great variety of igneous rocks, but only the most common are included in this simplified classification. Rock names used for both the rapidly cooled, finely crystalline or glassy volcanic rocks and the slowly cooled, coarsely crystalline plutonic rocks are included in Table 1.

For a complete review of igneous petrology, the reader is referred to the works of Carmichael, Turner, and Verhoogen (1974) and MacDonald (1972).

TABLE 1. SIMPLE CLASSIFICATION OF COMMON VOLCANIC ROCKS AND SOME PHYSICAL AND CHEMICAL PROPERTIES.

	Silicic	Intermediate		Mafic	
Main mineral phases	**K-feldspar, quartz oligoclase**	**Plagioclase biotite hornblende and pyroxenes, quartz**	**Plagioclase augite, orthopyroxene, hornblende**	**Plagioclase augite olivine**	**Feldspathoid, K-feldspar plagioclase, hornblende, biotite**
Rock name	**Rhyolite**	**Dacite**	**Andesite**	**Basalt**	**Phonolite trachyte**
	Average Chemical Composition* (Wt.%)				
SiO_2	73.66	63.58	57.7	50.83	56.90
TiO_2	0.22	0.64	0.6	2.03	0.59
Al_2O_3	13.45	16.67	17.0	14.07	20.17
Fe_2O_3	1.25	2.24	2.4	2.88	2.26
FeO	0.75	3.00	3.6	9.05	1.85
MnO	0.03	0.11	0.15	0.18	0.19
MgO	0.32	2.12	5.1	6.34	0.58
CaO	1.13	5.53	7.5	10.42	1.88
Na_2O	2.99	3.98	3.4	2.23	8.72
K_2O	5.35	1.40	1.6	0.82	5.42
Estimated extrusion temperatures (°C)**	800-900	900-1000	900-1000	1000-1200	1000
Viscosities at extrusion temperatures (Poise)†	10^{10}-10^{12}	10^{10}	10^5-10^7	10^2-10^4	10^5

*From Nuckolds (1954) and Williams (1932).
**From Carmichael et al. (1974).
†From MacDonald (1972).

REFERENCES FOR APPENDIX:

Carmichael, I. S. E., Turner, F. J., and Verhoogen, J., 1974. *Igneous Petrology*. McGraw-Hill, New York: 739 pp.

MacDonald, G. A., 1972. *Volcanoes*. Prentice-Hall, Inc., Englewood Cliffs: 510 pp.

Nuckolds, S. R., 1954. Average chemical compositions of some igneous rocks. Bull. Geol. Soc. Am., **65**: 1007-1032.

Williams, H., 1932. Geology of the Lassen Volcanic National Park. Calif. Univ. Pubs. in Geol. Sci., **21**: 195-385.

INDEX

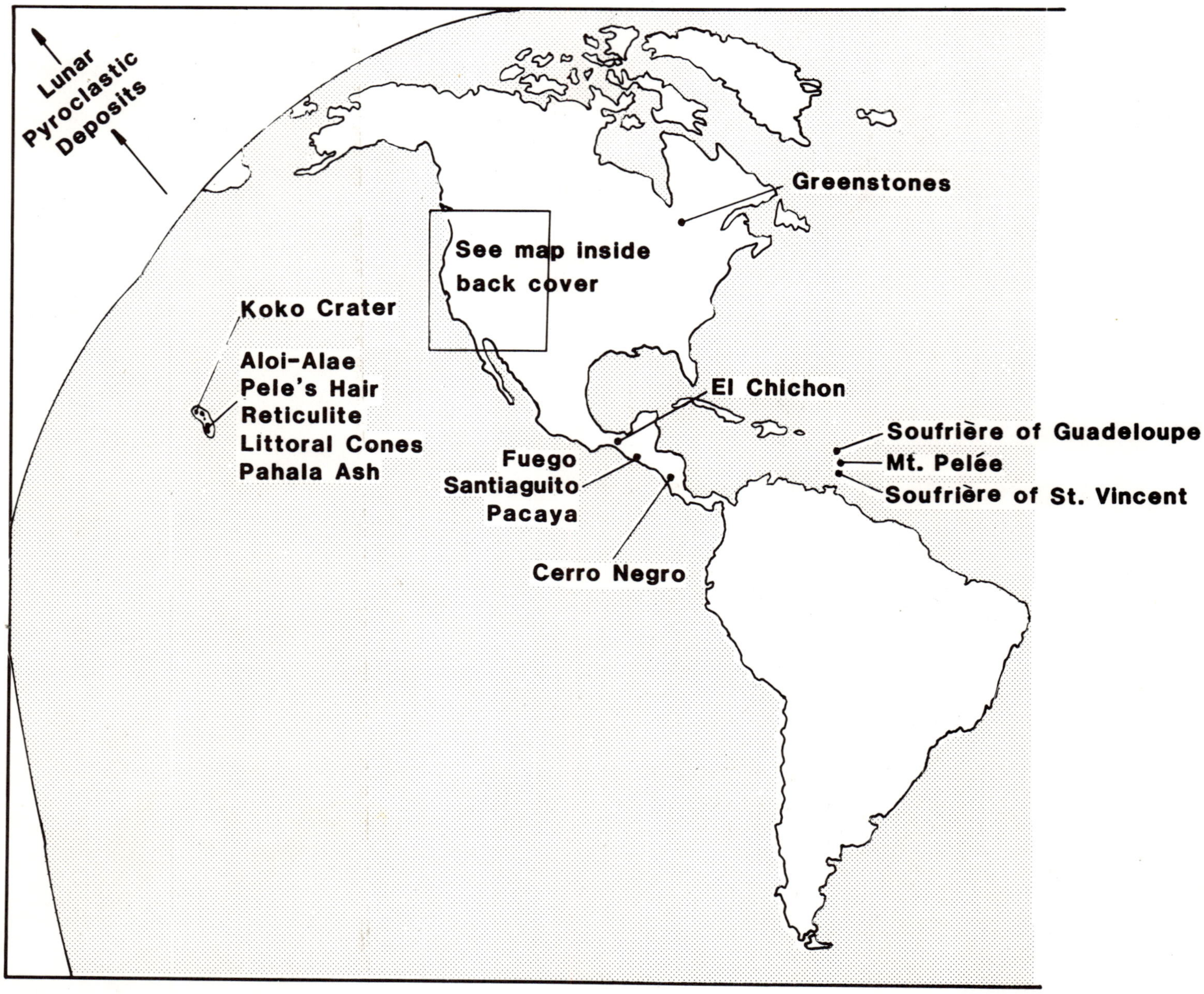
Lunar
Pyroclastic
Deposits
Greenstones
See map inside
back cover
Koko Crater
Aloi-Alae
Pele's Hair
Reticulite
Littoral Cones
Pahala Ash
El Chichon
Fuego
Santiaguito
Pacaya
Cerro Negro
Soufrière of Guadeloupe
Mt. Pelée
Soufrière of St. Vincent